Library of Congress Cataloging in Publication Data

Main entry under title:

Source book on materials for elevated-temperature applications.

Includes bibliographical references and index.

1. Heat resistant alloys.   I. Bradley, Elihu F.

TA485.S65   620.1'6'17   79-17831

ISBN 0-87170-081-6

II Series. American Society for Metals. ASM engineering bookshelf.

Contributors to This Source Book*

W. M. BOAM
Aerojet-General Corp.

E. F. BRADLEY
Pratt & Whitney Aircraft

J. Z. BRIGGS
Climax Molybdenum Co.

A. G. BUCKLIN
General Electric Co.

WILLIAM J. COLLINS
Corning Glass Works

HAROLD R. CONAWAY
Huntington Alloys, Inc.

R. H. COOK
Central Electricity Research
    Laboratories (England)

J. H. DAVIDSON
Creusot-Loire Département
    Études et Recherches Aciéries
    d'Imploy (France)

R. F. DECKER
International Nickel Co., Inc.

C. D. DESFORGES
Fulmer Research Institute Ltd.

M. J. DONACHIE, JR.
Pratt & Whitney Aircraft

W. T. EBIHARA
Union Carbide Corp.

G. A. FRITZLEN
Union Carbide Corp.

G. WILLIAM GOWARD
Pratt & Whitney Aircraft

N. J. GRANT
Massachusetts Institute of Technology

R. B. HERCHENROEDER
Union Carbide Corp.

W. HERDA
International Nickel Ltd.

R. I. JAFFEE
Battelle Columbus Laboratories

J. KEVERIAN
General Electric Co.

ROGER P. KING
Libbey-Owens-Ford Glass Co.

R. G. LYTHE
International Nickel Ltd.

G. N. MANIAR
Carpenter Technology Corp.

R. A. MILLER
American Brake Shoe Co.

L. A. MORRIS
Falconbridge Nickel Mines Ltd.

DONALD R. MUZYKA
Carpenter Technology Corp.

T. D. PARKER
Climax Molybdenum Co.

F. H. PENNELL
De Laval Steam Turbine Co.

*Affiliations given were applicable at date of contribution.

ROBERT M. PLATZ
Fairchild Engine and Airplane Corp.

C. D. PREUSCH
Crucible Steel Co. of America

A. J. RICKARD
International Nickel Ltd.

E. A. SCHOEFER
Alloy Casting Institute

C. W. SCHWARTZ
Misco Precision Casting Co.

F. A. SETTINO
Pittsburgh Plate Glass Co.

M. E. SHANK
Pratt & Whitney Aircraft

S. A. SHERIDAN
Ford Motor Aircraft Engine Div.
Ford Motor Co.

R. P. SKELTON
Central Electricity Research
   Laboratories (England)

R. A. SPRAGUE
Pratt & Whitney Aircraft

W. H. STRAUTMAN
Southwestern Portland Cement Co.

J. STRINGER
The University of Liverpool

C. P. SULLIVAN
Pratt & Whitney Aircraft

F. P. TALBOOM
Pratt & Whitney Aircraft

J. D. VARIN
Pratt & Whitney Aircraft

FRANCIS L. VERSNYDER
Pratt & Whitney Aircraft

E. T. VITCHA
TRW Inc.

R. M. WOODWARD
Owens-Corning Fiberglas Corp.

R. D. WYLIE
Babcock & Wilcox Co.

CONTENTS

Preface

In the era that preceded World War II, interest in materials for elevated-temperature technology was relatively limited and specialized. There were, among others, the petroleum-refinery and petrochemical applications, primary metals production and processing, and the ubiquitous steam boiler and steam engine in their diverse forms. Then, with the jet engine, there arose without warning an urgent need for dependable materials with combinations of high strength and corrosion resistance in a variety of hostile environments.

The early jet engines did, in fact, break more than the well-publicized "sound barrier." They shattered a mindless complacency regarding the true criticality of the need for new families of alloys and nonmetallics — materials for elevated-temperature applications. The "crisis" that followed — starting with the frantic pursuit of titanium and its alloys — was inordinately expensive and, as is characteristic of crisis-research, not a little wasteful.

Yet, the materials demands of that era were modest indeed compared to demands to come, demands that followed the memorable launching of Sputnik. That launching heralded a long series of events concerned with the exploration of outer space, followed in more recent years by the sobering realization that several conventional sources of energy, those upon which industrialized nations depended most, were by no means unlimited. Often, in the course of these events, major technical limitations — whether to the conquest of space or the development of substitute sources of energy — were identified as those imposed by the capabilities of currently available materials, especially materials for elevated-temperature service. Small wonder that reliable information on these materials is in greater demand now than ever before.

SOURCE BOOK ON MATERIALS FOR ELEVATED-TEMPERATURE APPLICATIONS responds to the technical needs of a community of designers and engineers whose numbers have grown and will continue to grow. Appropriately, its consulting editor, Elihu F. Bradley, has actively served this branch of technology throughout his entire professional career. That distinguished career as metallurgist and materials engineer has been prominently identified with the aircraft and aerospace industries, and has spanned the long, epoch-making trail from reciprocating engines to the sophisticated technology of space exploration. The contents of the book, which reflect his broad knowledge and longtime experience with materials, are presented in 13 major sections, a summary of which follows.

Survey Paper. This recent review by Desforges identifies the major types of elevated-temperature alloys in current use, with reference to their properties and applications. Due emphasis is given new developments in the processing of these alloys, which serve both to improve properties and ensure reliability in service. The problem areas that remain are also identified.

Turbines and Industrial Applications. The five articles in this section relate elevated-temperature alloys to applications in gas turbines; in industrial components, such as hot-working tools and furnace parts; in petrochemical process and refinery furnaces; and in chemical-plant processes employing aqueous nitric acid. Of outstanding value are the extensive data provided on strength and corrosion at elevated temperatures.

Properties and Environment. The important thrust of a comprehensive review by Cook and Skelton is that chemical and mechanical effects at elevated temperatures interact. Therefore, it is unsafe to assume that an alloy for a given elevated-temperature application can be chosen for its strength or corrosion resistance separately. These properties must be viewed as being functionally interdependent.

Super 12% Cr Steels. The exhaustive study of specially alloyed 12% Cr steels by Briggs and Parker, as published originally, ran to book length. A major portion of that work is presented here because of the continuing importance of these steels to elevated-temperature technology. Covered in depth are their applications and physical and mechanical properties, in the longest single article in this book.

Stainless Steels. The role of stainless steels as materials suitable for elevated-temperature service was established many years ago, primarily because of their resistance to corrosion and oxidation at high temperatures. In recent years, the strength of stainless steels at these temperatures has become a major design factor because of its importance in a majority of applications. T. D. Parker's contribution to this section provides authoritative data on the strength of stainless steels at elevated temperature and serves as a valuable guide in selection. Comparable coverage is afforded corrosion resistance at elevated temperature in an article by L. A. Morris.

Heat-Resistant Alloy Castings. This classic article on the cast alloys widely used in industrial elevated-temperature applications was originally prepared by E. A. Schoefer and was later revised by the ASM Committee for the 8th edition of *Metals Handbook*. It concludes with a series of data sheets tabulating the physical and mechanical properties of the cast alloys of principal industrial importance.

Iron-Base and Iron-Containing Superalloys. These superalloys are especially important not only because of their major contributions to aircraft and aerospace technology but also because iron was, and continues to be, a lower-cost substitute for nickel-base and cobalt-base superalloys in so-called "moderate strength" applications. Sullivan and Donachie, the authors of the comprehensive review presented here, have been acknowledged authorities in the field for many years.

Nickel-Base Superalloys. In this section, three articles have been selected to review nickel-base superalloys in a manner that broadens understanding of the fundamental relationship between their microstructures and resulting mechanical properties, as well as identifying alloys in widest use and their applications. Coverage is also given the role of heat treatment, effects of specific alloy additions, and effects of elevated-temperature exposure on mechanical properties.

Cobalt-Base Superalloys. Among the most important of the superalloys, the cobalt-base alloys are of particular interest in elevated-temperature applications above about 2000 °F and below 1500 °F. The intervening temperature range is dominated primarily by nickel-base superalloys. Many of the newer alloys are quite specialized, having been developed for a specific application. The two articles comprising this section review the properties of these alloys, microstructural-property relationships, and the in-process metallurgy of three of the leading wrought alloys.

Control of Structure and Properties. The contribution of this article by Muzyka is quite unique in that it does more than simply relate the properties of superalloys to their microstructures in a general sense. It reviews the fundamentals of processes that are needed to attain desired microstructures in wrought superalloys.

Role of Directional Solidification. "Something exciting has been happening during these last several years." With this modest observation, Versnyder and Shank introduce their review of the development and production of alloy single-crystal turbine blades for advanced aircraft jet engines — one of the most important metallurgical developments of the current century. The review considers casting and solidification, engineering properties, engine evaluation, and future developments.

Protective Coatings. G. W. Goward reports on protective coatings for high-temperature alloys, especially the coatings that are applied to blades and vanes in gas-turbine engines. Fortunately, these coatings are not restricted to turbine applications and might be considered for applications in many other industrial components.

Welding. The concluding article in this volume concentrates on a subject of fundamental importance to the performance and reliability of welded components in elevated-temperature service — the welding of dissimilar metals. An article by H. R. Conaway discusses the significance of dilution and establishes dilution limits for a range of superalloys. Also considered are design and fabrication for high-temperature service and the testing and inspection of welds.

The American Society for Metals extends its grateful appreciation to its President, Mr. Elihu F. Bradley, for his masterful contribution in selecting and organizing the articles in this book. Despite the pressure of his many duties, he brought to the editorial assignment both the expertise and thoughtful effort so important a subject deserves. It is fair to state that for his noble effort he, in common with many readers, can now reflect upon a noble and useful work. Finally, our most grateful acknowledgment is extended to the many authors whose work appears in this book and to their publishers.

PAUL M. UNTERWEISER
Staff Editor
Manager, Publications Development
American Society for Metals

WILLIAM H. CUBBERLY
Director of Reference Publications
American Society for Metals

Cover photo courtesy of Huntington Alloys, Inc.

Introduction

Elihu F. Bradley

Alloys used at elevated temperatures must also be able to withstand the deteriorating effect of the service atmosphere. In addition, of course, these materials must possess sufficient strength for the design conditions and have satisfactory stability to withstand damaging metallurgical structural changes at operating temperature. From the standpoint of resisting oxidation and high-temperature corrosion, the most important alloying element is chromium. It is not surprising, then, that corrosion-resistant steels, stainless steels, nickel-chromium alloys and superalloys, all of which contain significant amounts of chromium, are used extensively in high-temperature applications.

For use at moderate temperatures and moderate stress, the 12 Cr corrosion-resistant steels are satisfactory. Under conditions of somewhat higher stress at the same moderate temperatures, the so-called super 12 Cr steels, a versatile group containing in addition to chromium small amounts of molybdenum and/or other strong carbide formers and/or cobalt or nickel, have been used for some 40 years. As the operating temperature is increased but for conditions of low stress, higher-chromium steels, either the ferritic corrosion-resistant or the austenitic stainless steels containing nickel as well as chromium, or nickel-chromium alloys, are commonly selected. For very high temperatures there is increasing interest in the refractory metals of Groups V (vanadium, niobium, tantalum) and VI (chromium, molybdenum, tungsten) and ceramics. The refractory metals, however, exhibit very poor oxidation resistance, and their use is now restricted to nonoxidizing environments, and ceramics possess insufficient toughness for most structural applications. Elevated temperatures have been a disappointing limitation for titanium ever since its introduction into turbine engines in the middle 1950's. Two reasons for this limitation persist — namely, the affinity of titanium for interstitial elements, and inadequate creep strength at even the moderate temperature level of about 1000 °F. Therefore, for the most severe combination of stress and temperature, it remains for the remarkable superalloys to do the job.

The term "superalloy" was first used shortly after World War II to describe a group of alloys developed for use in turbosuperchargers and aircraft turbine engines, which required high performance at elevated temperatures. These alloys usually consist of various formulations made up from the following elements: iron, nickel, cobalt and chromium, as well as lesser amounts of tungsten, molybdenum, tantalum, niobium, titanium and aluminum. The most important properties of the superalloys are long-time strength at temperatures above 1200 °F and resistance to hot corrosion and erosion. Many types of alloys fall under the broad coverage of superalloys. These include iron-base alloys with chromium and nickel; complex iron-nickel-chromium-cobalt compositions; cobalt-base alloys, carbide strengthened; nickel-base alloys, solid-solution strengthened; and nickel-base alloys, precipitation or dispersion strengthened. The superalloys are used in both the wrought and the cast forms.

Generally, the strengths of the iron-base alloys, the complex iron-nickel-chromium-cobalt alloys and the nickel-base solid-solution-strengthened alloys are considerably lower than those of the nickel-base second-phase-strengthened and cobalt-base alloys at temperatures above 1200 °F. Early iron-base superalloys, such as 16-25-6 alloy containing 16% Cr, 25% Ni and 6% Mo, and complex iron-nickel-chromium-cobalt alloys, such as Fe – 20Ni – 20Cr – 20Co with small amounts of tungsten and molybdenum, are essentially solid-solution strengthened. Later iron-base alloys, containing small amounts (2 to 3%) of aluminum and titanium, achieved increased high-temperature strength through precipitation of an aluminum-titanium strengthening phase. An illustrative composition of the lower-strength nickel-base solid-solution superalloys that has been widely used in gas-turbine burner and combustor applications is Ni – 22Cr – 18Fe – 9Mo. Because of melting-point advantage, the cobalt alloys are usually stronger than the nickel alloys at temperatures of about 1800 °F and above. Cast cobalt-base alloys, characterized by a face-centered-cubic (austenitic) solid-solution matrix and containing complex carbides, have had a successful history as airfoils for gas-turbine engines (most turbine vanes and some turbine blades). One exception to this strength observation is the dispersion-strengthened nickel alloys, utilizing a dispersed-oxide strengthening phase, which show high strength at elevated temperatures but only moderate strength at intermediate temperatures. The secondary phase persists in the structure of these alloys as a strengthening mechanism throughout the solid state until melting occurs. In contrast, alloys strengthened by precipitation lose the strengthening phase by solution in the solid state at some temperature below the melting point. Dispersion-strengthened alloys are beginning to be used in some gas-turbine-engine burner applications.

Clearly, the nickel-base superalloys strengthened by a secondary precipitated phase are the most complex and, indeed, the most remarkable of all the superalloys. The physical metallurgy of these alloys is subtle, sophisticated and well understood. The structure consists of a face-centered-cubic, austenitic solid-solution matrix with a precipitated nickel-aluminum-titanium compound (gamma prime) as the principal strengthening phase. Various carbides, depending on the particular alloy composition and heat treatment, exist as second precipitated phases. These alloys are used in the most demanding applications relative to stress and temperature in gas-turbine engines. They have demonstrated remarkably useful strength at the highest fraction of the base metal melting point of any alloy system ever developed.

Castings are intrinsically stronger than forgings at elevated temperatures. The coarse grain size of castings, as compared to fine-grain forgings, favors strength at very high temperatures. In addition, casting compositions can be effectively tailored for high-temperature strength inasmuch as forgeability characteristics are not a factor. Higher elevated-temperature strength can be achieved in the nickel-base, gamma-prime-strengthened superalloys, for example, by lowering the chromium content, but at the expense of hot-corrosion resistance. Superalloys of this type, containing only 8 to 12% chromium, may require the use of coatings (such as diffused aluminum) to compensate for the loss in hot-corrosion resistance of the alloy. The addition of small amounts of hafnium (1 to 1.5%) causes a marked improvement in the intermediate-temperature ductility of these high-strength nickel superalloy castings. Because of high-temperature strength advantage, many aircraft gas-turbine engines use nickel-base, gamma-prime-strengthened, superalloy castings for the high-stress, high-temperature turbine-blade application. Innovation in the casting process has also resulted in improved high-temperature properties. The development of directional solidification involving controlled grain growth, whereby all crystals are aligned in the longitudinal direction, has provided increased high-temperature strength and, in particular, significant improvement in resistance to thermal fatigue.

Although materials technology has pushed the use of the superalloys to temperatures close to their melting points, materials scientists are studying methods for still further advances. Beyond today's directionally solidified castings is the production of single-crystal components, characterized by the complete absence of grain boundaries. Directionally solidified eutectics, wherein aligned whiskers grow from the eutectic phase within a ductile matrix, are emerging as fiber-reinforced alloys (in situ composites) of great strength and stability with the promise of even higher temperature capability. New approaches involving powder metallurgy and thermomechanical processing may significantly improve strength at intermediate temperatures. The use of powder metallurgy to disperse oxide particles in gamma-prime-strengthened superalloys has promise of improved high-temperature strength. The combination of composite technology with superalloy metallurgy may provide future turbine-blade materials.

The articles and technical papers contained in this Source Book have been selected to describe the metallurgical details of these various high-temperature alloys.

METALS AND ALLOYS
FOR HIGH TEMPERATURE APPLICATIONS
CURRENT STATUS AND FUTURE PROSPECTS

By

C. D. DESFORGES

FULMER RESEARCH INSTITUTE LTD.
(Stoke Poges) Slough, Bucks. (England)

INTRODUCTION

The task which has been given to me viz. a review of metals and alloys for high temperature applications is of such magnitude, both in the breadth of the topic and in the depth which is required to treat each constituent element of the spectrum of materials, that a few qualifying statements are necessary so that the limits of this paper can be clearly defined. Most of the papers presented at the Petten colloquium have been concerned with the specific requirements of a number of industrial sectors, both established and developing, which are of current prime importance in advanced industrial societies and will undoubtedly become important in the so-called emerging nations. I propose to identify the major alloy types which are in current use and to describe some of the basic principles which are believed to underly their properties and hence their applications. In particular I will describe recent developments in processing of these materials since control of manufacturing routes is leading to improved properties, cheaper products and greater quality control, which is so vital for components whose failure can have catastrophic consequences. An indication of problem areas and areas which merit further attention will also form a part of this paper.

1. PROPERTY REQUIREMENTS FOR HIGH TEMPERATURE APPLICATIONS

The large number of relevant properties, which are required in a high temperature material, deserve attention before describing the competitive metals and alloys of interest. If the refractory metals are included with the superalloys then service temperatures of 2,000° C can be considered as attainable. They must have creep resistance; fatigue resistance; thermal-shock resistance; good fracture properties and for b. c. c. metals, with a ductile to brittle transition, this temperature should be as low as possible; oxidation and hot-corrosion resistance; high stress to rupture values; ease of fabrication and joining; impact and erosion resistance; thermal properties, etc. By virtue of the wide range of engineering applications which require such materials, they must be available in a wide variety of forms: bar; castings, often of great intricacy; extrusions; forgings; sheet; tubings, and thus have to be processed economically by a number of production routes. Finally techniques are required so that the quality of the materials can be assessed and assured; their in-service behaviour has to be monitored and if at all possible this should be predictable from basic property data. This latter point is of some importance since materials are usually developed for service and our knowledge of their behaviour in operational service is significantly less than that of their « for service » properties, mainly as a result of the complex conditions in which a component functions at high temperature.

The large number of property requirements underlies the existence of materials problems, which are all the more difficult to solve when a combination of these factors is present. It is also worth considering how high temperature requirements can be met without having recourse to any specific material, since much of the existing technology of high temperatures makes use of such techniques to a greater or lesser degree. A component can be cooled either by a fluid or a gas, with a prime example being found in gas-turbine engines where blade cooling enables them to be used in gas temperatures exceeding those of the alloy melting point. Insulation can be used to limit the effect of high temperatures, as can physical phenomena such as ablation, radiation, etc., typically employed in rocket powered space vehicles. The drawbacks to these various approaches are numerous e. g. weight tends to be excessive and hence power/weight ratio decreases; very complex cooling passages are required in turbine blades and this demands very complex technical solutions; reliance on ablation is obviously limited by a time factor since the material wears away and has to be replaced.

Materials thus offer the major solution to these problems and provide the only answer when long term reliability is necessary e. g. electrical power generation equipment. Two classes of metallic materials can be identified for high temperature applications viz. superalloys and refractory alloys based on chromium, tantalum, niobium, molybdenum and tungsten. Their relative merits and demerits can be summarised in terms of their high temperature strength (fig. 1); their resistance to atmospheric attack at temperature; their ability to withstand shock loading i. e. their fracture properties. Processing routes for the fabrication of components from these materials must take into account these three factors and furthermore they must be designed so that these desirable properties are improved by the processing operation. Developments in this area have led to significant advances in the last ten years and a major theme of this paper will be a presentation and critique of liquid-phase and solid-phase processing routes which are being developed.

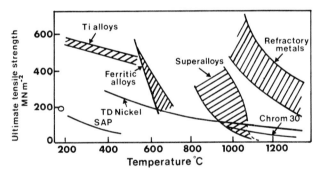

FIG. 1. — *Strength of high temperature alloys.*

Before discussing processing developments it is worth reviewing, albeit briefly, the salient features of superalloys and refractory alloys. Other papers at this colloquium have dealt in depth with specific properties and applications and there are several excellent publications to which reference can be made for more detailed information. What has to be borne in mind is the rapid pace of development of these materials and some sense of historical perspective is essential if the wide variety of developments are to be placed properly in context.

2. SUPERALLOYS

The name superalloys is widely used by metallurgists and engineers to cover a group of metallic alloys which have been developed, and indeed are now specifically designed, to function at high temperatures. More formally a superalloy is an alloy developed for elevated temperature service, usually based on group VIIIA elements (iron, cobalt and nickel) where relatively severe mechanical stressing is encountered and where high surface stability is frequently required.

Although there are only three basic types of superalloy based on the above definition, the technical and engineering literature abounds with a variety of materials, which have been developed for high temperature service. At the time of writing there are more than 100 superalloys commercially available and this quantity alone is a prospect which might daunt the most enthusiastic of designers concerned with the rational selection of the best alloy. An analysis of these alloys reveals complex compositions, forming methods from casting to forging, from controlled solidification procedures to powder metallurgy consolidation techniques using high pressures and temperatures.

The major driving force behind the development of superalloys has been the demand for more efficient power generation units, particularly gas turbine engines for military and civil applications in aerospace, marine and land based vehicles. The environment of a gas turbine component is one of the most demanding and aggressive situations which engineers have managed to devise and the ingenuity of the materials specialist has been taxed to the limit in attempting to satisfy these demands. It is not always appreciated that nickel and cobalt-base alloys have demonstrable useful strength at the highest fraction of the alloy melting point of any other alloy-systems, and if full use is to be made of this fact it is important that the basic factors controlling the properties of superalloys should be understood, even though at first sight they seem rather abstruse and complex.

This complexity stems from the interaction of several factors at high temperature and the difficulty of forming a strong solid to a precise form. Typical properties which have to be evaluated include strength, ductility, weldability, surface stability, formability, etc. The separate determination of each of these property groups has then to be followed up by an assessment of their relative change under the influence of changes in other parameters e. g. the effect of a heat affected weld zone on toughness. By defining what properties are of importance at high temperature the designer can then assess, albeit approximately, what emphasis has to be placed on chemical, physical or mechanical properties.

2.1. Property requirements and applications

It has been suggested that there are three types of property with which an engineer has to concern himself: quantifiable parameters, such as the stress rupture value at a given temperature; desirable but difficult to measure properties such as weldability and oxidation resistance; finally the desirable but unquantifiable properties, which include the rather unusual sounding term of « the forgiveness » of a material i. e. that the component will still function adequately even if fabrication errors have occurred. These three categories represent in order the decreasing ability of the designer to lay down specifications for a given component. Stress typifies a property which can be exactly prescribed and measured.

The design is usually arranged so that there is a uniformity of stress e. g. by altering geometry or operating conditions. As soon as stress levels vary or fluctuate the problem becomes more complicated and the onset of fatigue failure cannot be predicted with any great degree of certainty. Since fatigue is clearly related to surface damage it must be recognised that most of the lifetime of a component is spent in the crack propagation stage. The interaction of creep phenomena with fatigue crack propagation complicates still further the total stress situation, and since microstructural changes are occurring the problem becomes difficult to solve on the basis of present knowledge of materials behaviour. Although stress is a readily calculated parameter, there will inevitably be a scatter of values, related both to testing error and also to a lack of chemical homogeneity in any given component. Segregation produced during solidification leads to different parts of a nominally overall uniform composition ingot having variations in property. This is obviously increased as component size increases and the difficulties are compounded since a decision has to be made on the degree of scatter in properties which can be tolerated. High temperature situations are demanding because failure can bring about dramatic structural collapse and little imagination is required to foresee the result of such failure in for example, aircraft engines. Hence safety margins must be high, reliability is all important and inspection techniques both rigorous and comprehensive have to be employed. Weibull analysis a most useful design tool, which has been used for brittle materials, can play a role in solving this problem.

Properties such as weldability, surface stability, machinability are clearly important for high temperature materials but the establishment of absolute values has not been possible. Ranking schemes are usually adopted and relative comparisons made so that the suitability of two competitive materials can be measured. Interactions of such properties can lead to situations where no obvious solution can be found short of a decision based on experience, since a balance is being struck between parameters which themselves cannot be quantified. e. g. oxidation resistance, as has been clearly shown in a previous paper at this conference by Dr. Whittle. High temperature exposure usually leads to two, well recognised phenomena: the alloy surface is degraded because of reaction between the solid and the gaseous environment, which itself is a consequence of the innate tendency of the alloy to revert to a lower energy state e. g. oxide, sulphide, etc.; the alloy undergoes permanent plastic deformation under a constant stress, the amount of deformation being a function of time i. e. it is said to creep. The resistance of an alloy to both these factors basically determines its ability to function adequately at high temperature. Corrosion or oxidation resistance is vital since the gaseous surroundings of many high temperature components often contain sulphur, chlorine, sodium, which, even though present in small quantities, have profound effects on the surface stability of the alloy. Since this form of attack consumes metallic constituents, it is important

that the component retains adequate load-bearing properties in the reduced section thickness. Clearly, if a thin-walled tube is subjected to hot corrosion, the possibility exists that it can be perforated and design specifications must ensure that materials of high corrosion resistance are used in critical areas. Unfortunately this problem is complicated by three factors: the difficulty of translating laboratory measurements to full scale structures; the difficulty of simulating industrial conditions so that true performance data can be obtained; the large number of variables present in a high temperature corrosion reaction, which makes exact life prediction from simple data an almost impossible task.

Creep phenomena also present a considerable challenge since most of the empirical data obtained by testing cannot be used to predict eventual component life with any degree of exactitude. Until a direct reading indication of the exact state of a component with respect to its overall life is developed, great caution is required in extrapolating test results to longer times, higher temperatures or higher stress levels. This is a very real problem for the designer since the desirable goal of longer life can lead to lower maintenance costs, greater use of plant, higher availability, etc. But how can this be ensured in practice unless a test equal in duration to that of the expected service life is carried out ? No clear answer is available, although there are some indications that work at the National Physical Laboratory (England) to obtain microscopic information on the creep state of an alloy in service, is showing significant promise. The interplay of surface stability and creep resistance leads to the necessity for subjective value judgements, since it is almost inevitable that an improvement in one leads to a deterioration in the other. An excellent illustration of this can be found in the well known fact that, although chromium additions to steel improve oxidation resistance (through forming a protective oxide layer) the creep resistance decreases significantly, in some cases by an order of magnitude.

Having established a simple framework of properties which are essential for satisfactory high temperature service, the ways in which superalloys have been developed to satisfy these demands will be described in detail. An understanding of their basic metallurgy is essential if full advantage of these alloys is to be taken. They represent a significant technical achievement starting from the days of Whittle and the first jet engine to advanced engines powering *jumbo jets*, marine propulsion units, gas and oil pumping units, etc.

As an indication of the economic advantage which can be gained from running power plant at higher and higher temperatures, it should be pointed out that an increase from 900° C to 1,050° C in the inlet temperature of a gas-turbine engine increases the specific thrust by 10 %. If the temperature can reach 1,250° C then the increase is 30 %. Two approaches to this problem have been used: one has relied on the development of alloys with increasingly refractory properties; the other, an engineering design solution, has relied on the use of convection cooling of the components. In reality a combination of both is used, but the demands placed on materials are such that they are used in gas temperatures which are very close to the alloy melting point. The development of new alloys to satisfy these stringent requirements is a remarkable testament to the application of physical metallurgy principles to practical problems. Early gas-turbine engines contained only 10 % by weight of superalloys and operated at temperatures up to 815° C. The latest U. S. engines now contain up to 70 % by weight although UK engines contain significantly lower quantities of superalloy. Since aero-engine technology tends to run well ahead of industrial engine developments, it can be expected that the spectacular improvement in aero engine performance (halving specific fuel consumption; tripling specific thrust values; improving between-service life from 100 hours to 1,200 hours) will eventually be realised in land based units. The importance of energy conservation is receiving increasing attention and one obvious way of achieving this very desirable aim is to make the conversion of fuel oil to electrical power more efficient. In principle there is no reason why aero-engine efficiences of 55 % to 60 % cannot be achieved in industrial engines, in comparison with maximum, present day, levels of 40 %. The problems in many ways are less severe for a land-based unit since power to weight ratios are not as critical as for an aero engine. Furthermore there is far less thermal cycling in a power generating unit and hence thermal fatigue is almost eliminated. Oxidation difficulties are also reduced because thermally induced spalling of the protective oxide layers or of coatings is avoided. By using the superalloys, significant economies should prove possible in industries such as electrical power generation, land-based transport systems (rail and road); gas and oil pumping.

2.2. Physical Metallurgy background of Superalloys

Superalloys are based on the existence of stable, face centred cubic phases (austenite) based on iron, cobalt or nickel and their most useful property is the retention of very high strength levels at large values of the ratio-operating temperature/alloy melting point—known also as the homologous temperature. Equivalent body centred cubic alloys do not have this characteristic.

This property is also accompanied by high modulus values and good diffusivity of secondary alloying elements in the austenitic lattice. A vital factor in the development and retention of elevated temperature, mechanical properties has been the solubility of other elements in the austenitic matrix and the consequent possibility of forming strong intermetallic compounds based on nickel-aluminium (γ'), which are stable to relatively high temperatures.

The various elements present in these superalloys are specifically added to satisfy some service requirement and alloy development has been a series of compositional modifications designed to improve specific properties without adversely affecting the other. Table I shows in schematic form the role played by the various elements and indicates the complexity of the nickel-base alloys, whilst table II is a similar categorisation for cobalt-base alloys.

A schematic resume of the temperature ranges in which these various mechanisms operate and their effect on tensile strength and ductility values is shown in figure 2. Although this representation is not comprehensive in that dispersion effects both of strong fibres and stable particles have been left out, it gives a clear picture of the problems which new materials are called upon to solve.

Superalloys can be regarded as being a logical development step from stainless steels, which contain sufficient chromium to form a protective oxide layer upon exposure to high temperature. As an intermediate step there is a generic family of nickel-iron superalloys based on the formation of a stable austenitic phase through adding at least 25 % of nickel to the iron. The nickel not only stabilises the austenite but also promotes the formation of intermetallic phases containing either titanium or niobium. In this way the iron base superalloys can be used to higher temperatures (with a potential of about 800° C/825° C compared to 1,150° C for nickel base superalloys) possessing good stability but relatively poor oxidation resistance (although recent developments [1, 2] have indicated that iron-base alloys with excellent oxidation resistance properties can be produced for certain aerospace and gas-turbine applications). As a general rule the cost increases with increasing nickel content and certain components destined for high temperature service can be made from

TABLE I. — *Role of elements in nickel-base superalloys.*

Effect	Element						
	Co	Fe	Cr	Mo, W V	Nb, Ta, Ti	Al	C, B, Zr
Solid Solution strengthening. . .	X	X	X	X	X	X	
Intermetallic compounds formation.					X	X	
Grain boundary segregation. . .							X
Carbide formation.			X	X	X		X
Protective oxide formation. . .			X			X	
Typical alloys							
Nimonic 263	20	0.75	20	5.9--	--2.15	0.45	0.06 0.001 0.02
IN 100	15	—	10	3.0-1.0	--4.7	5.5	0.18 0.014 0.06
Hastelloy X	1.5	18.5	21.8	9.0 0.6-	---	—	0.10- -
Rene 41	11	5.0	19.0	10--	--3.1	1.5	0.09 0.002 -

TABLE II. — *Role of elements in cobalt-base superalloys.*

Effect	Ni	Cr	W	Ti, Zr, Nb, Ta	C
Solid Solution strengthening	X	X	X		
Intermetallic compound formation		X	X		
Carbide formation		X		X	X
Protective oxide formation		X			
Typical alloys					
X 40	10	25	7.5	—	0.45
L 605	10	20	15	—	0.10
HS 188	22	22	14	—	0.08

a low nickel content, iron base superalloy rather than the more expensive nickel base alloys. As well as being cheaper the high content iron alloy systems are more deformable and have high melting points and better machinability than the nickel-rich alloys. A particular drawback of the iron-base superalloys is their tendency to form embrittling phases both as a result of transformation of the strengthening Ni_3 (Al, Ti) and Ni_3Nb and also directly from the matrix e. g. Laves phases. Although increasing the application temperature will tend to rule out the wider use of iron-base alloys, they are still by far the largest volume of superalloy in service at temperature up to 800° C. One method of improving their properties at intermediate temperatures would be to develop a cheap precipitation process e. g. by adding manganese instead of nickel, and hence retaining the cost advantage of the iron-base alloy.

Applications, which could utilise such alloys, include small gas-turbine engines for land based transport units

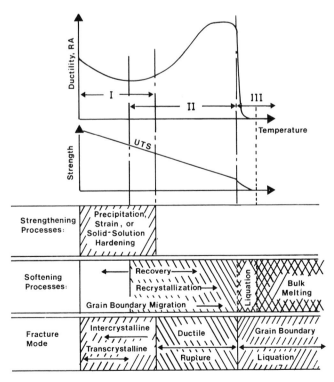

FIG. 2. — *Schematic portrayal of hot strength and ductility factors (Courtesy* of Welding Research Society).

e. g. lorries, buses and possibly automobiles. Powder metallurgy processing routes, which will be discussed later, may also lead to the use of iron base superalloys at much higher temperatures provided that the problem of a lack of tensile strength at intermediate temperatures (600° C to 750° C) can be overcome. This possibility exists because control and variation of the alloy composition will be greater using powders as a starting material in a processing route which uses consolidation by pressure.

2.2.1. Strengthening mechanisms. — The limitations outlined above are to a large extent overcome by both cobalt and nickel-base alloys, although each alloy group has specific advantages which allow them to be specified for a wide range of high temperature applications e. g. energy generation (gas turbines; jet engines; steam turbines; nuclear reactors); transportation systems (land, sea and air propulsion units); petrochemical plant. The largest potential is in gas-turbine engines and the desire to generate and use energy more efficiently will certainly lead to increasing usage despite their high cost. The state of-the-art position of alloy chemistry and alloying processing will now be discussed in detail before briefly reviewing some of the likely developents in superalloy technology. The major strengthening effect is achieved by adding to the austenitic matrix elements which, although they are totally soluble, distort the crystal lattice, so that dislocation motion, which is the prime cause of plastic flow in metals, is prevented. This effect is produced by additions of tungsten, molybdenum, tantalum, etc., although their efficiency is somewhat diminished by their high densities (table III). Cobalt and chromium, although added for other more important reasons, also increase the basic strength of the nickel matrix (as do nickel and chromium in a cobalt matrix). Improvements in mechanical strength at temperature of various grades of nickel-base superalloy have been obtained by increasing the amounts of alloying element as is shown in table IV.

The development of strength by the formation of a stable ordered, coherent intermetallic compound $Ni_3Al(\gamma')$ is of great importance in retaining strength at intermediate temperatures (about 750° C) in nickel-base superalloys. In general about 30 % by volume is formed during heat-treatment. In contrast cobalt-base super-

TABLE III. — *Elements present in superalloys.*

Metal	Melting Point (°C)	Density Kg m$^{-3} \times 10^3$
Iron	1,536	7.85
Nickel	1,455	8.9
Cobalt	1,492	8.71
Chromium	1,850	7.2
Niobium	2,468	8.55
Molybdenum	2,610	10.2
Tungsten	3,410	19.3
Tantalum	2,996	16.6
Aluminium	660	2.70
Titanium	1,668	4.5
Carbon	3,730	2.25

TABLE IV. — *Development of nickel base alloys.*

Alloy Type	Composition (wt%)						Property
	C	Cr	Co	Mo	Al	Ti	100hr stress rupture at 850° C MNm^{-2}
Nimonic 75	0.12	20	—	—	—	0.5	112
80A	0.08	20	—	—	1.5	2.4	130
90	0.10	20	17.5	—	1.6	2.4	172
95	0.12	20	17.5	—	2.0	3.0	228
100	0.20	11	20	5	5.0	1.3	239
105	0.15	15	20	5	4.7	1.2	285
115	0.15	15	15	3.5	5.0	4.0	320

alloys cannot be strengthened in this way because the precipitates dissolve at relatively low temperatures (about 800°/850° C) and they also possess an undesirable morphology. Research efforts to develop an equivalent cobalt-tantalum or cobalt-titanium compound are continuing and it is quite possible that in the future such phases will be produced in commercially available systems.

The properties of Ni$_3$Al are quite unique in that its strength increases with increasing temperature and it also possesses inherent ductility, unlike other strengthening phases such as carbides and borides, and in direct contrast to the severe loss in ductility which occurs when other intermetallic compounds e. g. sigma and Laves, form after heat-treatment or prolonged high temperature exposure.

The strengthening mechanism of γ' particles is now reasonably well understood. When the volume fraction is high and particles are fine in size, the particles are essentially stringed together and can be looked upon as a continuous array so that dislocation cutting of particles can occur. For the case of high strength nickel-base superalloys with a high volume fraction of γ' particles the strengthening theory of Copley and Kear [3] can be applied to explain their strength/temperature properties.

Regardless of γ' morphology, the yield strength of nickel-base superalloy drops fairly abruptly at about 760° C. This finding is consistent with a decrease in the degree of ordering in the γ' particles and an associated sharp decrease in Antiphase Boundary (APB) Energy which is the dominant restraining force to dislocation motion during yielding. Above 760° C the position of the strength-temperature change depends on strain rate, with the change moving to higher temperatures as strain rate is increased. This strain rate dependence strongly indicates that the disordering process in γ' is caused by the intruding dislocation and involves a kinetic process. The strength drop-off in nickel-base superalloys is a consequence of disordering about γ' by vacancy diffusion and a lowering of the APB. The suppression of such a diffusion process is impossible to accomplish by changes in γ' morphology or by processing means. Changes in γ' morphology will not affect the location of the strength drop and it is not affected by processing as results on thermomechanical treatments [4] and controlled solidification [5] of nickel-base superalloys have shown. In addition, recent results have confirmed that the strength loss- temperature of the current high strength nickel-base superalloys is about as high as one can achieve through alloying means.

Grain boundaries have always been the weak links in nickel-base superalloys and whilst they are strengthened by precipitates e. g. intermetallic compounds, carbides, or by elements e. g. boron, zirconium, hafnium, the strengthening effects will tend to be reduced during elevated temperature applications under load. This results from stress induced diffusion between grain boundaries where chemical changes can occur, resulting in the dissolution of strengthening phases at certain grain boundaries and a consequent deleterious loss in strength. By using coherent particles to strengthen a single crystal alloy it may be possible to avoid this adverse effect, but a gradual degradation of strength properties is inevitable in polycrystalline alloys operating at high temperatures under load. This means that the application of polycrystalline nickel-base superalloys is restricted to those temperatures where stress-induced diffusion is strictly limited [6]. Diffusion is thus the limiting factor controlling both the intergranular strength-loss temperature and also intergranular phase instability. This also suggests that improvements in mechanical properties will be obtained if phases that are immune to diffusion effects e. g. insoluble oxides, can be added to γ'-containing alloys. The first exploitation of this effect was in the production of ThO$_2$ dispersoids in nickel with the resultant alloy being known as T. D. Nickel [7] and its most recent success has been the use of high energy, dry, agitation milling of metallic powders with oxides in attritors [8] to produce a new type of superalloy. This process has been colloquially termed mechanical alloying [9] because of the intimate and homogeneous degree of mixing which is obtained by the milling action. This will be discussed later in this paper.

In addition to intermetallic compound strengthening, there is a further beneficial effect on creep resistance resulting from carbide formation in the grain boundaries. Cobalt-base superalloys depend on the formation of carbides for a large proportion of their strength, since intermetallic compound formation leads to severe embrittlement problems, and hence relatively high carbon contents (up to 1 %) are found compared to nickel superalloys (up to 0.2 %). The carbides tend also to form in the cobalt matrix rather than in the grain boundaries as is the case with nickel base superalloys.

The role of carbides is both complex and controversial, since opinion is divided as to whether or not nickel alloys should contain grain boundary carbides. Carbon free alloys show markedly lower creep life and it is now agreed that about 0.03 % to 0.05 % carbon is necessary for adequate rupture strength at high temperatures. The relative stability of the various carbides (from MC to M$_6$C where M is the metal) is a controlling factor in the long-term degeneration of superalloys at high temperatures. The most stable carbide (TiMo)C is a source of carbon for the formation of other carbides, particularly above 980° C, and it has been shown that for very long times (greater than 5,000 hours) the carbon content can decrease to such an extent that MC phase is totally absent. The intermediate carbide M$_{23}$C$_6$ (based on Cr$_{21}$(MoW)C$_6$) has a significant effect on properties, possibly by preventing grain boundary sliding and hence increasing rupture strength.

6

The cobalt alloys are strengthened to a large degree by MC carbides (Ta, Ti, Nb, Zr). These also furnish carbon for the formation of $M_{23}C_6$ phases which, if present as a fine dispersion, can act as a further strengthening phase.

2.2.2. Thermal effects.

— With such a wide variety of potential secondary phases in superalloys the influence of heat-treatment has to be considered in terms of optimising the microstructure; the effect of joining procedures; the long term degradation of the phases during service. Since there are many elements present, the first step is to solution treat the alloy to ensure that a homogeneous solid is formed, free from any segregation present as a consequence of casting. An aging treatment is then carried out at temperatures in the range 750° C to 850° C so as to form γ' and carbides. The amount of γ' which forms is a function of the relative amounts of aluminium and titanium and, as table IV shows for a simple alloy system based on nickel-chromium, the rupture strength is increased by almost a factor of three when aluminium and titanium are added. The presence of iron and cobalt will increase the amount of γ' formed at any given level of titanium and aluminium. The carbides are usually precipitated at grain boundaries during heat-treatment and care has to be taken to ensure that they do not form as a continuous film, since such a morphology will lead to brittleness at ambient temperatures. Optimum mechanical properties are obtained when a uniformly hardened matrix is accompanied by a fine dispersion of carbides in the grain boundary. The complex nature of the possible reactions in superalloys can also lead to grain size increases for both wrought and cast alloys. It should be recognised however that there is a limit to the fineness of the grains in that creep deformation tends to increase for grain diameters of about $1 \mu m$. As will be described later, practical use can be made of this relative ease of deformation of ultra-fine grained alloys so that a very strong material can be superplastically formed at very low loads. Creep deformation occurs in crystalline solids by three processes: grain boundary sliding, Herring-Nabarro or Coble diffusional creep, dislocation creep. If grain boundaries could be completely eliminated then the first two processes would be inoperative and creep resistance markedly improved. By controlling the directional solidification of an alloy so that grain boundaries are reduced or even eliminated, a practical application of this idea has been developed in the last ten years [10].

The presence of boron and zirconium in very small amounts has been found to be essential if high temperature creep resistance is to be obtained, although the exact mechanism, by which this improvement is brought about, is far from being understood. Since creep is essentially a diffusion controlled process, these elements may well interact with vacancies in the grain boundaries and reduce diffusion rates hence increasing creep resistance. A very recent development, which has markedly improved cast nickel-base alloys, has been the discovery that hafnium additions (up to 1.5 %) significantly improve ductility by modifying the carbide morphology.

Since superalloys have to withstand heavy loads for long times at elevated temperatures, components made from them have to be designed so that microstructural changes occurring during operation, which may result in a reduction in property levels, can be taken into account or minimised. Typical of these effects are an increase in precipitate size and a consequent reduction in strength, and also the formation of an embrittling phase (known as sigma) which causes very low ductility at room temperature. Similarly the grain size, which influences both tensile strength and creep strength, can alter with time at high temperatures, particularly when the initial grain size is not uniform throughout the component. Superalloys, which have been forged to final shape, will generally have a more uniform grain size than a cast alloy and at service temperatures lower than 0.5 Tm (where Tm = alloy melting point) forged alloys may prove to be superior to cast alloys. However above 0.5 Tm the as-cast alloys have superior creep properties, although this behaviour is usually attributed to the very high levels of hardening elements, which can be obtained in cast alloys, rather than to a grain size effect.

As mentioned earlier advantage can be taken of the fact that a component, which has most if not all of its grain boundaries parallel to the axis of stressing, will have improved ductility and thermal fatigue resistance plus greater rupture strengths, when compared to the identical alloy which has a random distribution of grain boundaries. The use of directional solidification to achieve this grain morphology will be discussed later when the influence of new processing technology on the properties of superalloys is described.

Cobalt-base alloys are distinguished from nickel-base alloys in several ways. They differ microstructurally because of the absence of hardening phase of γ' type, and they differ in composition having fewer alloying element additions. This latter point leads to little if any heat-treatment procedures being carried out on cobalt alloys, the carbide phases being formed during solidification. Only if recrystallisation or stress-relief are required are wrought alloys heat-treated, whilst cast alloys are sometimes given solution treatment followed by aging to reprecipitate carbides in a finer form, so that rupture strengths are increased. Although cobalt alloys have adequate ductility values at room temperatures especially after long term exposure to high temperatures.

As a consequence of microstructural differences the mechanical properties of nickel and cobalt superalloys vary as a function of temperature and the detailed manner in which this variation occurs dictates the applications which they can fulfill. In general the nickel base alloys have far superior tensile strengths up to 750° C; above this they tend to be of the same order of magnitude since the γ' phase becomes of less importance as it grows and eventually dissolves at high enough temperatures. At temperatures in the range 1,000° C to 1,200° C solid solution strengthening is the only source of mechanical resistance in conventional superalloys and hence nickel and cobalt alloys tend to be controlled more by their melting points with respect to the operating temperature. When rupture strength is considered the major difference depends on the lower rate at which it decreases with cobalt superalloys than for nickel superalloys. Thus at low temperatures the nickel base alloys are more resistant but this advantage is lost above 900° C, and cobalt alloys become competitive on strength grounds alone. As a further consequence of the relatively simple chemical compositions of cobalt alloys, higher melting points are measured than for nickel alloys so that at equal operating temperatures cobalt alloys are operating at lower ratios of service temperature to melting point than are the nickel-base alloys.

An important difference between nickel and cobalt superalloys is related to the superior hot corrosion resistance claimed for cobalt-base alloys in atmospheres containing sulphates, sodium salts, halides, vanadium and lead oxides, all of which can be found in fuel burning systems. Nickel forms a low melting point eutectic with nickel sulphide and hence in sulphur-bearing gases attack of the alloy surface is rapid and drastic. In contrast it is commonly believed that the oxidation resistance of nickel alloys is superior to that of cobalt based systems, because the relatively high aluminium contents of nickel alloys promote the formation of surface films of Al_2O_3 during oxidation in air, whereas the cobalt alloys rely on Cr_2O_3 formation for their high temperature oxidation resistance (in general Al_2O_3 films are more protective than Cr_2O_3 layers). The difficulty in selecting an alloy for high temperature service is compounded by the problem of relating laboratory data obtained by oxidation testing to in-service performance. For protection to be of any value the oxide layers must be adherent and in particular they have to remain adherent during thermal cycling (from high operational temperatures to ambient and *vice versa*). Loss of the oxide by spalling exposes a metallic surface which will then be oxidised in turn, thus consuming metal. This loss of metal can be critical for thin sheet material since perforation may well occur with catastrophic consequences for components such as combustion chambers, cracking chambers, etc. Even if perforation does not occur, oxidation will tend to reduce the load bearing capacity of a component since the metal cross section is reduced. A phenomenon of some consequence to designers is the penetration of the metallic alloy phase by oxides down grain boundaries. This is known as internal oxidation and assessment of this tendency is just as important as knowledge of the overall weight change resulting from oxidation. Recent developments in cobalt-base superalloys have tended to increase oxidation resistance by adding aluminium and silicon, whilst the addition of rare earth elements such as lanthanum [13] and cerium [14], and also yttrium [15] in very small quantities (less than 0.1 %) has been shown to give very marked improvements in the cyclic oxidation resistance of both cobalt and nickel-base alloys. Despite these improvements it has been found necessary to supplement their oxidation and corrosion resistant properties by the use of coatings which usually contain aluminium and chromium and act both as a protector of the alloy surface and also as a barrier to oxygen and other elements present in the atmosphere in which a given component has to operate. The limiting factor for the use of all superalloys is likely to prove to be their chemical resistance (both oxidation and hot-corrosion)

since it has now been shown that adequate mechanical strength can be obtained by adding inert phases, which remain stable up to the melting point of the metallic matrix. Coating technology, no matter how effective it now is or will be in the future, will not allow superalloys to be above 1,200° C for very long times.

The next step must be to use the refractory metals (niobium, tantalum, chromium, molybdenum, tungsten) or refractory, low density ceramic phases such as silicon carbide or nitride, with the distinct possibility that compounds based on silicon, aluminium, oxygen and nitrogen will become available through controlled reaction of powders with gaseous atmospheres. The attraction of the latter step lies in the wide abundance of cheap raw materials. Problems however will certainly arise because of the brittle behaviour of these compounds. The stress situation is simplified by virtue of the assumption that the component operates under elastic constraints. Surface defects or surface damage can markedly increase susceptibility to brittle failure, since these materials lack the ability to accomodate local stresses and strains from any number of sources. This will thus require very stringent analysis of the probability of failure and attempts will have to be made to reduce property scatter to a minimum so that Weibull techniques can be applied [17].

The importance of fatigue can be gauged from an estimate that about 90 % of engineering structures fail by fatigue [18]. Cyclic loads are difficult to predict in service and their effect on materials is dependent on microstructural characteristics and on macrostructural features, both of which are controlled by the processing route used to fabricate any given component. In contrast to low-temperature fatigue, which is characterised by transgranular initiation and propagation, high temperature fatigue usually results in intergranular fracture since slip-band cracking is suppressed as plastic deformation becomes more homogeneous and grain boundary sliding occurs. This problem is rendered more complex for high temperature materials, since there are the additional possibilities that thermal fatigue can occur as a result of non-uniform heating and cooling during thermal cycling and oxidation/hot corrosion phenomena will produce localised intergranular failure at the surface.

Nickel-base superalloys have relatively poor fatigue resistance, as can be seen from a comparison of their endurance ratio (endurance limit to yield strength) of approximately 0.25 with those of pure metals *e. g.* iron, nickel, aluminium with values of 1.0 and iron-base alloys with values varying from 0.5 to 1.0. Fortunately, because of the high yield strength of γ' and its maintenance at high temperatures, nickel base superalloys can still be used in fatigue. Casting defects can also be eliminated by a very novel process (H. I. P.) using a combination of high temperature and the simultaneous application of isostatic pressure [19]. The action is essentially a « healing » one and it is particularly effective for large, poor quality cast components of varying cross-sectional dimension with a consequent beneficial effect on fatigue properties.

Compositional control can also be used to improve fatigue properties. By lowering the carbon content so that the volume fraction of carbides is reduced a significant increase in the endurance ratio can be obtained. The production of a microstructure, which improves high temperature strength and maintains good ductility values, will materially assist situations as the low endurance ratio does not vary greatly with temperature.

Structural defects such as voids, casting porosity, cracked phases, massive heterogeneous grain boundary phases can all act as initiation sites for fatigue cracks. The importance of processing control so as to avoid or limit such defects has been realised and both directional solidification techniques and thermomechanical treatments can be used to achieve this objective. Gas porosity is significantly reduced both in size and density during directional solidification probably as a result of the absence of a dendritic microstructure which normally traps gas bubbles during solidification. Similarly a high degree of mechanical working can break up crack-nucleating brittle phases and close pores.

2.3. Processing

The processing of superalloys has tended to develop in terms of the choice of a certain alloy composition selected so as to provide specific mechanical or chemical properties e. g. a certain creep resistance or ability to withstand hot fuel gases without corroding significantly.

A useful framework within which processing developments can be described and classified, separates liquid phase technologies, involving primary melting and solidification in controlled atmospheres to produce specific microstructures in complex forms, from solid state processing, in which the ascast solid ingot is shaped by applying mechanical forces. Developments of specific processes occur for a variety of reasons and a complete discussion of them is beyond the scope of this paper. As an illustration of this point powder metallurgical technology has arisen in response to segregation problems which occur when a multiphase alloy solidifies. Each powder particle can be looked upon as being a micro-ingot and hence the scale of segregation is greatly reduced (almost to zero in some cases). Since the elimination of chemical heterogeneity is achieved by diffusion, a shortening of the diffusion path leads to shorter times to achieve homogeneity since the time is a power function of the distance. Powder metallurgical products are hence more homogeneous in composition compared with cast alloys and the flexibility of controlled compositional variations is greatly extended.

2.3.1. Processing effects on purity.
— A primary melting and casting operation would ideally produce a clean, uniform chemical composition, impurity and gas-free solid having a specific microstructure characterised by its uniformity, with a high material utilisation factor (low scrap levels) using a reliable, readily controlled inexpensive process. Reality is obviously some way from this ideal condition, but two specific processes: a primary melting in vacuum using induction heating (VIM) [21] and a secondary vacuum remelting opera-

tion [22] using high current low voltage electric arcs between a molten metal and a consumable electrode of the required alloy (VAR) have been developed for, and are widely used in, the production of nickel base superalloys which have to be made to very close compositional limits. As a complement to vacuum arc remelting an air melting process known as electroslag (flux) remelting (EFR) has been developed [23]. It depends for its action on the use of a consumable electrode of the final alloy composition which is submerged in a molten, chemically active, flux resistance heated by an electric current passing between electrode and molten pool beneath the flux (fig. 3) [24]. As the electrode melts droplets of alloy form and they are refined on passing through the flux. The refining rate is rapid because the droplets have a high surface area/volume ratio, resulting in the removal of most of the impurities and a significant reduction in inclusion content. This improvement in alloy chemistry is reflected in superior hot-workability and in improved mechanical properties compared to vacuum arc processed material.

Other melting techniques using electron beams (EBM) and plasma arcs (PAR) as heating sources have also been developed, especially for alloys where very strict impurity limits have been specified; but it is still too early to predict the future development of these novel technologies. Table VI summarises the various possible effects of some of these melting and refining processes.

The superalloys can be processed directly to shape via a casting route e. g. « lost-wax » investment methods or by shaping the cast ingot through the use of mechanical force e. g. rolling, forging or extrusion, the choice depending amongst other factors on the shape of the final component. Early compositions of nickel-base

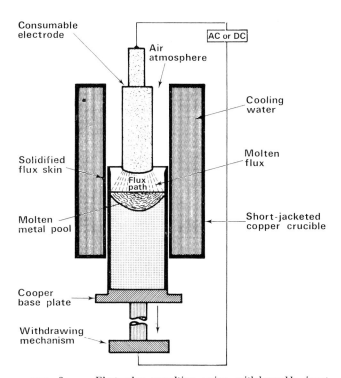

FIG. 3. — *Electroslag remelting using withdrawable ingot mechanism.*

alloys were developed so as to overcome the drawback of the relatively poor creep resistance of stainless steels, and the Nimonic series were all used in the forged condition mainly to make turbine blades and combustion chamber sheet. As strength requirements increased, alloy compositions were developed so as to satisfy this need, but the very fact that high temperature strength is produced leads inevitably to problems in hot working the alloy to shape. This difficulty can be overcome by using casting technology and hence producing very strong alloys, which can serve at higher temperatures than their forged counterparts. Processing developments have now taken over from compositional developments as the major route by which improved superalloys can be made and most of the present-day technology controls the properties of the final product through process variations, selected so as to give specific microstructures in complex shaped objects. This is an important point to recognise, since components such as turbine blades, etc. have their form controlled by engineering requirements e. g. aero-dynamic profile.

The pace of technical advance in the processing of superalloys can be measured in terms of a 1940 baseline when they were all without exception air melted, compared with 1976 when vacuum melting; electro-slag refining; directional solidification; superplastic forging; hot isostatic pressing of powders; dispersion strengthening are all in practical service as different parts of the overall processing route, whether it be through the liquid-solid transformation or solid-solid transformations.

The importance of these processing developments with respect to rupture strength is shown in figure 4 which demonstrates how the performance and temperature capability of gas-turbine blades have been improved in the last thirty-five years. The specific processes are detailed at the appropriate point on the graph.

Quality is greatly improved if impurities such as oxygen and nitrogen can be removed or at least minimised since they form inclusions which have a deleterious effect on mechanical properties such as fracture toughness. Vacuum melting and casting is able to achieve this result and spectacular increases in stress-rupture properties have been obtained in superalloys treated in this way, the effect being attributable to grain size and segregation control. Similarly investment casting, which had been carried out in air, is able to use the vacuum melted charge alloy and can also be performed in vacuum chambers (as the mould is easier to fill in the absence of air). Closely controlled compositions and lower metal pouring temperatures, grain size control to be achieved so that higher strength castings are possible. Recent developments in turbine technology have led to the use of internally cooled components and these intricate cooling passages can be made by using complex ceramic cores in an investment mould.

The production of these internal passages in high strength nickel-base alloys has been made possible by the application of electrochemical machining techniques. The external surfaces of components such as blades can be machined by conventional, well-established operations such as grinding and milling and also by electrochemical techniques. A complex air cooled blade can thus be made out of a solid blank although the cost

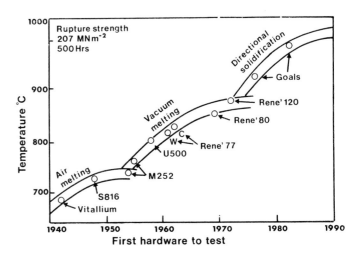

FIG. 4. — *Progress in turbine blade materials.*

of such an operation will be very high and the metal utilisation ratio will be very low leading to large amounts of discard material. The comparative rates of typical machining processes is shown in table V.

TABLE V

Machining Process	Superalloy Metal Removal Rate (in³/min)
Grinding	12
Turning	3
Milling	3
Electrochemical machining . .	1
Electrical discharge machining .	0.1

This represents a major advance since a higher operating temperature of the gas can be used without changing the alloy composition and the designer has been able to retain a material which has been fully characterised and with which he is thoroughly familiar.

2.3.2. **Processing effects on microstructure.** — An analysis of the mechanism, by which creep-rupture failure occurs, has revealed the importance of grain boundaries which are normal to the stress axis. By controlling the solidification process so that a steep temperature gradient is obtained and the solid/liquid interface moves in the direction of heat flow, these boundaries can be eliminated. This is now being exploited commercially for nickel base alloys (MAR-M200), with improvements in creep rupture strength at 982° C of about two orders of magnitude compared to a conventionally cast alloy (fig. 5). The ultimate limit of this technique is the production of a single crystal component which has no grain boundaries at all, and hence the elimination of grain boundary segregation. Practical exploitation of this has been demonstrated but the economic justification for the expense involved has not been fully established since there is a cost penalty of four times the conventional casting costs for uni-

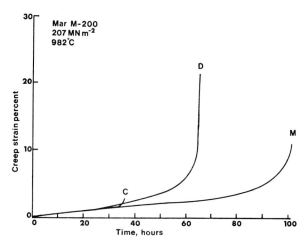

FIG. 5. — *Comparison of the creep properties at 982° C of conventional (C), directional (D), and monocrystal (M) mar-M200.*

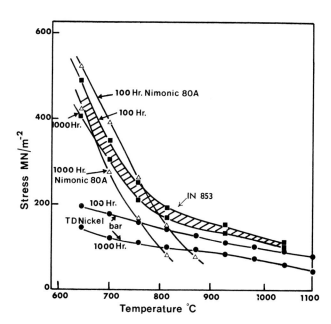

FIG. 6. — *100 hours and 1,000 hours rupture stress for dispersion strengthened superalloy compared with data for nimonic 80 A and TD nickel.*

directionally solidified alloys [25]. The need to improve strength at high temperature has also been partly fulfilled by processing developments which have taken two entirely different routes. As was pointed out earlier the strengthening phases in all superalloys dissolve at high temperatures leaving only solid solution strengthening, which is controlled by relatively heavy elements such as tungsten, molybdenum, tantalum, etc. What is needed is an inert dispersed phase which will be effective up to the alloy melting point. Prior to the introduction of powder metallurgy techniques this was not possible by conventional casting or forging routes and probably the best known material (and least used) utilising such an approach was thoria dispersed nickel (known as T. D. nickel), in which 2 % by volume of thoria was added by a rather complicated process. This alloy has excellent high temperature strength but is weak at low temperatures and also has little, if any, oxidation resistance. Precipitation hardening by Ni_3(Al, Ti) cannot be carried out by this method so that the development of an agitation milling process (using powders) known colloquially as « mechanical alloying », which produces both inert dispersoids and precipitation hardening, represents a major advance for superalloys [9]. A commercially available alloy IN853 with the strength of Nimonic 80A below 815° C and that of TD nickel above 815° C is now made in a variety of forms and may well be the pre-cursor for a new family of superalloys.

The importance of this development is clearly shown in figure 6. Similar results have been obtained for cobalt-base alloys and if the promise of the latter group is fulfilled a significant step will have been made in the production of more economic grades of superalloy. A specific and most important feature of these alloys is the need for an elongated grain structure with a high aspect-ratio if good high temperature properties are to be obtained [26]. Recent evidence has underlined the high degree of process control which is required for the production of this microstructure. A high temperature heat-treatment is needed plus a critical amount of « stored energy » which is induced by the effect of the dispersoid particles on the thermomechanical

treatment used to consolidate and shape the material. Since the milling process used to mix the metallic and non-metallic powders also « stores » energy in the powders, the subsequent heat-treatment requirements have to be clearly identified as a function of the milling time and the properties of the powder particles. A novel technique for producing this grain structure has been developed [27] using a solid state crystal growth process, known more colloquially as ZAP (zone aligned polycrystals). The non-recrystallised alloy is passed at a specific rate through a heated zone in which there is a steep temperature gradient. The alloy emerges from the temperature gradient with a strongly oriented recrystallised structure consisting of grains with aspect ratios of up to 15:1. In certain cases single crystals have been made. The stress-rupture properties at 1,100° C were doubled by this treatment. The use of powder metallurgy processes was initially introduced so as to overcome some of the segregation problems which are almost inevitable in large ingots of alloys which contain several elements. Each pre-alloyed powder particle is effectively a « micro-ingot » and hence upon consolidation by extrusion of forging a fully dense solid of uniform composition can be obtained. The design advantages of such powder processed alloys are significant: alloy melting temperature is increased; greatly improved hot workability; fine grain size which gives both strength and ductility at temperatures below about 900° C. Economic advantages which will undoubtedly arise from this approach include a better utilisation of material, since the use of a preform which has been compacted approximately to the forging shape will reduce forging costs, cut out expensive machining costs and conserve material. This is an important point since the material cost of a superalloy component is a very small fraction of the overall cost (an approximate figure for a 1 kg turbine blade costing £ 100 is £ 0.5 in material cost).

The fine-grained state of powder alloys not only gives improved properties but it can also be made use of during processing. At temperatures of the order of half the alloy melting point, stable fine grained (less than 10 μm) alloy can be strained slowly at very low loads and superplastic flow is said to occur. A commercial process for making turbine discs out of a nickel-base alloy IN-100, which is unforgeable by conventional routes, has been developed with the curious name of « gatorising ».

The pre-alloyed powders, prepared either by atomisation or rotating electrode processes, are consolidated using hot isostatic pressure technique or hot extrusion. This results in a very fine grained, homogeneous, microstructure which, when strained at rates of 10^{-2} min^{-1} at temperatures in the range 900° C to 1000 %, shows superplastic behaviour i. e. high elongations at low stresses. This can also be produced by compressive deformation in a forging process and by careful selection of the die material (molybdenum can be used) and design of the die shape, complex shapes can be made in single stage process. This is an interesting development, not only for its exploitation of unusual plastic properties but also because it illustrates the potential of powder routes for the direct production of shapes using alloys of closely controlled composition and conserving expensive material. The attraction of powder technology is further underlined by the possibility of designing new alloy compositions for powder processing routes and powder components. In this way specific properties such as corrosion-resistance, creep rupture strength, etc., can be obtained, even though they would be difficult, if not impossible, to produce in castings or by conventional forming processes. Dispersed non-metallic phases

can be added without any danger of producing large deleterious inclusions and there is evidence that high volume fractions of such phases can be added in this way [29]. Large components have been made in large quantities using this technique and it will certainly figure in the future development of powder routes which will conserve materials and energy as well as giving improved properties.

2.3.3. **Processing for composites.** — The use of strong, stiff fibres to reinforce weak metallic or plastic solids is now well-known and composite materials are widely used in engineering design especially for resins containing glass fibres. High temperature materials which could utilise this principle would be of enormous benefit but attempts by conventional routes e. g. putting strong metallic fibres (W) into superalloy matrices, have not been marked with success, because of reaction between fibre and matrix. Ceramic fibres have also been used but although the high temperature strength properties are outstanding, the oxidation resistance of such metallic matrix-ceramic fibre composites is poor [30], mainly because of the high surface area of metallic phase which is exposed to the oxygen and nitrogen, which can either diffuse in via the ceramic phase of via the ceramic-metallic interface. The production of oxide phases at these interfaces disrupts the mechanical rigidity of the composite and they are thus only likely to be useful if they can be used in inert atmospheres or if they can be protected by reliable coatings.

A far more attractive method with the potential of commercial application utilises the fact that liquid

TABLE VI. — *Comparison of the metallurgical possibilities of melting/refining processes for high temperature alloys.*

	Primary Melting Processes					Secondary Remelting Processes			
	AM	AM+VR	AM+AOD	VIM	PAM	VAR	EFR	EBM	PAR
Alloy Flexibility	−	+	+	++	++	++	+	+	++
Alloying	++	++	++	++	++	−	−	−	−
Superheating	+	+	+	++	++	−	−	++	++
Refractory Interaction	−	−	−	−	(−−)	++	++	++	++
Slag (Flux) Treatment	+	+	++	−	+	−−	++	−−	+
Composition Control	+	+	+	++	++	++	+	+	+
Removal of Gases	−−	+	+	++	(+)	++	−−	++	+
(Carbon) Deoxidation	−	++	+	++	(+)	+	−	++	(+)
Desulphurization	++	+	++	−	+	−	++	−	+
Decarburization	−	++	++	++	(+)	+	−	++	(+)
Volatilization (Impurities)	−−	+	+	++	(−)	+	−−	++	(+)
Microcleanliness	−−	+	+	++	++	++	+	++	++
Solidification Control	−−	−−	−−	−−	−−	+	+	++	++

Rating: + Good   − Poor   ( ) Probable effect
          ++ Better   −− Poorer

alloys undergoing a eutectic reaction can be solidified uni-directionally so that one phase can be caused to grow with a fibrous or lamellar morphology [31]. Because of the formation of fibres from the liquid state they are of low interfacial energy with respect to the matrix and hence are stable. An examination of the properties of the metallic elements shows that there are very few candidates for service at temperatures greater than 1,100° C and for this reason research is now being carried out not only on nickel-, iron- and cobalt-base matrices in *in situ* composites but also on intermetallic compounds which although brittle have much higher melting points [32]. As well as experimental investigations of potentially useful eutectic systems, there is now a considerable effort being made to calculate the range of existence of eutectic systems in multi-component alloy systems using a thermodynamic approach so that more rapid solutions may be found without unnecessary experimental work.

The mechanical properties of eutectic composites are controlled by the volume fraction of the reinforcing phase and the distance between the phases (in a manner analogous to that of the effect of grain size on yield strength as described by a Hall-Petch equation). The volume fraction is in turn controlled by the surface energy between the two solid phases and analysis [33] shows that a rod form occurs when the volume fraction is less than 32 % and a lamellar form at higher values. The fibres can withstand a higher elastic strain prior to fracture and are thus more efficient than lamellae. Against this however must be weighed the fact that fibres only form at low volume fractions and hence the overall composite strength will be lower than those measured in lamellae-reinforced composites. Stability considerations also tend to favour lamellae rather than fibres since they may withstand more severe thermal gradients and thermal cycling conditions.

Various methods have been employed to vary the

TABLE VII. — *Categories of* in situ *composites*
(*Lam* = Lamellar; *Fiber* = Fibre *and* M = Metallic component).

Ductile-Ductile

| Alloy | Second phase | | | Melting Point, °C |
	Composition	Form	v/o	
Ni-W	W	Fiber	6	1500
Ni,Co,Al-Cr,W	Cr,W	Lam		
Ni$_3$Al-Mo	Mo	Fiber	26	1306
Ni$_3$Al-Ni$_7$Zr$_2$	Ni$_7$Zr$_2$	Lam	42	1192

Ductile-Semiductile

γ-δ				
Ni-Ni$_3$Nb	Ni$_3$Nb,δ	Lam	26	1270
γ'-δ				
Ni$_3$Al-Ni$_3$Nb	Ni$_3$Nb,δ	Lam	44	1280
Hypoeutectic			32	1280
γ/γ'-δ				
Ni,Al-Ni$_3$Nb	Ni$_3$Nb,δ	Lam	∿35	1272 to 1274
Ni,Cr,Al-Ni$_3$Nb	Ni$_3$Nb,δ	Lam	∿33	1244 to 1257

Ductile-Brittle

Co-Co$_3$Nb	Co$_3$Nb	Lam	---	1270
Ni-Cr	Cr	Lam	23	1345
Ni-NiBe	NiBe	Lam	38 to 40	1157
Ni-Ni$_3$Ti	Ni$_3$Ti	Lam	29	∿1300
M-MC				
Co-TaC	TaC	Fiber	16	1402
Ni-TaC	TaC	Fiber	∿10	----
M-M$_7$C$_3$				
(Co,Cr)-(Cr,Co)$_7$C$_3$	(Cr,Co)$_7$C$_3$	Fiber	30	1300
(Ni,Cr)-Cr$_7$C$_3$	Cr$_7$C$_3$	Fiber	30	1305

Table VII (continued).

Brittle-Ductile

Alloy	Second phase			Melting Point, °C
	Composition	Form	v/o	
MO-M				
$(ZrO_2, Y_2O_3)-W$	W	Fiber	∿6 w/o	
$(HfO_2, Y_2O_3)-W$	W		∿12 w/o	
MgO-W	W			
Cr_2O_3-Re	Re			
-Mo	Mo		8.6	1760
-W	W			
Ni_3Ta-Ni_3Al	Ni_3Al	Fiber	35	∿1360
CoAl-Co	Co	Fiber	35	∿1400

Brittle-Brittle

Alloy	Second phase			Melting Point, °C
	Composition	Form	v/o	
$Al_2O_3-ZrO_2$		Lam		1870
$Al_2O_3-(ZrO_2, Y_2O_3)$		Fiber		1890
$Al_6Si_2O_{13}-Al_2O_3$				1840
NiO-CaO		Lam		1720
$MgAl_2O_4-MgO_2$	MgO	Fiber	>50 w/o	1995
$ZrO_2-Y_2O_3$	ZrO_2	Lam	6.38 w/o	
Fe,Cr-Fe,Cr,Nb	Fe,Cr,Nb	Fiber	22•1	1275±10

volume fraction of the phases [34] present in eutectics e. g. coupled growth of off-eutectic systems: monovariant eutectics: multi-variant and multi-component systems. In this way considerable compositional latitude is obtained and a satisfactory balance struck between properties such as high temperature strength and high temperature oxidation resistance, etc. Table VII [34] illustrates some of the compositions and categories (based on mechanical properties) of some *in situ* composites studied to data.

The equipment used to make these eutectics essentially requires a controlled solidification in a high temperature gradient (G) at a sufficiently high speed (R) for economic production routes, without losing plane-front solidification conditions and breakdown to a colony structure.

By utilising liquid metal cooling to improve the thermal gradient and hence increase the ratio of G/R faster production rates are possible than for simple radiation cooling in air.

In order that these materials should be cost effective it is important that the cycle time of a unit should be as short as possible but compatible with satisfactory structures being produced in a reproducible manner [35]. This means that high values of G will be required if the value of R is to be high i. e. fast processing rates. The gradient can be increased either by more efficient cooling e. g. liquid metal coolants or by increasing the liquid alloy temperatures. This latter possibility

will create problems with the refractories and particularly with cores used to make hollow parts.

Two types of eutectic system with commercial potential have been developed in recent years: one based on the $Ni_3Al(\gamma')Ni_3Nb$ (δ) lamellar system in which the Ni_3Nb is the reinforcing phase [36]; the other using carbide fibres e. g. TaC in both cobalt-base and nickel-base alloys [37]. Their rupture properties are superior to dispersion-strengthened and precipitation hardened alloys as can be seen in figure 7.

As well as economic problems there are potentially more serious problems related to the thermal stability of eutectic systems with respect to thermal cycling and to temperature gradients, which may be set up in a cooled component [38]. Degradation of the structure and loss of properties can occur and there is a great deal of effort being injected into research programmes so that the response of *in situ* composites to these constraints can be assessed. It will also be essential for rigorous quality assurance procedures to be established, if reliance on these new composite materials is to be ensured for critical applications.

The oxidation behaviour of the composites developed to date have been dependent on the amount of chromium and aluminium in the alloy and the importance of an adequate level of protection at 1,000° C is shown in figure 8. It is almost certain however that some form of protective coating will be required for the first generation of *in situ* composites.

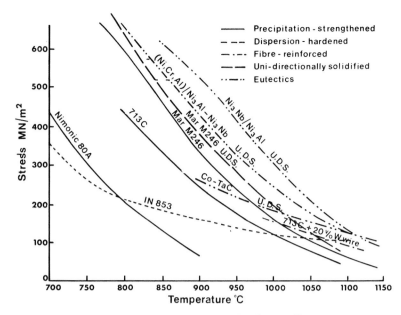

FIG. 7. — *100 hour rupture strengths of superalloys.*

At the time of writing trial quantities of nickel and cobalt-base alloys reinforced with carbides and intermetallic phases are being assessed and they may play a significant role in gas-turbine technology.

All of these process developments have taken place within the last twenty years and although they have occurred mainly in the field of aerospace technology, their impact outside this field may well be more extensive in the future. All of them cost money and the potential benefits, if such alloys are to be specified and hence prepared by these relatively novel processes, have to be assessed.

2.3.4. Fabrication problems.

— Apart from these primary processing developments what else can the designer expect from these materials? More efficient utilisation of material is essential if greater economy is

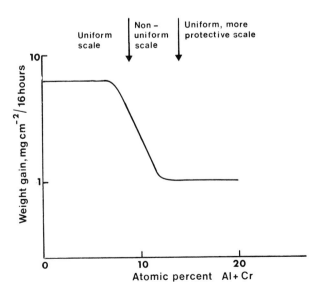

FIG. 8. — *Oxidation of composites at 1,000° C as function of chromium and aluminium content.*

to be achieved and this involves processes which do not waste the alloys early in their forming. For this reason powder routes will certainly increase in importance as will the use of hot isostatic compaction to produce a shape as close as possible to the desired final form. An idea of the degree of utilisation of superalloy material during processing can be gauged from a figure published for aeroengines: out of 25,000 kg of starting material only 3,000 kg are found in the final product, taking into account the primary casting yield, forming and machining losses. The cost of machining is by far the most expensive item in the cost accounts for using superalloys. Machining is a difficult process since by definition superalloys are strong and tough and they can only be cut at speeds of about 5 % to 10 % of those used for common steels, and tool bit changes have to be frequently made [39]. As heat-treatment processes further improve strength levels, machinability deteriorates still further (machining in the un-heat-treated state is not possible because of the difficulty of maintaining dimensional tolerances) and with the trend to even higher performance levels this property will become an object of further attention from machine tool engineers. The most widely used processes depend on chip-making operations since they remove more metal at higher rates than un-conventional methods such as electrochemical machining. As can be seen from table V grinding is the fastest process and involves very high energy/unit volume of material removed (about thirty times more than for turning). Since the superalloys cannot be modified so as to render them more machineable (adding lead or a sulphide phase as is done for steels, is not practicable because of the high temperature service for which have been designed) two possibilities are available: develop new cutting tool materials or develop new processes, which will make the final object in a single operation. The cobalt-containing high speed steels have to be used for difficult machining operations e. g. broaching, drilling, as the superalloys are still hard when the conventional high speed steels start to soften *i. e.* about 600º C. The carbides are used for finishing operations such as face milling. Non-conventional processes have been developed for superalloy machining for several reasons: greater economy can be obtained; the workpiece does not degrade during machining (surface work hardening will occur with normal processes); unique shape requirements. The common feature of processes such as chemical machining, electropolishing, discharge machining, etc., is the fact that metal removed is independent of the mechanical properties of the superalloy. Developments at present being evaluated include the use of ultra fine-grained carbides in conventional compositions and newly designed compositions in which the cutting phase and matrix are selected and processed by powder routes. If inert oxides and nitrides with high hot hardness levels can be incorporated into a refractory metal binder then a revolutionary advance in cutting behaviour may well occur.

The drive to conserve energy, materials and hence reduce financial burdens will lead to more intensive effort in this area of final processing of a component.

The joining of superalloy components is not without problems, especially when the dispersion strengthened alloys are considered, since fusion techniques cannot be used as they destroy the alloy properties in the weld zone. Although welding has many drawbacks *e. g.* weld zone cracking; local mechanical property changes in the weld zone, in particular reduction in the ductility levels and fatigue limit; reduction in corrosion and oxidation resistance because of local compositional changes; it is still the most widely used fabrication technique for superalloys. Several processes are used including gas-tungten arc, resistance and electron beam and although they differ in detail their application always involves the minimisation of the heat input to the weld zone, so that cracking tendencies are reduced. The advantages of welding techniques are related to their low cost, low weight and retention of high strength in the weld zone and for these reasons welding will remain the most important joining method for superalloys, despite some of the drawbacks mentioned above.

3. REFRACTORY METALS

The superalloys have unique properties, which have, been vital to the development of high temperature processes. By controlling composition and developing fabrication routes, which make more economic use of the alloys, a wide range of applications have been.

The future will certainly see the further development of processes with as few steps as possible between raw material and final finished component, but the ultimate limit of superalloys is related to their melting point and to their ability to resist atmospheric attack above 1,200º C. Above this range two avenues can be explored: the use of refractory metals and the application of ceramic phases, and it is the former category which will be briefly analysed before considering some of the future developments which may occur in high temperature materials. Table VIII shows how the refractory

alloys compare with superalloys, provided that their surfaces are protected from oxidative degradation. Their main advantages include much higher melting points, high moduli of elasticity and hence an ability to withstand much higher loads than superalloys.

TABLE VIII. — *U. T. S.* (MNm^{-2}) *at temperature.*

Alloy	1,100º C	1,320º C	1,540º C	1,760º C
Superalloys . . .	245/350	—	—	—
Niobium alloys . .	350	168	119	—
Molybdenum alloys.	630	385	252	182
Tantalum alloys. .	560	364	210	105
Tungsten alloys. .	700	420	280	210
Chromium alloys .	315	119	—	—

Unfortunately with the exception of chromium they have very high densities, which makes their use for rotating parts difficult to envisage particularly for tungsten and tantalum. In contrast to the superalloys the refractory metals have b. c. c. structures and hence have a ductile to brittle fracture transition at a specific temperature which is in excess of ambient. This presents an immediate problem to the engineer who has to design with a potentially brittle material even though they have adequate high temperature ductility values. Furthermore they tend to be embrittled by interstitial elements such as oxygen, carbon and nitrogen, which poses serious problems in obtaining high purity materials for commercially viable processing routes.

Chromium-base alloys are of potential interest since they have good oxidation resistance because of the formation of a protective Cr_2O_3 scale in a manner similar to that of the superalloys. They can be solid-solution strengthened with other refractory elements and further high temperature strength can be obtained by using dispersed phases such as carbides or nitrides. The most telling factor against chromium alloys is the effect of nitrogen on ductility. Amounts of nitrogen as low as 0.01 wt % can raise the temperature by several hundred degrees centigrade and despite several efforts to use rare earth elements to fix the nitrogen, it has not yet proved possible to protect chromium from embrittlement when exposed to air and thus give adequate design ductility. Unless a dramatic new development occurs it is rather unlikely that chromium alloys will be used in power generation units.

Molybdenum alloys were initially strengthened by dispersion of TiC and ZrC the alloys being known as TZM and TZC, but the realisation that HfC was more efficient and gave improved ductility values (10 % tensile elongation) in the swaged condition led to an increasing interest in Mo — Hf — C alloys. As with superalloys, significant improvements in high temperature strength have been obtained by using thermomechanical processing developments. By extruding the alloys at high temperatures (2,000º C) and then swaging at lower temperatures (1,400º C) an increase of 40 % can be obtained in the tensile strength (as measured at 1,350º C) to levels as high as 600 MNm^{-2}. In a similar manner, the tensile properties of tungsten can be improved although there are indications that minor compositional variations can lead to rapid coarsening of the HfC phase and loss of creep properties.

Despite these advances in strengthening (provided that oxidation is prevented) by processing developments, the poor low-temperature ductility of these alloys prevents them from being used on a wider scale. Purity levels have to be high *i. e.* low interstitial contents. Inert dispersoids are needed if a stable fine-grained microstructure is to be produced, and an expensive element, rhenium, has to be added if the so called « ductilising » effect is to be obtained.

A recent paper [40] has analysed the respective merits and demerits of the refractory metals and their alloys and it is suggested that there is still some realistic hope of improving their high-temperature strength and perhaps their ductility values. The most striking improvement is to be found in chromium alloys containing Ta, B and C which have strengths at 0.65 Tm (Tm = melting point) greater by a factor of 45 than that of high purity chromium. By optimising thermomechanical processing routes so that the strong dispersed phases are maintained in a fine and stable state, it has been suggested that further strength increases will become possible. This still leaves the thorny problem of ductility and it is evident that the effect of elements such as rhenium, iron, cobalt, etc., is worth exploring in greater depth particularly in terms of the possibility that the effect is related to the number of « s and d » electrons in the solute. All of this will come to nothing if the components are exposed to an oxidising environment, so applications will have to be limited to inert or reducing environments *e. g.* helium-cycle turbines or totally reliable coatings must be developed.

4. FUTURE TRENDS AND PROBLEMS

From a discussion of what has been achieved to the uncertain aspects of future trends presents certain difficulties, not the least of which is the reasonable likelihood that one will be proved wrong by future events. If the future is to be looked at in terms of both future demands and solutions to present problems then a sensible analysis can be presented.

Superalloys will most certainly be limited by their relatively low melting points, compared with the refractory alloys, to operational temperatures no greater than 1,200° C/1,250° C. They present problems of phase stability in isothermal and thermal-cycling conditions, which are likely to be all the more complex as long-term, maintenance-free, operation is required for high temperature components. This thus suggests that a useful line of research should investigate phase stability problems both in conventional alloys and in the newer directionally solidified eutectics and powder metallurgical products.

The problem of stability will involve diffusion information, phase analysis and the further development of phase computational techniques, *e. g.* (Phacomp) [41] based on the concept that bonding in electron compounds is promoted by electron vacancies in the 3d sub shell of transition metals. The technique has been used to predict the likelihood of certain embrittling phases being formed, based on an analysis of the composition of the matrix after major phases have been taken into account. The physical chemistry of multi-component, multi-phase alloys is not well understood and improvements in theory could well prove of great value. The interaction of fatigue and creep phenomena also requires further study and this should be carried out in conjunction with phase stability investigations so that the effect of microstructural changes on these failure mechanisms can be fully assessed. In addition to phase stability, the surface stability of materials must be studied in greater depth, for apart from some elementary guide lines, knowledge of the behaviour of complex, multi phase surface layers (formed either by oxidation or hot corrosion) is not very profound. In particular problems of oxide adhesion, hot-corrosion attack and the development of protective coatings with long lifetimes are worthy of attention. This applies with equal force both to superalloys and to refractory metals.

Processing developments have been at the centre of most of the recent technical improvements in power generation materials and it is in this area that further progress is likely to be made. By using both liquid phase processes and powder metallurgical routes it should be possible to design new alloys which have their compositions selected on the basis of a desired specific microstructure necessary for specific properties to be obtained.

The development of laser technology and its application to fabrication techniques *e. g.* cutting and welding and to surface alloying and heat-treatment will certainly be beneficial to high temperature materials. Their potential is significant and provided that multidiscipline programmes to develop systems can be successfully executed, applications will be found.

Similarly, processing developments can bring about the application of casting techniques to the refractory alloys which are fabricated at present by forging methods. The ductility problem of the refractory metals still requires an enduring solution which will allow their excellent high temperature strength to be usefully employed.

Protective coatings, if they can be relied upon, will also improve the chances for the practical exploitation of these alloys. These major problems represent a challenge to both the physical metallurgist and the engineer.

If this topic is reviewed at the 10th annual Petten Colloquium, then an assessment of the degree of success, which has been obtained, can be made. Or rather will we be assessing the advances made in producing ceramic engines as part of the next major advance in high temperature materials?

*
* *

REFERENCES

[1] ALLEN (R. E.). — *Procs.* of the 2nd International Conference on Superalloys Processing, XI-X10, *M. C. I. C.*, Ohio. 97°.

[2] DESFORGES (C. D.), LAWN (R. E.), WILSON (F. G.). — *Fulmer Research Institute* (U. K.), 1975 *(unpublished results)*.

[3] COPLEY (S. M.), KEAR (B. H.). — *Trans. A. I. M. E.*, 1967, *239*, 984.

[4] OBLAK (J. M.), OWCZARSKI (W. A.). — *Met. Trans.*, 1972, *3*, 617.

[5] TIEN (J. K.), GAMBLE (R. P.). — *Mater. Sci. Engineer.*, 1971, *8*, 152.

[6] TIEN (J. K.). — *Procs.* of the 2nd Int. Conf. on Superalloys Processing, W1-W12, *M. C. I. C.*, Ohio, 1972.

[7] ANDERS (F. J.), ALEXANDER (G. B.), WARTEL (W. S.). — *Met. Progr.*, 1962, 88.

[8] WADHAM (H.). — *J. Oil Col. Chem. Ass.*, 1964, 728.

[9] BENJAMIN (J. S.), CAIRNS (R. L.). — Modern Developments in Powder Metallurgy 5, Procs. of the Int. Powder Metallurgy *Conf.*, 47, *J. Hausner*, New York, 1970.

[10] VERSNYDER (F. L.), SHANK (M. E.). — *Mater. Sci. Engineer.*, 1970, *6*, 213.

[11] VARIN (J. D.). — *The Superalloys*, 1972, *8*, 231-257 (ed. by C. T. SIMS and W. C. HAGEL), *J. Wiley*, New York.

[12] VERSNYDER (F. L.), GUARD (R. W.). — *Trans. Amer. Soc. Metals*, 1960, *52*, 485.

[13] HATWELL (H.), DESFORGES (C. D.), MOENTACK (P. L.). — *U. S. Patent 3, 591, 371*, July 6th, Oxidation-resistant cobalt-base alloys. 1971,

[14] RICHARDS (E. G.), TWIGG (P. L.). — *Mem. Sci. Rev. Met.*, 1971, *68*, 167-177.

[15] FELTEN (E. J.). — *J. Electrochem. Soc.*, 1961, *108*, 490-495.

[16] JACKSON (C. M.), HALL (A. M.). — *NASA* TMX-53448, 1966.

[17] DAVIES (D. G. S.). — *Proc. Brit. Ceram. Soc.*, 1973, *22*, 429-452.

[18] DIETER (G.). — *Mechanic. Metallurgy*, 1961, 296.

[19] WASIELEWSKI (G. E.), LINDBLAD (N. R.). — *Procs.* of the 2nd Int. *Conf.* on Superalloys, 1972. Processing D1-D24 *M. C. I. C.*, Ohio.

[20] JAHNKE (J. P.), BRUCH (C. A.), 1974, Agard-CP-156 (ed. by E. R. THOMPSON and P. R. SAHM).

[21] DARMARA (F. N.). — *J. Met.*, 1967, *19*, 42-48.

[22] PETER (W.), SPITZER (H.). — *Stahl und Eisen*, 1966, *86*, 1383-1393.

[23] DUCKWORTH (W. E.), HOYLE (G.). — Electroslag Refining Chapman and Hall Ltd., London, 1969.

[24] KLEIN (H. J.), PRIDGEON (J. W.). — *Procs* of the 2nd Int. *Conf.* on Superalloys. Processing, B1-B26, *M. C. I. C.*, Ohio, 1972.

[25] COLE (G. S.), CREMISCO (R. S.). — The *Superalloys*, 1972, *17*, 479-508 (ed. by C. T. SIMS and W. C. HAGEL) 1972, *J. Wiley*, New York.

[26] WILCOX (B. A.), CLAUER (A. A.). — *Acta Met.*, 1972, *20*, 743.

[27] ALLEN (R. E.). — *Procs.* of the 2nd Int. *Conf.* on Superalloys Processing, X1-X10. *M. C. I. C.*, Ohio, 1972.

[28] MOORE (J. B.), TEQUESTA (J.), ATHEY (R. L.). — *U. S. Patent 3, 519, 503*, July 7th, Fabrication method for the high temperature alloys, 1970.

[29] DESFORGES (C. D.), WILSON (F. G.). — Fulmer Research Institute *(unpublished work)*, 1975.

[30] DESFORGES (C. D.). — Practical Metallic Composites, B43-B46, March, 701-74-Y, *Institution of Metallurgists*, London, 1974.

[31] LEMKEY (F. D.), BAYLES (B. J.), SALKIND (M. J.). — UACRL, D910261-4 July, *United Aircraft Research Laboratories*, 1965, U. S. A.

[32] *Conference* on *in situ* Composites, 1975, Bolton Landing U. S. A. Procs. *(to be published)*.

[33] HUNT (J. D.), CHILTON (J. P.). — *J. Inst. Metals*, 1963, *92*, 21.

[34] LEMKEY (F. D.). — Solidification Technology (ed. by J. T. BURKE), *Syracuse University Press*, 1974.

[35] ALEXANDER (J. A.), GRAHAM (L. D.). — Agard CP-156, 67, *Thompson and P. R. Sahm*, 1974.

[36] LEMKEY (F. D.), THOMPSON (E. R.). — *Procs.* of Lakeville *Conf.* on *in situ* composites, 1972, 105-120.

[37] BIBRING (H.), SEIBEL (G.), RABINOVITCH (M.). — *Mem. Sci. Rev. Met.*, 1972, *69*, 343-358.

[38] DAVIES (G. J.). — Practical Metallic Composites, March, 701-74-Y, D1-D15, *Institution of Metallurgists*, London, 1974.

[39] *The Superalloys*, 1972, *534* (ed. by C. T. SIMS and W. C. HAGEL).

[40] KLOPP (W. D.). — *J. Less-Common Metals*, 1975, *42*, 261-278.

[41] WOODYATT (L. R.), SIMS (C. T.), BEATTIE Jr. (H. J.). — *Trans. A. I. M. E.*, 1966, *236*, 519.

SECTION II:
Turbines and Industrial Applications

High-temperature oxidation and corrosion of superalloys in the gas turbine (*A review*)

R I Jaffee, J Stringer¶
Metal Science Group, Battelle Columbus Laboratories, Columbus, Ohio 43201
Received 21 July 1971

1 Introduction

Of the various applications for high-temperature structural materials the gas turbine appears to offer the most challenging set of requirements. To gain efficiency the turbine inlet temperatures of aircraft gas turbines have been increasing, and it is not out of reason to contemplate the use of temperatures as high as 1300°C, although the more usual range for advanced gas turbines is 1000–1100°C. The major limitation on turbine inlet temperature has been and remains the strength of the turbine rotor blade and stator vane materials. There have been major improvements in the high-temperature strength of these materials, primarily through the use of strong casting alloys hardened by a substantial volume fraction of the coherent γ' precipitation $(Ni, Co)_3(Al, Ti)$. However, the major means for achieving higher gas temperatures is through the use of bypass air cooling, whereby a small percentage of the inlet air is diverted through cooling passages in the hot turbine blading and vanes, lowering their temperature about 200–260°C below the gas temperature to the point where their strength capability is sufficient.

Up to the present time corrosion and oxidation resistance has not been a design consideration in the selection of aircraft gas turbine materials other than that the materials resist the environment, which superalloys containing about 20% chromium do adequately up to the temperature at which strength is maintained (~900°C). However, the development of the stronger gas turbine alloys was facilitated by a reduction in chromium content to about 10 $^w/_o$Cr to permit an increased amount of γ' hardening. This caused a reduction in both high-temperature oxidation resistance and, more importantly, a reduction in hot-corrosion resistance. Hot corrosion, in general terms, means the accelerated attack of the hot parts of the turbine due to presence of certain compounds in the combustion products. This has been a serious problem for many years in land-based turbines operating on impure fuels and in marine gas turbines due to seal-salt ingestion; in addition, the use of higher strength alloys of lower chromium content has resulted in recent years in hot-corrosion problems in gas turbines in aircraft operating in marine environments. The reduced oxidation and hot-corrosion resistance has been combatted by coating the hot components with aluminide-type coatings. This entails additional expense and a loss of reliability compared with the former situation where uncoated alloys with adequate oxidation and corrosion resistance were operated at lower gas temperatures. Furthermore, the coatings may be speedily eroded in land-based or marine turbines due to the ingestion of solid particles in the air or in the use of fuels with relatively high ash contents.

Despite the fact that the present turbines are coping with the problem of higher gas temperatures, there is an undercurrent of dissatisfaction with this solution by the producers and users of gas turbines. There is a desire for alloys which would be adequately strong, yet corrosion resistant enough for use without protective coatings.

¶On leave from The University of Liverpool, Liverpool, England.

There are unanswered questions dealing with the long-time chemical stability of cooled blades and vanes operating with high turbine inlet temperatures. Furthermore, there is an appalling lack of basic information on which to base better solutions to the problems of service in this severe operating environment.

In part this is because the alloys used are complex, often containing as many as ten components; and have been developed in an empirical fashion over many years. The environment is also complex, and as a consequence the correlation between performance in the engine and simple laboratory tests has been poor, so that many of those concerned with the development of practical alloys feel that laboratory studies on simple easily controlled systems are of very limited value. As a result, the emphasis has moved from basic research in favour of studies on complex alloys in conditions which attempt to simulate as closely as possible those in the engine, but this has unfortunately resulted in a lack of understanding of many of the corrosion and oxidation problems.

2 The conditions in the engine

In a gas turbine air is ingested from the atmosphere and compressed. The intake air is filtered in the case of land-based or marine turbines, but not in the case of aircraft engines. The compression ratio varies with the type of engine: in a high-performance aircraft engine current compression ratios are as high as 20:1. The compressed air is heated during compression to approximately 400–500°C.

Fuel is injected into the combustion chamber, mixed with part of the air from the compressor, and burnt; the remaining air is mixed with the combustion gases downstream from the combustion zone, thus lowering the temperature and increasing the volume of the hot gases. Gas turbines can operate with a wide range of fuels— powdered coal, peat, natural gas, and various grades of oil—but nearly always have a very high air:fuel ratio, so the combustion gas is highly oxidising. However, local reducing conditions may result from unburnt fuel leaving the combustion chamber or may exist beneath ash deposits.

The flame temperature depends on the fuel, but is commonly about 2000°C; the maximum possible temperature of the gases leaving the combustion chamber and impinging on the first stationary row of blades in the turbine (the nozzle guide vanes) is determined largely by the properties of the vanes and first-row rotor blade materials. The stationary vanes are subjected to aerodynamic stresses of the order of $3 \cdot 5 - 7$ kg mm^{-2}, while the rotor blades have much higher stresses imposed by the rotation which are of the order of $14 - 18$ kg mm^{-2}; however, the gas cools some 100°C between the vanes and the rotor. In addition, the vanes are subjected to severe thermal shock. Both the vanes and the blades are cooled in modern high-performance gas turbines to permit higher gas temperatures to be used; the cooling imposes considerable thermal stresses on the blade during steady-state running, but may reduce the magnitude of transient thermal stresses.

In a land-based turbine operating with residual fuel and filtered air, the main corrosive impurities come from the fuel, and there are problems with the formation of ash. Ideally, the ash should be fine and solid: there is then some problem with mechanical erosion of the blading, but no real corrosion. However, the presence of low-melting point phases in the ash can result in the build-up of adherent deposits in the engine, and these not only cause the aerodynamic efficiency to fall off, but allow corrosion to take place beneath them. The most serious of the molten ash-forming constituents of the fuel is vanadium, but sodium and potassium salts may also be harmful; sulphur is also an important impurity which can lead to severe corrosion problems.

The fuel burned in aviation engines is a highly refined distillate fuel and is relatively free of contaminants, containing a maximum of 0·01% ash and 0·5% sulphur. The major source of impurities is thus the intake air, and of those contained in it sodium chloride and sodium sulphate, usually originating from sea salt, are the most important. Because a gas turbine may consume 50 times more air than fuel, 1 ppm sea salt in air would be equivalent to 50 ppm in fuel, and there thus appears to be little benefit in trying to reduce further the impurity content of jet engine fuel.

The composition of sea salt is as follows:

	g l^{-1}	%
NaCl	23	52
MgCl$_2$.6H$_2$O	11	25
Na$_2$SO$_4$.10H$_2$O	8	18
CaCl$_2$.2H$_2$O	1·4	3
KBr, KCl	1·1	2
	44·5	100

The sodium chloride will react with sulphur in the fuel according to the reaction

$$2NaCl + SO_2 + O_2 \rightarrow Na_2SO_4 + Cl_2.$$

The reaction in detail may well differ from this, since the chlorine will react with some other species to produce a chloride.

The reaction from left to right is highly favoured, and under equilibrium conditions virtually all the sodium chloride will be converted to sodium sulphate. However, the gas velocity is high, of the order of 300 m s^{-1}, and the length between the combustion zone and the guide vanes is of the order of 0·3–0·6 m. The total time available for reaction is therefore 10^{-3} s or so for the air injected at the burner and less for the air injected further along the combustion chamber. It may well be that this is insufficient time for equilibrium to be attained. Nevertheless, the deposits on blades from gas turbines operated in marine environments always show the presence of sodium sulphate, and almost never show the presence of chloride.

It is possible to calculate the 'dew point'—the temperature at which salt species will condense on an inert solid surface—as a function of the concentration in the gas phase, and this has been done for NaCl and for Na$_2$SO$_4$ by DeCresente and Bornstein (1968) as a function of the sodium chloride content of the intake air and of the compression ratio of the turbine. Figure 1 shows their results: Figure 1a is the dew point for sodium chloride assuming no reaction with sulphur, while Figure 1b assumes complete conversion to sodium sulphate. Haryslak and Pollini (1967) have measured directly the salt concentration on the deck of a destroyer: their results are shown in Figure 2. This suggests a maximum of about 1 ppm salt in the intake air, of which about 90% will be removed by efficient filtration. Other estimates have suggested higher concentrations and from a design point of view it would probably be better to regard 10 ppm as a possible maximum. With a compression ratio of 20:1, 1 ppm salt would give an Na$_2$SO$_4$ dew point of about 860°C; the higher value would correspond to a dew point of closer to 1000°C. The melting point of pure Na$_2$SO$_4$ is 884°C, but this is reduced by various additions. In particular, a simple eutectic is formed with NaCl with a eutectic temperature of 625°C at 65 mol % Na$_2$SO$_4$. The metal surface is not, however, inert; and this may well invalidate the simple calculations. Burner rig tests show condensation of Na$_2$SO$_4$ at 900°C, 0·1 ppm synthetic sea salt, 15 atm although the calculations give a dew point of 710°C (DeCresente and Bornstein, 1968) for these conditions.

Various additives are present in aviation fuel to improve viscosity, control visible smoke emissions, and so forth. Little research has yet been done on the effects of these on corrosion, but it seems probable that they will be minor in comparison to the effects of ingested salt.

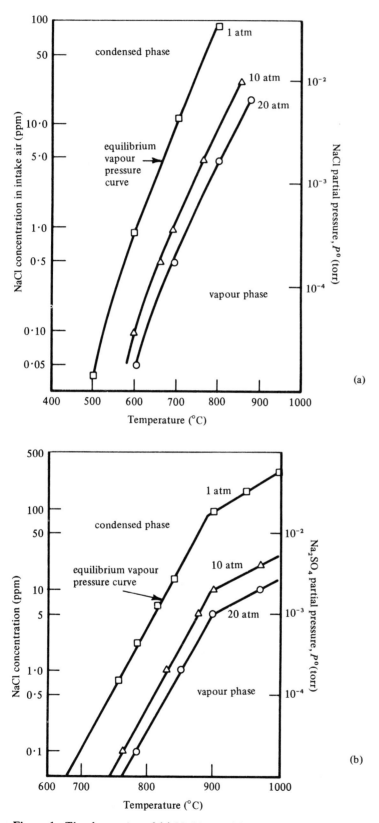

Figure 1. The dew point of (a) NaCl, and (b) Na_2SO_4 as a function of NaCl concentration and pressure (DeCrescente and Bornstein, 1968).

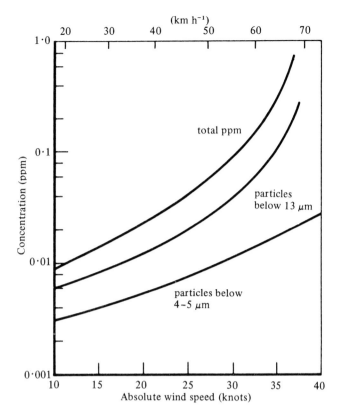

Figure 2. Averaged sea salt aerosol concentrations (Haryslak and Pollini, 1967).

3 Gas turbine alloys and coatings

3.1 *Alloys*

The development of gas turbine alloys in the early 1940's was based on existing high-temperature materials, notably the 80 $^w/_0$Ni–20 $^w/_0$Cr heater alloys and the 70 $^w/_0$Co–30 $^w/_0$Cr stellites. It was discovered that additions of relatively small amounts of aluminium and titanium to the nickel-based alloys produced a considerable increase in strength due to the formation of a precipitate based on the ordered face centred cubic phase γ' Ni$_3$(Al, Ti). This precipitate formed essentially coherent boundaries with the face centred cubic primary solid solution. The solubility of aluminium and titanium in the solid solution decreased with temperature allowing heat-treatment of the alloys. In general, the aim of much of the alloy development has been (a) to increase the maximum solubility of aluminium and titanium, (b) to diminish the low temperature solubility, and (c) to increase the temperature of the solvus line. In the United Kingdom most turbine blades have been forged, and the need to have a hot-working range has restricted the degree of strengthening that could be achieved. In the United States most blades are investment cast, and higher concentrations of aluminium and titanium can be tolerated. In general, the strength of nickel-base alloys increases with increasing (Al+ Ti) content.

Decreasing the chromium content permitted a greater maximum solubility of aluminium and titanium, and so alloys with progressively lower chromium contents were developed; this led, however, to a decline in oxidation resistance and, in particular, to a very sharp decline in hot-corrosion resistance. Nowadays, low chromium alloys are used where hot-corrosion problems are not anticipated, while for corrosive environments alloys generally contain at least 15 $^w/_0$Cr. Table 1 shows the composition of some typical modern nickel-base alloys.

Cobalt-base alloys are strengthened by a completely different method. The equivalent ordered phase Co$_3$Al is close-packed hexagonal and does not produce significant strengthening; instead, the alloys rely on solid-solution strengthening of the matrix and a carbide distribution. The alloys are therefore simpler, containing about 10 $^w/_o$ tungsten, $0\cdot2$ $^w/_o$ carbon, and 25 $^w/_o$ chromium, with a carbide former such as zirconium. In addition, approximately 10 $^w/_o$ nickel is usually added to suppress the face-centred-cubic to close-packed-hexagonal transformation.

The nickel-base superalloys are in general stronger than the cobalt-base superalloys at temperatures below about 1100°C, but at higher temperatures the nickel alloys become weaker as a result of the resolution of the γ'. Accordingly, nickel alloys are almost invariably used for the highly stressed rotor blades, while cobalt alloys are frequently used for the nozzle guide vanes. Since cobalt alloys do not rely on γ' strengthening, there is no need to reduce the Cr content, and this may explain in part the better performance of cobalt-base alloys in hot corrosion conditions.

In 1964 the U.S Navy established a programme to develop more 'sulfidation' (hot-corrosion) resistant turbine blade and nozzle guide vane alloys for marine turbines, with the following objectives:

(i) Develop a turbine blade alloy combining the hot corrosion resistance of Udimet 500[1] in 1% sulphur diesel fuel combustion products with the strength and ductility of alloy 713C at 870°C (100 h rupture life at 30 kg mm^{-2}).

(ii) Develop a cobalt-base turbine nozzle with the 1040°C strength of WI-52 (100 h rupture life at 7 kg mm^{-2}) but with improved hot corrosion resistance.

Several alloys were developed including MAR-M-421 (Martin-Metals), Udimet 710 (Special Metals), IN-738 (International Nickel Co.), and more recently MAR-M-432 (Martin Metals), which came close to meeting the above targets for blades. Several experimental alloys developed by the General Electric Company for the Marine Engineering Laboratory meet the targets for a cobalt-base vane material. (Sims *et al.*, 1969).

From alloy development studies conducted it appears that chromium, and to a lesser extent titanium and cobalt, are the only elements which contribute beneficially to hot-corrosion resistance of nickel-base alloys. Some additional elements, such as tantalum, were found beneficial solely to oxidation resistance. Some, such as molybdenum and aluminium, were found detrimental to hot-corrosion resistance, and

Table 1. Composition of alloys.

Alloy designation	Nominal chemical composition ($^w/_o$)										Temperature capability[b] (°C)	
	C	Cr	Ni	Co	Mo	W	Cb	Ti	Al	Other[a]	100 h	1000 h
MAR-M-421		15·5	bal.	10	1·75	3·5	1·75	1·75	4·25	–	975	920
Udimet 710	0·07	18	bal.	15	3	1·5	–	5·0	2·5	–	965	915
IN-738	0·17	16	bal.	8·5	1·75	2·6	0·9	3·4	3·4	1·75 Ta	990	930
MAR-M-432	0·15	15·5	bal.	20	–	3·0	2·0	4·3	2·8	2·0 Ta	990	930
IN-792	0·21	12·7	bal.	9	2	3·9	–	4·2	3·2	3·9 Ta	1005	950
Udimet 500	0·08	19	bal.	19	4·2	–	–	3	3	–	925	860
Alloy 713C	0·12	12·5	bal.	–	4·2	–	2	0·8	6·1	–	990	915
B 1900	0·10	8·0	bal.	10	6·0	0·1	0·1	1·0	6·0	4·3 Ta	1000	950
WI-52	0·45	21	1·0	bal.	–	11·0	2·0	–	–	2·0 Fe	905	870
X-40	0·50	25·5	10·5	bal.	–	7·5	–	–	–	1·0 Mn 0·5 Si	870	815

[a] Most of these alloys contain $0\cdot01$–$0\cdot02$ B and $0\cdot10$–$0\cdot20$ Zr.
[b] Temperature to produce rupture in indicated time at 14 kg mm^{-2}.

[1] The compositions of this and other alloys mentioned in this section are given in Table 1.

other alloy elements were generally found neutral or ineffective. In recent years it has been found that a proper balance of alloying elements, particularly the refractory metals, can substantially improve hot-corrosion resistance even at relatively low chromium levels. For example, preliminary results indicate that a new alloy IN-792 possesses nearly the corrosion resistance of Udimet 500, but with roughly 30°C increased temperature capability over alloy 713C, based on rupture strength.

The above description of the Navy programme fairly well demonstrates that the alloy-producing industry will respond to well-defined targets. More guidance of this kind is needed. This will provide the alloy producers with meaningful targets, whose achievement—they may be reasonably assured—will result in an increased market for their products.

It is also apparent that the alloy properties achieved by conventional means are close to the maximum possible. Some new approaches and/or mechanisms which might be considered in the future are: (i) dispersion hardening, (ii) fibre strengthening, (iii) rare earth alloying, (iv) new precipitation mechanisms, and (v) intermetallic compounds.

Dispersion hardening. Oxide dispersions result in very little strengthening in the low-temperature range, and need to be made more effective or combined with other strengthening mechanisms.

Fibre strengthening. Refractory metal fibres have been found to be effective, but at a sacrifice in density and oxidation resistance. Work is needed on non-metallic fibre strengthening, such as with Al_2O_3 fibres.

Rare earth alloying. Rare earths in oxide form have been found to improve hot corrosion resistance. They also improve oxidation resistance by favouring the preferential oxidation of chromium to form a protective Cr_2O_3 layer. Thus, the rare earth oxide finely dispersed in the alloy matrix could improve hot corrosion, oxidation resistance, and strength simultaneously.

New precipitation mechanisms. Cobalt-base alloys are strengthened by solid solution strengthening or carbide dispersions. There is no known precipitation hardening mechanism for cobalt analogous to γ' hardening in nickel-base alloys. A new approach to strengthen cobalt alloys is needed.

Intermetallic compounds. Intermetallic compounds have excellent oxidation resistance but lack ductility. Their potential for high-temperature strength is a moot point. Some intermetallics like NiAl exhibit low strength at elevated temperatures. Development of structural intermetallics to provide a good combination of strength, ductility, and oxidation resistance is needed.

3.2 Coatings

Requirements for superalloy coatings for gas turbine use have been stated as the following:

  (i) they must be resistant in the thermal stress environment;
 (ii) they must be metallurgically bonded to the substrate;
(iii) they must be thin and uniform;
 (iv) they must have self-healing characteristics;
  (v) they must be ductile enough to withstand substrate deformation without cracking;
 (vi) they must not degrade the mechanical properties of the substrate; and
(vii) they must have diffusional stability.

All current superalloy coatings are based on the use of aluminium as the primary coating constituent. Processes used are: (a) pack cementation, (b) hot dipping, (c) slurry, and (d) electrophoresis. Regardless of the technique, the coatings are predominantly NiAl (nickel-base alloys) or CoAl (cobalt-base alloys), with minor additions of an alloying element, usually chromium. Although no coating satisfies all performance requirements, the aluminide-type coatings have successfully extended the life expectancies of gas turbine engine blades and vanes, and have been used in many military and commercial engines. Future coatings for superalloy hot section components will have to exhibit performance capabilities superior to those of current coatings. Improved coatings for blades and vanes are expected to have life expectancies in excess of 12000 h at metal temperatures of 925°C and 1010°C. respectively. Research and development programmes of the type listed below have been recommended to promote the development of the improved coating systems for future engines:

(i) Conduct studies to evaluate thoroughly the potential (i.e., determine advantages and limitations) of newer coating techniques for depositing advanced coatings on superalloy components used in gas turbine engines. Processes to be considered should include: (a) cladding, (b) electrodeposition from fused salts, (c) pyrolitic deposition from organometallic vapours or solutions, and (d) physical vapour deposition, especially as applied to alloys. Successful research results in developing new coating processes will probably be necessary in order to permit deposition of advantageous coating alloys.

(ii) Develop a recoating capacity that will minimise (or eliminate) substrate-metal loss.

(iii) Develop non-destructive testing methods for process control and for predicting the useful remaining life of coated turbine-engine components. Techniques should measure or identify defects, corrosion damage, coating thickness, extent of coating–substrate interdiffusion, fatigue damage, and wall thicknesses of cooled components.

(iv) Develop manufacturing techniques for the reliable coating of internal cooling passages of extremely small diameter used in convection and transpiration cooling.

(v) Develop automated manufacturing processes in order to improve coating reproducibility and reliability.

4 Oxidation of gas turbine materials

It has been well known for many years that alloys of iron, nickel, or cobalt with chromium contents of the order of 20 $^w/_0$ have excellent oxidation resistance, and it is therefore not surprising that there has been a great deal of work on the oxidation behaviour of these alloys. Recently Giggins and Pettit (1969) have studied the oxidation of Ni–Cr alloys at temperatures in the range 800–1200°C, Kofstad and Hed (1969) have studied the oxidation of Co–Cr alloys, and Wood et al. (1970) have compared the oxidation behaviour of Fe–Cr, Ni–Cr, and Co–Cr alloys.

Giggins and Pettit (1969) distinguish three groups of Ni–Cr alloys: those containing less than approximately 10% Cr, those containing between 10% and 25% Cr, and those containing over 25% Cr. Small amounts of chromium increase the rate of oxidation of nickel, and these dilute alloys oxidise according to a parabolic rate law. The scale consists of two layers: an outer compact layer of NiO and a porous inner layer composed of a mixture of NiO and $NiCr_2O_4$. In addition, there is internal oxidation in the metal with the formation of Cr_2O_3 particles.

In the intermediate range the overall oxidation rate decreases rapidly with increasing chromium content. The rate law is parabolic only after long times, of the order of 20 h at 1100°C. The oxide growth on the surface appears to be irregular: in some places a thin, apparently single-layered, oxide forms; in other places one finds a thick oxide, apparently consisting of three distinct layers. Electron microprobe and X-ray diffraction studies indicate that the thin oxide and the layer adjacent to the metal in the thicker oxide are essentially Cr_2O_3. The next layer is a mixture of NiO and $NiCr_2O_4$, and the oxide furthest from the metal is NiO. Chromium is depleted from the metal beneath the scale, but there is no internal oxidation. After long times, the Cr_2O_3 layer is continuous over the specimen surface and is of fairly uniform thickness, and the parabolic oxidation rate is apparently controlled by diffusion through it. At short times, before the establishment of the parabolic rate law, the Cr_2O_3 layer is not continuous, and locally the chromium in the metal is internally oxidised, forming Cr_2O_3 particles in a chromium-depleted matrix. Apparently, the continuous layer of Cr_2O_3 develops in these regions only after a significant amount of the alloy has been oxidised.

The more concentrated alloys follow a parabolic rate law, the rate being independent of alloy content. The external scale is predominantly Cr_2O_3, with a small amount of $NiCr_2O_4$. In general there is no internal oxidation, but at long times some internal oxide particles may appear.

The variation of oxidation rate of cobalt–chromium alloys with chromium content has the same general form, but the turning points are displaced to rather higher chromium contents. In addition, the behaviour seems to be much more sensitive to oxygen pressure. Below about 15% Cr the rate is more rapid than for pure cobalt; Kofstad and Hed (1969) have discussed the oxidation of a Co–10% Cr alloy in the temperature range 800–1300°C for oxygen pressures in the range 0·2–760 torr. The scale consists of two layers: an outer layer of columnar CoO crystals, and an inner layer of mixed CoO, $CoCr_2O_4$, and Cr_2O_3. A small amount of internal oxidation of chromium is also observed. The overall oxidation rate follows a parabolic rate law, with the rate constant depending on $p_{O_2}^{1/3}$.

Above 15% Cr the rate diminishes with increasing chromium content, but below 25% Cr the structure of the scale is much the same. Kofstad and Hed suggest that the rate controlling process is diffusion of cobalt outwards through the CoO, with the spinel effectively reducing the cross-sectional area for diffusion. In alloys containing more than approximately 25% Cr a continuous layer of Cr_2O_3 forms, but the transition to this behaviour is very sensitive to oxygen pressure: at 760 torr oxygen a continuous Cr_2O_3 layer is not formed until the alloy contains over 30 $^w/_o$Cr while at 2 torr oxygen a continuous Cr_2O_3 layer may develop with as little as 20 $^w/_o$Cr.

In general, protective oxidation involves the diffusional transport of one or more of of the reactants through the oxide layer, and it is plainly desirable that an oxide with a slow transport rate should be formed. There is relatively little difference in the rate of transport of nickel through NiO, chromium through Cr_2O_3, and chromium through $NiCr_2O_4$: a factor of about five difference, with the first being most rapid. Nickel diffusion through $NiCr_2O_4$ is rather slower than chromium diffusion, but again the difference is not great. Cobalt diffusion through CoO is significantly more rapid, some two orders of magnitude faster than that of nickel through NiO (Birchenall, 1968). From this it might seem most desirable to form a spinel layer, but in fact a continuous protective spinel is not often observed, and it is doubtful if such a layer would be thermodynamically stable.

In practice therefore it is more common to aim for a continuous Cr_2O_3 layer, and the slow rates of oxidation in the binaries correspond to the establishment of such a layer. However, Cr_2O_3 is removed by the formation of the volatile oxide CrO_3:

$$Cr_2O_{3(s)} + 1\tfrac{1}{2}O_2 \rightarrow 2CrO_{3(g)},$$

and the rate of volatilisation starts to become significant at about 1100°C in still air. However, in high velocity gas streams the rate of oxidation/volatilisation may increase enormously, since it is controlled by diffusion through a gaseous boundary layer. (Graham and Davis, 1971).

From both the transport and tha volatilisation points of view, it would be desirable to form an Al_2O_3 oxide. The transport of aluminium through Al_2O_3 is between two and three orders of magnitude slower than the transport of the metal ions in the spinels, and the oxidation/volatilisation is negligible at normal temperatures. For this reason there has been some investigation of Ni(Co)–Cr–Al ternary alloys,

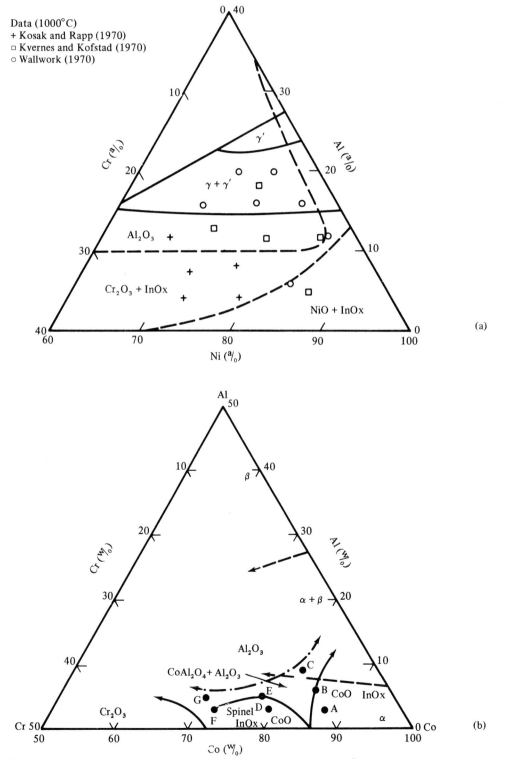

Figure 3. Oxide map for (a) the ternary system Ni–Co–Al at 1000°C, and (b) the ternary system Co–Cr–Al at 1100°C and 760 torr O_2 indicating the types of scale which form over the various compositions (Wallwork and Hed, 1971).

notably by Kvernes and Kofstad (1970), Kosak (1969), Kosak and Rapp (1970), Stott *et al.* (1971), Wallwork and Hed (1971) and Giggins and Pettit (1970). Wallwork (1970), and Wallwork and Hed (1971) have represented the oxides formed on a ternary diagram, a technique apparently first used by Scheil and Schulz (1932). Figure 3 shows the diagrams for the Ni–Cr–Al and Co–Cr–Al system, from which it can be seen that alloys containing relatively low chromium and relatively high aluminium will form predominantly Al_2O_3 scales.

In agreement with this the low-chromium high-aluminium superalloys Alloy 713C and B 1900 do both from predominantly Al_2O_3 scales and, of course, Al_2O_3 scales are also formed on the high aluminium coatings based on NiAl and CoAl used to protect all the alloys.

Scale adherence is also a matter of importance, since the blades and vanes are subjected to large numbers of severe thermal shocks in operation. Unfortunately, Al_2O_3 scales have relatively poor thermal shock resistance; whereas Cr_2O_3 scales are quite good, and are particularly adherent when the metal contains minor amounts of rare earths or oxide dispersions[2]. The effect of oxide dispersion on the adhesion of the Al_2O_3 scales has not yet been tested, but there are some indications that it may be improved considerably (M.S.Seltzer, private communication).

5 Hot corrosion of superalloys

It has been known for many years that at high temperatures in the presence of salt heat-resisting alloys undergo rapid corrosion. This problem became important in the gas turbine in the late 1950's, originally in land-based turbines burning impure fuels and later in marine turbines burning aviation kerosene. With the use of the low-chromium high-strength nickel-base superalloys and the increase in engine temperature the problem became of importance in aircraft gas turbines, particularly those operating near the sea. The maximum rate of attack appears to be in the temperature range 800–1000°C, although this varies from alloy to alloy.

Sykes and Shirley (1951) concluded, from a study of corroded samples, that sodium sulphate was the principal corrosive agent, although laboratory tests showed that as little as 0.3% NaCl increased the rate of attack considerably. Shirley (1956) showed that the main corrosion product on nickel-based alloys was a voluminous porous oxide, principally NiO containing Cr_2O_3 dispersed in it. The oxide penetrates deeply into the metal, isolating regions of metal, and ahead of this there is a thin band of small grey globules, which appear to be chromium sulphide.

Simons *et al.* (1955) suggested that two quite distinct processes were involved:
(i) a 'triggering' stage, involving some reducing agent R:

$$Na_2SO_4 + 3R = Na_2O + 3RO + S,$$
$$M + S = MS,$$

where M is the metal and MS is its sulphide; and
(ii) an autocatalytic destruction of the metal:

$$Na_2SO_4 + 3MS = 4S + 3MO + Na_2O,$$
$$4M + 4S = 4MS.$$

The triggering stage was thought to be sporadic and unpredictable, and the auto-catalytic reaction rapid and violent.

The idea of an incubation period prior to the onset of attack has been noted a number of times, and it is usually regarded as due to the reduction of the initial protective air-formed Cr_2O_3 oxide.

[2] The early literature is discussed by Kubaschewskii and Hopkins (1967); for a later paper on the effects of a dispersed phase see Giggins and Pettit (1971).

Hancock (1961) appears to have been the first to note that the penetration of oxide into the metal, and the morphology of the chromium sulphide, suggested the presence of a liquid phase at temperature, although chromium sulphide itself has a very high melting point. He suggested that in fact a liquid nickel–sulphur phase is formed, which decomposes on solidification to nickel and a nickel/nickel sulphide eutectic at a eutectic temperature of 637°C. On further slow cooling an exchange reaction takes place with chromium displacing the nickel to form the thermo-dynamically more stable chromium sulphide. Recently, Goebel and Pettit (1970) and Seybolt (1970) have supported this view, although several other investigators have suggested that nickel sulphide will only be formed in atypically high sulphur activities.

Danek (1965) remarked that there were three schools of thought concerning the mechanism of the attack:

(a) sulphur reacts with chromium to form chromium sulphide, depleting the matrix. The depleted matrix oxidises rapidly, the oxide carrying sulphide globules with it, the latter being themselves slowly oxidised, releasing sulphur;

(b) again, chromium sulphide is formed, but this itself then oxidises rapidly, liberating sulphur and leading to the development of an oxide containing chromium-poor islands of metal which are relatively slowly consumed.

(c) the sulphide formed is nickel sulphide which as a liquid penetrates rapidly into the metal; it oxidises rapidly liberating more sulphur.

DeCrescente and Bornstein (1968) first applied thermochemical reasoning to the reaction and concluded that the sulphur activities produced by the dissociation of sodium sulphate in the vapour phase in the oxygen-rich combustion gases were too low to produce sulphidation, and that therefore hot corrosion required the condensation of sodium sulphate on the metal surface; this view is generally accepted although corrosion can take place in Na_2SO_4 vapour if the oxygen activity is low enough (Seybolt, 1970). Quets and Dresher (1969) have used the Pourbaix method of presentation to describe the thermochemistry of the reaction and concluded that in order to obtain sulphidation it is necessary to reduce the activity of the sodium oxide in the fluid salt by the formation of complexes of the type Na_2CrO_4. However, Goebel and Pettit (1970) point out that this treatment does not take account of the variation of sulphur and oxygen activity possible within the liquid sulphate, and that when this is considered there is no need to postulate the formation of complexes, although there is little doubt that their formation will assist in the generation of a self-sustaining reaction.

It has been remarked that the presence of small amounts of sodium chloride enhance the reaction rate very considerably, and indeed some authors have suggested that pure sodium sulphate is not corrosive at all. Waddams et al. (1959) and Gray (1961) have suggested that the chlorine forms volatile chromium chloride or oxychloride molecules which are then oxidised to form non-protective chromium oxide.

Bornstein and DeCrescente (1969) suggest that sulphur is not essential to the hot corrosion, and have shown that rapid corrosion and a substantially similar morphology can be obtained with sodium oxide with no sulphur present; however, at the present time this probably represents a minority view.

In general, the higher the chromium content the better the hot-corrosion resistance. Some other elements appear to be beneficial, notably aluminium and titanium; Lewis and Smith (1961) first showed that these elements could be regarded as 'chromium equivalents': their data are shown in Figure 4. Rentz (1966) has attempted to extend this idea to other elements; his chromium equivalent is

$$^w/_oCr + 3 \cdot 8 \ (^w/_oAl-5) + 2 \cdot 0 \ (^w/_oW) - 12 \cdot 5 \ (^w/_oC) - 1 \cdot 4 \ (^w/_oMo) - 1.$$

While there may be some doubt about the detailed validity of this formula, it expresses the general view among investigators that aluminium and tungsten are beneficial, while molybdenum is harmful. The reasons for this are not well understood. In general, cobalt-base alloys resist hot corrosion rather better than nickel-base alloys. This may be because commercial cobalt alloys contain more chromium than most nickel-base superalloys, but Seybolt (1970) suggests that in part it is because the cobalt-cobalt sulphide eutectic temperature is significantly higher (872°C), and in part because the diffusion of sulphur in cobalt is approximately two orders of magnitude slower than the diffusion of sulphur in nickel at 1000°C.

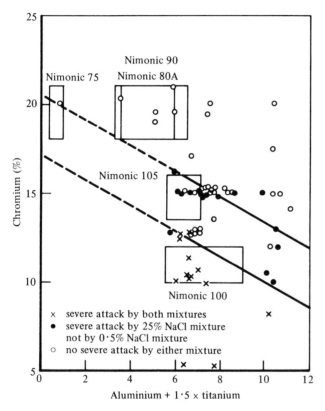

Figure 4. Effect of chromium, titanium, and aluminium on corrosion of nickel-base alloys (0·3% Si) by 0·5 and 25% NaCl mixtures showing relation of specification compositions to limits for severe attack at 900°C (Lewis and Smith, 1961).

5.1 *Methods of evaluating hot corrosion*

The preceding description of hot corrosion has been entirely qualitative, because there is very wide disagreement on how to evaluate the reaction. In his early studies Shirley (1956) developed the so-called crucible or half-immersion test, in which a rod of the specimen material is immersed up to its midpoint in a molten salt mixture held in a crucible, the entire assembly being heated in a resistance furnace in air. Originally, a salt mixture consisting of 99% Na_2SO_4, 1% NaCl was used, but it is now more common to use the much more aggressive 90% Na_2SO_4, 10% NaCl mixture. This test is extremely severe: most alloys are heavily corroded after 24 hours, and designers of alloys feel that the correlation with experience in the engine is poor. A second widely-used test was introduced by Simons *et al.* (1955), in which the sample is initially coated with the salt mixture, and is then oxidised—either in air or oxygen—in a gravimetric apparatus. In principle, this method is more attractive in that it permits the generation of quantitative data, and furthermore probably resembles the sort of process involved in the engine. It too is a fairly severe test, and again there are some doubts about its correlation with practical experience.

Finally, so far as laboratory studies are concerned, there are attempts to expose specimens to well-defined sulphur and oxygen activities, usually by means of SO_2/O_2 mixtures. This gives by far the best controlled experiment and the most meaningful data in a formal sense, but without a full understanding of the complex hot corrosion reactions the translation to the practical situation is very difficult.

The remaining tests are basically attempts to simulate, in a more-or-less controlled fashion, the conditions in the engine. These may use a synthetic combustion gas blowing on a heated specimen, or involve the actual combustion of fuel in a combustion chamber, the hot gases playing on specimens which are not otherwise heated. Salt can be sprayed in with the air or mixed with the fuel. The specimens are sometimes stationary, sometimes rotated; some techniques use cylindrical specimens while others use wedge specimens designed to simulate aerofoil shapes.

There are two methods of evaluating the degree of damage: the first measures the weight loss after descaling, while the second uses metallographic measurement of the affected metal. The majority of investigators now agree that the metallographic method, in spite of its rather subjective nature, is the better.

With all these test methods there is an alarmingly wide scatter of results, and different investigators will even rank materials in a different order. In part this is due to an inherent irreproducibility in the reaction associated with the incubation period, which many test procedures do not take sufficient account of; but in part it is due to inadequate control over the various reaction parameters and a lack of appreciation of the sensitivity of the results to this kind of control. A more general agreement on a standard 'practical' test procedure, supplemented by a more careful basic study of reaction mechanisms, would help. It is clear that the reaction is so complex that, without a much greater degree of understanding which can only be achieved by careful research, a proper control of the practical tests and a sensible appreciation of their results will not be possible.

Acknowledgements. This paper is based on a pilot paper for discussion presented at the AGARD Materials Group Meeting, Istanbul, Turkey, in September, 1969. One of the authors (J. Stringer) would like to thank Battelle's Columbus Laboratories for the award of a Visiting Fellowship, during the tenure of which this paper was written.

References

Birchenall, C. E., 1968, in *Mass Transport in Oxides* (Eds. Wachtman and Franklin, N.B.S. Publ. 296, August, 1971), p.119.

Bornstein, N. S., DeCrescente, M. A., 1969, *Trans. AIME*, **245**, 1947.

Danek, G. J., Jnr., 1965, *Naval Engineers J.*, **77**, 859.

DeCrescente, M. A., Bornstein, N. S., 1968, *Corrosion*, **24**, 127.

Giggins, C. S., Pettit, F. S., 1969, *Trans. AIME*, **245**, 2495.

Giggins, C. S., Pettit, F. S., 1970, Commemorative Symposium on the Oxidation of Metals, Electrochemical Society Fall Meeting, Atlantic City, N.J., October.

Giggins, G. S. , Pettit, F. S., 1971, *Met. Trans.*, **2**, 1071.

Graham, H. C., Davis, H. H., 1971, *J. Am. Ceram. Soc.*, **54**, 89.

Gray, P. S., 1961, in discussion to Hancock (1961), p.213.

Goebel, J. A., Pettit, F. S., 1970, *Met. Trans.*, **1**, 1943; 3421.

Hancock, P., 1961, in *Proceedings of the First International Congress on Metallic Corrosion*, London, April (Butterworths, London, 1962), p.193.

Haryslak, L. W., Pollini, R. J., 1967, in *Hot Corrosion Problems Associated with Gas Turbines*, ASTM Special Technical Publication No.421, p.146.

Kofstad, P. K., Hed, A. Z., 1969, "Mechanisms of oxidation in the Co–Cr system", 4th International Congress on Metallic Corrosion, September, Amsterdam, Holland.

Kosak, R., Jnr., 1969, Ph.D. Thesis, Ohio State University.

Kosak, R., Jnr., Rapp, R. A., 1970, Commemorative Symposium on the Oxidation of Metals, Electrochemical Society Fall Meeting, Atlantic City, N.J., October.

Kubaschewski, O., Hopkins, B. E., 1967, *Oxidation of Metals and Alloys,* 2nd Ed. (Butterworths, London), pp.247–8.

Kvernes, I., Kofstad, P., 1970, Final Report on Contract F61052-67-C0057, European Office of Aerospace Research, OAR-USAF, January, 31.

Lewis, H., Smith, R. A., 1961, in *Proceedings of the First International Congress on Metallic Corrosion*, London, April (Butterworths, London, 1962), p.202.

Quets, J. M., Dresher, W. H., 1969, *J. Materials*, **4**, 583.

Rentz, W., 1966, Thesis, Rensselaer Polytechnic Institute [quoted by M. J. Donachie, Jnr., R. A. Sprague, R. N. Russell, K. G. Ball, and E. F. Bradley, 1961. in *Hot Corrosion Problems Associated with Gas Turbines*, ASTM Special Technical Publication No.421, p.85].

Scheil, E., Schulz, E. H., 1932, *Arch. Eisenhüttenw.*, **6**, 155.

Seybolt, A. U., 1970, G.E. Research and Development Center, Technical Information Report No.70-C-189, June.

Shirley, H. T., 1956, *J. Iron and Steel Inst.*, **182**, 144.

Simons, E. L., Browning, G. V., Liebhafsky, H. A., 1955, *Corrosion*, **11**, 505.

Sims, C. T., Bergman, P. A., Beltran, A. N., 1969, "Progress in the development of hot corrosion resistant alloys for marine applications", paper No.69-GT-16 presented at ASME Gas Turbine Conference, March 9.

Stott, F. H., Wood, G. C., Hobby, M. G., 1971 (to be published).

Sykes, C., Shirley, H. T., 1951, in *Symposium on High Temperature Steels*, Special Report No.43, (The Iron and Steel Institute, London), pp.153–169.

Waddams, J. A., Wright, J. C., Gray, P. S., 1959, *J.Inst.Fuel*, **32**, 246.

Wallwork, G. R., 1970, 2nd International Conference on Strengthening of Alloys, Asilomar, California.

Wallwork, G. R., Hed, A. Z., 1971, *J. Oxidation of Metals*, **3**, 229.

Wood, G. C., Wright, I. G., Hodgkiess, T., Whittle, D. P., 1970, *Werkstoffe und Korrosion*, **21**, 900.

Effects of Turbine Atmospheres on Sulfidation Corrosion

M. J. DONACHIE, JR.   E. F. BRADLEY   R. A. SPRAGUE   F. P. TALBOOM

Recent applications of gas turbine engines have required that they pack substantially more horsepower in less space than does any other propulsion system. Consequently, several Coast Guard and naval forces are using or planning to design, build, and commission various types of craft, including high-speed antisubmarine-warfare ships, propelled by gas turbines. There are special turbine problems associated with such marine operations and these are primarily associated with salt atmospheres. Salt ingestion in the presence of sulfur-containing fuels can lead to the formation of a sulfide and accelerated hot corrosion (sulfidation) attack. Furthermore, this attack need not be limited to marine sea-level operation. Sulfidation has occurred in aircraft gas turbines in instances where sulfate-contaminated water has been used for water-injection engines. Sulfidation damage may also occur in engines where the air has been contaminated by salt. Experience has shown that the former, marine, type of sulfidation is the same as sulfidation which may be encountered by aircraft gas turbines which are exposed to heavy concentrations of salt, even though basic differences in sulfur concentration exist from one fuel (Marine Diesel -- 1 percent S) to another (JP5R -- 0.35 percent S). Consequently it is feasible to describe a standard type of sulfidation attack for all cases by consideration, principally, of the sulfidation effects produced in burning JP5R fuel under sulfidizing conditions.

HISTORICAL REVIEW

Sulfur adsorption on pure nickel promotes the formation of a low-melting-point nickel-nickel sulfide eutectic. This effect does not exist on nichrome (80 percent nickel-20 percent chromium). Furthermore, nickel-base alloys derived from the basic nichrome composition are inherently highly oxidation resistant owing to a combination of chromium, aluminum, and other alloy elements. As alloy-element changes produced a lowered chromium content of the base metal, a form of accelerated hot corrosion, known eventually as sulfidation, appeared. The external appearance of turbine

Fig. 1   Nimonic 100 first-stage turbine blade from turboprop engine. Note deterioration due to sulfidation attack

hardware which suffered from sulfidation is shown in Figs. 1 and 2 for Nimonic 100 and INCO 713 alloys. The macroscopic appearance of this sulfidation attack is characterized by severe exfoliation in many cases and by distinct color changes (greenish hue) in the region of accelerated attack. Even a highly resistant material such as Waspaloy has shown attack under severe conditions in instances where sulfate-contaminated water has been used inadvertently for water-injection engine operation, Figs. 3 and 4. The metallographic exam-

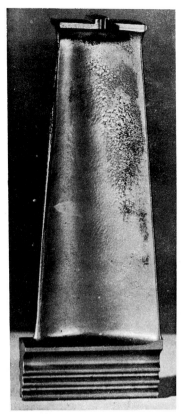

Fig. 2   INCO 713 turbine blade after
approximately 130 hr of experimental
engine testing.   Note heavy scaling
of airfoil surface

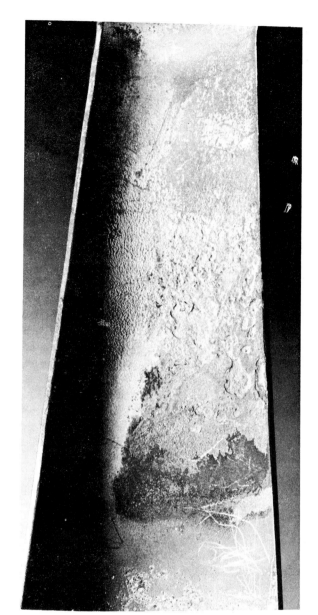

Fig. 3   Waspaloy turbine blade after engine operation
showing sulfidation attack on airfoil

inations of a Nimonic 100 and an INCO 713 turbine
blade confirmed that in each blade severe oxidation had been preceded by a thin zone characterized by the formation of a light grey sulfide
phase which was chromium rich, Figs. 5 and 6.

Although few instances of sulfidation other
than those reported in the foregoing were known
initially, it was apparent by 1960 that studies
of the phenomenon of sulfidation were required.
Consequently, by controlled experimental engine
operation and the development of laboratory rig
testing procedures at the authors' company, it
was possible for these sulfidation phenomena to
be reproduced in the laboratory for study.   From
these studies a series of laboratory testing procedures have become relatively standard tools in
the study of sulfidation behavior of alloys (1-
5).[1]   These methods, which were pioneered by the
authors' company, together with test results on
a variety of metals, coatings, and operating conditions, are described in the following section.
For a further review of the subject, see References (5-9).

―――――――――

[1] Numbers in parentheses designate References
at the end of the paper.

SULFIDATION CORROSION STUDIES

Apparatus

Crucible tests have been advocated (10,11)
for the study of new high-strength, low-chromium-
content nickel-base alloys, but have not been successful (12).   The authors' company therefore dis-
continued crucible tests at an early stage and
concentrated upon the development of a new type
of burner test which would more closely simulate
engine operating conditions (13).   The burner test
rig developed to establish sulfidation effects is
shown in Fig. 7.   Corrosive elements are ingested
at a point in the burner nozzle.   These elements
may be either sodium chloride or synthetic sea

Source: ASME Publication 67-GT-2

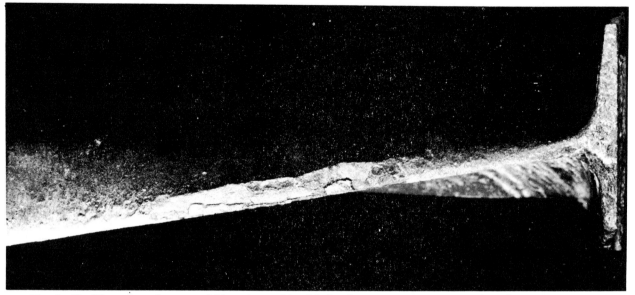

Fig. 4   Trailing edge of uncoated Waspaloy turbine blade after engine operation with sulfate-contaminated injection water.   Note severe exfoliation

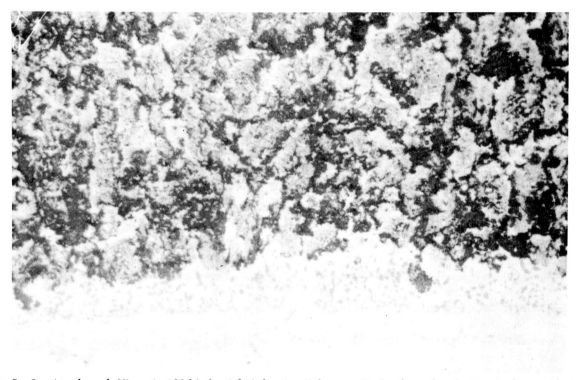

Fig. 5   Section through Nimonic 100 blade airfoil showing light grey Cr_2S_3 phase (bracket) preceding oxidation (unetched)   (X1000) reduced for reproduction

salt in aqueous solution. The salts can be vaporized and added to the combustion stream through the port (tube) in the minimum-diameter section of the convergent-divergent nozzle. The combustor nozzle is shown schematically in Fig. 8. Specimens have a simulated (wedge shaped) airfoil section, Fig. 7(a). Fig. 7(b) shows the sulfidation rig in operation. During operation, the rig relies upon the sulfur content of the appropriate fuel (Marine Diesel, JP5R) for its sulfur environment. The air-fuel ratio is held on the oxidizing side and salt usually is ingested at 3.5 to 35 ppm salt/ air ratio. Automatic temperature control and monitoring are used wherever feasible, and provi-

Fig. 6   Section through INCO 713 blade airfoil showing light grey Cr_2S_3 phase (bracket) preceding oxidation. Etchant: mixed acids.   (X1300) reduced for reproduction

sions for temperature cycling are incorporated.

Metallurgical and External Chemical Effects

The metallographic appearance of a sulfidized area on nickel-base turbine hardware has been shown. Furthermore, the temperature ranges of interest in sulfidation have been delineated (12). Evaluation and analysis of burner test data are of interest. In burner rig testing, present procedures require weight-change and metal-loss curves to be generated by the periodic examination of specimens, normally at about 20-hr intervals, in the temperature range of interest (1450-2150F). These curves are established by weight-change measurements prior to a descaling operation and by micrometer evaluation of metal loss where feasible. At the end of testing, metallographic examinations are performed on transverse sections at intervals along the wedge (airfoil) specimen. Weight loss eventually reported in sulfidation testing is the total weight loss per specimen. The metal loss per specimen diameter is the metallographic measurement at the most heavily attacked and corroded region of the specimen and represents the metal actually lost due to erosion, depletion,

and internal (intergranular) attack. It is a measurement of the total mechanical and metallurgical loss of material per cross-sectional area.

Tests on a wide variety of nickel and cobalt-base alloys have been made in the burner rigs, and the visual appearance of sulfidation test specimens is shown in Fig.9. The results of such tests may be interpreted in terms of the effects of salt, fuel sulfur content, cycle, temperature, and alloy composition (including the effects of coatings). Alloy compositions are given in Table 1.

Effects of Salt and Sulfur on Corrosion. The effects of salt and sulfur on sulfidation corrosion are not completely separable. It appears confirmed (5,13) that sulfur is required for the observed sulfidation attack, probably in the form of sodium sulfate which is produced during combustion or is present in natural sea water. In addition, the presence of a halide ion (Cl^-) seems to be required for the process of sulfidation to take place; however, the halide ion itself does not attack or degrade the base alloy. In order to separate the effects of salt and sulfur, tests may be carried out at the same temperature and salt concentration while sulfur concentration is

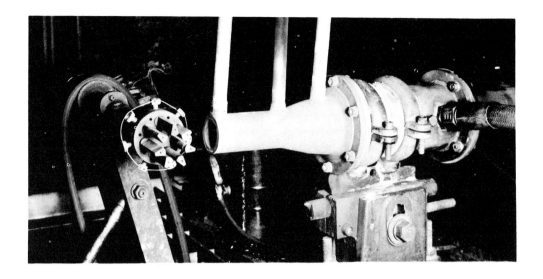

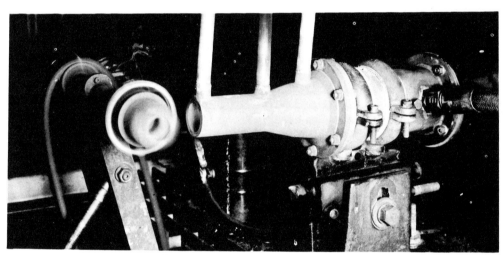

Fig. 7   Burner test rig showing (a) specimens and test configuration, (b) operation

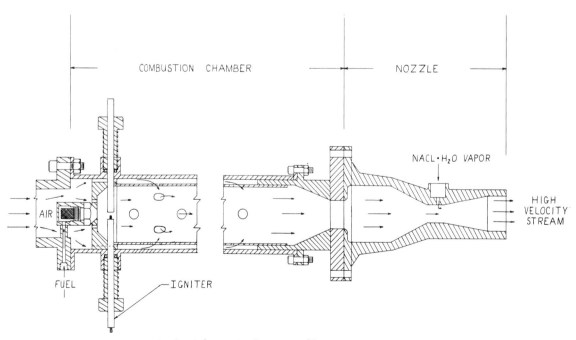

Fig. 8   Schematic diagram of burner nozzle system

Fig. 9   Uncoated nickel-base alloy test bars after 1650 F - 120 hr isothermal sulfidation test using JP-5R fuel and 3.5 ppm synthetic sea salt: air ratio.   Left to right: Cast Udimet 700, wrought Udimet 700, Waspaloy, IN-100, B-1900, MAR-M 246, INCO 728, and MC 102

varied. When this is done, it is found that most alloys show essentially similar weight losses in JP5R fuel (0.35 percent S) to their weight loss in marine diesel fuel (1.0 percent S), with lesser percentage differences due to sulfur as test time increases; Table 2. The fact that fuel sulfur content did not have a significant effect on sulfidation is in agreement with the results of other investigators (13) who have used a pressurized burner rig to evaluate sulfidation. On the other hand, examination of weight-change data obtained after short-time high-salt-concentration (5.0 hr -- 35.0 ppm salt/air) runs and after long-time low-salt-concentration (500 hr -- 3.5 ppm salt/air) runs indicates, as shown in Figs.10 and 11, that when either test time or salt concentration is increased at a fixed temperature, the separation between alloys becomes more apparent (12). The importance of salt is thus demonstrated. Consequently it is frequently suggested that high-salt-concentration runs be substituted for lower-salt-concentration runs as a means of expediting testing. This amounts to a change in cycle, and will be discussed again in a following section. It is interesting to note, however, that "the addition of greater amounts of salt in an attempt

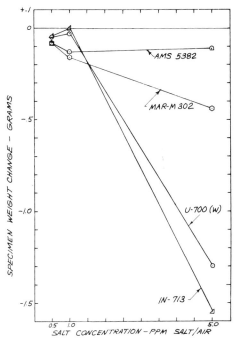

Fig. 10   Effect of salt concentration on weight loss after sulfidation rig testing

TABLE 1

Composition of Turbine Alloys

Cobalt-base Alloys

Alloy	C	Cr	W	Ni	Ta	Other		
AMS 5382	0.50%	25.5	7.5	10.5	-	-		
MAR-M 302	0.85	21.5	10.0	-	9.0	B 0.005	Zr 0.20	
WI-52	0.45	21.0	11.0	-	Cb 2.0	Mn 0.25	Si 0.25	Fe 2.0

Nickel-Base Alloys

Alloy	C	B	Zr	Cr	Ti	Al	Co	Mo	Cb	Ta	W
Udimet 700	0.08%	0.03	-	15.0	3.5	4.3	18.5	5.2	-	-	-
Waspaloy	0.08	0.005	0.06	19.5	3.0	1.3	13.5	4.3	-	-	-
MAR-M 21D4	0.15	0.015	0.05	15.5	1.75	4.25	10.0	1.75	1.75	-	3.5
MAR-M 246	0.15	0.015	0.05	9.0	1.50	5.5	10.0	2.5	-	1.5	10.0
B-1900	0.10	0.015	0.10	8.0	1.0	6.0	10.0	6.0	-	4.0	-
INCO 728	0.05	0.02	0.10	15.0	-	6.0	10.0	2.0	1.0	2.0	2.0
Nimonic 100	0.03	-	-	11.0	1.5	5.0	20.0	5.0	-	-	-
INCO 713	0.15	0.01	0.10	13.0	0.75	6.0	-	4.5	Cb+Ta 2.3		-
MC 102	0.04	-	-	20.0	-	-	-	6.0	6.5	2.5	Mn 0.30 Si 0.25
IN 100	0.18	0.014	0.06	10.0	4.7	5.5	15.0	3.0			V 1.0

Fig. 11 Comparison of airfoil wedge specimens after 1650 F - 500 hr isothermal sulfidation test using 1 percent sulfur marine diesel fuel and 3.5 ppm salt: air ratio. Left to right: INCO 713, Udimet 700, AMS 5382, and MAR-M 302

to produce alloy discrimination in a shorter time period may lead to conservative test results" (12). Fig.12, modified from that in Reference (12), illustrates this point.

From the preceding discussion, it may be concluded that alloy evaluation may be carried out in either JP5R or marine diesel fuel as long as appropriate attention is given to a uniform salt concentration when multiple tests are to be compared in an evaluation of alloy behavior.

Effects of Alloy Composition. The visual appearance of sulfidized test specimens has been shown, Fig.9 and 11. Metallographic examination of such specimens confirms the similarity of the sulfidation attack to that produced in turbine airfoils. It is apparent from a cursory inspection of Figs.9 and 11, however, that different alloy compositions may be expected to vary in their sulfidation resistance. Variations also may be noted within a given alloy composition and it has been noted that wrought alloys show slight inter-

TABLE 2

Effect of Sulfur Concentration on Weight Loss After 50 and 100 Hour
Tests at 0.5 ppm Nominal Salt/Air Ratio at 1650F

Alloy	50 Hours		100 Hours	
	MD-1%S	JP-5R-0.4%S	MD-1%S	JP-5R-0.4%S
MAR-M 302	-0.10 gms	-0.10 gms	-0.20 gms	-0.25 gms
Waspaloy	-0.10	-0.03	-0.17	-0.06
AMS 5382	-0.15	-0.04	-0.20	-0.10
WI-52	-0.30	-0.25	-0.80	-0.65
IN-100	-1.35	-0.85	-3.25	-2.70
IN-713	-0.04	-0.10		
U-700 (W)	-0.05	-0.03		

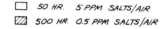

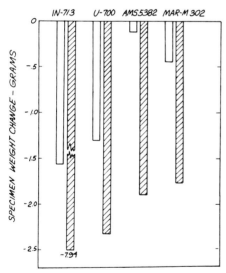

Fig. 12   Effect of salt concentration and
time on weight loss after 1650 F rig testing
using 1 percent sulfur marine diesel fuel

sible inversions in the relative sulfidation resistance of alloys, as previously suggested (12).

In general, such data as presented suggest
the most obvious and logical effect is an increased sulfidation-corrosion resistance with increased alloy chromium content. The effects of
other elements are less obvious. Based on recent
tests (14,15), molybdenum is probably detrimental
to sulfidation resistance, while cobalt, aluminum,
and titanium additions are probably beneficial.
The degree to which these elements are effective
in promoting improved sulfidation-corrosion resistance is obviously dependent upon the test cycle conditions.

Unfortunately, the elemental combinations
which provide the best sulfidation-corrosion resistance are not usually conducive to the highest
strength in nickel-base superalloys. Present
high-strength alloys have relatively poor sulfidation resistance when compared to such older
wrought alloys as Waspaloy and M-252. On the
other hand, the strengths of the newer, lower-
chromium alloys, such as IN 100, are so much superior to the strengths of the wrought alloys that
the cast alloys find extensive and increasing use
in gas turbine technology. Furthermore, these alloys when properly protected by coatings show substantial sulfidation resistance; e.g., a factor of
10:1 improvement at 1800F. Coatings are not the
only means of improved sulfidation resistance,
and it has been shown by a detailed development
program at the authors' company that alloy compo-

granular sulfidation when compared to cast alloys
of the same nominal composition, Fig.13. The effects of composition on the 120-hr isothermal sulfidation resistance of several nickel-base alloys
are shown in Fig.14. Weight-loss and metal-loss
data are given in Fig.15. When similar alloys
were run a total of about 184 hr, the normalized
weight and metal-loss data, Fig.16, indicate pos-

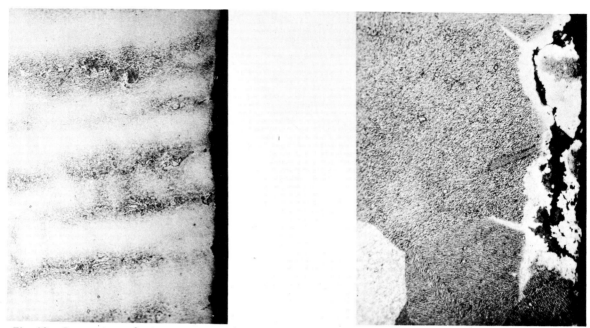

Fig. 13 Comparison of microstructures of cast (left) and wrought (right) Udimet-700 alloy after identical sulfidation testing. Note intergranular sulfidation attack on wrought Udimet-700. Etchant: Kalling's. (X250) reduced for reproduction

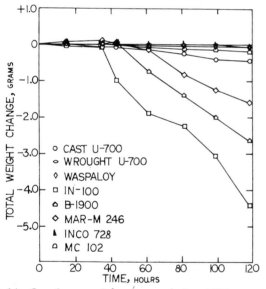

Fig. 14 Specimen weight change during 1650 F - 120 hr isothermal sulfidation test using JP-5R fuel (0.4 percent S) and 3.5 ppm salt: air ratio

- ○ CAST U-700
- ○ WROUGHT U-700
- ◇ WASPALOY
- □ IN-100
- △ B-1900
- ◇ MAR-M 246
- ▲ INCO 728
- □ MC 102

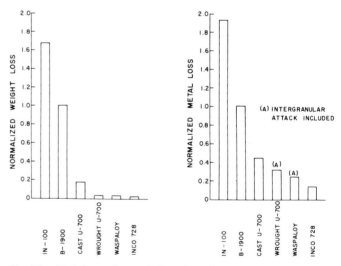

Fig. 15 Weight and metal-loss data for certain alloys shown in Fig. 14, normalized to B-1900

sitions with strength equal to B-1900 and three times greater sulfidation resistance can be produced.

One other observation which is worthy of note is that isothermal sulfidation testing invariably has shown plane front sulfidation attack in cast alloys, but has shown plane front plus intergranular attack in wrought alloys. The reasons for intergranular attack may be in the chemical homo-geneity of the wrought products. In cyclic sulfidation testing, to be described, it was found for one alloy, U-700, that the cyclic testing enhanced planar front attack at the expense of intergranular penetration.

It should be clear from the preceding metallurgical arguments that no one simple picture can be used to explain the sulfidation resistance, or lack thereof, in nickel-base turbine alloys. Cobalt-base alloys are in a similar state of knowledge and study. Such alloys are inherently more sulfidation resistant than nickel-base alloys pri-

TABLE 3

Effects of Coating on Sulfidation-Erosion Resistance of WI-52

	Uncoated, 1450F-100 Hours 3.5 ppm Salt:Air	Coated, 1650F-500 Hours 3.5 ppm Salt:Air
Weight Change	-0.25 gms	-0.02 gms

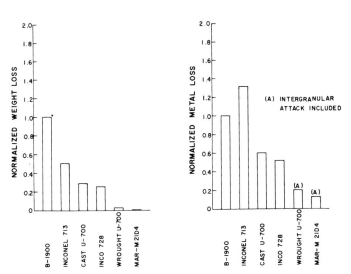

Fig. 16   Weight and metal-loss data after 1650 F - 184 hr
sulfidation test using JP-5R fuel and 3.5 ppm synthetic
sea salt: air ratio.   Data normalized to B-1900

marily because of their high chromium contents.
Under appropriate operating conditions, however,
they do sulfidize. The metallographic appearance
of sulfidation in cobalt alloys is similar to that
in nickel-base alloys. The effects of coating on
cobalt-base alloys, which are not as oxidation resistant as nickel-base alloys, are shown in Table
3. This table shows that both weight change and
metal loss were negligible for coated MAR-M 302
after 50 hr/3.5 ppm salt/air at 2000F when compared to coated MAR-M 302 in isothermal sulfidation (admittedly a somewhat high temperature for
sulfidation attack) using marine diesel fuel.
Furthermore, uncoated WI-52 showed greater weight
loss (-0.25 gm) after 100 hr at 1450F in marine
diesel fuel, 3.5 ppm salt/air than coated WI-52
showed after a similar test for 500 hr at 1650F.
The beneficial effects of coatings are obvious.
Cycle and Temperature. The microstructural
characteristics of sulfidation corrosion are affected by variations in component metal temperatures (16). Conflicting data are available.
Burner tests indicate that erosion bars are always
attacked most rapidly in the hot spot at all temperatures during oxidation testing. However, in
isothermal sulfidation testing, sulfidation at-

tack is not seen until metal temperatures of about
1500F are reached. As test temperatures are
raised, the hot spot eventually does not show sulfidation attack and accelerated attack seems to
become most severe at the erosion bar extremities.
From these observations, it appears that sulfidation does not occur above about 1900F and is most
rapid at 1700-1750F. It would appear, therefore,
that two conditions of test, namely temperature
and salt, will have a significant effect upon sulfidation attack. Since the aim of any dynamic
sulfidation test is the reproduction of experimentally observed conditions in gas turbines, it
is essential that a given test be reproducible
and correlate with engine experience. Bare (uncoated) alloy testing in isothermal sulfidation
has correlated well with such engine data as are
available. Relative rankings are substantiated.
However, it was not always evident that relative
rankings by isothermal testing discriminated between coatings as clearly as would be desired.
This undoubtedly is due in part to the finite
thickness of a coating compared to the infinite
thickness of a base alloy. In other words, the
coating cannot replenish itself from an infinite
reservoir and in addition, it undoubtedly tends
to local spalling of oxide layers in engine testing due to the thermal cycles involved. Consequently, two cyclic sulfidation-oxidation testing
procedures were evolved: (a) Blade cycle, 1550F
(3min) + 1850F(2 min) + cool(2 min); (b) Vane cycle, 1750F(3min) + 2050F(2 min) + cool(2 min).
The term sulfidation-oxidation is used to describe
the testing since both processes are involved,
particularly in the vane cycle. These cycles were
then applied to coated and uncoated alloys; however, coated alloys were tested using 35 ppm salt/
air while uncoated alloys were tested using 3.5
ppm salt/air. Table 4 compares the results of
vane and isothermal sulfidation testing with engine testing of coatings on the same nickel-base
alloy substrate. The correlation between cyclic
and engine tests is apparent. A reverse sulfidation cycle, 2050F(2 min) + 1750F(3 min) + cool
(2 min), also confirmed the relative coating rankings, but cycles to failure did not correlate as
well with engine experience.

TABLE 4

Comparison of Isothermal, Vane Cyclic and Experimental Engine
Sulfidation Testing of Aluminide Coatings on A Nickel-Base Superalloy

Test	Coating (Best - Left, Poorest - Right)		
Isothermal	A	B	C
Vane Cyclic*	B	C	A
Engine*	B	C	A

* By thickening coating C to thickness of coating A, order
may be changed to CBA.

The success of cyclic testing on coatings led to cyclic testing on bare uncoated alloys. However, no change in relative rankings of alloys has occurred due to thermal cycling. This probably obtains from the fact that the uncoated alloys act as an essentially infinite reservoir for the elements depleted at the surface by sulfidation. Thus, any enhancement in attack on alloys is due to the effects of cyclic thermal fluctuations on surface oxide formation and retention. Furthermore, it is not clear that cyclic sulfidation-oxidation really enhances attack on cast nickel-base alloys. For example, if the temperature of maximum sulfidation attack (1750F) is used and time at this temperature is ratioed against total cycle time, it appears that time to one gram weight loss versus metal temperature follows a common line for cast nickel-base alloys whether time (weight loss) is accumulated in isothermal testing or cyclic testing.

One feature of cyclic testing on uncoated alloys is worthy of attention. Wrought alloys degrade more rapidly in cyclic than in isothermal testing. Consequently a wrought alloy of the same composition as a cast alloy will now exhibit the same sulfidation characteristics (weight loss, metal loss) as the cast alloy when tested cyclically. The reasons for this behavior apparently lie in the nature of the intergranular attack on wrought alloys. Apparently grains or portions of grains can be spalled by cyclic testing, whereas no attack other than intergranular attack is observed in isothermal testing. The result of cyclic testing seems to bring uncoated wrought and cast alloys more in line with engine experience.

DISCUSSION AND SUMMARY

The future use of gas turbine engines will bring turbine hardware more in contact with salt atmospheres and provide the opportunities for sulfidation to occur. Much effort has been devoted to theoretical and experimental evaluations of the sulfidation process (1-5,12). However, the essence of present and near-future engine requirements necessitates the use of dynamic tests which simulate expected operating conditions. Consequently, burner tests frequently are used to assess the sulfidation resistance of alloys. From the original isothermal tests on nickel-base alloys, there has been developed an understanding of the effects of chromium (beneficial), molybdenum (detrimental), and other elements on sulfidation-corrosion resistance. Furthermore, salt, not sulfur, is seen as the primary variable in the production of enhanced sulfidation attack. This sulfidation attack takes place at temperatures between about 1500 and 1900F, but is apparently most pronounced between 1550 and 1800F.

Consequently, since the test conditions of sulfidation may be varied, it is apparent that sulfidation probably ought to be studied at 1550 and about 1750F (peak temperature for attack). Furthermore, salt/air and cycle ought to be varied to provide the best correlation of rig data with experimental engine data. This has been done and has been shown to be a very successful procedure. The results of cyclic testing tend to bring coated and uncoated nickel-base alloys more in line with engine experience.

REFERENCES

1 "Final Report on FT4A-2 Laboratory Sulfidation Testing, Laboratory Component Testing," Pratt and Whitney Aircraft Report No. 2218, 1963.

2 W. Rentz, "Sulfidation (Hot Corrosion)," paper presented at Annual Meeting of the Metal-

lurgical Society of the AIME, New York, February 1966.

3 "Final Summary Report on Development of Hot-Corrosion-Resistant Alloys for Marine Gas Turbine Service," General Electric Company, U. S. Navy Marine Engineering Laboratory Contract N600 (61533) 63218, January 1966.

4 H. Quigg, et al., "Effect of JP-5 Sulfur Content on Hot Corrosion of Superalloys in Marine Environment," Phillips Petroleum Company, Progress Reports 1-4, U. S. Navy Bureau of Naval Weapons Contract NOw 64-0443-d, 1964-1966.

5 W. Rentz and M. Donachie, "Oxidation and Sulfidation Corrosion of Nickel Base Superalloys," paper presented at ASM National Metal Congress, Chicago, November 1966.

6 G. Danek, "State-of-the-Art Survey on Hot Corrosion in Marine Gas Turbines," Naval Engineers Journal, vol. 77, 1965, p. 859.

7 P. Bergman and I. Kalikow, "Corrosion and Erosion Tests," presented at 6th Annual National Conference on Environmental Effects on Aircraft and Propulsion Systems, Princeton, N. J., September, 1966.

8 G. Danek, "Hot-Corrosion in Marine Gas Turbines," presented at ASM National Metal Congress, Chicago, November 1966.

9 W. Rentz, "Sulfidation Corrosion of Nickel-Base Superalloys," submitted to Rensselaer Polytechnic Institute in partial fulfillment of the requirements for the MS degree, January 1966.

10 L. Graham, J. Gadd, and R. Quigg, "Effect of Grain Orientation on the Hot Corrosion Behavior of Superalloys," presented at ASTM Annual Meeting, Atlantic City, N. J., June 1966.

11 C. Sykes and H. Shirley, "Scaling of Heat-Resisting Steels, Influence of Combustible Sulfur and Oil-Fuel Ash Constituent," in Special Report No. 43, Iron and Steel Institute, 1951, p. 153.

12 M. Donachie, et al., "Sulfidation of Hot Section Alloys in Gas Turbine Engines," presented at ASTM Annual Meeting, Atlantic City, N. J., June 1966.

13 R. Schirmer and H. Quigg, "Effects of JP-5 Sulfur Content on Hot Corrosion of Superalloys in Marine Environment," presented at ASTM Annual Meeting, Atlantic City, N. J., June 1966.

14 P. Bergman, C. Sims, and A. Beltran, "Development of Hot-Corrosion-Resistant Alloys for Marine Gas Turbine Service," presented at ASTM Annual Meeting, Atlantic City, N. J., June 1966.

15 M. Shepard, W. Danesi, and N. Olson, unpublished research, Pratt and Whitney Aircraft, East Hartford, Connecticut.

16 A. Liu, "Malfunction of TF2036-B1A Engine S/N 1," AVCO Lycoming Div. Rept. No. TN 0194, July 1964.

HIGH TEMPERATURE MATERIALS REQUIREMENTS OF THE METALLURGICAL INDUSTRIES

By

J. H. DAVIDSON

Creusot-Loire
Département. Études et Recherches
Aciéries d'Imphy (France)

INTRODUCTION

There is a considerable overlap in the materials requirements of the metallurgical and other industries. Taking materials in the broader sense, there is one major field which concerns particularly, if not exclusively, the metallurgical industry, namely that of refractory ceramics; furnace and ladle linings etc. However, this subject being outside the author's range of competence, the discussion will be limited to metallic materials.

Some materials exposed to high temperatures are chosen chiefly for economic reasons; when time lost in replacement is insignificant, low life can be accepted on price grounds, for instance in the case of ingot moulds. It is proposed to consider here three areas of application in which, while the price factor remains important, materials properties limit process performance: hot-working tools, furnace parts, electric heating elements.

Generally speaking the principal requirements for high temperature materials are high mechanical strength and good oxidation/corrosion resistance. In the case of hot working tools, the most important factor is high temperature mechanical strength, whereas oxidation/corrosion resistance is the major prerequisite for electric heating elements and a combination of both is essential in materials for furnace parts.

HOT WORKING TOOLS

Materials for hot-working tools (forging anvils and dies, extrusion dies, hot shear blades, etc.). require high hot hardness, ability to withstand impact stresses and thermal shock and adequate abrasion resistance. The most commonly used materials are the hot-work tool steels with around 0.4 % C and various W, Cr, V, Mo or Co additions. These steels maintain a high resistance to deformation up to about 550 to 600º C. The important factors are the temperature of the tool, the time at temperature and the strength of the work material. Even in cases where tool steels give satisfactory performance, advantage can often be gained by using superalloy tools whose longer life offsets the increased cost. Figure 1 compares the hot tensile strength of tool steels and certain superalloys. The latter are employed for extrusion dies and for hot-working anvils and dies for forging superalloys, and this is the area in which progress is most likely to be situated.

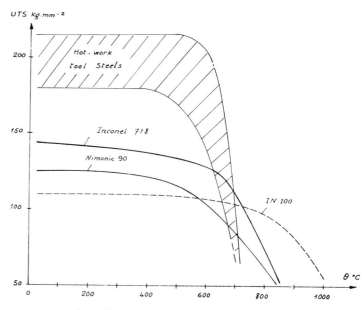

FIG. 1. — *Hot tensile strength of hot-work tool steels and of some nickel-base superalloys.*

The problem in hot-working superalloys is obviously that they have been designed to resist high temperature deformation. This resistance is derived from a combination of solid solution strengthening and precipitation hardening by intermetallics. Figure 2 shows that, to a first approximation, temperature capability increases with the total hardener content. As regards solution hardening, the most important aspect in creep is the reduction in diffusion rate.

This is greater the higher the melting point of the element added, and for the elements considered, the melting point increases with atomic weight. For this reason, the solid solution hardeners are plotted on a weight percent basis, whereas atom percentages are used for elements contributing to γ' precipitation hardening, since volume fraction is of major importance in this case. Niobium and hafnium are considered as solid solution hardeners in the diagram and ought

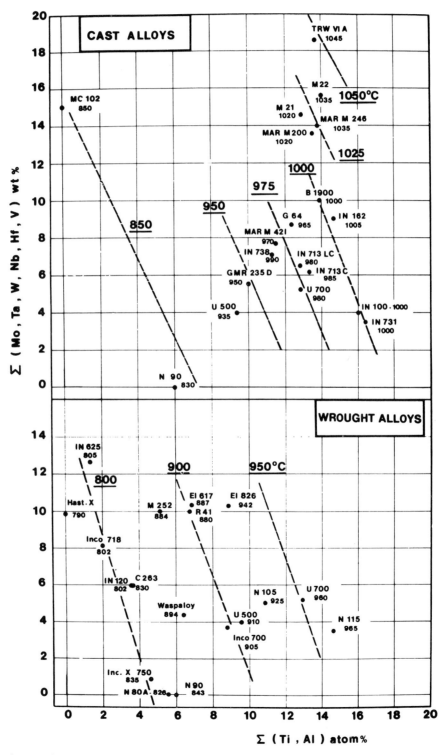

FIG. 2. — *Influence of composition on creep strength of nickel-base superalloys. Temperature for 100 hours rupture life at 14 kg. mm⁻².*

Source: *Revue Internationale des Hautes Températures et Réfractaires*, Vol 13, 1976

perhaps more appropriately to be included as precipitation strengtheners, but in fact this produces little change in the overall result. Increasing volume fraction of intermetallic precipitates is accompanied by a continuous rise in solution temperature, figure 3, and leads to lower incipient melting temperatures and eventually to the formation of eutectic γ'. The relatively low flow stress region between complete γ' solution and the incipient melting temperature thus becomes progressively smaller. This is illustrated in figure 4 in which the behaviour of a purely solution hardened alloy (Hastelloy B) is included for comparison.

It would clearly be an advantage in the case of alloys with narrow single phase temperature regions to be able to maintain the work-piece temperature as long as possible. Furthermore, slow deformation rates lead

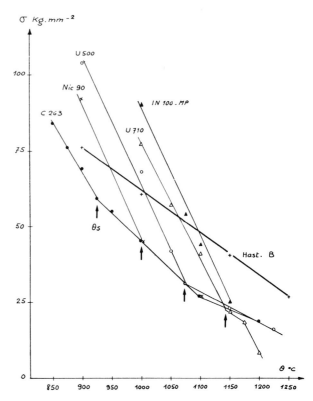

to lower flow stresses, figure 5, and even to superplasticity in the case of very fine-grained materials, such as can be obtained by powder-metallurgy techniques, at high temperatures. Similarly, slow strain rates lead to early recrystallisation, either dynamic (fig. 6) or static, with a consequent-increase in ductility (fig. 7).

It can be seen that isothermal or hot-die forging could open up considerable scope for the hot-working of superalloys and is particularly interesting now that powder metallurgy has made it possible to obtain for-

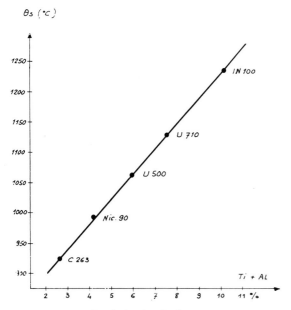

FIG. 3. — *Variation in γ'-solvus temperature with total Ti + Al content.*

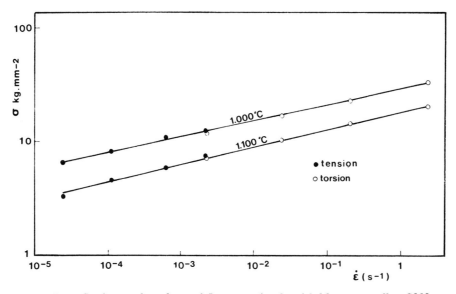

FIG. 5. — *Strain-rate dependence of flow-stress for the nickel-base superalloy C263. Data from tensile and torsion tests. Correlation using Von Mises criterion.*

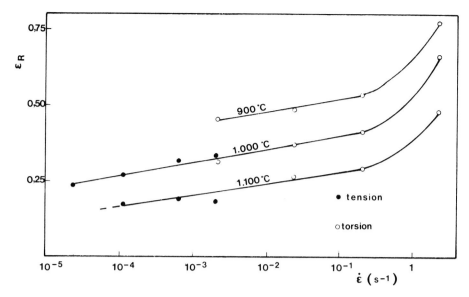

FIG. 6. — *Strain-rate dependence of strain to onset of dynamic recrystallisation in alloy C263. Data from tensile and torsion tests.*

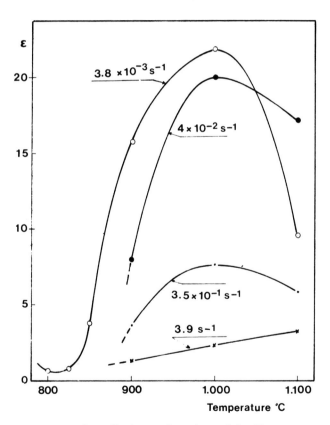

FIG. 7. — *Strain-rate dependence of ductility in torsion for alloy C263.*

geable alloys of higher strength. Apart from the need for special die-heating arrangements, the tool material presents the major obstacle. Molybdenum-based TZM alloy has been used for superplastic forming of powder metallurgy IN 100 [1], but as with other refractory alloys, oxidation resistance is extremely poor. Ceramic materials could probably be employed if care is taken to avoid excessive mechanical and thermal shock.

Carbide and/or solution strengthened Ni or Co-base alloys with flatter flow-stress temperature curves might provide an answer, but little work seems to have been done in this field.

FURNACE PARTS

Furnace parts-muffles, rollers, rails, belts, baskets, etc. usually need both high temperature strength and oxidation/corrosion resistance to a greater or lesser degree depending on the application. Good thermal shock and wear resistance are also often required. The alloys used are generally solid-solution or carbide hardened. They are known as « heat-resisting » alloys and are capable of supporting lower stresses than the « superalloys », but can usually operate at higher temperatures.

Oxidation/corrosion resistance

Good oxidation resistance is obtained by the formation of a stable oxide film, impermeable to ionic diffusion. Most heat-resisting alloys owe their protection to a continuous surface film of Cr_2O_3, being based on the Fe — Ni — Cr system. Figure 8, plotted using the data reported by Brasunas *et al.* [2], shows the effect of composition on the oxidation resistance in this system. The optimum behaviour is obtained when the chromium content is sufficient for the formation of a continuous Cr_2O_3 layer. Less chromium is needed in alloys with higher nickel contents and the minimum rate of attack is lower. However, the results shown were obtained in short-time tests ($\sim$ 24 hours). In practice higher chromium levels are preferable, in order to replace losses due to spalling or evaporation in the form of CrO_3 or CrO_2.

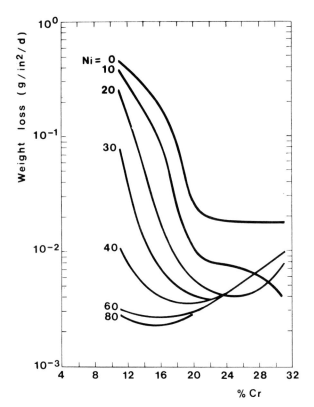

FIG. 8. — *Oxidation at 980° C of* Fe — Ni — Cr *alloys containing ~ 0.4 % C and ~ 1.2 % Si. Results taken from* Brasunas *et al.* [2].

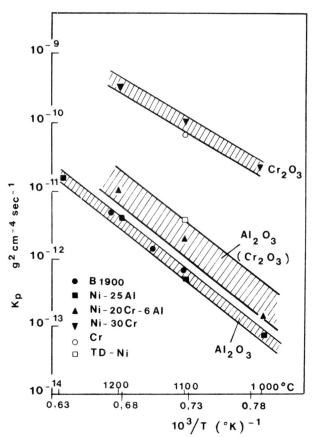

FIG. 9. — *Effect of oxide type on the oxidation kinetics of various alloys. After* Goward [3].

Al_2O_3 affords better protection than Cr_2O_3 as far as oxidation is concerned, due to a smaller concentration of ionic effects. This is illustrated in figure 9, after Goward [3]. Roughly 5 % by weight of aluminium is required to form a continuous film. However, apart from certain superalloys and Fe — Cr — Al heating element materials, which will be discussed in the next section, little use has been made of Al_2O_3 protection for furnace components.

Oxidation, and particularly spalling resistance, can be markedly improved by minor additions of « active » elements such as the rare-earths, but their use in commercial alloys is not yet widespread.

Although oxidation is the most common form of attack, corrosion by other species is often at least as important. In carburising atmospheres, an impermeable oxide is an advantage, since carbon penetration is limited. Nickel, which has little affinity for carbon and which improves thermal shock resistance due to higher ductility and lower thermal expansion, is generally favourable. However, the composition Ni-20 % Cr is especially vulnerable in strongly carburising conditions, since the ternary eutectic in the Ni — Cr — C system, which melts at 1,045° C, occurs at Ni-20 % Cr-3.5 % C. The liquidus temperature rises along the eutectic valley, reaching 1,305° C at 32 % Cr-2.2 % C [4]. In general, the chromium content should be sufficient to allow for losses due to carbide formation.

Because of its relatively high solid solubility and the slow kinetics of nitride formation, attack by nitrogen is not normally a problem, but can become more severe when it is present in the nascent state. In this case,

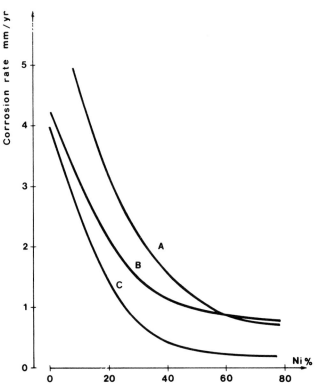

FIG. 10. — *Effect of* Ni *content on the corrosion rate of* Fe — Ni — Cr *alloys in three different nitriding atmosphères :* A. *586 hours in a nitriding furnace at ~ 540° C,* B. *1 011 hours in an ammonia heater at 460° C,* C. *1 540 hours in an ammonia circuit at 500° C. After* Moran *et al.* [5].

as with carbon pickup, a high nickel content is preferable, as shown in figure 10, after Moran *et al.* [5].

Corrosion by sulphur-containing gases can be extremely deleterious, particularly under reducing conditions. This is due to the exceedingly low solid solubility of sulphur, together with the formation of low melting-point eutectics (645° C in the Ni — S system, 877° C for Co — S, 988° C for Fe — S and 1,350° C for the Cr — S system). Chromium, which forms a high melting-point stable sulphide is therefore a favourable element, whereas high nickel contents are deleterious. Sulfidation resistance can be considerably improved by fairly small additions of aluminium, as shown by Schultz *et al.* [6], figure 11.

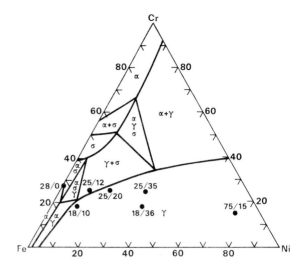

FIG. 12. — *Some common heat-resisting alloy matrices based on the* Fe — Ni — Cr *system (650° C section. Metals Handbook 1948).*

At medium temperatures, low stacking fault energy (compositions near the austenite phase boundary) is probably important for good creep strength, whereas high nickel content (low diffusivity) seems to be more important at higher temperatures, figure 13.

Considerable hardening is obtained by carbon additions, figure 14, and many heat resisting alloys produced as castings contain about 0.4 % C. Solid solution hardening is often enhanced by alloying with elements such as tungsten, molybdenum and niobium. Cobalt probably strengthens mainly by lowering the stacking fault energy. Total additions are limited by sigma

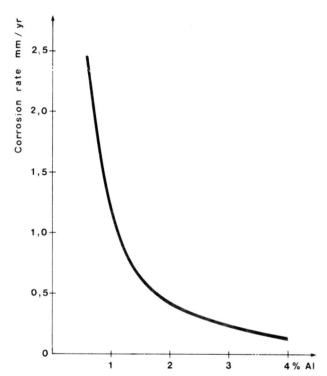

FIG. 11. — *Effect of aluminium on the resistance to sulphur attack of an alloy* Fe — 32Ni — 20 Cr *930 hours exposure at 593° C in an atmosphere containing 9 to 33 %* H_2S, *18 to 65 % light hydrocarbons and 2 to 72 % nitrogen. After Schultz* et al. [6].

In the case of corrosion by hot salts, of which the most severe are the halides, the main requirement for satisfactory resistance is a high chromium content, as has also been found for the combined attack by sulphur and NaCl, frequently encountered in marine turbines [7, 8], and the same is true for resistance to vanadium containing ashes.

Creep resistance

The mechanical strength required of furnace parts is in general fairly moderate, in many cases being limited to the ability of the component to support its own weight.

Figure 12 shows the location of some of the most commonly used matrices in the Fe — Ni — Cr system.

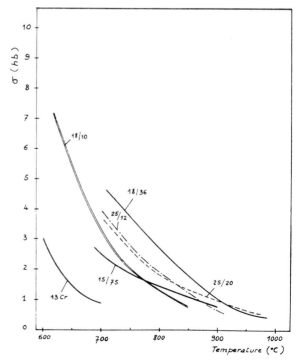

FIG. 13. — *Creep strength of some simple heat-resisting alloy matrices. Stress for 1 % elongation in 10,000 hours. Figures indicate* Cr *and* Ni *contents, balance* Fe.

phase formation. They can be higher in more nickel-rich matrices, but the price factor is often the major obstacle. Figure 15 shows the effect of various additions to an Fe — 35Ni — 25Cr matrix, while figure 16 compares these same materials with a number of more special commercial alloys, whose nominal compositions are

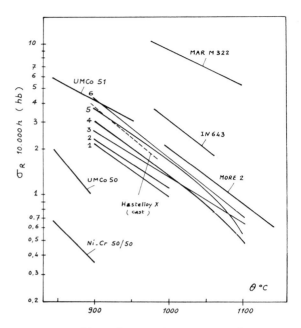

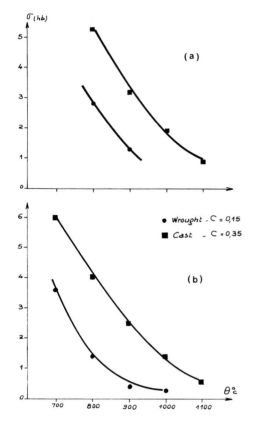

FIG. 14. — *Effect of carbon content on the creep strength of a 25 % Cr — 20 % Ni steel; (a) stress for rupture in 1,000 hours; (b) stress for 1 % elongation in 10,000 hours.*

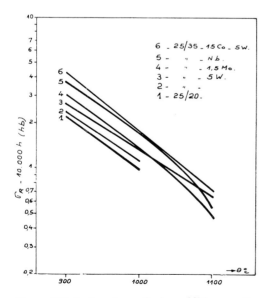

FIG. 15. — *Effect of various alloying additions on the creep strength of cast* Fe — Ni — Cr *heat-resisting alloys containing 0.4 % C.*

given in table I. Among these, the cobalt-base super-alloy MAR M 322 represents the upper performance limit of its category. The poor creep resistance of the Ni — 50Cr composition, developed for use in conditions of severe vanadium corrosion, has been greatly improved by an addition of 1.5 % Nb in the alloy designated IN 657, whose creep strength is comparable to that of a 25 % Cr-20 % Ni steel (curve 1 in fig. 16). The UMCo-50 and UMCo-51 grades, the latter of which contains 2 % Nb in the Fe — 50Co — 28Cr matrix, are extremely insensitive to attack by slags and molten salts and have excellent wear and thermal shock resistance. The alloy IN 643 is an alloy developed for the manufacture of cast tubes for the petrochemical industry, and is a good illustration of the problem which faces the alloy-designer in this field, namely that of obtaining a maximum combination of creep strength and corrosion resistance while avoiding the formation of embrittling sigma phase.

ELECTRIC HEATING ELEMENTS

Electric heating elements can be considered as a special type of furnace component. Except in the case of the ferritic Fe — Cr — Al alloys, the main problem is that of oxidation/corrosion resistance, but there are also certain physical property requirements: high electrical resistivity, low temperature coefficient of resistivity and high thermal emissivity of the oxide layer. The mechanical strength level needed is low; an element must be able to support its own weight and must be sufficiently ductile to absorb thermal strains. However, this aspect becomes important if resistance to environmental degradation makes it possible to work at very high homologous temperatures.

TABLE I. — *Nominal compositions of alloys figuring in the text (wt %).*

Alloy	C	Fe	Ni	Co	Cr	W	Mo	Mn	Ti	Al	Mn	Si	Zr	Ta	V
T 63W, Mo-RE 1.	0.4	bal	33	—	25	5	—	—	—	—	1.0	1.5	—	—	—
Manaurite 35 D.	0.45	bal	35	—	25	—	2	—	—	—	1.2	1.2	—	—	—
Manaurite 36 X.	0.4	bal	33	—	25	—	—	1	—	—	1.0	1.5	—	—	—
Supertherm . .	0.4	bal	35	15	25	5	—	—	—	—	1.0	1.5	—	—	—
MO-RE 2. . .	0.4	—	bal	—	33	17	—	—	—	—	—	—	—	—	—
IN 643. . . .	0.5	3	bal	12	25	9	0.5	2	0.15	—	< 0.3	< 0.3	0.08	—	—
Hastelloy X . .	0.1	18.5	bal	1.5	22	0.6	9	—	—	—	⩽ 1	⩽ 1	—	—	—
IN 657. . . .	0.1	⩽ 1	bal	—	50	—	—	1.5	—	0.15	—	—	—	—	—
Umco-50 . . .	0.08	bal	—	50	28	—	—	—	—	—	0.8	0.8	—	—	—
Umco-51 . . .	0.35	bal	—	50	28	—	—	2.1	—	—	0.8	0.8	—	—	—
Mar M 322 . .	1.0	—	—	bal	21.5	9	—	—	0.75	—	—	—	2.25	4.5	—
Nimonic 90 . .	0.07	—	bal	19	19	—	—	—	2.50	1.50	—	—	—	—	—
C 263	0.06	—	bal	20	20	—	5.9	—	2.15	0.45	—	—	—	—	—
U 500	0.10	—	bal	18	18	—	4.0	—	3.0	3.0	—	—	—	—	—
U 710	0.07	—	bal	15	18	1.5	3.0	—	5.0	2.5	—	—	—	—	—
IN 100 . . .	0.18	—	bal	15	10	—	3.0	—	4.7	5.5	—	—	—	—	1.0

Except in cases of forced convection, the heat, W, lost from an element is almost exclusively due to radiation, so that

$$W = \varepsilon\sigma S(T_E - T_F)^4$$

where ε is total emissivity, σ is Stefan's constant, S is the total radiating surface area, $T_É$ is the temperature at the element surface and T_F is the furnace temperature. This heat is produced by the current in the heater elements, and is also given by

$$W = \frac{V^2 A}{J\rho L}$$

where V is the applied voltage, J is the mechanical equivalent of heat, ρ is the resistivity, A is the cross-sectional area of the elements and L their total length.

Because dimensions are limited by a need for a compromise between mechanical strength and environmental resistance on the one hand and radiating-surface-to-volume ratio on the other hand, and since convenient operation demands reasonable current and voltage levels, this places requirements on the resistivity. Thus, in practice, values ranging from $\sim 10^{-5}$ to $\sim 50\,\Omega.\text{cm}$ are acceptable, although $10^{-4}\,\Omega.\text{cm}$ is a more usual lower limit.

While more exotic metallic materials can be used in some special small-scale furnaces, and although certain refractory compounds are beginning to find industrial applications at very high temperatures, by far the major part of all heating elements are made from alloys based on the Fe — Ni — Cr and Fe — Cr — Al systems, and the present discussion will be limited to this range of products.

Approximate isoresistivity contours are plotted in figure 17 for the Fe — Ni — Cr system and in figure 18 for the Fe — Cr — Al system. The beneficial effects of chromium and aluminium, and of nickel in iron-rich alloys, are clearly visible.

Since a resistivity of about 80 to $100 \times 10^{-6}\,\Omega.\text{cm}$ is generally considered to be a minimum value for convenient operation, the principal alloys in the Fe — Ni — Cr system are austenitic. Because of this, creep strength is not usually a problem, life being limited chiefly by oxidation-corrosion. As has been discussed in the section dealing with materials for furnace parts, good

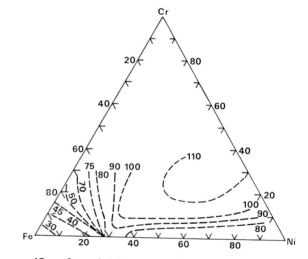

FIG. 17. — *Iso-resistivity contours in the* Fe — Ni — Cr *system* $(\mu\Omega.\text{cm at } 20^o\,C)$.

environmental resistance requires at least 15 % Cr and, except in the case of sulphur-containing atmospheres, performance improves with increasing nickel content. Many of these alloys also contain 1 to 2 % Si and this increases both oxidation resistance and electrical resistivity.

Alloys based on the Fe — Cr — Al system, with 20 to 30 % Cr and 4 to 6 % Al, whose electrical resistivity is high and varies little with temperature, offer exceptional oxidation-corrosion resistance at elevated temperatures, due principally to the formation of a continuous Al_2O_3 surface film. Furthermore, the solidus temperature is in the region of 1,500° C, compared with values between 1,355 and 1,410° C for the austenitic Ni — Cr and Fe — Ni — Cr grades. However, the ferritic structure leads to mechanical property limitations, namely low creep strength and lack of ductility. The ductile to brittle transition characteristic of body-centered cubic materials occurs at around 100 to 200° C in these alloys, so that it is advisable to carry out final forming operations between 200 and 300° C for thick sections. Rapid grain-growth, which sets in above about 850 to 900° C, enhances the embrittlement,

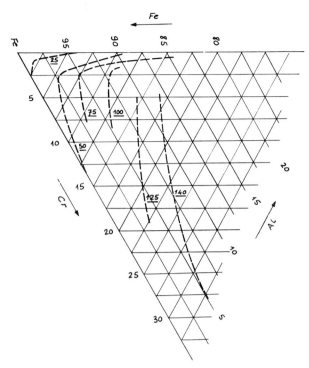

FIG. 18. — *Iso-resistivity contours in the* Fe — Cr — Al *system* (μΩ.cm at 20º C).

leading to an almost glass-like behaviour at room-temperature after service in this range. Ductility can also be decreased by the phenomenon known as « 475º C embrittlement » which occurs on exposure at temperatures between 400 and 550º C, and which is generally attributed to either an order-disorder reaction or to a very fine precipitation. However, this temperature region can usually be largely avoided in practice and the embrittling reaction is reversible, the initial condition being restored by annealing for a few minutes at 600º C. While the lack of ductility makes the Fe — Cr — Al alloys difficult to manufacture and delicate to use, the most common cause of failure at high temperatures is excessive creep deformation, entraining local thinning or short-circuiting. In spite of these drawbacks, the ferritic alloys are widely used since they represent the highest temperature capability of the common metallic materials and because of their relatively low price.

Future trends will probably see more generalised use of rare-earth additions for increased oxidation/corrosion resistance. However, the ferritic alloys have by far the best inherent temperature capability, due to their high melting points, and would be assured of more widespread use if the problems of creep strength and embrittlement could be solved.

CONCLUSIONS

Three fields of particular interest to the metallurgical industries have been described, in which materials requirements are especially demanding. While at the forefront of technological advance appreciable further progress in performance will probably require the introduction of ceramics, in the majority of cases cost is a primordial factor, and much can still be done to improve the price: service-life ratio in metallic alloys. This paper, which has only treated metallic metarials, has attempted to outline the present situation and to point out areas in which improvement is desirable, without necessarily pretending to provide the answers.

* * *

REFERENCES

1] Moore (J. B.), Tequesta (J.), Athey (R. L.). — *U. S. Patent*, 1970, 3, 519, 503.
[2] Brasunas (A. de S.), Gow (J. T.), Harder (O. E.). — *Proc. ASTM*, 1946, 46, 870.
[3] Goward (G. W.). — *J. Metals*, 1970, 31.
[4] Köster (W.), Kabermann (S.). — *Archiv f. d. Eisenhüttenwesen*, 1955, 26, 627.
[5] Moran (J. J.), Mihalisin (J. R.), Skinner (E. N.). — *Corrosion*, 1961, 17, nº 4.
[6] Schultz (J. W.), Hulsizer (W. R.), Abbott (W. K.). — (Paper nº 18), NACE *Conference « Corrosion 72 »*, 1972.
[7] Waddams (J. W.), Wright (J. C.), Gray (P. S.). — *J. Inst. Fuel*, 1959, 32, 246.
[8] Beltran (A. M.), Shores (D. A.). — The superalloys, (Sims, C. T., Hagel, W. C., ed.), *Wiley*, New York, 1972, p. 317.

Higher efficiency, improvement in reliability, and economics, have led to the need for greater consideration to be paid to the selection of metallurgical materials for use in petrochemical furnace equipment. Developments likely to meet these and future, requirements are described.

Recent Developments in Materials for High-Temperature Service in Petrochemical Process and Refinery Furnaces

W. Herda and A.J. Rickard F.I.M.

INTRODUCTION

THE CONTINUING aims for increased efficiency in petrochemical furnace plant, particularly ethylene furnaces and steam-hydrocarbon reformers, has led to greater sophistication in design, increase in the size of individual plant and in operating temperatures, pressures etc. These trends, together with a desire to improve reliability and overall economics, place increasingly stringent demands on the furnace tube materials in terms of their ability to meet design and operation requirements.

This paper describes some of the salient metallurgical considerations governing the selection of materials for petrochemical furnace tubes. Some new material developments to meet current and future requirements are outlined.

ETHYLENE PYROLYSIS FURNACES

Ethylene is one of the most important of the basic raw materials used by the plastics and synthetic fibre industries. Production capacity in Western Europe has been rising at an unprecedented rate: this is shown in Table I which presents estimates of ethylene requirements up to 1980. To meet this demand, arising principally from the United States, Japan, and Western and Eastern Europe, about 120 large plants will need to be built before 1980. Over the last decade individual plant size has increased from 10 000 to 450 000 t.p.a. and, indeed, some of 750 000 t.p.a. capacity are being considered.

In most instances ethylene is produced by a process of thermal cracking of hydrocarbon-steam mixtures (steam cracking). The essential reaction takes place in tube-coils in furnaces that are heated to 800–1 000°C. Since the cracking furnaces, the tubes, and associated equipment, represent approximately 20% of the overall cost of the plant, the careful specification of tube materials is an important factor in controlling both the initial cost of the plant and the ultimate running costs, linked to maintenance and replacements.

As a factor in improving the economy of the process the general trend for furnaces has been towards operation of higher cracking temperatures and to have shorter residence time of the reactants in the tube, requiring high heat flux density. This leads to a better distribution of desirable products and particularly improved olefine yields. Material selection must, of course, follow this trend to operation at higher temperatures.

The earlier built "medium severity" plants were designed with tube banks horizontally disposed within the furnace and these were maintained in position by suitable supports that also needed to withstand the operating temperatures.

However, as operating conditions became more severe it became difficult to find tube support materials able to withstand the higher temperatures required in the radiant section of the furnace, in order to maintain the high heat flux in the tube and reactant. To overcome this fundamental difficulty, modern "high severity" furnaces are designed with the reactant tube coils installed vertically in the furnace. This allows tube supports to be placed in the relatively cooler part of the furnace.

To achieve the optimum heat-flux into the reactants there is now a firm trend to increase the number and reduce the diameter of tubes. Since they are usually produced as centrifugal castings there is a practical difficulty in reducing the internal diameter of the tubes; the current limitation in this respect is of the order of 50 mm in diameter.

Oxidation/Carburisation Characteristics

The tube-material has to be sufficiently strong and resistant to corrosion by oxidation/carburisation mechanisms and, insofar as these characteristics are affected by the temperature of operation of the process, the selection of the tube material becomes a key factor for economic operation. Both wrought and cast alloys are well-established for use in "low" and "medium" severity plants. The wrought Alloy 800 and cast 25 Cr/20 Ni steel (see Table II for compositional details) are typical examples.

In the initial years of operation of the pyrolysis process the plant operators reported problems with pyrolysis tubes arising from carburisation/oxidation reactions. In particular diffusion of carbon into tubes caused differential coefficients of expansion that led to high internal stress values in the tubes and consequent failure by stress-rupture. Other features of some of the early failures were associated with hot spots, arising during the decoking operation, and also uneven temperature profiles along or across the tubes during service. Both of these phenomena led to extremely severe oxidation/carburisation reactions. Examples of this type of failure are shown in Fig. 1.

One early failure, Fig. 1(b), shows a creep fracture in a 25 Cr/20 Ni cast steel pyrolysis tube. In this instance poor welding technique caused deposition of coke and con-

Table I Ethylene demand 1970–1980 in M t/a

Country	1970	1975	1980
U.S.A.	9·7	9·3	17·0
Western Europe	5·7	9·3	13·8
Japan	2·5	4·5	7·0

Mr Rickard is Senior Marketing Engineer in the U.K. Marketing Office of International Nickel Ltd., London, and Mr Herda is Senior Project Engineer in the Commercial Development Department.

Table II Nominal Alloy Compositions

Designation	Composition %												Other Elements
	C	Si	Mn	Cr	Ni	Co	Fe	Mo	Ti	Al	Nb	W	
Wrought Alloys													
Alloy 600	0·15	—	1·0	15	Bal	—	8	–	–	–	–	–	
Alloy 625	0·10	0·5	0·5	22	Bal	—	5	9	–	0·4	–	–	
Alloy 800	0·10	1·0	1·5	21	32	—	Bal	–	0·3	0·3	–	–	
Alloy 807	0·10	0·7	1·0	21	40	8	Bal	–	0·3	0·3	1·0	5	
IN-793	0·05	0·35	–	20	34	–	Bal	–	0·3	1·8	–	–	Cu 0·3
Cast Alloys													
35/25NiCrW	0·5	1·7	0·6	25	35	–	Bal	With W & Co additions depending on grades					
35/25NiCrNb	0·5	1·7	0·6	25	35	–	Bal	With Nb addition					
35/25NiCrNbW	0·5	1·7	0·6	25	35	–	Bal	With Nb abd W additions					
19/9 CrNi	0·2–0·4	2·0	2·0	21	10	–	Bal	–	–	–	–	–	
25/12CrNi	0·2–0·5	2·0	2·0	26	12	–	Bal	–	–	–	–	–	
25/20CrNi	0·2–0·6	2·0	2·0	26	20	–	Bal	–	–	–	–	–	
35/15NiCr	0·35–0·75	2·5	2·0	15	35	–	Bal	–	–	–	–	–	
60/12NiCr	0·35–0·75	2·5	2·0	12	60	–	Bal	–	–	–	–	–	
66/17NiCr	0·35–0·75	2·5	2·5	17	66	–	Bal	–	–	–	–	–	
IN-643	0·5	0·5	0·5	25	48	12	3	0·5	0·1	–	2	9	Zr 0·1
IN-519	0·3	1·0	0·7	24	24	–	–	–	–	–	1·5	–	
IN-657	—	—	–	50	48·5	–	–	–	–	–	1·5	–	

Note: Several of the above alloys are the subject of patent rights.

sequently most severe carburisation at this point. Considerable distortion can arise when irregular carburisation occurs, this leading to local differences in coefficients of expansion, to differential thermal stresses, and ultimately to fracture (Fig. 1(c)).

These examples serve to demonstrate that local overheating of tubes, even in plants operating under relatively moderate conditions, can cause serious damage to individual tubes.

An extremely important finding from world-wide field experience was that tubes with an internally machined surface gave substantial improvement in resistance to oxidation/carburisation phenomena. In practice this has led to a very substantial overall improvement of tube life, and the installation of internally machined tube is now an established practice.

Hubert[1] has recently published the results of 7 000 h trials of machined and unmachined tubes in the field. A range of materials were tested at temperatures from 870–1 060°C. Some of the results of his tests are summarised in Table III, which clearly illustrates the merit of internally machining the tubes. Nominal compositions of these alloys, as well as others mentioned later in this paper, are all summarised in Table II.

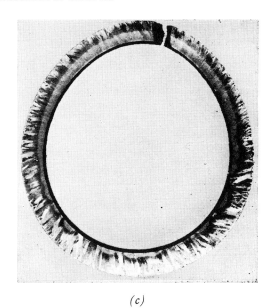

(c)

(a)

(b)

Fig. 1—Tube failure in 25/20 Cr Ni Cast Tubing ascribed to carburisation

    (a) Typical localised carburisation at a hot spot on the internal surface of an ethylene pyrolysis tube.

    (b) Local carburisation in a pyrolysis tube due to bad welding technique—crevices in the root run resulted in carbon sedimentation.

    (c) Irregular carburisation in a pyrolysis tube due to inhomogeneous temperature profile, resulting in bulging and creep failures.

Table III   Carburisation Trends in a Series of Cast Heat Resistant Ni Cr Alloy Steel Tubes. (After Hubert)

Distance from Internal Surface (mm)	Increase in Carbon Content (%)							
	25/20 Cr Ni		35/25 Ni Cr W		35/25 Ni Cr Nb		35/25 Ni Cr Nb W	
	as-cast	machined-bore	as-cast	machined-bore	as-cast	machined-bore	as-cast	machined-bore
1	1·75	0·11	1·95	0·19	0·6	0·07	0·2	0·04
2	1·6	0·13	1·9	0·21	0·5	0·11	0·15	0·05
4	1·1	0·10	1·3	0·16	0·15	0·06	0·1	0·03
6	0·65	0·07	0·3	0·09	<0·05	0·04	<0·05	0·02
8	0·3	0·05	0·15	0·05	<0·05	0·03	<0·05	0·01
10	0·1	0·04	0·05	0·02	<0·05	0·02	<0·05	nil

All cast tubes exposed for 7 000 h in a carburising atmosphere at 1 000°C.

In "high severity" plants corrosion at the higher operating temperatures becomes much more difficult to control, and hot spots or maldistribution of temperature cause stress differences that can have even more serious consequences.

Influhnce of Cast Structure

Hubert's[1] work suggests that the crystal structure of an alloy may be important in the terms of resistance to carburisation. In particular, he suggests that the resistance of columnar structures to carburisation is greater than that of equiaxed structures, as is illustrated in Table IV, for experi-

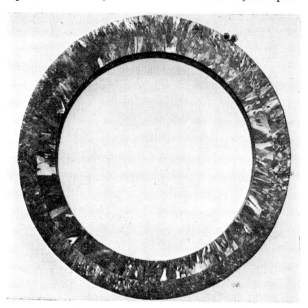

Fig. 2—Macro-structure of weld-formed tube.

Table IV   Influence of Cast Grain Structure on Carburisation Characteristics of 25/20 Cr Ni Steel. (After Hubert)

	Increase in Carbon Content (%)							
	Equiaxed Grains				Columnar Grains			
	Distance from Internal Surface (mm)							
Days at 1 050°C	0·1	0·2	0·4	0·8	0·1	0·2	0·4	0·8
2	0·5	0·5	0·5	0·45	0·65	0·7	0·6	0·55
4	1·0	1·1	0·8	0·7	1·2	1·3	1·15	0·8
8	1·2	1·4	1·3	1·0	*	1·4	1·3	1·1
16	1·5	1·75	1·7	1·4	*	*	1·5	1·3
32	2·1	2·5	2·8	2·7	*	1·6	1·8	1·6

* not measured.

ments extending over a period of 32 days. The difference in behaviour, however, appears to be less at temperatures above 1 050°C. Some foundries aim to control structure for ethylene furnace tubes.

"Weld-formed" tubing involving basically continuous electroslag casting has recently been developed in Japan, and one claim being made for such material is that the structure can be readily controlled to give a columnar structure in the bore. A section of such a tube is shown in Fig. 2.

Recent Trends in Alloy Development

The traditionally established materials have limitations for operation in the hottest parts of the radiant section of "high severity" furnaces and this has led to the development of improved pyrolysis tube materials. Work in a number of laboratories has indicated the compositional changes necessary to provide an alloy having improved carburisation/oxidation resistance. Chromium, nickel and silicon in increased amounts can improve this property dramatically. Figure 3 after Avery[3] shows the depth of carburisation of a range of alloys with increasing nickel and silicon content and it is seen that as nickel content increases the carbon gradient across the wall becomes less steep. This

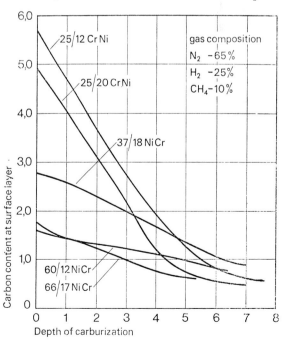

Fig. 3—Carburisation characteristics of various Cr–Ni–Fe alloys after exposure of 1 000 h, (after Avery).

Table V Comparison of Derived Stress-Rupture Properties of 25/20 Cr Ni and IN-519 Cast Steels with IN-643

Test Temperature (°C)	*Rupture Stress N/mm²*								
	25Cr Ni (0·35/0·44% C)			*IN-519*			*IN-643*		
	1 000 h	*10 000 h*	*100 000 h*	*1 000 h*	*10 000 h*	*100 000 h*	*1 000 h*	*10 000 h*	*100 000 h*
800 …	72	47	(23)	93	50	(32)			
900	37	22	(12)	44	30	(21)	76	63–68	(59)
950	26	14	(9)	33	19	(13)	58	48	(36)
1000	17·5	—	(5)	24	12	(7·0)	44	37–39	(30)
1050	12	—	—	—	(10)	—	29–33	22–29	—
1100	(9·5)	—	—	—	—	—	(27)	(20)	—

(—) extrapolated figures.

results in a lower stress-gradient during operation and reduces the chance of stress-rupture failure.

Lewis[2], and more recently Hubert[1], have also demonstrated that carbide-stabilising elements, and particularly niobium and tantalum, improve the resistance to carburisation/oxidation. A comparison of the behaviour of the commercial niobium-containing (35 Ni/25 Cr, Nb and 35 Ni/25 Cr, Nb, W) and niobium-free (35 Ni/25 Cr, W and 25 Cr/20 Ni) alloys made in Table III illustrates this benefit.

A cast nickel-base alloy designated IN-643 has recently been introduced in which good carburisation/oxidation resistance and high stress-rupture strength and ductility are combined. Its high temperature corrosion behaviour is compared with that of cast 25 Cr/20 Ni steel, Fig. 4; under cyclic oxidation and carburising test conditions where 25 Cr/20 Ni steel shows a severe weight loss, the IN-643 exhibits a small rate of weight gain, suggesting that a stable protective surface oxide film continues to be developed during service.

The effects of composition upon corrosion have been studied by workers in a number of laboratories. Even though laboratory test conditions cannot exactly reproduce service conditions they have afforded useful pointers for the development of improved alloys. Silicon has been found to delay the carburisation reaction significantly and this is reflected in a preference by some operators for higher silicon contents in centrifugally-cast tubing.

Table V compares the rupture strength of 25 Cr/20 Ni steel and IN-643 obtained in tests on centrifugally-cast tube material. The good rupture ductility exhibited by IN-643 (10–17% after exposure at 1 000°C) may well be used to advantage. By virtue of its high rupture properties combined with relatively good carburisation resistance, IN-643 is already being evaluated for service in high severity ethylene crackers by major chemical companies. One such trial has already been in operation for a year.

In horizontally tubed ethylene furnaces the tube supports are usually of cast 25 Cr 12 Ni steel. However, with operating temperatures increasing up to 1 050°C and tendencies towards failures by creep and oxidation, the cast 25 Cr/20 Ni alloy steel became used more frequently.

The stringent service required in many "high severity" furnaces requires castings in stronger material, and more highly alloyed steels such as 45 Ni 25 Cr with 5% W and 35 Ni 25 Cr with 5% W and/or 1·5% Nb are now being used at metal temperatures up to 1 150°C due to their high creep strength and oxidation resistance. In view of its outstanding high temperature strength the previously mentioned alloy IN-643 may be of interest in the severest applications.

Fuel-Ash Corrosion

Another example, where a change of material has enabled a practical problem to be circumvented, is illustrated by an

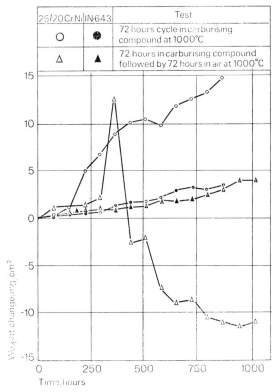

	25/20CrNi	IN-643	Test
	O	◉	72 hours cycle in carburising compound at 1000°C
	△	▲	72 hours in carburising compound followed by 72 hours in air at 1000°C

Fig. 4—Carburisation and carburisation/oxidation resistance of 25/20 Cr Ni and IN-643 at 1 000°C.

Table VI Hot Corrosion Resistance of IN-657 with other Cast Heat Resistant Alloys

Alloy	Test Temp. °C	Exposure Time h	*Environment and Weight Loss*	
			75% Na₂SO₄ + 25 NaCl mg/cm²	80% V₂O₅ + 20% Na₂SO₄ mg/cm²
IN–657	800	300	47	130
50/50 Cr Ni		300	40	170
25/20 CrNi Steel		16	540	300
IN–657	900	300	17	350
50/50 Cr Ni		300	40	380
25/20 CrNi Steel		16	640	1 000

Table VII Derived Rupture Properties for 25/20 Cr Ni Cast Steel Compared with 50/50 Cr Ni and IN-657 Cast Heat Resistant Alloys

Temperature °C	25/20 Cr Ni Steel			50/50 Cr Ni			IN-657		
	Stress (N/mm^2) to rupture in								
	1 000 h	10 000 h	50 000 h	1 000 h	10 000 h	50 000 h	1 000 h	10 000 h	50 000 h
700	130	93	72*	65	30*	—‡	150	107	85*
800	72	47	35*	38	20*	—‡	74	485	35*
900	37	22	6·0*	20·5	10·0*	6·0*	33	20	13·6*
1000	17·5	9·5	6·1*	9·2	42*	—‡	13·5	7·5	4·5

*—extrapolated value.

‡—insufficient data.

experience some years ago, with a small "low severity" furnace fired by heavy residual oils. The 25 Cr/20 Ni and 18/8 Cr Ni steel tubes showed corrosion due to oil ash on the fire-side and excessive carburisation arising from hot spots on the product side. Although the creep strength of a simple 50 Cr/50 Ni alloy is hardly adequate at temperatures above 850°C, it was decided to install a trial tube in this material. After four years' operation no corrosion problems of either kind had been experienced.

Whilst the use of heavy fuel-oil for firing petrochemical furnaces is not common for various reasons, the need for an improved 50 Cr/50 Ni alloy having stress-rupture properties at least comparable to 25 Cr/20 Ni cast steels was apparent. Thus a 50 Cr/50 Ni alloy containing niobium, designated IN-657, was developed which has a stress-rupture strength at 900°C similar to that of the cast 25 Cr/20 Ni steels, but a much improved resistance to corrosion by oil ashes (Table VI). This material could be useful in smaller plants that are occasionally designed for versatile use with a range of fuels, including heavy residual oils. The properties of the new alloy IN-657 are compared with those of the simple 50 Cr/50 Ni alloy and of cast 25Cr/20 Ni steel in Table VII.

Heat Exchangers and Associated Items

After the cracking reaction the resulting gas mixtures need to be rapidly and uniformly cooled to prevent a further secondary reaction occurring. A considerable amount of heat is extracted from the gases in this cooling process which, incidentally, is used to generate high-pressure steam for use in the plant. For example, a 450 000 t.p.a. capacity ethylene plant produces 300 t/h of steam in about 20 heat exchangers.

In this "quenching operation" the various components of the equipment, i.e. the transfer piping, various pipe fittings and other components of the heat exchangers are exposed to extremely severe cyclic thermal conditions. To resist these thermally-induced fatigue conditions, the transfer pipe materials must have good stress-rupture properties combined with good ductility, even after extended exposure to the service temperatures. In several plants use has been made of welded fabrications of wrought and cast stainless steels of the 25 Cr/20 Ni type. In many cases, failures occurred due to carbide and sigma-phase embrittlement even after quite short periods of service. Indeed, in those parts of the equipment where start-up and shut-down operations induced stress peaks in some parts of the design, flanges have been recorded as having fractured in service. Alloy 800 is now frequently specified for these applications because of the exceptionally severe demands upon material properties. Another wrought alloy, designated Alloy 807, has been developed for similar applications and is already gaining service experience in one plant. This alloy has a significantly greater strength than Alloy 800, combined with very good ductility even after exposure to service temperatures (Table IX).

The cast 32 Ni/20 Cr steel with 1% niobium has also been recently used extensively for transfer piping in some installations in Western Europe. Past field and research experience suggest that the wrought alloys 800 and 807, as well as the above-mentioned cast alloy, can offer the best alternative to meet the specific mechanical-strength problems that arise in transfer piping systems.

REFORMER APPLICATIONS

Catalytic steam-reforming of hydrocarbon feedstocks finds extensive world-wide application for the production of such basic chemicals as ammonia, methanol, hydrogen, oxo-gas, and reducing gas.

In recent years, plant capacities have increased in size considerably, very often accompanied by design improvements that necessitate higher operating pressures and temperatures. The proper selection and use of materials in this reforming equipment is, of course, an important requirement to ensure both reliability and profitable operation of the plant.

The reforming operation is carried out by passing the reactants over a catalyst contained in heated reformer tubes. These tubes comprise a substantial proportion of the overall investment in the plant.

Centrifugally-cast tubes in 25 Cr/20 Ni steel have been

Table VIII High Temperature Oxidation Resistance of Alloy 519 in Comparison with Other Cast Heat Resistant Alloys

Time (h)	IN-519			25/20 Cr Ni Steel			37/18 Ni Cr Steel		
	Weight gain in air—mg/cm^2								
	900°C	1 000°C	1 100°C	900°C	1 100°C	1 100°C	900°C	1 000°C	1 100°C
4	0·21	0·64	2·12	0·11	0·38	1·05	0·10	0·24	0·89
16	0·45	1·05	5·29	0·23	0·92	1·93	0·23	0·65	1·58
25	0·60	1·24	6·64	0·31	1·11	2·40	0·32	0·79	2·04
49	0·79	1·59	7·64	0·49	1·51	3·25	0·46	1·05	3·10
81	0·91	1·88	8·34	0·63	1·83	4·07	0·60	1·44	4·09
100	1·00	1·98	8·73	0·68	1·96	4·48	0·69	1·64	4·63

successfully operated in many hundreds of plants for many years and this alloy is still the standard material for the majority of current reformer applications. Some tube failures, however, indicate the possibility of limitations of the use of reformer catalyst tubes in 25 Cr/20 Ni steel as more severe requirements develop in modern plants.

The most common types of failure take the form of longitudinal cracks through the tubes. They are usually remote from welds, without sign of catastrophic oxidation and are generally consistent with some form of overheating; an example is illustrated in Fig. 5 for a tube section which failed after 30 000 h at an operating temperature of 870°C and at a pressure of 24 atmospheres. Catalyst deterioration can result in decreased permeability of the catalyst bed and consequent severe local overheating. In other instances, overheating has been caused by incorrect adjustment of burners leading to irregular tube temperatures throughout the furnace. In view of the harmful effect of temperature cycling on the stress-rupture properties of tube materials it is not surprising that the tube failures were particularly prevalent in those plants which reported a large number of shut-down periods during the initial years of operation. Although longitudinal tube stresses are generally lower than circumferential stresses, there have been a number of instances of circumferential cracking of welds. One explanation for this is suggested by Fig. 6 which shows that the stress-rupture strength of the weldments can be substantially lower than that of the parent material. Some tests suggest that shielded gas-welding processes produce a weld strength somewhat higher than deposits from coated metal-arc electrodes; but this is a controversial topic. Another aspect of this type of failure is the fact that heavy-walled reformer catalyst tubes (above 20 mm wall thickness) have shown a higher probability of failure than expected with thinner-walled tubes. This suggests that thermal gradients that induce thermal stresses across the tube wall are another important factor that can influence the life-expectation. Figure 7 shows an example of a 25 Cr/20 Ni steel reformer tube that failed at a weld after 12 months' service during which it was subjected to more than 20 shut-downs. The wall thickness was 28 mm.

Improved Reformer Tube Alloys

In recent years an alloy, designated IN-519, has been developed which could be regarded as an "optimised 25 Cr/20 Ni steel": the composition is given in Table II. The addition of niobium in this material has resulted essentially in a substantial improvement of rupture strength ($\geqslant 25\%$) over that of 25 Cr/20 Ni steel (see Fig. 6). A carbon content of 0·25–0·35% has been chosen for this alloy since it results in a good compromise between creep properties, weldability and the maintenance of ductility even after prolonged service.

This alloy, generally shows an elongation of 8–15% at room temperature even after exposure for 1 000 h at 800°C; this may be compared with typical comparative values for cast 25 Cr/20 Ni steel of 3–5%. The creep ductility of IN-519 after 10 000 h at 900°C is 8–10% compared with 2–3% for 25 Cr/20 Ni steel.

Figure 8 demonstrates the importance of the niobium/carbon ratio in determining the weldability of the alloy.

The oxidation resistance of IN-519 is comparable with that for the 25 Cr/20 Ni cast steels (Table VIII) for temperatures up to at least 1 000°C.

It would appear that IN-519 will be a most economical material for tubes having wall temperatures in the temperature range up to 1 000°C and where tubes in 25 Cr/20 Ni steel are presently used with wall thicknesses above about 16 mm.

In some countries, e.g. Austria, fired process vessels are subject to Code Approval by authorities, and in consequence the weld-joint factor may need to be considered when cal-culating tube-wall thickness values. Alloy IN-519 offers the advantage of a very high rupture strength even in the weldments. Consequently, higher weld-joint factors can be realistically considered. Figure 6 illustrates this point. It is seen that the strength of the welded IN-519 samples at 900°C is significantly greater than that of unwelded samples in 25 Cr/20 Ni steel. This factor suggests an attractive possibility for using welding consumables of IN-519 for joining 25 Cr/20 Ni steel.

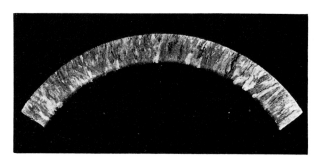

Fig. 5—*Creep failure in a cast 25/20 Cr Ni reformer tube after 30 000 h at 870°C wall temperature and 24 ats. internal pressure.*

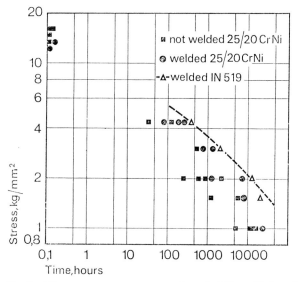

Fig. 6—*The stress rupture properties of (1) welded and unwelded 25/20 Cr Ni cast steel (2) welded IN-519 (24/24 Cr–Ni–Nb) steel.*

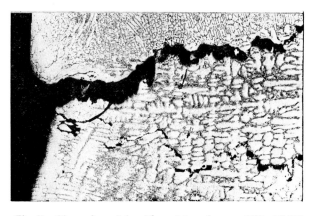

Fig. 7—*Circumferential weld cracking of a cast 25Cr 20 Ni. steel reformer tube after twelve months of operation Wall temperature 930°C.*

Field trials of two tubes of IN-519 were initiated some four years ago in a town's gas reformer, when very severe conditions in terms of temperature cycling resulted in many creep failures of 25/20 Cr Ni steel. The IN-519 tubes are still performing very well after over 30 000 h exposure to the field conditions.

The first commercial applications for IN-519 involved replacement tubes in an ammonia synthesis reformer which had been converted to use natural gas as a feedstock rather than naphtha; a change which needed the tubes to operate at a higher wall temperature. This IN-519 tube-set has been operating successfully for two years. Five other commercial applications for IN-519 reformer catalyst tubes have now been developed involving a total of 700 tubes.

Some recent failures of reformer tubes in 25 Cr/20 Ni steel have been reported due to catastrophic oxidation on the product side. In some instances the abnormally severe oxidation/carburisation that had occurred could be associated with a rather high lead content in the tube material, but in some other instances no definite reason for failure could be recognised. Suggestions are that there may have been some unidentified contamination of the feed steam and that porosity in the bore of non-machined tubes might at least be contributory factors. Taking a direct analogy with pyrolysis furnace tubes it seems reasonable to conjecture that machining of the tube surface could improve the resistance to carburisation/oxidation.

Occasional cracking has occurred in some reformers due to stress-corrosion, particularly in the cold end pieces of tubes, mainly, but not exclusively, in the vicinity of the weld-seams. This type of failure, for example that illustrated

Table IX Stress-rupture properties in 'as-extruded' tube of Alloy 807

Test Temperature (°C)	Stress (N/mm²) to produce rupture in			
	100 h	1 000 h	10 000 h	100 000 h
700	219	161	(123)	(83)
800	115	77	(54)	(43)
900	57	36	(21)	(12)
1 000	26	15	(7·5)	(5·5)

(–) data outside stress limits of Larson-Miller diagram from which these figures were computed.

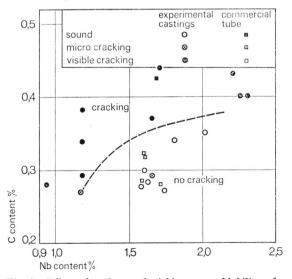

Fig. 8—Effect of carbon and niobium on weldability of cast IN-519 (24/24 Cr–Ni–Nb).

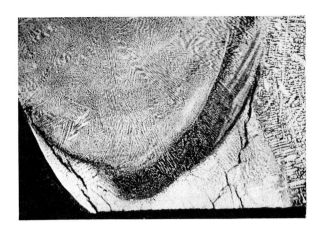

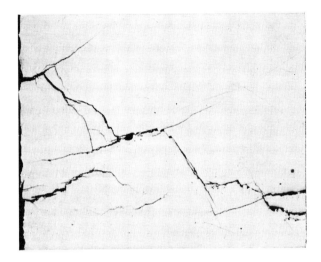

Fig. 9—Transcrystalline stress-corrosion cracking at the weldment and the heat-affected zone of a cast 25/20 Cr Ni steel reformer tube (top end).

in Fig. 9, commonly occurred in designs where inlet and outlet pigtails were at an angle of 90° to the reformer tube access. This feature gives rise to dead zones where condensation can occur. Stress-corrosion cracking failures at the top end of the tube have occurred both in low and high-

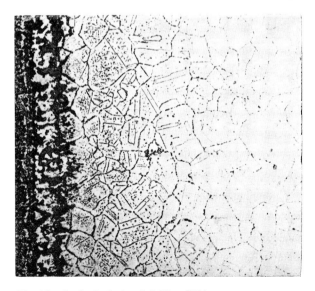

Fig. 10—Carburised pigtail (Alloy 800).

pressure reformer plants. The failures were generally caused by chloride-containing condensates. In one or two instances polythionic acid could be deemed responsible for this corrosion in the sensitized zone of the tube material. Stress-corrosion cracking in the bottom ends of the reformer tubing was found largely in high-pressure plants and was associated with alkali-containing condensates emanating from the catalyst in naphtha reformers. Design changes introducing axial pigtail entry and exit, eliminating "dead zones" have largely obviated failures of this kind.

For production of some chemicals, particularly methanol and oxo-synthetic gas as well as reducing gas, there is a tendency to increase operating temperatures; for producing reducing gas in some projects a design temperature above 1 100°C is envisaged. It is expected that the earlier described alloy IN-643 should help meet the severe requirements of strength and resistance to corrosion at this high temperature.

Table V shows the improvement in strength of -IN643 over that for the cast 25 Cr/20 Ni steels. This high rupture-strength is obtained in combination with a very high rupture-ductility and retained ductility after prolonged service. A number of proprietary alloys containing nickel contents in the range 35–45% and with about 25% chromium and strengthening additions have also been developed for this application. It would appear, however, that the alloy IN-643 has higher rupture strength and ductility which can be utilised in meeting the most stringent conditions in future reformer developments.

Pigtails

The pigtail connecting the reformer catalyst tubes and manifolds are another critical part of reformer furnace construction. Despite some trends towards pigtail-less designs the majority of reformer plants are still equipped with them. The alloy which has given good service for many years is Alloy 800 as specified particularly in ASTM B407. Failures of pigtails which have been reported as due to creep were generally associated with:

(a) Deviations of chemical composition from the specification.
(b) Material not used in the proper condition of solution annealing as set out in ASME Code 1325 Grade 2. In some of these instances the fact was also overlooked that after cold bending the pigtails may be required to be re-solution annealed if the service temperature is to be above the recrystallisation temperature of the materials.
(c) Bad pigtail design. Thermally induced strain can sometimes cause tubes to twist and impose stresses upon the pigtails.

In some instances it has been known for carburisation to cause problems, particularly when it results in embrittlement. Such embrittlement of tubes makes it difficult to isolate tubes by the "nipping technique", often used to isolate individual tubes.

The dependence of carburisation rate on the process-parameters is not fully understood. In addition to the metal temperature, the carbon/steam ratio of the feedstock may be significant. An example of a carburised pigtail in Alloy 800 is illustrated in Fig. 10. In a number of recent plants alloy 807 has been used as an alternative. The stress rupture properties of this alloy are summarised by Table IX. In the temperature range 800–950°C Alloy 807 is 80–100% stronger than Alloy 800. Furthermore, after 5 000 h exposure at 800°C the material still exhibits elongation values of 30–40%. Since Alloy 807 has also slightly improved carburisation resistance its use is to be considered for pigtail applications in the temperature range above 850°C.

A number of failures in the middle 60's, with cast 32 Ni 18 Cr headers, that were mainly due to thermal shock or emergency shut-down conditions, resulted in a reappraisal of the specifications for header material. It was discovered that most failures were associated with secondary carbide embrittlement in conjunction with adverse design and operational factors. In consequence both designers and operators looked for more ductile wrought and cast materials than the high carbon cast 37 Ni/18 Cr steels previously used, and which would also have sufficient creep strength. To fulfil the needs set by the size of the headers and operating conditions, two particular materials were suggested for this component, viz. wrought Alloy 800 or cast 32 Ni/20 Cr/1 Nb steel.

The latter alloy offers good weldability as well as good retained ductility (15% after 10 000 h exposure at 800°C; compared with only 2–3% for the corresponding niobium-free alloy). The stress-rupture properties of this material also showed substantial improvement when compared with high carbon cast 37 Ni/18 Cr steel.

Although Alloy 800 is well-established for use in headers, increase in metal wall temperatures has also led to a consideration of the higher strength Alloy 807 for this application.

The tubes are the components deserving primary concern in reformer furnace design but there are also needs for tube-sheets in waste-heat boilers subject to severe conditions of temperature and stress. Various proprietary alloys are very commonly used for this application but IN-643 might offer some advantages for the most severe conditions.

REFINERY FURNACE APPLICATIONS

Wrought stainless steels are an established choice for refinery furnace tubes in thermal-crackers, HDS units, etc., but it is not so well appreciated that centrifugally-cast products have performed successfully in these refinery furnace applications. Centrifugally-cast products are often

Table X High-temperature Corrosion of Various Heat Resistant Alloys in Sulphur Containing Atmospheres

Alloy	$3\% \, SO_2$–Air				$1 \cdot 5 \, Vol\% \, H_2S$–Hydrogen			
	650°C	900°C	900°C	1 000°C	500°C	600°C	200°C	800°C
	Wt. loss–mg/cm²		Depth of Corrosion penetration—mm					
Alloy 800+Al (IN-793)	0·10	2·2	0·13	0·03	0·05	0·13	0·15	0·56
Alloy 800	0·5	6·8	0·64	0·23	0·05	0·18	0·71	2·10
60Cr40Ni		6·8			0·03	0·43	1·30	4·0
50Cr50Ni		6·9			0·10	0·31	1·27	4·0
35Cr65Ni		4·2			0·13	0·28	1·07	4·0
25Cr20Ni Steel		1·9			0·05	0·23	0·81	4·0
25Cr12Ni Steel					0·05	0·10	0·96	4·0
27Cr iron					0·05	0·23	0·83	4·0

considered on the basis of costs, particularly if the design wall thickness of the wrought tubing is above 6 mm. Additionally, cast products can offer more than just a price advantage over wrought stainless and straight chromium steels. For instance, by using cast 20 Cr/15 Ni steel appreciable advantages in oxidation resistance and creep strength can be obtained. Moreover, the tube material is less sensitive to the effects of occasional overheating periods of the kind that would lead to severe oxidation or creep-failure of wrought products. Additionally, there is some indication that internally machined cast-tubes may not be as susceptible to coke deposition as wrought steels.

Treatment of high sulphur-containing hydrocarbons is becoming very important in the light of air pollution legislation. The desulphurisation plants are expected to require tubular products with improved resistance against both oxidation and sulphidation, and this will pose a serious problem where tube operating temperatures exceed about 550°C; in the range where the choice of wall thickness is based on creep strength values.

To minimise the effects of sulphidation some companies have experimented with parts coated by diffusion of aluminium, but success has been variable. An attempt to provide a solution to this problem has been made by developing an aluminium-modified alloy of the Alloy 800 type, designated IN-793. The analysis is summarised in Table II, and resistance to sulphur-bearing atmospheres as compared with that of some conventional heat-resisting alloys, is given in Table X. These preliminary results are encouraging and further opportunities for field evaluations are required.

Occasional reports are received of failures of 25 Cr/20 Ni alloy steel castings for such components as the tube supports. In instances such as these, where failure has been due to inadequate creep strength, one could suggest that consideration be given to use of Alloy IN-519 before changing to more sophisticated and possibly more expensive materials.

OTHER APPLICATIONS

The topics discussed represent the major petrochemical applications of heat-resisting alloys. There are some other applications, however, that merit some mention particularly where reasonable process operation has only been realised by the use of nickel-containing alloys.

For example in the manufacture of styrene good oxidation resistance and creep strength are required in the material for steam superheater tubing. The superheated steam is used as a diluent in the dehydrogenation of ethylbenzene. The steam is superheated to the temperature range 600–650°C and the metal wall temperature of the furnace tubing is approximately 800°C. In most instances 19 Cr/9 Ni steel is satisfactory but if the above temperatures are to be exceeded 25 Cr/20 Ni steel is occasionally specified.

Furnace tubes are needed in reactors used for the synthesis of carbon disulphide from natural gas or refinery sulphur. Cast 25 Cr/20 Ni steel is again a well-established choice for the tubes, but this steel, although less resistant to overall general corrosion when compared with 28% chromium steel, is less subject to severe carburisation and embrittlement than the wrought chromium steels.

Techniques have been developed recently for the manufacture of heavy-water using a petrochemical feedstock. As part of one process, ammonia needs to be cracked in the temperature range 700–800°C, and the resistance of the furnace tubes towards nitriding is clearly significant. Alloy 800 and Alloy 600 have been first considered but in one instance Alloy 625 has been specified since it offers an optimum combination between resistance to nitriding and high-temperature strength.

CONCLUSION

Some of the trends in material selection for petrochemical furnace tubes, tube supports etc., have been outlined in the context of current and potential problems. Whilst established cast and wrought furnace tube materials, for example cast Cr Ni steels, Alloy 800 and the 300 series of stainless steels, will continue to satisfy a large proportion of petrochemical furnace requirements for the foreseeable future, several newer proprietary and non-proprietary alloys are now in moderately extensive service, and others are being evaluated to meet some of the more stringent material requirements inherent in the more advanced boundary areas of ever-developing high-temperature petrochemical furnace technology.

REFERENCES

1. R. Hubert, J. Thuillier. "Carburisation in Heat-Resisting Alloys Used in Pyrolysis Furnaces". Acieries du Manoir Pompey 1971.
2. H. Lewis. "Factors Affecting Corrosion Resistance of High-Temperature Alloys in Carburising Gases". British Corrosion Journal, 1968, **3**, 166–175.
3. N. S. Avery. "Cast Heat-Resisting Alloys for High Temperature Service". A.R.C. Bulletin 143, 1969.

Alloys for Nitric acid at high temperatures

Corrosion tests were performed on alloys containing Fe, Ni and Cr in 10–70% nitric acid at 100–130°C. Order of merit was found to generally correlate with increasing Cr content, but the high-nickel alloys proved to be superior to austenitic steels

Dr R. G. Lythe, International Nickel Ltd, Clydach, Glam, UK

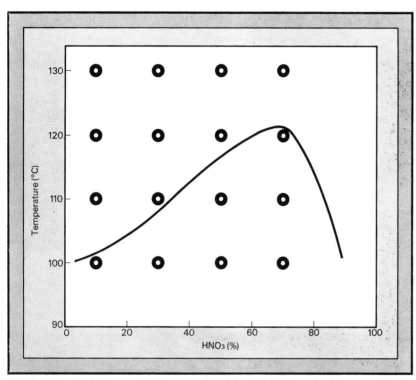

Fig. 1. Boiling point/concentration curve for nitric acid at atmospheric pressure. The 16 test conditions are indicated by circles

Table 1. Materials tested for corrosion behavior in nitric acid

Alloy	Composition wt%									
	C	Mn	Si	Cr	Ni	Fe	Mo	Nb	Ti	Others
304 stainless steel	0·065	1·0	0·38	18·4	9·8	Bal.	c. 0·2	<0·05	<0·05	
304L stainless steel	0·025	0·75	0·62	17·8	10·0	Bal.	c. 0·2	<0·05	0·05	
321 stainless steel	0·065	0·77	0·70	18·1	9·6	Bal.	c. 0·2	<0·05	0·48	
347 stainless steel	0·065	0·88	0·51	18·4	9·9	Bal.	c. 0·2	0·88	<0·05	
316 stainless steel	0·045	1·6	0·59	17·0	11·2	Bal.	2·8	<0·05	<0·05	
316L stainless steel	0·030	1·40	0·39	17·3	12·7	Bal.	2·6	<0·05	<0·05	
18Cr/14Ni/4Si	0·020	0·83	4·74	16·8	17·2	Bal.	<0·1	<0·05	c. 0·2	
Low-carbon 25Cr/20Ni	0·025	0·72	c. 0·35	25·5	19·7	Bal.	<0·1	0·28	<0·05	
Incoloy 800	0·035	0·80	0·38	20·4	32·0	Bal.	c. 0·1	<0·05	0·5	
Incoloy 825	0·040	0·10	0·33	18·8	39·0	Bal.	2·7	<0·05	0·82	Cu 2·0
Corronel 230	0·025	0·10	c. 0·3	32·7	Bal.	2·6	<0·1	<0·05	c. 0·2	
50/50 Ni/Cr	0·025	—	—	49·6	Bal.	—	—	—	—	Al <0·1, Mg <0·002

CHEMICAL plants are increasingly required to function at higher temperatures and pressures. As many processes employ aqueous acids, including HNO_3, it is envisaged that alloys will be required for use under progressively more onerous corrosive conditions in plants containing nitric acid at elevated temperatures and pressures. Certain stainless steels and nickel alloys are known to possess good resistance to corrosion by HNO_3 at temperatures up to atmospheric boiling point but there are only limited data on the corrosion susceptibility of these materials at higher temperatures.[1-4] Accordingly, a study has been made of the corrosion behaviour of nickel alloys and stainless steels in both the solution-treated and sensitised conditions in liquid-phase nitric acid in the range 10–70% HNO_3 and 100–130°C in order to provide a guide for selection of materials. The technique employed represents a considerable 'overtest' owing to the retention of corrosion products within the test vessel and thus the results require careful interpretation in the light of practical experience.

The experimental procedure used a modification of the method of Miller *et al.*[5] Intergranular penetration of each specimen which had been exposed to HNO_3 was assessed by focusing a binocular microscope on the surface of the corroded material. The results obtained, taken in conjunction with those of weight loss, gave a reasonable indication of the extent of such localised corrosion. In addition, the depth of penetration derived by sectioning and examining a few specimens in the conventional manner showed good correlation with the above. Fig. 1 shows the temperatures and concentrations employed in the tests.

Most of the alloys (Table 1) were commercially produced and supplied in the softened condition as 20 s.w.g. sheet. Types *304L*, *316L* and 18Cr/14Ni/4Si were only available as 10 s.w.g.

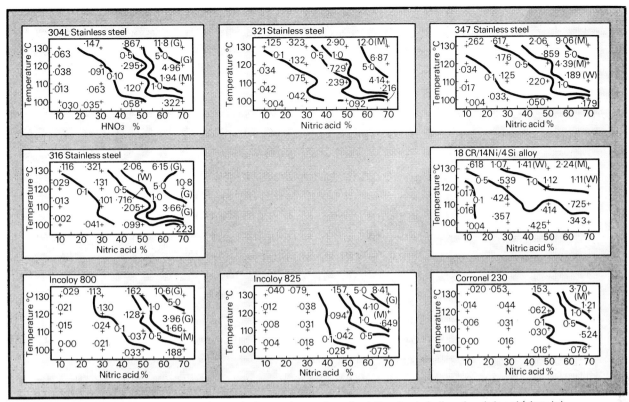

Figs. 2–9 (reading left to right and top to bottom). *Rates of corrosion of solution-treated alloys by nitric acid (mm/a).* *Letters W, M and G indicate weak, medium and great penetration, respectively*

sheet; the first two were cold-rolled to 20 s.w.g. and subsequently heated at 1,050°C for 30 min followed by a water quench. The high-silicon steel was hot-rolled at 1,100°C to 20 s.w.g. and given the same solution-treatment. The 50Cr/50Ni alloy was melted and cast in the laboratory, part of the material was left as cast and part extruded at 1,120°C in a mild steel can to 16-mm rectangular bar; the mild steel was subsequently pickled off. Some of each of the above materials was sensitised by heat-treatment in air at 650°C for 5 hr. Specimens measuring 50 × 10 mm were machined from each of the alloys and polished prior to testing.

Results

Tables 2 and 3 and the isocorrosion charts, Figs. 2–17, show the results obtained by testing the alloys. The tables refer specifically to materials examined at only a few points in the temperature-concentration range and were thus not fully studied. The contour plots were drawn from the data points given on each chart assuming that the corrosion rate is a linear function of concentration and a logarithmic function of temperature—both the calculation and the drawing were performed using an IBM *1130* computer. The letters in parenthesis represent the presence and extent of intercrystalline penetration: *(I)* indicates incipient intercrystalline attack, while *(W)*, *(M)* and *(G)* designate weak, medium and great penetration, respectively. Sectioning a few specimens showed that these corresponded to <1, 1–10, 10–20 and >20 mm/a.

Examination of Figs. 2–17 shows that there is no inflection of the isocorrosion curves at the boiling point curve. The results are thus similar to those of Zitter[1] and differ from those reported by Fontana.[2,3] Tests conducted by the latter at temperatures below the boiling point were made in flasks open to the atmosphere. Hence corrosion products, such as nitrous acid, which accelerates the corrosion reaction,[3] were oxidised by ambient oxygen, and corrosion rates were low, unlike the present experiments where such products were confined within the reaction vessel. Some results are thus higher than would be expected in plant practice and should be interpreted in conjunction with operating experience. Overall, however, the results

are in agreement with those obtained by both of the above workers, Zitter especially showing the superior corrosion resistance of 25Cr/20Ni steels over that of 18Cr/10Ni,[1] which has also been found in this study.

Solution treatment

A comparison of the corrosion behaviour of all solution-treated materials that were fully studied is illustrated in Fig. 18; the 0·5 mm/a line has been adopted as it probably represents the maximum tolerable corrosion rate. Above 110°C the materials fall into four groups which are, in order of decreasing area of possible usage: (1) *Corronel 230*; (2) *304L* stainless steel and *Incoloy 800*

Table 2. Overall corrosion rates of solution-treated materials after 48-hr test (mm/year). *I* indicates intercrystalline attack, *W* and *M* indicate weak and medium penetration respectively

Test condition		304 stainless steel	316L stainless steel	Cast 50/50 Ni/Cr	Extruded 50/50 Ni/Cr	Low-carbon 25Cr/ 20Ni
Acid concentration	Temperature (°C)					
10	100	0·000	0·041	0·000	0·000	0·012
10	110	0·013				
10	120	0·021				
10	130	0·079	0·030	0·013(I)	0·000	0·008
30	100	0·023				
30	110	0·060				
30	120	0·091	0·074	0·042(I)	0·000	0·027
30	130	0·220				
50	100	—				
50	110	0·129				
50	120	—				
50	130	—				
70	100	0·141	0·319(I)	0·086(I)	0·077	0·158
70	110	1·22(W)	1·34(W)	0·238(M)	0·408	0·274
70	120	2·87(W)				
70	130	7·32(M)	9·00(M)	2·250(M)	2·250(W)	3·620

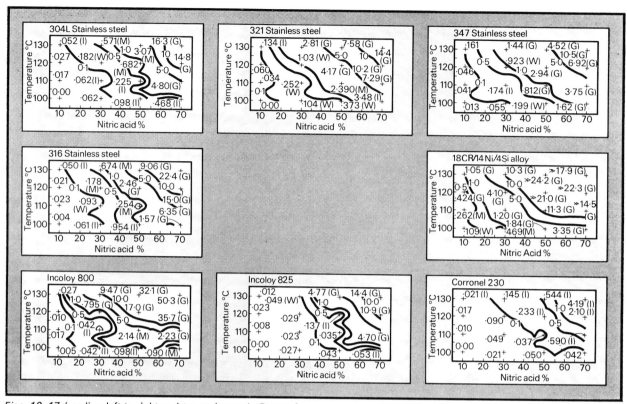

Figs. 10–17 (reading left to right and top to bottom). Rates of corrosion of sensitised alloys by nitric acid (mm/a). Letters W, M and G indicate weak, medium and great penetration, respectively

and *825*; (3) *316, 321* and *347* stainless steels; and (4) 18Cr/14Ni/4Si stainless steel. Below 110°C the materials cannot be similarly grouped, although the superiority of *Corronel 230* remains clearly apparent.

It may be seen from Figs. 2–6 and Table 2 that, in the solution-treated condition and with only two exceptions, low-carbon stainless steels *304L* and 25Cr/20Ni are superior in corrosion resistance to all the other steels tested throughout the temperature concentration range; Type *304L* is slightly more susceptible to corrosion under the most severe conditions than 25Cr/20Ni stainless steel as would be expected from the higher chromium and nickel contents of the latter.[1,3] The two exceptions are type *304* in the range 0·1–1 mm/a and 18Cr/14Ni/4Si at high temperatures and concentrations. Type *304* stainless steel is next in order of merit having been less corroded than the other materials except under the more aggressive conditions; its higher carbon content *vis-à-vis 304L* has only a relatively small deleterious effect when the steel is in

the solution-treated condition.[6] The high-silicon steel (18Cr/14Ni/4Si) has the lowest susceptibility to corrosion at acid concentrations of greater than 50%, i.e. it is less corroded the more oxidising the environment;[7,8] the effect is thought to be caused by the presence on the surface of the steel of a colloidal silicon acid.[7,9] The remaining steels containing molybdenum (types *316* and *316L*), titanium (type *321*) or niobium (type *347*) are essentially similar in behaviour though the last is marginally inferior at the lower end of the temperature-concentration range; the greater susceptibility of these steels may be correlated with the above additions in accordance with published data.[1,10–13]

It is apparent from Figs. 7–9 and Table 2 that high-nickel alloys are in general superior to austenitic stainless steels and may be placed in order of decreasing corrosion resistance: 50Cr/50Ni alloy, *Corronel 230*, *Incoloy 825* and *Incoloy 800*. This coincides with decreasing chromium content as would be expected,[1] the increased nickel content of *Incoloy 825* also possibly account-

ing for its slight superiority over *Incoloy 800*.[3] The similarity in behaviour of cast and extruded 50Cr/50Ni indicates that hot-working this alloy has little effect on corrosion resistance.

The incidence of localised attack on each of the materials under the different test conditions is shown in Figs. 2–9 and Table 2. All the stainless steels and nickel alloys suffered intercrystalline corrosion under the more severe conditions, except for high-silicon stainless steel and low-carbon 25Cr/20Ni; the latter alloy was uniformly corroded under all circumstances. The 18Cr/14Ni/4Si alloy showed very finely-dispersed shallow pitting. These pits, which were round-bottomed, tended to overlap laterally and would thus eventually give rise to uniform attack if reaction were faster. Within the susceptible group, the higher-chromium alloys, *Corronel 230* and 50Cr/50Ni alloy suffered the least intercrystalline attack.

Sensitised materials

Table 3 and the isocorrosion charts in Figs. 10–14 illustrate the effects on the stainless steels of a sensitising heat-treatment of 5 hr at 650°C. Their comparative behaviour may be explained in terms of the grain boundary chromium impoverishment theory,[14] the three principal factors being the chromium and carbon contents and the presence of stabilising elements such as titanium and niobium. It is apparent that the high-chromium low-carbon 25Cr/20Ni alloy is superior in corrosion resistance throughout the temperature-concentration range to all the other steels tested,

Table 3. Overall corrosion rates of sensitised materials after 48-hr test (mm/year). *I* indicates intercrystalline attack, *W, M* and *G* indicate weak, medium and great penetration respectively

Test condition		304 stainless steel	316L stainless steel	Cast 50/50 Ni/Cr	Extruded 50/50 Ni/Cr	Low-carbon 25Cr/ 20Ni
Acid concentration	Temperature (°C)					
10	100	0·133	0·010	0·004	0·010	0·016
10	130	1·33(*I*)	0·067	0·008(*W*)	0·004	0·006
30	120	4·60(*G*)	1·46	0·017	0·008	0·033
70	100	12·1(*G*)	5·37	0·115(*I*)	0·210	0·578(*I*)
70	110	20·1(*G*)	10·8(*G*)	0·415(*I*)	0·764	1·529(*I*)
70	130	46·9(*G*)	15·5(*G*)	1·89(*W*)	3·03	9·678(*G*)

as might be predicted from published data.[15,16] The low-carbon type *304L* may be placed next in order of corrosion resistance except for the better behaviour of titanium-stabilised type *321* and niobium-stabilised type *347* in 70% acid at 130°C. The molybdenum-containing type *316* and its low-carbon variant, type *316L*, are generally superior to types *321* and *347* except at the highest temperatures and concentrations where the latter two alloys show marginally better behaviour. Types *347* and *321* exhibit similar corrosion resistance, although the titanium-stabilised *321* is somewhat inferior under the more severe test conditions, as would be expected.[1,10] The deleterious effect of molybdenum is not particularly distinct in the sensitised state, although comparison of the behaviour of *316L* and

304L does indicate the superiority of the molybdenum-free *304L*.[9] The relatively poor performance of *18Cr/14Ni/4Si* may have been caused by insufficiently oxidising conditions for which this composition has been optimised.[7,8]

The results obtained on the similarly heat-treated nickel alloys are given in Figs. 15–17 and Table 3. With the exception of *Incoloy 800*, this group again displays generally better corrosion resistance than the austenitic stainless steels and a decreasing order of merit is apparent: 50Cr/50Ni alloy, *Corronel 230*, *Incoloy 825* and *Incoloy 800*. Comparison of the results on cast and extruded 50Cr/50Ni alloy in Table 3 shows that hot-working prior to sensitisation seems to have only a relatively small adverse effect on the corrosion resistance. The order is largely coinci-

dent with decreasing chromium content and also the presence of free carbon; the greater level of stabilising elements in *Incoloy 825* over that in *Incoloy 800* may explain the superiority of the former.

The fully-tested materials are compared in Fig. 19, the 0·5 mm/a isocorrosion line having once again been adopted. At temperatures above 110°C, the materials may be placed in order of decreasing area of possible usage: (1) *Corronel 230*; (2) *Incoloy 825*; (3) *304L* stainless steel; (4) *316* stainless steel; (5) *347* stainless steel; (6) *321* stainless steel; (7) *Incoloy 800*; (8) 18Cr/14Ni/4Si stainless steel. The superiority of *Corronel 230* is once again apparent below 110°C although the gradation of materials behaviour is more complex at the lower temperatures.

Localised attack

The incidence of localised attack is also shown in Figs. 10–17 and Table 3. All materials, apart from extruded 50Cr/50Ni alloy, suffered intercrystalline corrosion, the depth of penetration increasing with both temperature and concentration. The order of merit obtained is much the same as that based on weight loss, given above, except that type *347* is less susceptible to this kind of corrosion than type *316*. At acid concentrations above about 50%, a thick (0·5-mm) adherent film was formed on the surface of the high-silicon 18Cr/14Ni/4Si, which was later found to be virtually all silica. The occurrence of intergranular corrosion on cast 50Cr/50Ni alloy reflects primarily the large grain size of this material (6 mm) as compared with crystallite diameters of 0·1 mm present in the extruded sample.

Comparison of these data with those given on the solution-treated materials shows that corrosion rates were increased upon sensitisation, the differences being more marked with *304* stainless steel and *Incoloy 800* (up to two orders of magnitude) than with low-carbon or stabilised alloys.

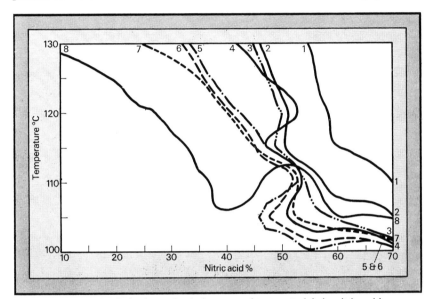

Fig. 18. Comparative behaviour of solution-treated test materials in nitric acid (0·5 mm/a isocorrosion line). Curve numbers refer to: 1 – Corronel 230, 2 – Incoloy 825, 3 – Incoloy 800, 4 – 304L stainless steel, 5 – 316 stainless steel, 6 – 321 stainless steel, 7 – 347 stainless steel, 8 – 18Cr/14Ni/4Si alloy

Fig. 19 (below). Comparative behaviour of sensitised test materials in nitric acid (0·5 mm/a ioscorrosion line). Curve numbers refer to: 1 – Corronel 230, 2 – Incoloy 825, 3 – 304L stainless steel 4 – 316 stainless steel, 5 – 347 stainless steel, 6 – 321 stainless steel, 7 – Incoloy 800, 8 – 18Cr/14Ni/4Si alloy

Acknowledgment

The author is indebted to International Nickel Ltd for permission to publish this article.

References

1. Zitter, H., Dechema Monograph 1962, **45**, 129.
2. Fontana, M. G., *Ind. Eng. Chem.*, 1953, **45**, 93A.
3. Beck, F. H., and Fontana, M. G., *Corrosion*, 1953, **9**, 287.
4. Bishop, C. R., *ibid*, 1963, **19**, 308.
5. Miller, R. F., *et al, ibid*, 1954, **10** (1), 7.
6. Brown, M. H., *et al*, ASTM Spec. Publ. no. 93, 1949, p. 103.
7. Coriou, H., *et al, Mem. Scient. Rev. Met.*, 1964, **61** (3), 177.
8. Armijo, J. S., and Wilde, B. E., *Corrosion Sci.*, 1968, **8**, 649.
9. Desestret, A., *et al, Electrochim. Acta*, 1963, **8**, 433.
10. Ternes, H., and Schwenk, W., *Archiv für das Eisenhüttenwesen*, 1965, **36** (2), 99.
11. LaQue, F. L., ASTM Spec. Publ. no. 93 (1949).
12. Heeley, E. J., and Little, A. T., *JISI*, 1956, **182** (1), 241.
13. Schwenk, W., and Ternes, H., *Archiv für das Eisenhüttenwesen*, 1965, **36** (2), 109.
14. Wiester, H. J., *et al, ibid*, 1959, **30** (5), 299.
15. Bourrat, J., Paper presented at 15th International Conference of Société Chimie Industrie, Paris, May 1968.
16. Edstrom, J. O., and Skold, B., *JISI*, 1969, **207** (7), 954.

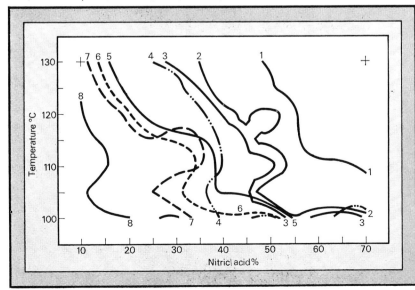

Properties and Environment

Environment-Dependence of the Mechanical Properties of Metals at High Temperature

by R. H. Cook and R. P. Skelton

Premature failure of engineering components can result from the combined influences of corrosion and residual or applied stresses to produce the effects found in corrosion fatigue and stress corrosion cracking.[1,2] In choosing alloys for operation in the creep range it has usually been assumed that their strength and corrosion resistance are separately acceptable. However, there is now a growing awareness that chemical and mechanical effects at elevated temperatures will interact; this review summarizes experimental studies under externally applied stresses at temperatures above about 0·4 of the absolute melting point. We shall not be concerned with details of stresses and strains at oxide/metal interfaces since these have been reviewed elsewhere.[3,4] Early work on environment-dependent deformation has been reviewed[5] and a conference on environment-sensitive mechanical properties, mostly dealing with effects at ambient temperature, was held in 1965.[6] Board[7] has discussed surface chemistry, creep, and fatigue in impure helium environments.†

Types of Metal/Environment Interaction

Chemical changes near a specimen surface will affect its mechanical properties to an extent depending on the type and rate of the reaction and on the specimen dimensions. The rate of reaction may itself depend on stress and strain distribution, so the resulting interaction can be complex. Table I lists possible reactions which will be discussed in turn.

'Inert Atmospheres'

No atmosphere is fully inert although attempts have been made to reduce corrosion by testing in vacuum or 'inert' gases. A vacuum may promote evaporative loss of alloying elements (e.g. manganese and chromium from austenitic steels[8]) and corrosion by residual gases. At 1·3 mN/m², a clean metal surface will be saturated with a layer of adsorbed gas in 2–3 seconds at 925 K.[9] Also, the relative proportions of residual gases in a vacuum will differ greatly from their ratio in air.[10]

Inert gases will not readily adsorb on or dissolve in metals[11] but it is difficult to purify them to a level below about 0·2 vpm total impurities. Conclusive interpretation of test results from vacuum and inert gases thus requires a knowledge of impurity levels which, unfortunately, is usually lacking.

R. H. Cook, BMet, PhD, and R. P. Skelton, BSc, are in the Materials Division, Central Electricity Research Laboratories, Leatherhead, Surrey.
† Fidler and Collins[174] have recently reviewed the effect of sodium on mechanical properties.

Formation of an Adsorbed Layer

A model of the equilibrium between adsorption (implying absence of chemical combination) and desorption predicts a linear dependence of adsorption coverage on pressure at low pressure and a saturation of the surface at high pressure.[12] However, during adsorption of oxygen on nickel the extent of surface coverage and the 'sticking probability' varied with surface orientation.[13] Typically, saturation occurred at about quarter coverage and the bond formed was ionic with a heat of adsorption of 110 kcal/mol, suggesting an associated surface energy change of about 2 J/m² (2000 ergs/cm²).

Such large changes in surface energy should affect mechanical properties, a mechanism invoked by Forestier and Clauss,[14] who measured the fracture strengths of fine wires of copper, gold, platinum, silver, and tungsten at room temperature. Strength was greatest in vacuum and in other environments was reduced progressively as the boiling point of the gas surrounding the specimen increased (Fig. 1). The effect of atmosphere became less as wire diameter was increased from 30 to 200 μm.

Formation of a Surface Layer of Second Phase: Oxidation

Oxidation, e.g. by the reaction $2\frac{x}{y}M + O_2 \rightleftarrows \frac{2}{y}M_xO_y$, is accompanied by release of standard free energy ΔG° when the reactants are in their standard states (oxygen at atmospheric pressure). Free energy data for oxidation are summarized in Fig. 2.[15] When $\Delta G^\circ = 0$ the oxide is in equilibrium with oxygen at atmospheric pressure and is at its standard dissociation temperature. Dissociation temperature T varies with oxygen partial pressure according to the relation:

$$\Delta G^\circ = RT \ln p_{O2}$$

Thus at 1000K Fe_3O_4 and Cr_2O_3 become unstable at ~ 1 fN/m² and 10^{-7} aN/m², respectively.

Two important parameters controlling rate of oxidation are *oxygen availability* and *solid-state diffusion rates*. The effect of the former is summarized by:

$$\text{Oxidation rate} \propto (p_{O2})^{1/n}$$

where n is between 2 and 8 depending on the details of the reaction.[16] The second arises when the reaction, occurring at the oxide/metal interface, is controlled by diffusion of oxygen through the oxide (anion diffusion), producing a stress at the interface. Alternatively when the reaction occurs at the oxide/gas interface it requires an outward flow of metal atoms (cation diffusion) causing vacancy accumulation near or below the oxide/metal interface. This may lead to oxide spalling or secondary defects in the metal[17] which may modify mechanical properties.

The magnitude and sign of the stresses set up during oxide growth may be estimated from the Pilling–Bedworth ratio,[18] i.e. the ratio between the molecular volume of oxide and the atomic volume of metal. This concept must be applied with caution, especially where cation diffusion is controlling.[19]

During oxidation of alloys, reaction rate may be determined by selective oxidation of reactive elements.[20] At low oxygen levels oxidation of AISI 304* steel in argon/oxygen mixtures at 1090 K proceeded by formation of Cr_2O_3 but with above 70 vpm of oxygen the oxide type changed to a spinel (M_3O_4) as iron became oxidized.[20] This was associated with a change in creep behaviour.

Chromium-rich oxides (along grain boundaries) formed in austenitic steels at ~ 1000 K in helium with a low p_{O_2} whereas iron was present in the oxides formed in air.[21] Selective oxidation of aluminium and titanium has been reported in γ' strengthened nickel-base alloys exposed to helium with a low p_{O_2}.[22]

Formation of Sub-Surface Particles of Second Phase

This occurs in systems where a minor constituent in the alloy has a much greater affinity for gas atoms than has the major constituent. Examples are found in the internal oxidation of Cu–Si alloys (forming SiO_2) and in internal carburization of Fe–Ni–Cr alloys (forming $Cr_{23}C_6$). The sub-surface second phase so formed may occur randomly within grains or preferentially at grain boundaries.

Carburization from a gaseous environment requires first the breakdown of carboniferous gases, e.g.

$$2CO \rightleftarrows C + CO_2 \text{ (Boudouard reaction)},$$

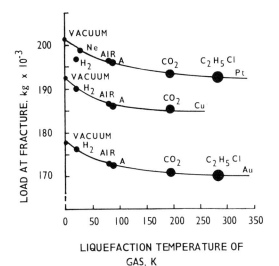

Fig. 1 The effect of the liquefaction temperature of various gases upon the fracture strength of wires of gold, copper, and platinum (Forestier and Clauss, 1955).[14]

* Compositions of commercial alloys are given in Table IX.

a reaction catalysed by iron and nickel. Carbon deposition is favoured by high CO/CO_2 ratio and low temperature.[7] Carburization in AISI 304 steel occurred only when the ratio $(p_{CO})^2/(p_{CO_2})$ exceeded 0.10^{23} and it has been found in austenitic steels and nickel-base alloys in helium with more than 5000 vpm CO.[24] Methane may also cause carburization.[25]

TABLE I
Possible Types of Metal/Environment Interaction

Direction of Mass Transport	Examples of Effect
None	Atmosphere inert to metal
From atmosphere to metal	(i) Formation of a thin adsorbed layer on the specimen.
	(ii) Formation of an adherent surface layer of second phase (e.g. oxidation)
	(iii) Formation of second phase particles or grain-boundary films in the sub-surface layer.
	(iv) Formation of a second phase in surface cracks in the specimen.
From metal to atmosphere	(i) Evaporative loss or chemical dissolution of elements from the specimen to the environment.

McCoy[26] noted carburization during creep of AISI 304 steel in CO and in CO_2 at 1090 K. The observation in CO_2 appears to conflict with the above results but CO_2 is unstable above 1000 K, dissociating to CO and O_2. From Fig. 2 the equilibrium ratio of $(p_{CO})/(p_{CO_2})$ at 1090 K in the presence of iron and Fe_3O_4 is ~ 3 which could be carburizing. Whether this ratio is reached will depend upon gas flow rate, &c.

Litton and Morris[27] reported carbon transport via sodium at 725–925K from an Fe–0.87%C source to a sample of AISI 316 steel which was carburized to depths of ~ 100 μm. Carburization may raise the proof stress of austenitic steels but it seriously impairs tensile ductility.[28]

The rate of internal nitridation[29,30] of an austenitic steel increased with increasing applied stress. Stress also affected orientation of hydride platelet stringers formed in zirconium tubing at 575 K.[31,32] In general, oxidation or precipitation is likely to be assisted by applied stress when the reaction reduces the overall energy of the system.

Formation of a Second Phase in Cracks

Oxidation in surface cracks has been observed on numerous occasions. If oxide completely bridges a crack, it may increase the load bearing capacity of a specimen.[33]

Loss of Alloying Elements to the Environment

Decarburization occurs most readily in alloys of high carbon activity. Mild steels were severely decarburized in sodium at 750 K, whilst low alloy steels were little affected.[34] At 925 K $2\frac{1}{4}$Cr/1Mo steel was decarburized whilst9 Cr/1Mo steel was only slightly attacked. Decarburization caused loss of room-temperature tensile strength and it increased ductility.

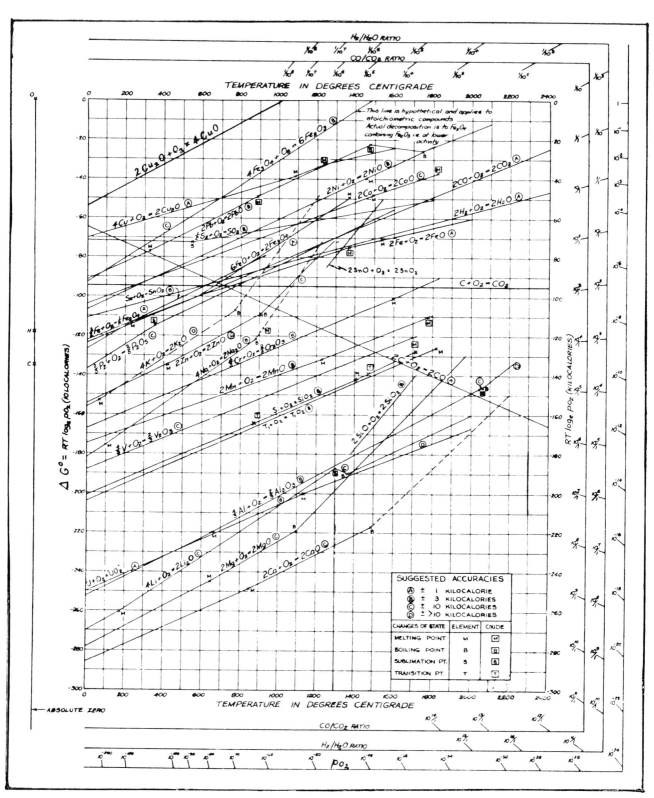

Fig. 2   *Free-energy data for oxide formation (after Richardson and Jeffes, 1948).*[15]

1 kcal = 4·1868 kJ; 2·303 R = 19·1546 J/mol degC;
1 atm = 101·325 kN/m²; 1 V equiv. = 96·606 kJ.

TABLE II

Summary of Test Conditions Used in Studies of the Effect of Environment upon Creep of Nickel

Temperature, K	Environment	Test Duration, h	Remarks	Ref.
925	Air, vacuum	<4000	6 mm-dia. bar specimens	38, 39
1090	Air	<1000	Ni/NiO/Ni specimens	33
925	Air, vacuum	< 80	6 mm-dia. specimens. Vacuum of 1·3 mN/m² (10⁻⁵ torr)	43
785	Range of p_{O_2} and vacuum	< 300	4·5 mm-dia. specimens. p_{O_2} varied from 0·013 mN/m² to 13·3 kN/m² (10⁻⁷ torr–10² torr)	42
925 1090	Air, vacuum		Effect of notches	45
925 1090	Air, vacuum		'Hour-Glass' specimens	46
925 1090	Air, vacuum, Ar, H₂,N₂	<8400	Sheet and bar specimens	40
975–1050	Air, N₂, H₂	~ 100	Studied stress and temperature dependence of secondary creep rate	47
755–1090	N₂, air, vacuum, He–2% O₂	< 800	2 purities of nickel (99·99% and 99·8%)	48
1200	Ar, H₂	~ 800	Atmosphere cycling	26
1090	Ar, air, NaOH	<2000	Tube specimens	49
875	Air	< 200	Specimens pre-oxidized at 1473 K	44

Sodium may also remove nickel, chromium, niobium, and molybdenum from steels,[35] nickel removal causing a ferrite layer to form on austenitic steels.

Alloying elements may also be lost through evaporation.[8] Ryabchenkov and Maksimov[36] argued that the oxide free surface of argon-tested specimens allowed selective evaporation of alloying elements from steels and nickel-base alloys, so leading to impairment of creep and fatigue resistance. There are, however, other possible mechanisms (cf. selective oxidation in impure helium). Evaporative loss will be less in an inert gas than in vacuum because of the shorter mean free path of the evaporating species.[37]

The Influence of Environment on Creep Behaviour

This section first describes work on nickel which clearly reflects the importance of oxidation. Similar effects obtain in nickel alloys but other reactions and structural variations arise. Creep of austenitic steels reveals conflicting rôles of oxygen which are complicated by simultaneous nitridation in air. Finally, ferritic steels and other materials are discussed.

Nickel

Important experimental work is summarized in Table II.

Effects of Air and Vacuum

Air increases the rupture life of nickel in long term tests at high temperature whilst vacuum produces the higher rupture life in short tests at low temperature.[38,39,40] The resulting crossover behaviour is illustrated in Fig. 3 for two temperatures. The effect is independent of grain size.[38,39,40] The crossover roughly coincides with a change from transgranular failure in short tests to intergranular failure in long tests.[40]

The observed air weakening has been ascribed to a lowering of surface energy, and hence of the work to propagate a crack, by gaseous adsorption on fracture faces.[41,42] Stegman et al.[42] tested nickel in the air-weakening regime for a range of values of p_{O_2} and again found the lowest creep rates and

longest rupture lives in the highest vacuum. Rupture life decreased with increasing p_{O_2}, as shown in Fig. 4(a), until at a pressure of about 13 mN/m² it became insensitive to further increase in p_{O_2}. The authors supposed this to be a critical pressure required to maintain complete coverage of a growing fracture surface by adsorbed gas. However, Fig. 4(b) shows that when oxygen was admitted to a specimen under vacuum during steady state creep, the new creep rate only became steady after about 10h. If oxygen adsorption were controlling, this should have been achieved immediately.[9] This suggests that the sub-surface oxygen content may have an important effect on crack growth and creep rate.

Air-Strengthening

Extension of rupture life during long-term tests in air (Fig. 3) was associated with oxide in intergranular cracks,[38,39,40] reducing their propagation rate. Cracks were formed during early secondary creep, coinciding with a slight increase in creep rate relative to behaviour in vacuum[41,43] but subsequently the oxide apparently supported part of the applied load, so delaying fracture. This often happened during tertiary creep and was followed by a marked reduction in creep rate (indicated by the broken line in Fig. 3).

To estimate the strength of an oxide filled crack, Cass and Achter[33] tested sintered sandwiches of nickel oxide in nickel at 1090K and observed rupture stresses similar to those at which air-strengthening was apparent in nickel. Thus delayed rupture can be reasonably attributed to strengthening by oxide bridging. This mechanism depends on the conditions of oxide formation, however, since Douglass[44] showed that pre-oxidation of nickel at 1475 K produced intergranular oxide which *impaired* creep properties at 875 K.

Effects of Specimen Geometry

Douglas[40] found that air-strengthening in nickel sheet ~ 1·3 mm thick was much greater than in 13 mm-dia. bar at 1090 K.

Notch-strengthening of nickel at 1090 K can occur.[39,45] A notch prolonged life in air by a factor of ~ 8:1 which was greater than that (~ 2:1) in vacuum. Tests on specimens of non-uniform diameter showed that air-strengthening was more pronounced in thin sections than thick[46] since oxidation caused the specimens to fail at regions away from the minimum cross section whereas similar specimens tested in vacuum failed at the minimum section. These effects suggest that oxidation rate is sensitive to stress or strain rate.

Effects of Other Environments

Generally, tests on nickel in hydrogen, nitrogen, argon, and vacuum have produced similar creep behaviour.[40,47,48] Notable exceptions are:
  (i) Creep behaviour of 99·8% Ni was impaired in nitrogen, particularly around 750 K, due to intergranular embrittlement,[48] and
  (ii) creep rate in hydrogen at 1200 K was about twice that in argon.[26]
Interchanging these gases produced reversible changes in rate.

Activation energies for creep in air and hydrogen were similar although creep rate at a given stress in air was about half that in hydrogen.[47]

McHenry and Probst[49] tested tubes and bars of 99·3% nickel in argon, air, and sodium hydroxide at 1090K. Air produced the greatest rupture resistance, more so in tubes (1·5 mm wall thickness with air inside and outside) than in 6 mm-dia. bar. Sodium hydroxide reduced rupture resistance compared with both air and argon owing, in part, to mass transfer corrosion, i.e. transfer of material from a hot zone to a cooler region.

Nickel-Base Alloys

Experimental work on these alloys is summarized in Table III.

Effects of Air and Vacuum or Inert Gases

The effect of air, relative to vacuum or inert gases, is similar to that described for nickel. At low stresses, oxidation increased the rupture life of Inconel,[40] Hastelloy C,[50] Inconel X,[51] Ni/20Cr alloys,[45,52] and single crystals of Mar-M200.[53] It also reduced secondary creep rate and enhanced the ductility of Inconel.[26,30] On the other hand at high stresses air impaired the rupture resistance of Inconel 550,[54] Ni/19Cr and Ni/19Cr/4Al alloys.[55,56] Thus at intermediate stress and temperature a crossover from air weakening to oxidation strengthening can again be expected. Figs. 5 and 6 illustrate this for Ni/19Cr alloys.[56]

There are exceptions, however. Udimet 500 was always more rupture-resistant in vacuum than in air[45] whilst Hastelloy X[57] and Inco 713C[58] displayed no consistent differences between rupture life in air and in helium. Microstructure (or heat-treatment) affected the response to environment of Inconel X[45] and René 41[59] (see below) whilst silicon level and melting history affected oxidation-strengthening of Ni/20Cr alloys[52] which displayed internal oxidation.

Oxidation-strengthening again occurred when oxide formed in cracks nucleated during secondary creep[40,41,52] and, in some cases, this led to a stage of low creep rate after tertiary creep (Fig. 6).

Effects of Specimen Geometry

At 1090K, the rupture life in air of Inconel specimens 0·5 mm thick was over three times that of material 1·5 mm thick at the same stress.[40]

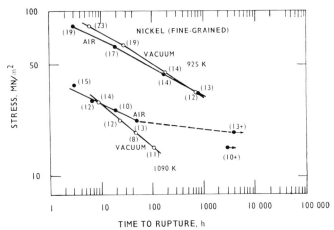

Fig. 3 *Comparison of the effects of vacuum and air on the creep-rupture behaviour of nickel Figures in parentheses denote % elongations at fracture (Shahinian and Achter, 1959).*[39]

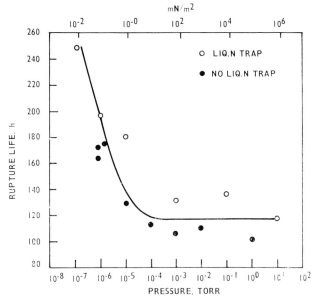

Fig. 4(a) *Effect of oxygen partial pressure upon rupture life of nickel at 875 K, 58·5 N/mm² (Stegman, Shahinian, and Achter, 1969).*[42]

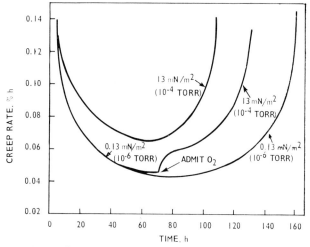

Fig. 4(b) *Effect of oxygen partial pressure and oxygen admission on the creep rate of nickel at 875 K, 58·5 N/mm² (Stegman, Shahinian, and Achter, 1969).*[42]

Source: *International Metallurgical Reviews*, Vol 19, 1974

TABLE III

Summary of Test Conditions Used in Studies of the Effect of Environment upon Creep of Nickel-Base Alloys

Alloy	Temperature, K	Environment	Test Duration, h	Remarks	Ref.
Hastelloy C	1145	Air, vacuum, H_2, He, N_2	200–2000	Vacuum 100 mN/m^2 (0·7 torr)	50
Inconel X	1090	Air, vacuum, O_2, N_2, He	100–400	Plain and notched specimens	51
Ni/19Cr/4Al	975–1310	Air, vacuum, O_2, N_2			55
Ni/19Cr	865–1310	Air, vacuum	2–1600	Studied stress and temperature dependence of creep rate	56
Nichrome V Udimet 500 Inconel X	975–1200	Air, vacuum		Two heat-treatments on Inconel X	65
Inconel 550	1090	Air, vacuum	⩽ 300		54
Mar-M200 single crystals	1035	Air, vacuum	150	Vacuum 13 $\mu N/m^2$ (10^{-7} torr)	53
Ni/20Cr Ni/20Cr/3·6Al Inconel X	975–1090	Air, O_2, N_2, He, He + 2% O_2, vacuum	< 200	Plain and notched specimens	41
Inco 713C	975–1090	Air, He, impure He	<1100	Cast test pieces	69, 58
Hastelloy X	1145	Air, He	200–1900	All data below alloy manufacturers specifications	57
Inconel	975–1175	Air, Ar, N_2	7–7500	Several specimen thicknesses	40
Inconel	1090	Air, Ar, O_2, N_2, CO, CO_2, H_2	⩽2000		26, 30
Ni/20Cr	925–1255	Air, Ar	0·001–1000	Effects of minor constituents	52
René 41	1025–1175	Air, Ar	Tensile tests	Effect of varying precipitate particle size	59
Ni/6W	1175	Ar, vacuum	< 20		60

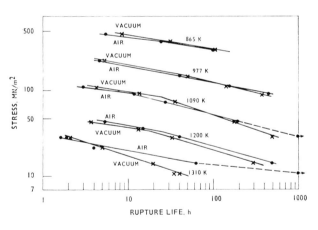

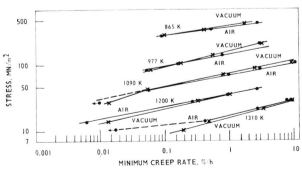

Fig. 5 The influence of stress upon the rupture life and minimum creep rate of Ni–19Cr in air and vacuum (Shahinian and Achter, 1959).[56]

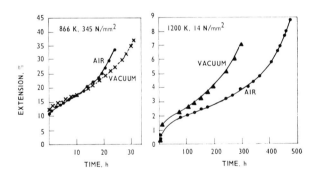

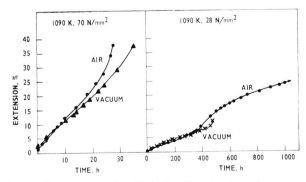

Fig. 6 Creep curves of a Ni–19Cr alloy in air and vacuum (Shahinian and Achter, 1959)[56]

Notches have a complicating effect. Inconel X was notch strengthened in air but severely notch weakened in vacuum and notches generally tended to increase oxidation strengthening.[41,45,51]

Effects of Microstructure

Normally heat-treated Inconel X, containing a fine precipitate of γ' ($Ni_3(Al,Ti)$), had generally about 10% greater rupture life in vacuum than in air.[45] However, when the γ' precipitate was coarsened or partly re-dissolved, rupture lives in vacuum exceeded those in air by 50%. This was found in both notched and plain specimens.

In tensile tests on René 41 treated to give a range of γ' particle sizes, the ratio minimum ductility in air: minimum ductility in argon at $\sim$ 1075–1175 K varied from $\sim$ 0·15 for a particle size of < 20 nm (200 Å) to $\sim$ 0·95 for a particle size of 0·25 μm (2500 Å).[59] Minimum elongation in argon was 10–20% and that in air was 2–17%, depending on heat treatment.

Effects of Other Gases

In Inconel a reduction in secondary creep rate in nitrogen relative to the rate in argon was attributed to the formation of nitride particles.[26] Nitrides did not form in air tests and creep behaviour was indistinguishable from that in oxygen, suggesting that in air an oxide film inhibited nitrogen dissolution.[30] The creep rate of Inconel was also reduced during tests in carbon monoxide and carbon dioxide by carbide formation.

Hydrogen increased creep rate (relative to that in argon) and reduced rupture life of Inconel, possibly by assisting growth of internal creep cracks.[30]

Johnson et al.[60] tested a Ni/6W alloy and found lower ductilities in argon ($p_{O_2} \sim$ 0·1 N/m^2) than in vacuum ($p_{O_2} \sim$ 0·1 mN/m^2). The effect, particularly evident at coarse grain sizes, was associated with increased grain-boundary cavitation during the argon tests. This observation is discussed further below.

Austenitic Steels

Important experimental work is summarized in Table IV. Fig. 7 illustrates the creep behaviour of sheet specimens of AISI 304 steel in various gases.[61] Several reactions can affect creep rate and these will be discussed individually.

Oxidation: General Observations

There is conflicting evidence for oxidation strengthening in austenitic steels. Its occurrence was noted in AISI 304 at 1005 K[41] and at 1200 K[30] but not at the intermediate temperature, 1090 K.[61] In AISI 347 steel, tests in air produced generally lower creep resistance than did similar tests in argon, nitrogen, and vacuum at 905–1065 K.[62] In attempting to reconcile these observations, we may note that Francis and Hodgson[19] found oxidation strengthening in some stress ranges but not in others. They pre-oxidized 0·2 mm-dia. wires of 20Cr/25Ni/Nb steel in CO_2 and compared the creep rates in subsequent tests at 1125 K in vacuum with those of unoxidized specimens. Oxide growth stresses reduced creep rate at high applied stresses (20–35 N/mm^2) but increased it at low stresses.

Effects of Varying Oxygen Partial Pressure

Varying p_{O_2} in argon/oxygen mixtures caused changes in the secondary creep rate of AISI 304 steel at 1090 K (Fig. 8). These were correlated with a change in oxide type from adherent Cr_2O_3 at partial pressures < 70 vpm oxygen to a less

---

TABLE IV

Summary of Test Conditions Used in Studies of the Effect of Environment upon Creep of Austenitic Steels

Alloy	Temperature, K	Environment	Test Duration, h	Remarks	Ref.
AISI 304	975–1200	Air, O_2, N_2, Ar, He, H_2, CO, CO_2	100–1000	Sheet specimens	30, 61, 26
AISI 304	1005	Air, O_2, He	$\sim$ 100	6 mm-dia. bar specimens	41
20Cr/25Ni/Nb	1125	CO_2	$\leqslant$ 400	Wire	19
AISI 304	1090	Ar–O_2 mixtures	$\leqslant$ 1300	Sheet	20
Pure Fe	1340–1370	Ar–O_2 mixtures	0–4		63
18Cr/8Ni/Ti	1075	Air, vacuum	$\leqslant$ 1000		64
18Cr/8Ni/Nb	905–1065	Air, vacuum, N_2, Ar	0·5–1000		62
AISI 304	925	He + 2% H_2, air, vacuum	$\leqslant$ 100		66
AISI 321	,,				
AISI 316	975–1090	He, air	$\leqslant$ 5000		68 67,
AISI 316	925	He, air, Na	$\leqslant$ 4000		77, 70
AISI 316	1090	He, air	$\leqslant$ 1000		69
AISI 316	1075	Air, impure He	$\leqslant$ 7600		21
Incoloy 800	925				
AISI 304	810–1035	Na, He	< 1000		71
AISI 316	,,				
AISI 316	810–1035	Na	$\leqslant$ 4000		74,75
AISI 304	,,				
AISI 316	975	Na, Ar, air	< 3000	Bar specimens	73
AISI 304,	975	Na, Ar	1000–1500		72
AISI 316	,,				
18Cr/8Ni/Nb	875–1075	Air, steam	0–1200		170

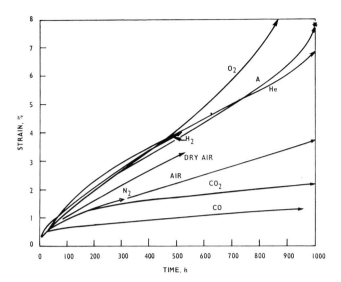

Fig. 7 Effect of environment upon creep behaviour of AISI 304 steel at 1090 K, 23·5 N/mm² (McCoy and Douglas, 1960).[61]

adherent spinel oxide at above this level.[20] The Cr_2O_3 was thought to act as a dislocation barrier, its effectiveness increasing with thickness and hence with p_{O_2}. The spinel oxide in contrast permitted an increase in creep rate. At partial pressures > 250 vpm O_2 the decrease in creep rate was ascribed to a reduction in crack growth rate by oxygen (Fig. 8(c)).

Short term creep of pure austenitic iron at ~ 1350 K was studied in argon with up to 1000 vpm O_2[63] (Table IV). Increasing the degree of oxidation, either before or during creep, caused an increase in secondary creep rate and reduced rupture life, failure occurring by growth of surface- and wedge-cracks. Oxygen promoted the formation of intergranular cavities, as was also found in an 18Cr/8Ni/Nb steel[64] at 1075 K tested in air and vacuum. Here, also, the creep rate in air was greater than in vacuum and failure at low stresses (< 90 N/mm²) resulted from void coalescence. Cavitation and wedge cracking was also seen in AISI 316 steel tested in air and in helium of low p_{O_2}[65] at 1025 K and the mechanism explains the higher creep rate and earlier onset of tertiary creep in AISI 304 steel in oxygen compared with other gases (Fig. 7).

The effects of oxygen on creep of austenitic steels can be summarized as follows:

(i) Oxide growth stresses may assist or retard creep

(ii) Oxidation may retard crack growth

(iii) Oxygen assists the formation of sub-surface intergranular cracks. This may be caused by vacancy injection during oxidation or by an effect of dissolved oxygen.

Nitriding

Nitriding of an 18Cr/8Ni steel during creep, first noted by Dulis and Smith[29] at 1090 K and subsequently studied by McCoy et al.,[26,30] can account for the strengthening seen in air and nitrogen compared with helium and argon, Fig. 7. Its occurrence was critically dependent on stress since at 1200 K nitrides did not form in a test at 7 N/mm² (duration 524h) whereas they did at 17·5 N/mm² (duration 52h).[26]

In contrast, there was no evidence of strengthening by nitriding in AISI 347 tested in air, nitrogen, argon, and vacuum at 1065 K and below.[62] Air appeared to be more important as a source of oxygen than of nitrogen.

Carburizing

The low creep rates seen in AISI 304 steel during exposure to CO and CO_2 (Fig. 7) were ascribed to carburizing. This has been found at 875 to 1090 K.[26] The amount of carbon absorbed by 0·5 mm-thick sheet during 1000 h exposure to CO_2 raised the average level from 0·02 to 0·25 wt.-% and tensile ductility was reduced to about half that of specimens given the same thermal treatment in argon.[26] Carbide distribution was sensitive to material condition. In material deformed by 10% cold work, the carbides produced during creep in CO_2 at 1090 K appeared on slip planes whereas in annealed specimens they took the form of grain-boundary networks.

Copper plating of AISI 304 steel before creep testing in CO prevented carburizing and the creep behaviour of a specimen given this treatment was similar to that of an unplated specimen tested in argon.[61]

Tests in Helium (Reactor Coolant)

Comparisons between the stress rupture life in air and in nominally pure helium have been conducted on several austenitic steels (Table IV) but no consistent effect was found.[66-70] Kirschler and Andrews[70] observed higher secondary creep rates in helium than in air (Fig. 9). There was more intergranular failure in air than in helium,[68] consistent with oxygen-enhanced void growth as discussed earlier.

To simulate the environment in a helium cooled reactor, Wood et al.[21] have tested 2·5 mm thick specimens of AISI 316 and Incoloy 800 in helium containing CO 1000, H_2 1000, and H_2O 100 vpm approximately. In both steels the helium-base atmosphere induced selective oxidation along grain boundaries. This nucleated cracks, so leading to an increase in creep rate and a reduction in rupture life by factors of 2–3, relative to similar tests in air (in which this form of corrosion was absent). Recent work[65] suggests that intergranular oxide cracking in helium occurs mainly at high strain rate. At low strain rates (e.g. below ~ 2 × 10⁻⁵ h⁻¹ at 1025 K) cracking did not occur and there was no impairment of creep-resistance.

Tests in Sodium (Reactor Coolant)

Creep rates in helium and in flowing sodium (30 ppm O_2) were similar in AISI 316 at 925 K but were greater than those in air (Fig. 9). Rupture lives were similar in the three environments. Atkins[71] found that creep rates in helium and static sodium were similar during biaxial tests on AISI 304 and AISI 316. Collins and Zebroski[72] conducted similar tests on these steels in flowing sodium, and control tests in argon, at 975 K. In both steels sodium reduced rupture life, relative to argon, at given hoop stresses. In AISI 304, the life reduction was ~ 10% at 1000 h increasing to 50% at 10 000 h, probably owing to decarburizing. In other work,[73] little difference was found in rupture life of variants of AISI 316 tested in flowing and static sodium, argon, and air at 975 K. Surface attack, absent in static sodium, caused a surface layer of ferrite to form in flowing sodium.

High-carbon sodium (~ 7000 vpm) caused carburizing of AISI 316 at 925K. This reduced creep rate and increased rupture life although ductility was impaired.[74] Grain boundary ferrite was seen in association with M_7C_3 carbides.[75]

High oxygen sodium caused formation of a chromite layer on AISI 316 and AISI 304 steels and intergranular penetration of ferrite also occurred. This had little effect on creep rate but marginally reduced rupture life of AISI 304.[75]

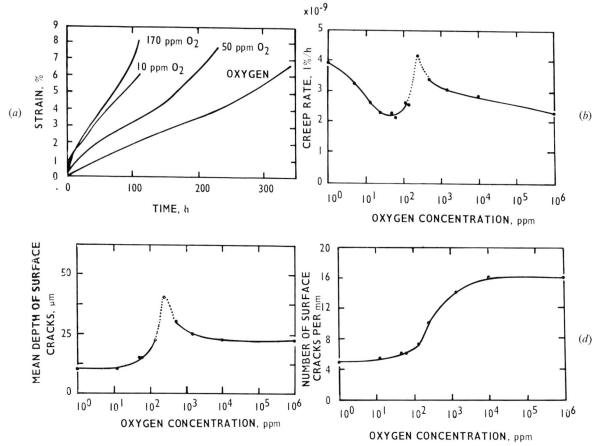

Fig. 8   *Effect of the concentration of oxygen in argon upon creep characteristics of AISI 304 steel at 1090 K, 23·5 N/mm².*
(a) Creep curves; (b) secondary creep rate; (c) mean crack depth; (d) crack density (Rodriguez, 1967)[20]

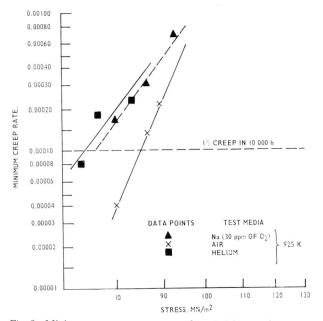

Fig. 9   *Minimum creep rate vs. stress for AISI 316 steel at 925 K in helium, sodium, and air (Andrews and Barker, 1964).[77]*

Ferritic Steels

Experimental conditions for these steels are summarized in Table V.

Gaseous Atmospheres

Trace impurities in otherwise 'inert' atmospheres can increase the creep- and rupture-resistance of plain carbon steels. Rupture lives in 0·15%C steel specimens tested in 99·8% argon were greater than in those tested in purified argon at 1035 K.[50] Similarly, in tests on a 0·03% steel in argon of two purities and in vacuum of 130–0·13 nM/m², rupture lives were greater in the atmospheres containing the larger amounts of impurity.[76] Such impurity-strengthening diminished as test temperature increased from 675 to 1075 K.

Tests on $1Cr/\frac{1}{2}Mo$ and $2\frac{1}{4}Cr/1Mo$ steels indicated negligible differences in the creep behaviour in air and nominally pure helium[68,69,77] (Figs. 10 and 11), vacuum and $He + 2\%$ H_2.[66] Shahinian,[51] using short term tests, found greater rupture lives in $1\frac{1}{4}Cr/\frac{1}{2}Mo$ and 12Cr/1V steels tested in air and oxygen than for the same steels tested in helium, nitrogen, and vacuum.

Hydrogen at pressures in the range 2·8–10 MN/m² can seriously reduce rupture lives of some steels at 810 K.[78,79] This was attributed to rapid decarburizing of the steels by

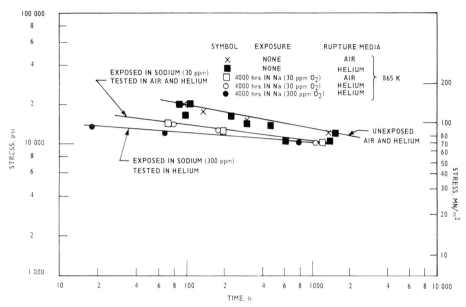

Fig. 10 Stress-rupture data for $2\frac{1}{4}Cr$ $1Mo$ steel in air and helium at 825 K showing effect of exposure to sodium (Andrews and Barker, 1964).[77]

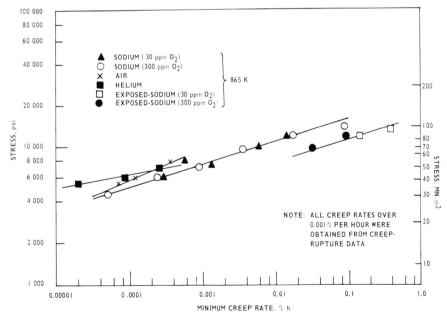

Fig 11. Minimum creep rate vs. stress for $2\frac{1}{4}Cr$ $1Mo$ steel at 865 K in air and sodium, showing effect of exposure to sodium (Andrews and Barker, 1964).[77]

TABLE V

Summary of Test Conditions Used in Studies of the Effect of Environment upon the Creep of Ferritic Steels

Alloy	Temperature, K	Environment	Test Duration, h	Ref.
Fe/0·15C	1035	He, air, Ar	< 500	50
Fe/0·19C				66
Fe/1Cr/$\frac{1}{2}$Mo	810–925	Air, vacuum, He + 2% H_2	<6000	,,
Fe/0·03C	675–1075	Ar (2 purities), vacuum	<1100	76
Fe/1Cr/$\frac{1}{2}$Mo	865–925	Air, He	<2000	68
Fe/1Cr/$\frac{1}{2}$Mo	810, 825	Air, He	<3000	67
Fe/1Cr/$\frac{1}{2}$Mo	925	Air, He	<2000	69
Fe/1$\frac{1}{4}$Cr/$\frac{1}{2}$Mo	865	Air, vacuum, O_2	< 200	51
Fe/12Cr/1V	925	He, N_2		,,
Fe/2$\frac{1}{4}$Cr/1Mo	865	Air, He, Na	<2000	77
Fe/0·2C Fe/$\frac{1}{2}$Mo Ferrovac E Fe/12 Cr	700–865	H_2 (high pressure)	< 500	78, 79

combination of hydrogen and carbon, resulting in methane-filled intergranular fissures growing under the action of stress. Plain carbon steels (e.g. 0·2%C) and those with about $\frac{1}{2}$%Mo were particularly prone; but as the carbon activity decreased (e.g. by reducing its concentration, or introducing elements forming stable carbides) the susceptibility to attack diminished.

Sodium Environments

Exposure to sodium increased secondary creep rate and reduced rupture life of 2$\frac{1}{4}$Cr/1Mo steel at 865 K[77] (Figs. 10 and 11). This was due to decarburizing,[34] which was also noted during tests in air and helium after pre-exposure to sodium.

Other Materials

Cadmium, Zinc, and Magnesium

Early work on single crystals of zinc and cadmium, reviewed elsewhere,[5] has shown the potential importance of a surface layer as a dislocation barrier. Electrolytic dissolution of an oxide film on cadmium caused an increase in creep rate[80] whilst alternate copper plating and de-plating of zinc undergoing creep at 475 K caused reversible changes in creep rate[81] of single crystals but not of polycrystals.[47]

During oxidation of zinc and fine-grained magnesium[17,82] the introduction of a surface supersaturation of vacancies opposed the vacancy flux of Herring–Nabarro creep, so reducing the creep rate in air to about one third of that in an 'inert' nitrogen–hydrogen mixture.[82] Although the Pilling–Bedworth ratio of magnesium is negative, the oxide strength was insufficient to account for the effect.

Aluminium, Copper, and Silver

Sweetland and Parker[83] tested commercial-purity aluminium and copper at 475 K in helium after mechanically removing the oxide film. During secondary creep, admission of air to the specimen chamber (allowing oxide re-formation) caused a 60% reduction in creep rate.

Price[84] examined the influence of dissolved oxygen ($\sim$ 0·01 at.-%) on the creep of silver in air at 525–825 K in tests lasting up to 1200 h. Oxygen prolonged secondary creep by curtailing grain-boundary sliding but shortened tertiary creep by promoting cavitation.

Summary of Environmental Effects during Creep

Studies of the influence of environment upon creep properties have covered several materials and atmospheres. Whilst there is some confusion and apparent contradiction in the results, some important features can be summarized as in Table I.

1. Formation of an adsorbed layer on crack faces has been invoked by some workers in models explaining weakening, but there is no direct support for this.

2. Formation of a surface layer of oxide cannot be distinguished entirely from oxide formation in cracks but it appears to reduce creep rate and increase rupture life of nickel and its alloys, aluminium, and copper. In austenitic steels oxidation can either increase or decrease creep rate.[19,20,62] Point defects, induced by oxidation, can oppose Herring–Nabarro creep in magnesium[82] but may assist cavity formation in other materials.

3. Formation of sub-surface second phase particles generally leads to a reduced creep rate and an increased creep life with possible impairment of ductility. The commonest reactions are internal oxidation, nitriding, and carburizing. Reaction rates may depend on applied stress.

4. Development of oxide in surface cracks improves rupture resistance of nickel and its alloys and possibly retards crack growth in austenitic steels.[20,62]

5. Loss of alloying elements to the environment, pronounced in low carbon steels exposed to sodium, generally leads to decreased rupture life and increased ductility and creep rate.

6. There is evidence also that environmental effects will depend on the nature of the deformation processes,[45,59,68] specimen geometry,[40] and the presence of notches.[39,45]

The Influence of Environment on Fatigue Behaviour

Adverse effects of air and water vapour during room temperature fatigue are well documented.[85–90] There are several reviews of environmental effects during fatigue at room temperature[91,92] and at varying air pressure.[93,94] This section is concerned with fatigue at elevated temperatures where most work has been based on 'total endurance' tests as summarized in Table VI. There is also current interest in a mechanistic study of fatigue crack growth.[95–99]

High-Frequency Fatigue

It is necessary to distinguish between 'high' and 'low' frequency fatigue tests since in the former stress is usually controlled over a small cyclic strain range whilst the latter are conducted using strain control. For most cases the transition between the two types of test occurs at 0·1–1 Hz.

Endurances in Air and Vacuum or 'Inert' Environments

In reversed bend tests on lead at 500 cycles/min Snowden[100] demonstrated a difference of two orders of magnitude in fatigue life between vacuum, air, and pure oxygen. At all strain levels vacuum endurances exceeded those in air, which exceeded those in oxygen. Intermittent stress-free exposure to air had no effect on the lifetime in vacuum. At high temperature (1090 K) vacuum also improved endurance, relative to air, of the Co-base alloy S-816 and the Ni-base alloy Inconel 550[54] although the effect was much smaller than that seen in lead. Endurances for the nickel-base alloy converged at low stresses indicating a possible strengthening effect of air (Fig. 12). Similar convergence of air and vacuum data was noted for AISI 316 steel at 1090 K[101] and a crossover of the air and vacuum curves occurred for nickel, where it was suggested that oxide in cracks could prolong life in air at low stresses. Crossovers have also been seen in a ferritic stainless steel and a Ni/Cr alloy in the range 875–1025 K[36] where tests in purified argon gave shorter endurances than those in air, impure argon, or sulphur dioxide. Also, in single crystals of the alloy Mar-M200, air endurances were less than those in vacuum at room temperature whilst the reverse was true at high temperature[96] (Fig. 13). A thin oxide film, formed during testing, suppressed surface crack initiation but oxide formed during pre-exposure did not.

During vacuum tests on AISI 316 steel at 775 K, oxygen admission (at 10–15% of the expected life) caused reversible changes in the rate of fatigue hardening.[102] Since a previous oxide layer did not affect the subsequent vacuum hardening rate, it was proposed that oxygen reacted with freshly exposed slip steps.

Fatigue lives of silver at about 600 K were only slightly greater in vacuum (1·3 $\mu N/m^2$) than in oxygen at 1 atm,[103] possibly because silver oxide is unstable above about 475 K so that oxygen is virtually an inert gas at 600 K. However, some effects of oxygen persisted during creep at above 475 K.[84]

Endurances in Other Environments

Trace amounts of oxygen in nitrogen and hydrogen substantially impaired the fatigue-resistance of lead[100] although when these gases were purified the values of endurance approached those found in vacuum.

Some high-frequency tests have been conducted in reactor coolants but generally the observed endurances were similar to those in air. (This contrasts with low frequency testing.) Endurance at 10^7 cycles was marginally improved in carbon dioxide, relative to air, for an austenitic steel tested at 925 and 1025 K;[104,105] and Magnox Al 80 displayed little difference in endurance when tested in air, carbon dioxide,[106–108] and argon[109] at 675–775 K.

Endurances of AISI 347 steel in static sodium and in air at 775 K were similar but the endurance of mild steel was impaired by sodium, particularly when the initial carbon level of the steel was low. No decarburizing was detected as would be expected in flowing sodium.[34,72]

TABLE VI

Summary of Principal (Total Endurance) Experiments Referred to in Text

Material	Environment	Type of Test*	Temperature, K	Frequency	Semi-Stress Ranges, MN/m²	Strain Ranges, %	Cycles to Failure	Remarks (See also Text)	Ref.
Lead	Vacuum, air, O₂, N₂, H₂	R.B.	295	8·3 Hz		0·14–0·02 (total)	10^5–2×10^7	N_f vac. > N_f air > N_f O₂	100
S-816	Vacuum, air	P.-P. + Mean Tension	10·0	33 Hz	300–150	See Fig. 12	10^5–10^8	No crossover for S-816	54
Inconel 550									
Inconel X	Vacuum, air	R.B.	1090	5 Hz		3–0·9 (Total)	2×10^4–10^7	N_f vac > N_f air. No crossover	101
AISI 316				2·5 Hz		2–0·2 (Plastic)	10^4–10^7		
Nickel				12 Hz		1·5–0·6 (Plastic)	10^4–2×10^6	As above but crossover at 4×10^5 cycles Thicker specimens. No crossover	
						1–0·2 (Plastic)	2×10^4–10^7		
11Cr Ferritic s.s.	Air + 6% H₂O, 4·5% CO₂ 0·3% SO₂, argon	ROT. B.	875 925, 1025	47 Hz	380–140		10^4–10^9	Crossover for 11Cr at 260 MN/m² (10^6 cycles)	36
35Ni/15Cr steel									

TABLE VI (continued)

Material	Environment	Type of Test*	Temperature, K	Frequency	Semi-Stress Ranges, MN/m²	Strain Ranges, %	Cycles to Failure	Remarks (See also Text)	Ref.
Mar-M200	Vacuum, air	P.-P.	295–1200	10 Hz	See Fig. 13				96
20Cr/25Ni/Nb s.s.	Air, CO₂ + 1% CO	R.B.	1025	400 Hz	250–185		10^5–10^7	CO₂ endurance marginally greater	104
20Cr/25Ni/Nb s.s.	Air, CO₂	ROT. B.	925	83 Hz	250–185		10^4–10^8	As above, after 10^5 cycles	105
0.17C Steel	Air, town gas	ROT. B.	825	45 Hz	200–160		2×10^1–10^7	N_f town gas $< N_f$ air above crossover (5×10^6 cycles 0·17C, 2×10^6 cycles 0·39C)	111
0.39C Steel		R.T.		30 Hz	150–100		10^5–2×10^7		
Inconel X, Inconel	Air, ammonia	P.-P.	811	20–30 Hz	690–275		10^2–10^7	No effect. Tests on Inconel, Inconel X sheet, and Inconel butt welds	112
Mild Steel AISI 347	Air, Na (Internal) Air (External)	P.-P.	755	Not given	190–80		10^4–10^7	Hollow specimens, static Na Cracks started internally	110
½ Mo steel	Vacuum, air	R.B.	775	1 c/min (+ Hold)	See Fig. 16			825K tests + hold times ~ 300 min also undertaken	123
1Cr/Mo/V steel	Air, steam	R.B.	825	1 c/min		2·0–0·4 (Total)	2×10^2–5×10^3	Steam had < 1 ppm O₂	128
A286	Air, vacuum	P.-P.	866	0·01–10 c/min		0·09–9 (Plastic)	10^2–3×10^3 (Air) 20–3×10^4 (Vac.)	Transgranular cracks in vacuum Intergranular in air	125
Magnox Al 80	Air, CO₂, argon	T.C.†	675, 725			See Table VII		Except data in last line (Tube bend tests)	108
20Cr/25Ni/Nb	Vacuum, air, CO₂	R.B.	975	$\dot{\varepsilon}_t \sim 10^4$/s		1·2–0·5 (Plastic)	5×10^1–10^3	Tests on thin (0·5 mm) section material	133
2¼Cr–1Mo steel	Air, sodium, helium	R.B.	866	~ 3 c/min	See Figs. 17, 18				70, 77
AISI 316			925						
Nimonic 75, 90, 95	Air, argon	T.C.	1145, 1295		See Table VIII			Intergranular surface oxidation	137

*R.B. - Reversed Bending; ROT. B. - Rotating Bending; P.-P. - Push-Pull;
T.C. - Thermal Cycling; R.T. - Reversed Torsion.
†See Remarks

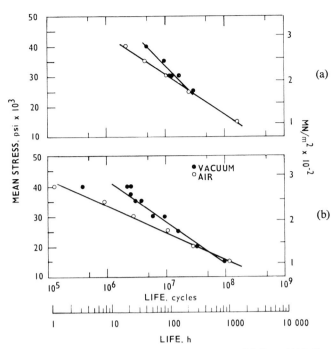

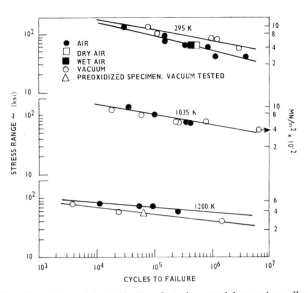

Fig. 13 *Fatigue life at 10 Hz of single-crystal low-carbon alloy Mar-M200 at 295–1200 K (Gell and Leverant, 1973).*[96]

Fig. 12 *Axial tensile fatigue properties of Inconel 550 at 1090 K in air and vacuum.* (a) *Ratio of cyclic to mean stress* = 0·125; (b) *ratio of cyclic to mean stress* = 0·667 Testing frequency 33 Hz (Nachtigall et al., 1965).[54]

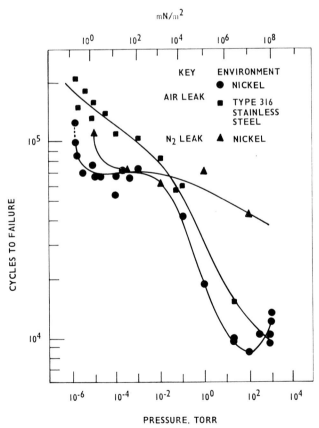

Fig. 14 *Effect of pressure and environment on endurances of nickel and AISI 316 steel at 1090 K in 5 Hz reversed bending tests (Achter et al., 1963).*[113]

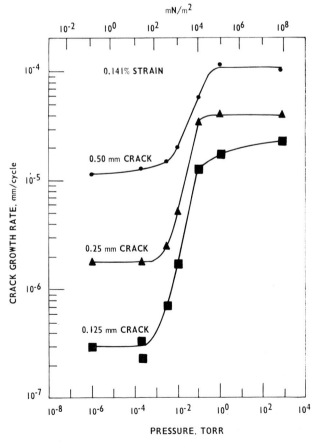

Fig. 15 *Effect of oxygen pressure on rate of fatigue crack growth in AISI 316 steel at 775 K and 10 Hz (Smith et al., 1969).*[115]

Crack initiation could be assisted by surface oxide break-up (e.g. at high strains) but delayed if it remained intact. Endo[111] attributed improved endurance of mild steel tested in town gas (compared with air) at 825 K to the appearance of an adherent grey film. However, Weisman[112] found that despite a tightly adherent scale formed during testing of Inconel in ammonia at 775–875 K, lifetimes in ammonia and air were similar.

Endurances at Low Gas Pressure

Residual reactive gases may adversely affect endurance. Fatigue life of nickel was markedly increased at 1090 K as air or nitrogen pressure was reduced from 0.13 MN/m^2 to 0.13 mN/m^2.[113] A plateau occurred at a transition pressure (T.P.) of ~ 133 mN/m^2, Fig. 14. Above this pressure, the rate of decrease of endurance was smaller in nitrogen than in air, presumably due to slight oxygen impurity in the nitrogen. By varying temperature and p_{O2} with known levels of residual gases (argon, hydrogen, nitrogen, &c.) it was found[114] that T.P. increased with temperature in the range 575–825 K. It was supposed that higher temperatures increased the oxidation of crack faces so reducing oxygen concentration at the crack tip and increasing T.P. Experiments on an austenitic steel[115] at 775 K and 1075 K correlated the T.P. with a sharp increase in crack growth rate for given crack length (Fig. 15). Also it was found that

(i) most of the lifetime was spent in the growth of small cracks (as cracks became longer, the effect of pressure was reduced) and
(ii) on changing the atmosphere the crack growth rate was a function only of the immediate environment.

However, a period of about one hour was necessary to establish a new steady growth rate after reducing p_{O2} from 133 N/m^2–0.13 mN/m^2. During this time, the crack extended by $\sim 80\,\mu$m, suggesting that there was an oxygen-affected zone this deep ahead of the crack when pressure was reduced. This observation implies that propagation rate is controlled by oxygen content in the material *ahead* of the crack.

In experiments on nickel at 575 K, fatigue lives at 0.13 μN/m^2 in water vapour and oxygen were identical, but there was no sharp T.P. for water vapour.[10] Oxygen had a much greater effect than carbon monoxide, carbon dioxide, nitrogen, or hydrogen since endurances in these at 13 mN/m^2 were only slightly less than at 0.13 μN/m^2.

A well-defined T.P. has also been found in lead[102] where it was independent of strain amplitude and the pressure-dependence was similar in single- and poly-crystals.[116] However, at high oxygen pressure, where intergranular cracking predominated in polycrystals, the endurance in single crystals was much greater. Increasing oxygen partial pressure also increased the contribution of intergranular cracking in nickel.[102,113]

Metallography

Most of the observations in this section are confined to failed specimens, i.e. to crack propagation rather than initiation.

During fatigue tests on an austenitic steel at 775–875 K, Coffin[117] observed preferential oxidation along grain boundaries, the width of the heavily oxidised zone increasing as strain range was reduced. This suggests that cyclic strain caused repeated cracking of the oxide along boundaries, accelerating subsequent oxidation.

Mar-M200 tested in air at 1200 K and 10 Hz displayed surface cracks filled with oxide.[118] Oxide was thought to reduce crack resharpening in compression, so retarding crack growth. Internal cracks were unoxidized and narrow and they propagated faster. Nimonic alloys tested at 975 K and above displayed precipitate-free zones at the surface and along and ahead of intergranular cracks.[119,120] These were caused by oxidation of Ti, Al, and Cr; and crack propagation could be expected to be assisted by oxidizing atmospheres.[120] Similarly, chromium transfer from the alloy to the scale was found in S-816 and Inconel 550 tested in air[54] whilst there was evidence for evaporative loss of chromium from S-816 in vacuum.

Atmosphere affected the distribution of slip bands in Inconel X750 at 775 K,[121] fine slip being seen at an oxygen pressure of 133 N/m^2 and coarse slip in a vacuum of 80 μN/m^2. Contrary to experience in other materials, fine slip in this case was associated with much shorter ($\times 40$) times for crack initiation.

Crack distribution in nickel at 575 K was also sensitive to atmosphere[113] since cracks in vacuum-tested specimens were wide whereas those in specimens tested at higher p_{O2} were narrow and more numerous. This suggests that oxygen assisted crack initiation[113] as was found for creep.[42] However, at 1075 K the main effect of oxygen was on crack propagation since many small intergranular cracks were seen in vacuum whereas a single long crack appeared at higher p_{O2}. Similar results were noted for AISI 316 steel at 775 and 1075 K[115] and diffusion of oxygen ahead of the growing crack may be important.[113] Studies on nickel at 575 K[10] showed that water vapour was less effective than oxygen in promoting cracking.

In summary, oxygen appears to assist crack growth (see, for example, Fig. 15) although exceptions may arise when oxide forms along a crack.[118] There are still no convincing data on effects of atmosphere on crack initiation.

Low-Frequency Fatigue

The high strain amplitudes involved in low frequency fatigue tests may produce effects quite different from those seen at high frequency. For example, oxide may continually crack or spall[122] and possibly affect endurance. Furthermore, many tests have incorporated hold times[123] which can introduce creep damage, thus reducing endurance (number of cycles to failure) but always increasing total time to failure. Studies of the interaction of creep and fatigue have been conducted to identify effects of environment at certain parts of a strain cycle.

Comparison of Air and Vacuum or Inert Environments

There has been no systematic study of low-frequency fatigue in controlled partial pressures of active impurities.

During reversed bend tests on a low-alloy steel at 875 K and 1 cycle/min, endurances in argon were twice those in air.[124] Introduction of a hold time of 30 min at zero stress in air (stress-free oxidation) did not impair endurance whilst hold times of 30–300 min in tension progressively reduced it. This suggested a component of damage due to creep and/or oxidation under stress. The contribution of each was assessed by testing a $\frac{1}{2}\%$ Mo steel in vacuum (1.3 mN/m^2) and air at 775 K.[123] A 30 min tension hold in vacuum reduced endurance relative to continuously cycled (vacuum) specimens so there was a creep component of damage (Fig. 16). However, the specimens continuously cycled in air gave even lower endurances and air specimens with 30 min hold time lower still. Thus, it was concluded that (*a*) both oxidation and creep

Source: *International Metallurgical Reviews*, Vol 19, 1974

TABLE VII

The Effect of Test Atmosphere on the Fatigue Endurance of Magnox Al 80[108]

Strain Range	Frequency, c/h	Temperature, K	Test Atmosphere					
			Carbon Dioxide		Argon		Air	
			Endurance		Endurance		Endurance	
			Cycles	Hours	Cycles	Hours	Cycles	Hours
1.2×10^{-4}	12	675	Undamaged at 15 000–21 000	Undamaged at 1500–2000	10 000–25 000	800–2000	—	—
4.8×10^{-4}	3	675	10 000–20 000	3000–7000	1000–3000	700–1000	—	—
2.4×10^{-4}	8	725	9000–18 000	1000–2000	3000–5000	400–700	—	—
4.8×10^{-4}	3	725	4000	1200	1500	500	—	—
1.76×10^{-4}	1020	675	>1 000 000	1000	—	—	40 000	40

TABLE VIII

The Effect of the Nature of the Environment on the Thermal Endurance of Nimonic Alloys[137]

Nature of Cycle	Material	Mean Number of Cycles to Failure			
		$T_2 = 1145$ K		$T_2 = 1295$ K	
		Air	Argon	Air	Argon
Slow heating to T_2, followed by rapid (2 min) cooling to 295 K	Nimonic 75	—	—	415	548
	Nimonic 90	—	—	240	410
	Nimonic 95	503	707	65	110

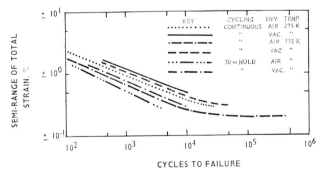

Fig. 16   Effect of hold time in air and vacuum upon the fatigue endurances of a $\frac{1}{2}$Mo steel at 275 and 775 K (White, 1969).[123]

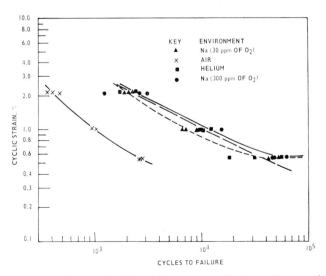

Fig. 17   Influence of cyclic strain range upon fatigue endurance of $2\frac{1}{4}$Cr 1Mo steel in sodium, air, and helium at 865 K. (Cycle used was approximately up 5 s, hold 5 s, down 5 s, hold 5 s.) (Andrews and Barker, 1964).[77]

during a hold can lower endurance, (b) that oxidation in the presence of a stress is much more damaging than oxidation alone, and (c) (from Fig. 16) that oxidation had a more serious effect on endurance than did creep.

Metallography revealed transgranular cracking in continuous vacuum and air tests but intergranular cracks (surface and internal) predominated on the tension side of hold-test specimens. Cracks on the compression side were transgranular, blunt, and probably non-propagating and those formed in air were oxide filled. Thus intergranular crack propagation was a likely outcome of creep damage during a hold time.

Oxidation-induced damage was suggested by Coffin[125] from tests on A286 steel in vacuum ($1.3 \ \mu N/m^2$) and air. Endurances in vacuum were frequency independent, whilst in air the (much lower) endurances decreased as frequency decreased. In Udimet 500 oxidation along grain boundaries ahead of a crack was considered an important prelude to propagation since denudation of γ' and a layered oxide were seen there.[126]

Thus the resolution of creep and environmental components of damage appears helpful. It is not always possible to separate a true creep component of damage although that occurring remote from a surface may offer a good approximation. At the surface some elements (e.g. oxygen) can assist intergranular cracking; it has been suggested that this may reduce the period normally associated with fatigue crack initiation.[127]

Effects of Other Environments

The endurance in steam of a low alloy steel at 825 K differed little from that in air both for continuous and hold-type tests.[128] Surface oxide spalled in air but an adherent film was produced in steam. However, the behaviour of oxide films near a crack tip is probably more significant than on a surface. During fatigue of a low carbon-manganese steel Coffin[129] noted that oxide which began to form along secondary crack edges completely filled the crack as temperature increased up to 775 K. Introduction of a nitrogen–hydrogen mixture suppressed surface oxidation. Coffin proposed that the 'packing' of cracks with low density oxide during the tensile cycle can lead to a wedging action as the crack is closed, thus increasing the propagation rate, but this contradicts other work where oxide in cracks reduced their growth rate.[118]

Reactor Environments

(a) Carbon dioxide. Contrary to its behaviour at high frequency,[106–109] Magnox Al 80 was environment-sensitive during low-frequency fatigue, endurances in carbon dioxide being three to ten times those in air and argon (Table VII).[108]

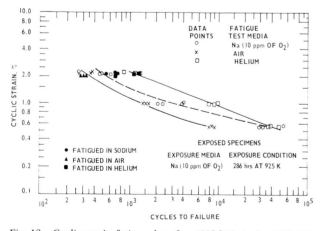

Fig. 18 Cyclic strain fatigue data for AISI 316 steel at 925 K in sodium, air, and helium (cycle used as for Fig. 17) (Kirschler and Andrews, 1969).[70]

Bending tests on 20Cr/25Ni/Nb steel at 1025–1125 K and 0.1 and 1 cycles/min revealed excessive laminar oxidation in cracks,[130] suggesting that successive oxide layers had spalled during reversed strain. Further work[131] revealed intergranular cracks completely filled with oxide whilst only a thin oxide formed on the specimen surface. Endurance in vacuum is increased relative to that in carbon dioxide although this trend may reverse at low strain rates.[132,133] Stress-free exposure to carbon dioxide at 1125 K for 500 h reduced subsequent fatigue endurance in that gas but did not affect endurance in vacuum, suggesting that oxidation affected crack nucleation rather than propagation.[131]

Metallography of 20Cr/25Ni/Nb specimens tested in CO_2 indicated decarburization in the strained areas, preferential oxidation in the grain boundary/surface region, and no σ phase near surfaces or cracks (although present elsewhere).[134] This indicated rapid chromium diffusion along grain boundaries.

(b) Helium. Bend tests in helium ($0.5–2$ vpm O_2) have been conducted on $2\frac{1}{4}$Cr/1Mo steel at 866 K[77] and on AISI 316 steel at 925 K.[70] For both materials (Figs. 17 and 18) endurance was greater in helium than in air. The ferritic steel displayed oxide-filled transgranular cracks in air tests but not in helium. In the austenitic steel the cracks formed in helium were oxide-free. Pre-exposure to sodium was effective only when the sodium contained high levels of carbon when it caused carburizing to a depth of 200–260 μm, so impairing subsequent fatigue resistance in helium.

(c) Sodium. Compared with air, flowing sodium markedly improved endurance of $2\frac{1}{4}$Cr/1Mo steel at 865 K (Fig. 17) and of AISI 316 steel at 925 K (Fig. 18). The bend tests did not cause surface cracking of the austenitic steel. Liquid sodium would probably penetrate cracks since zero contact angle was observed in nickel after surface oxide was reduced.[135]

Thermal Cycling

Isothermal tests[130,131] and temperature cycling over a range of 50–200 K about a mean of 1125 K in CO_2 produced intergranular oxidation and spalling in a 20Cr/25Ni/Nb steel.[136] Crack nucleation during cycling was caused by chromium depletion and intergranular corrosion.[137,138] Decarburization may improve thermal fatigue resistance through increased ductility.[139]

Though more difficult to perform than isothermal tests, thermal fatigue experiments more closely reproduce the cycles experienced in service. Glenny and Taylor[137] rapidly cooled Nimonic alloys from 1145 to 295 K in purified argon after stages of slow heating and found greater endurances than in air tests (Table VIII). In Nimonic 90 cracks < 75 μm deep developed in argon, but highly oxidized grain boundary cracks up to 250 μm deep formed in air. Tests on this material[140] showed that endurances in air at upper cycling temperatures of 1175 and 1275 K were similar. Corresponding endurances in vacuum were greater, and by a larger factor at the higher temperature.

During thermal fatigue of EN-25 steel between 315 and 875 K in hydrogen, cracks followed the shear planes.[141] Although the number of cycles to initiation was the same as for air, cracks (transcrystalline in both cases) were sharper in hydrogen, thus lowering endurance. Tests in argon reduced the rate of cracking relative to air.

Spalling

Spalling, due to the difference in expansion coefficients between oxide and metal, can occur during large temperature changes and high rates of change. It may also occur isothermally where growth stresses in oxides, oxide thickness, interface void nucleation, and the relative creep behaviour of metal and oxide become important.[4]

Spalling usually causes a net weight loss as seen for a cobalt-base alloy, Fig. 19,[142] and in the Ni/Cr/Co/Mo alloy ZhS-6 K, Fig. 20,[143] both cooled from 1375 K. Clearly the effect is more marked the higher the cycling rate. Spalling sometimes reveals cracks which are oxide free[144] so, again, surface observations cannot give a reliable indication of either the state of oxidation or the behaviour of oxide in a crack.

Coatings which reduce oxidation improve both spalling-resistance and thermal fatigue-resistance, as seen for aluminide coatings on nickel-base alloys.[145]

Fatigue Crack Initiation and Propagation

Total endurance results do not readily yield data on crack initiation. There are few instances in the literature where the effect of environment on initiation or propagation can be separately and explicitly stated. However, oxidation clearly affects crack propagation, and data as in Fig. 15 are required for thermal fatigue and combined creep/fatigue. Since there is evidence that the fraction of total life spent in initiation increases with decreasing strain range,[127] information is also required on the effect of environment upon cycles to initiate a crack.

Summary of Environmental Effects in Fatigue

1. In high-frequency tests there is conclusive evidence that increasing oxygen pressure in the range 1.3 $\mu N/m^2$ to 133 N/m^2 reduces endurance in lead,[100] nickel,[113,114] stainless steel,[115] and Inconel.[121] A possible contribution is from enhanced diffusion during vibration,[146] e.g. of oxygen ahead of a crack tip. There is little difference in endurance of Magnox Al 80 and 20Cr/25Ni/Nb steel respectively in air and CO_2.[104,105,108]

2. In low-frequency tests there is for Magnox Al 80 and 20Cr/25Ni/Nb steel evidence of an increased endurance in CO_2.[108,133] Ferritic and austenitic steels also have a longer life in helium and sodium than in air.[70,77] Because data are sparse at very low pressure,[125] there is as yet no evidence of a T.P. at low frequency.

3. Repeated spalling of surface oxide will clearly reduce cycles to crack initiation in fatigue but cannot indicate oxide behaviour at the tip of a propagating crack.

4. At low stresses 'crossovers' in endurance between air and inert environments can occur at high frequency.[36,54,101,111] This is probably caused by oxide in cracks.

5. In thermal fatigue tests intergranular corrosion and spalling reduce the resistance of austenitic steels in CO_2. Conversely, a vacuum or an inert gas increases the thermal fatigue endurance of Nimonic alloys relative to air.

Effects of Trace Impurities on Crack Propagation

This section is concerned with (i) the mechanism of gas flow to a crack tip and (ii) possible effects there of active elements. Bulk oxide restrictions will be ignored for the present.

Gas Access to a Crack Tip

In both creep and fatigue at low pressure, increasing p_{O_2} can reduce time to fracture (Figs. 4 and 14.) There is usually a plateau in endurance above the T.P., where it has been proposed[9,42,113,115,147] that the rate of arrival of gas molecules

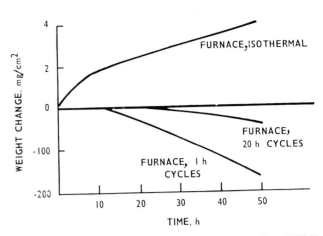

Fig. 19 *Effect of thermal cycling to room temperature from 1375 K upon the weight change of alloy WI-52. (Note change of weight-change scale.) (Probst, 1970).*[142]

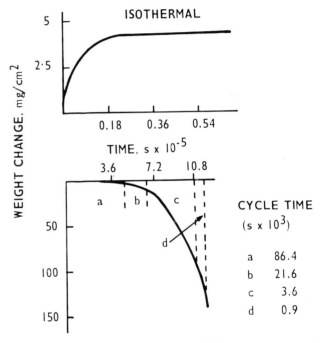

Fig. 20 *Effect of thermal cycling from 1375 K on weight change in alloy ZhS-6K. (Note weight-change scales.) (Gertsriken* et al., *1963).*[143]

at a crack tip just balances the rate of creation of new surface atomic sites (general adsorption theory). In the capillary reduction theory the arrival rate is considered to be attenuated by crack dimensions.[100,148]

General Adsorption Theory

Using simple kinetic theory,[149] Stegman *et al.*[42] considered a creep crack of length *l* in material of atomic spacing Ψ. Each atomic row at the crack tip will be exposed for an interval $\dfrac{\Psi}{dl/dt}$ before the crack moves on to expose the next row. Adsorption was regarded as fully effective only if each atom in the row was covered by a gas atom during this interval. Neglecting the possible effect of crack geometry, the time required to saturate the surface with a diatomic gas is S/v, where S is the number of surface sites per unit area and ν is frequency of molecular impacts. Equating these two times it was shown that:

$$\text{T.P.} = \frac{4\left(\frac{dl}{dt}\right)\sqrt{MRT}}{L\,\Omega\sqrt{3}} \qquad \dots (1a)$$

where Ω is the atomic volume of the metal, M is the gramme molecular weight of the gas, R is the gas constant, T is absolute temperature, and L is Avogadro's number. The corresponding equation for fatigue, assuming that gas access is possible only during a tensile cycle, is[115]

$$\text{T.P.} = \frac{8f\left(\frac{dl}{dN}\right)\sqrt{MRT}}{L\,\Omega\sqrt{3}} \qquad \dots (1b)$$

where f is the frequency of fatigue testing.

Capillary Reduction Theory

Crack geometry may restrict gas flow to a crack tip. For example, fatigue damage in lead[100] changed from intercrystalline to transcrystalline as cracks lengthened and impeded oxygen access. The model[148] considers that, at low pressures, gas flow along a crack of length l and width α is molecular rather than viscous (Poiseuille type). If a uni-molecular layer forms on crack walls during flow, the characteristic time τ for complete crack coverage is[150] $\tau \sim \dfrac{3l^2}{8\alpha^2\delta^2\nu}$ where δ^2 is the molecular cross-sectional area. Snowden[100] found that at the T.P. for lead, τ was much greater than the time for half a cycle so the model does not provide a quantitative understanding of the T.P.

Appraisal of the Gas Access Models

There are several shortcomings to the above models. The predicted values of T.P. underestimate those observed in fatigue.[115] For example, using data from Fig. 15, equation. 1(b), predicts T.P. as 1.3 N/mm^2 which is considerably below that observed. The discrepancy was still greater at 1075 K. However, reasonable agreement between observed and calculated T.P. (equation 1(a)) was found in creep[42] although crack growth rate was not measured accurately.

The predicted dependence of T.P. on crack growth rate and frequency (equation 1(b)) has not been found experimentally. Fig 15 shows that T.P. was independent of dl/dN for AISI 316 whilst in lead[151] T.P. at 64 cycles/min was similar to that at 500 cycles/min (Fig. 21). There are also indications[42,94,114] that temperature-dependence of T.P. is much greater than $(T)^{\frac{1}{2}}$, equations 1(a) and 1(b).[94]

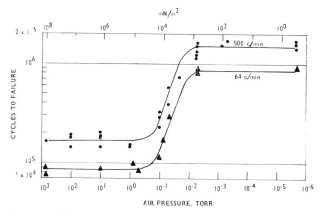

Fig. 21 *Effect of air pressure upon the room-temperature fatigue life of lead at different frequencies (Snowden, 1964).*[151]

TABLE IX

Nominal Compositions, wt.-%, of Alloys Mentioned in Text

Alloy	Fe	Ni	Cr	C	Ti	Al	Nb	Mo	Mn	Other
AISI 304	Bal	9·5	19·5	<0·08	—	—	—	—	1·5	
AISI 316	Bal	10·5	19·5	<0·08	—	—	—	2·5	1·5	
AISI 321	Bal	10·5	19·5	<0·08	1·0	—	—	—	1·5	
AISI 347	Bal	10·5	19·5	<0·08	—	—	1·0	—	1·5	
A286	Bal	26·0	15·0	<0·5	2·0	0·2	—	1·3	1·3	0·015 B, 0·5 Si.
Incoloy 800	Bal	35	20	<0·08	0·3	0·3	—	—	<1·0	
Inconel	<9	Bal	13	<0·15	—	—	—	—	<1·0	
Inconel 550	7	Bal	15	0·05	2·5	1·2	0·9	—	0·6	
Inconel X Inconel X750	9	Bal	15	0·04	2·5	0·7	1·0	—	0·5	
Inco 713C	—	Bal	12·5	0·12	0·8	6·0	2·0	4·2	—	
Nichrome V	—	Bal	20	—	—	—	—	—	—	
Hastelloy C	6	Bal	15	—	—	—	—	17	—	4·5 W
Hastelloy X	18	Bal	22	0·1	—	—	—	9	0·5	
Udimet 500	—	Bal	18	0·08	2·9	2·9	—	4	—	18 Co
René 41	—	Bal	19	0·09	3·1	1·5	—	10	—	11 Co
Mar-M200	—	Bal	9	0·15	2·0	5·0	1·0	—	—	10 Co, 12·5 W
S—816	4	19	20	0·4	—	—	4	4	1·3	42 Co, 4 W
Nimonic 75	<5	Bal	20	0·1	0·4	—	—	—	<1·0	
Nimonic 90	—	Bal	19·5	0·07	2·4	1·4	—	—	0·7	18 Co
Nimonic 95	5	Bal	19·5	<0·15	2·9	1·9	—	—	<1·0	18 Co
WI 52	2	—	21	0·45	—	—	2	—	—	Bal Co, 11 W
En 25	Bal	2·5	0·7	0·3	—	—	—	0·6	0·6	
Magnox Al 80	—	—	—	—	—	0·8	—	—	—	0·005 Be, Bal Mg
ZhS–6K	—	Bal	11·5	0·17	2·7	5·5	—	4	—	5 Co, 5 W

Some of the reasons for these difficulties are as follows:

(i) The models neglect the complex effects of pressure, temperature, etc. on the rates of adsorption[12] (or dissolution) of gas at the crack tip.

(ii) Combination of reactive gases with crack faces (gettering) will tend to reduce impurity partial pressure at the crack tip.

(iii) Crack propagation occurring by breaking of successive bonds at the crack tip is an oversimplification since void coalescence ahead of a crack is known to be important in creep and high temperature fatigue fracture.

(iv) Metallography of failed specimens[113,115] suggests that gas access down a fatigue crack may not be restricted during compression, equation 1(b), since crack widths were much greater than applied strain amplitude.

The Effect of Trace Impurities in Vacuum and Inert Gases

We will briefly examine the relative effects on crack propagation of the following alternatives:

(i) A given partial pressure of impurity in an inert gas,

(ii) A similar impurity pressure in a vacuum.

The T.P. for given partial water vapour pressures in nitrogen or argon at 1 atmosphere was an order of magnitude above that for the same (total) water vapour pressures alone during fatigue of aluminium at room temperature,[152] suggesting that inert gases may prevent water vapour from reaching the crack tip. Hydrogen was much less effective in screening water vapour.

From kinetic theory of gases[149] the collision frequency of impurity molecules with a specimen surface will be unaffected by collision with other gas molecules (cases (i) and (ii) above). However impurity arrival rates at a crack tip are different in the two cases. If, for given crack dimensions, impurity arrival rate $n_{1\text{imp}}$ in an inert gas pressure p_1 is by viscous flow while a lower pressure, p_2, of impurity alone results in molecular flow of $n_{2\text{imp}}$, then assuming zero pressure at the tip of a cylindrical crack it may be shown from the equation of gas flow [153] that:

$$\frac{n_{1\text{imp}}}{n_{2\text{imp}}} \sim \frac{\alpha}{4\lambda_1}$$ where λ_1 is the molecular mean free path (m.f.p.)

at pressure p_1. If the crack is so narrow that flow is molecular in both cases (i) and (ii), we find that $\frac{n_{1\text{imp}}}{n_{2\text{imp}}} = 1$, i.e. impurity flow rates become identical.

For typical values of α, however, the rate of arrival at the crack tip is greater in the viscous case*, conflicting with experiment.[152] This suggests that the reactive gas loss at an advancing crack tip is not large enough to cause net flow down a crack but that impurity molecules arrive instead by general diffusion.

The coefficient of interdiffusion D_{ab} between two gas species a and b is given by $D_{ab} = \lambda \frac{\sqrt{6kT}}{3} \sqrt{\frac{1}{m_a} + \frac{1}{m_b}}$. For Knudsen flow ('streaming'), λ is replaced by the crack width α. D_{ab} thus decreases as the m.f.p. decreases. Further, the rate of diffusion of, for example, O_2 (for which $m = 32$) depends on the molecular weight of the inert gas. The ratio of diffusion rates for O_2 in helium ($m = 4$) and in argon ($m = 40$) is thus 2·2:1. Diffusion thus seems likely to govern the arrival of impurity molecules down narrow cracks.

---

* The Knudsen Number α/λ_1 (> 100 for viscous or < 1 for molecular flow) determines a critical crack size defining the type of flow. Taking $\lambda_1 = 5 \times 10^{-5}$ mm at atmospheric pressure, then $\alpha > 5 \times 10^{-3}$ mm for viscous flow and $\alpha < 5 \times 10^{-5}$ mm for molecular flow. For a crack width of 5×10^{-3} mm, $n_{1\text{imp}}/n_{2\text{imp}} = 25$.

Surface Energy Considerations

Environmental/Metal Interface

We can now consider the reactions occurring at a crack tip. Gas adsorption can lower surface energy and hence reduce the work required in fracture. This mechanism has been used[42,101,154] to account for a 'weakening' effect of air relative to vacuum at high stresses and 'low' temperatures. Jaffe,[155] however, has suggested that the energy reduction is far too small to be of importance.

The discrepancy between true surface energy and the energy to propagate unit area of fracture (effective fracture surface energy, EFSE) may be seen as follows. As p_{O_2} was increased, the surface tension of clean silver at 1145–1220 K decreased from about 1·15 N/m.[156] Similarly oxygen adsorption reduced the surface energy of δ-ferrite at 1685 K from 2·4 J/m² ($p_{O_2} = 10$ a N/m²) to a 'saturation' value ~ 1·7 J/m² ($p_{O_2} > 1$ N/m²).[157] In contrast, EFSE is generally 10–10⁵ J/m² for the intergranular creep failure of several brittle materials,[157] EFSE increasing with increasing temperature. Similar values are expected in fatigue, but EFSE would be considerably higher for ductile materials such as nickel.

Thus the reduction ($\Delta\gamma$) in true surface energy represents only about 10^{-5} of the EFSE. Differentiation of the Griffith crack growth equation gives $\Delta\sigma/\sigma \simeq \frac{1}{2}\Delta\gamma/\gamma$ so, for $\Delta\gamma/\gamma \simeq 10^{-5}$, the change in fracture stress will be immeasurably small even if curvature and strain at a crack tip modify γ.

Prager and Sines[59] proposed a surface energy model for high-temperature ductility of René 41. They considered that gas adsorption could occur on Petch–Stroh cracks nucleated at dislocation pile-ups. True surface energy γ was related to strain at fracture ε_f by $\gamma = \frac{K^2\pi(1-\mu)\varepsilon_f^2}{2aGd}$, where K is the work-hardening coefficient (assumed independent of strain), μ is Poisson's ratio, G is shear modulus, d is grain diameter, and a is a constant (~ 5).

Assuming a 50% reduction in γ due to adsorption, the calculated ductility reduction is $\sim 25\%$ compared with observed decreases $\sim 80\%$. The approach fails to indicate why cracks in less environment-sensitive material (coarse γ' precipitate) were not prone to adsorption effects and also to distinguish between surface energy and EFSE.

A further objection to the surface energy/adsorption model is that the change in creep rate or fatigue crack growth after a change in environment (Fig. 4(b)) is not instantaneous.

Sub-Surface Interfaces

It is possible that crack propagation rate is controlled by adsorption on interfaces ahead of a crack.

(i) Rodriguez,[20] in attempting to account for oxygen-retarded crack growth in AISI 304 steel, suggested that the Griffith equation for an intergranular crack of length l may be written

$$\text{Fracture stress} = \sqrt{\frac{2E(2\gamma - \gamma_{GB})}{\pi l}}$$

where E is Young's modulus, γ is the surface energy and γ_{GB} is the grain-boundary energy. Reduction in γ_{GB} by, for example, oxygen diffusion and adsorption on boundaries ahead of a crack[159] could raise the stress for propagation. Again, γ is not distinguished from EFSE and γ will also probably be reduced by adsorption of the same impurity species as that reducing γ_{GB} (see below).

(ii) Tipler and McLean[160] have conducted creep tests in vacuum on a dilute Cu–Sb and Cu–O alloy. There was an increase in grain boundary cavitation and creep rate, and a decrease in rupture life and ductility relative to pure copper. This was attributed to solute segregation, lowering the overall energy for grain-boundary cavitation by reducing the term $(2\gamma - \gamma_{GB})$. It was estimated that the decrease in energy for cavitation $= 100\ C/C_s$ J/m^2, where C is solute concentration and C_s is its maximum solubility. This work clearly has far reaching implications in environmental studies, since many important elements (e.g. oxygen) have limited solubility in metals. Furthermore, this is the only approach so far in which surface energy can be separated from EFSE. Such effects[160] may have caused oxygen-enhanced cavitation in an austenitic steel,[64] in γ–Fe,[63] in a nickel-tungsten alloy[60] and in silver.[84] The results are also consistent with oxygen weakening observed in nickel and its alloys.[41,42]

In summary, it is clear that changes in surface energy due to adsorption at a crack tip are far too small compared with EFSE to alter significantly the stress for crack propagation. However, adsorption may be important in controlling structure ahead of a crack. Measurements of oxygen concentrations there and of EFSE in controlled environments would give a further guide to mechanisms involved.

A Suggested Model of Trace Impurity Effects

It is proposed that an impurity (e.g. oxygen) controls creep or fatigue crack growth in a metal (e.g. nickel) as follows:

(i) The impurity diffuses through the gas phase to the tip of a crack. The concentration there will depend on total gas pressure, crack width, the accompanying gas, and on the amount lost to the crack walls.

(ii) An equilibrium is set up between the impurity in the gas at the tip and that dissolved in the metal, surface adsorption acting as an intermediate step. For a diatomic gas, the concentration in solution is proportional to $\sqrt{p}$.

(iii) The impurity will then diffuse into the metal ahead of the crack, e.g. oxygen will segregate to the surface of grain boundary cavities,[160] reducing the radius at which voids become stable. Growth will then occur by vacancies which may arise from surface oxidation condensing at the void. Subsequent coalescence will aid crack growth.

Oxygen thus has a unique rôle in assisting crack propagation since it promotes both nucleation and growth of voids. The solubility in metals of hydrogen and of nitrogen is generally similar to that of oxygen (although rates of adsorption may be different) so both these elements will stabilize voids[160] but not assist subsequent growth.

The rate-controlling factor will be the slowest of the following: (a) gaseous diffusion, (b) adsorption, (c) solid state diffusion, and (d) void growth by vacancy condensation. If crack growth in creep or fatigue is faster than the rate of growth of the environmentally affected (cavitated) zone, then it will not be assisted by environment. If the reverse is true, then crack growth will be assisted.[96] There will thus be a rapid change in intermediate behaviour, e.g. the T.P. effect (Figs. 15, 21). From the above, this need not be caused by pressure alone. The depth of the affected zone was estimated at $\sim 80\ \mu$m during fatigue of AISI 316 steel at 1075 K. This yields a diffusion coefficient of $\sim 10^{-8}$ cm^2/s, which is reasonable for an interstitial element at that temperature, suggesting that solid-state diffusion was rate-controlling.

Effect of Bulk Oxide on Mechanical Properties
Stresses Induced by Surface Oxide

The rôle of surface oxide as a barrier to dislocation egress

is well substantiated for single crystals. However in polycrystals the main effect of surface oxide may be to affect creep, especially of thin specimens, through stresses arising in this layer. Oxidation can cause an extension $\sim 2\%$ in some austenitic steels[161] during stress-free exposure of 2 500h at 1025–1175 K. Growth stresses were estimated as ~ 5 N/mm^2 during oxidation of 20Cr/25Ni/Nb steel wire at 1125 K in carbon dioxide[19] and ~ 10 N/mm^2 during oxidation of nickel in air at 1275 K[162] although these can either assist or oppose applied stresses.[19]

Oxide Formation under Stress

If oxide-growth stresses cannot deform the underlying metal microcracking occurs in the oxide itself at a critical oxide thickness e.g. in 20Cr/25Ni/Nb steel.[163] Alternatively the oxide may spall. Such damage may alter the subsequent oxidation rate since (i) oxide cracks will enhance oxygen access to the metal surface, e.g. by molecular flow,[164] or (ii) a Cr-depleted zone beneath spalled oxide may form a non-protective (e.g. Fe-rich) layer[163] giving a large re-oxidation rate.[165]

In tests on iron[166] and on 20Cr/25Ni/Nb steel,[167] oxidation rate was faster for cold-worked surfaces, the oxide layer being non-porous and adherent. Thus, if a region is continuously deformed, as in fatigue, this may result in a thicker oxide layer than on a strain-free surface. Further, the spalling of oxide at surfaces will not necessarily reflect its behaviour at a fatigue crack tip. Data from stress-free corrosion experiments must therefore be applied with caution to materials likely to be deformed in service, where oxidation rates are likely to be enhanced.

Oxide in Cracks

Strain-enhanced oxide formation near a crack tip can modify crack propagation as follows:

(i) Vacuum rewelding upon crack closure[168]
(ii) Oxide bridging and load bearing[33,101] } Decreasing propagation rate.
(iii) Oxide filling preventing compression resharpening[118] (Crack blunting due to increased tip radius.)

(iv) Oxide wedging at crack mouth[169] or tip[129]
(v) Oxide formation and subsequent rupture contributing directly to increase in crack length } Increasing propagation rate

None of these mechanisms pertains to crack initiation.

Creep Crack Propagation

There is clear evidence for (ii) as discussed above. In tests on several carbon steels between 625 and 825 K there was some evidence of (iii) and (iv)[169] but much more work is required to establish these as important mechanisms.

Intergranular oxide may either increase or decrease the rupture resistance of nickel depending on the temperature and stress but, generally, oxide in creep cracks reduces growth rate.[43] For example in 18Cr/8Ni/Nb steel in steam,[170] oxide retarded crack growth compared with a thinner oxide formed in air. Steam, however, assisted crack nucleation.

Since oxide behaviour depends so much on conditions of formation, results from pre-exposure tests must be treated with caution.

Fatigue Crack Propagation

Oxide in fatigue cracks is much thicker than oxide formed at an unstressed surface, emphasizing again the importance of strain-enhanced formation. Vacuum rewelding (i) is doubtful because after a tensile cycle the region ahead of the crack has suffered slip which is unlikely to be reversible.

The high-frequency work of Achter *et al.* (e.g. Fig. 14) shows that endurance increases with improvement of the vacuum, and the weakening effect of oxygen may be attributed to oxygen diffusion ahead of an intergranular crack.[93] Oxide-strengthening at atmospheric pressures and low fatigue stresses observed for alloys of high nickel content[36,54,113] is most satisfactorily explained by an oxide-sealed crack restricting further oxygen access.

Thus it is necessary to determine when oxide in cracks is least disturbed. Large strain rates in high-strain fatigue are likely to lead to repeated oxide cracking. Low deformation rates presumably produce a lower volume of oxide per unit time, minimizing wedging effects in compression and allowing more time for relaxation in the surrounding metal. A stable oxide is thus envisaged as forming when the growth rate is equal to the crack opening displacement rate.

Sub-Surface Voids

Vacancies injected into a metal by cation-diffusion controlled oxidation may promote nucleation and growth of intergranular cavities. We may estimate the resulting void volume by considering a cylindrical specimen of unit length and diameter Λ. If all vacancies from the oxide/metal interface diffuse to grain-boundary cavities, then for oxide thickness x, it may be shown that the number of voids per unit grain-boundary area $n_v = \dfrac{x\,\Omega d}{\pi \bar{r}^3 \Omega_o \Lambda}$ where Ω_o and Ω are the atomic volumes of metal in the oxide and the pure metal, respectively, d is grain size, and $\bar{r}$ is the mean void radius. The number of voids per arbitrary length y of grain boundary is thus $y\sqrt{n_v}$.

After creep of an austenitic steel at 1075K in air the number of voids per unit boundary (0·16 mm) length was between 10 and 40, or twice the number seen in vacuum.[64] In this work $\Lambda \sim 5$ mm, $d \sim 0.5$ mm, and $r \sim 5$ μm, so taking $x \sim 10$ μm as a typical oxide thickness in air the number of voids per unit length calculated from above is ~ 9. Vacancies generated from oxidation thus contribute significantly to extra cavity volume seen in air specimens.

Structure Changes ahead of Cracks

Carburization (e.g. in austenitic steels) raises the proof stress but reduces tensile ductility[28] whereas decarburization (e.g. in Nimonic 75) leads to a loss in strength[171,172]. Whilst room-temperature experience suggests that changes in bulk yield stress have little effect on fatigue crack propagation rates, compared with changes in mechanisms of crack extension,[173] structural changes ahead of the crack (especially cavitation) and oxide identification behind it will still be important at high temperature.

Concluding Remarks

The work summarized shows that the mechanical properties of metals are complex functions of environment and of material composition and geometry. Much experimental work has been done to establish the effects of, for example, reactor atmospheres, but its value has often been impaired by failure to record the environment composition and to conduct careful metallographic examination after test.

Many valuable studies however have sought to investigate fundamental effects of environment in tests conducted at several partial pressures of an active impurity (e.g. oxygen). It is accepted that there is often a critical pressure above which time to fracture is markedly reduced. This does not arise from reduction of fracture surface energy by adsorption but rather from traces of active dissolved impurity which assist in cavity nucleation ahead of the main crack front. The depth of the zone affected depends on the gas impurity concentration, on rate of impurity dissolution in the metal, and on solid state diffusion. Above a certain crack growth rate or below a critical pressure one or more of these processes will be too slow to cause weakening.

Vacancies are also required for an increase in void density. Many will be generated by deformation within the metal but supply can be increased (e.g. by a factor of 2) during cation-diffusion controlled oxidation, as in austenitic steels where cavitation in air occurs faster than in vacuum. Oxygen thus assists in weakening because of its rôle in both the nucleation and growth of voids.

Surface Energy Consideration

It is different once substantial quantities of oxide form. Oxide in cracks can retard crack growth by stifling oxygen access to the tip and by supporting the applied load; this is well documented for nickel. There is as yet little evidence that oxide in a crack can assist propagation by wedging. The interplay between oxygen adsorption (weakening) and oxide formation (strengthening) can cause crossovers in lifetimes for tests at several strain rates in oxidizing gases when compared with those in inert environments.

Stresses induced by oxidation can either oppose or assist an applied stress. The rate of oxide formation too can depend on stress and strain. For example, it is much greater in a fatigue crack than on a free surface. This may be due to repeated cracking of the oxide under strain, leading to lack of protection and enhanced oxidation in the underlying metal. Clearly oxidation data from stress-free experiments cannot strictly be applied to components undergoing strain.

Other demonstrable effects of environment are that point defects induced by oxidation alter diffusion creep rate (as in magnesium) and that discrete particles of a second phase (e.g. carbide and nitride) forming in steels generally reduce creep rate at the expense of ductility.

There is obviously great scope for further study of environmental effects during high temperature deformation, particularly to describe more fully the processes occurring ahead of a growing crack and to establish the reasons for the differences in the properties of oxides formed under different conditions.

Acknowledgements

The authors' thanks are due to Drs. J. Barford, I. L. Mogford, R. T. Pascoe, C. E. Richards, and R. D. Townsend for discussions during the preparation of this review, which is published by permission of the Central Electricity Generating Board.

References

1. P. T. Gilbert, *Met. Rev.*, 1956, **1**, 379.
2. D. deG. Jones and I. L. Mogford, 'Corrosion Fatigue', p. 30. **1972**: Houston (Nat. Assoc. Corr. Eng.).
3. J. Stringer, *Corrosion Sci.*, 1970, **10**, 513.
4. D. R. Holmes and R. T. Pascoe, *Werkstoffe u. Korrosion*, 1972, **10**, 859.
5. I. Kramer and L. J. Demer, *Progress Mat. Sci.*, 1961, **9**, 133.
6. A. R. C. Westwood and N. S. Stoloff (ed.), 'Environment Sensitive Mechanical Properties'. **1965**: New York (Gordon and Breach).
7. J. Board, *J. Brit. Nuclear Energy Soc.*, 1970, **9**, 101.
8. T. D. Atterbury, *Metallurgist*, 1973, **5**, 238.
9. M. R. Achter and H. W. Fox, *Trans. Met. Soc. AIME*, 1959, **215**, 295.
10. H. H. Smith and P. Shahinian, *Amer. Soc. Test Mat. Spec. Tech. Publ.* **462**, p. 217, 1970.
11. R. Blackburn, *Met. Rev.*, 1966, **11**, 159.
12. A. R. Ubbelohde, 'An Introduction to Modern Thermodynamic Principles'. **1952**: Oxford (Clarendon Press).
13. A. U. McRae, *Surface Sci.*, 1964, **1**, 319.
14. H. Forestier and A. Clauss, *Rev. Mét.*, 1955, **52**, 961.
15. F. D. Richardson and J. H. E. Jeffes, *J. Iron Steel Inst.*, 1948, **160**, 261.
16. R. E. Smallman, 'Modern Physical Metallurgy'. **1970**: London (Butterworth).
17. R. Hales, P. S. Dobson, and R. E. Smallman, *Metal Sci. J.*, 1968, **2**, 224.
18. N. B. Pilling and R. E. Bedworth, *J. Inst. Metals*, 1923, **29**, 529.
19. J. M. Francis and K. E. Hodgson, *Mater. Sci. Eng.*, 1970, **6**, 313.
20. P. Rodriguez, *Trans. Indian Inst. Metals*, 1967, **20**, 213.
21. D. S. Wood, M. Farrow, and W. T. Burke, *J. Brit. Nuclear Energy Soc.*, [*Conference on Effect of Environment on Material Properties in Nuclear Systems*], p.213, **1971**.
22. R. A. U. Huddle, *ibid.*, p. 203.
23. H. Inouye, 'Proceedings of IAEA Conference on Corrosion of Reactor Materials, Vienna,' p. 319, **1962**.
24. J. C. Bokros and H. E. Schoemaker, *U.S. Atomic Energy Commission Rep.* (**GA-1508**), 1961.
25. J. E. Antill, K. A. Peakall, and J. B. Warburton, Ref. 21, p. 187.
26. H. E. McCoy, Ref. 23, p. 263.
27. F. B. Litton and A. E. Morris, *J. Less-Common Metals*, 1970, **22**, 71.
28. A. Thorley and C. Tyzack, Ref. 21, p. 143.
29. E. J. Dulis and G. V. Smith, *Trans. Met. Soc. AIME*, 1952, **194**, 1083.
30. H. E. McCoy, W. R. Martin and J. R. Wier, *Proc. Inst. Environment* [*Conference on Global and Space Environments*], p. 163, **1961**.
31. E. D. Hindle and G. F. Slattery, Ref. 21, p. 1.
32. P. Hurst and C. Tyzack, *ibid.*, p. 37.
33. T. R. Cass and M. R. Achter, *Trans. Met. Soc. AIME*, 1962, **224**, 1115.
34. J. Sannier, O. Konovaltschikoff, D. Leclercq and R. Darras, Ref. 21, p. 155.
35. M. J. Fevery, N. Wieling, F. Casteels and P. Libotte, *ibid.*, p. 167.
36. A. V. Ryabchenkov and A. I. Maksimov, *Metallovedenie i Term. Obrabot. Metallov*, **1968**, (12), 21.
37. P. Kofstad, 'High Temperature Oxidation of Metals'. **1966**: New York, (John Wiley).
38. P. Shahinian and M. R. Achter, *Trans. Met. Soc. AIME*, 1959, **215**, 37.
39. P. Shahinian and M. R. Achter, 'High-Temperature Materials' (edited by R. F. Hehemann and G. F. Ault), p. 448. **1959**: New York (John Wiley).
40. D. A. Douglas, *ibid.*, p. 429.
41. P. Shahinian and M. R. Achter, 'Proceedings of the Symposium on Crack Propagation, College of Aeronautics, Cranfield', p. 29, **1962**.
42. R. L. Stegman, P. Shahinian and M. R. Achter, *Trans. Met. Soc. AIME*, 1969, **245**, 1759.
43. R. J. Sherman and M. R. Achter, *ibid.*, 1962, **224**, 144.
44. D. L. Douglass, *Mater. Sci. Eng.*, 1968, **3**, 255.
45. P. Shahinian, *Trans. Amer. Soc. Mech. Eng.* [D], 1965, **87**, 344.
46. T. C. Reuther, P. Shahinian and M. R. Achter, *Proc. Amer. Soc. Test Mater.*, 1961, **61**, 956.
47. M. R. Pickus and E. R. Parker, *Amer. Soc. Test Mat. Spec. Tech. Publ.* **108**, p. 26, 1951.
48. P. Shahinian and M. R. Achter, 'Inst. Mech. Eng. Joint International Conference on Creep, London,' p. 7, **1963**.
49. H. T. McHenry and H. B. Probst, 'High Temperature Materials' (edited by R. F. Hehemann and G. F. Ault), p. 466, **1959**: New York (John Wiley).
50. O. C. Shepard and W. Schalliol, *Amer. Soc. Testing Mat. Spec. Tech. Publ.* **108**, p. 34, 1951.
51. P. Shahinian, *Trans. Amer. Soc. Met.*, 1957, **49**, 862.
52. R. Widmer and N. J. Grant, *Trans. Amer. Soc. Mech. Eng.*, [D], 1960, **82**, 882.
53. D. J. Duquette, *Scr. Metall.*, 1970, **4**, 633.
54. A. J. Nachtigall, S. J. Klima, J. C. Freche, and C. A. Hoffman, N.A.S.A. (**TN D-2898**), 1965.
55. P. Shahinian and M. R. Achter, *Proc. Amer. Soc. Test. Mat.*, 1958, **58**, 761.
56. P. Shahinian and M. R. Achter, *Trans. Amer. Soc. Met.*, 1959, **51**, 244.
57. A. F. Weinberg, R. J. Coburn and R. Wallace, *U.S. Atomic Energy Commission Rep.*, (**GA-2372**), p. 97, 1961.
58. J. C. Bokros, *ibid.*, (**GA-1629**), 1960.
59. M. Prager and G. Sines, *Trans. Amer. Soc. Mech. Eng.*, [D], 1971, **93**, 225.
60. W. R. Johnson, C. R. Barrett and W. D. Nix, *Metall. Trans.*, 1972, **3**, 695.
61. H. E. McCoy and D. A. Douglas, *U.S. Atomic Energy Commission Rep.* (**TID-7597**), p. 748, 1960.
62. I. LeMay, K. J. Truss, and P. S. Sethi, *Trans. Amer. Soc. Mech. Eng.*, [D], 1969, **91**, 575.
63. R. R. Hough and R. Rolls, *Metal Sci. J.*, 1971, **5**, 206.
64. E. C. Scaife and P. L. James, *ibid.*, 1968, **2**, 217.
65. R. H. Cook, *CEGB Rep.*, (**RD/L/N267/73**), 1973.
66. F. Garofalo, *Proc. Amer. Soc. Test. Mat.*, 1959, **59**, 973.
67. D. W. Carreau, M. J. Donachie, B. Lund, N. Pompilio and R. Shepheard, *U.S. Atomic Energy Commission Rep.* (**GA-1532**), p. 123, 1960.
68. R. G. Shepheard, and M. J. Donachie, *Trans. Amer. Soc. Met.*, 1962, **55**, 45.
69. J. C. Bokros, W. H. Ellis, and D. G. Guggisberg, *U.S. Atomic Energy Commission Rep.*, (**GA-1195**), p. 134, 1959.
70. L. H. Kirschler and R. C. Andrews, *Trans. Amer. Soc. Mech. Eng.* [D], 1969, **91**, 785.
71. D. F. Atkins, *Rep.* (**AI-AEC 12721**), p. 165, 1969.
72. G. D. Collins and E. L. Zebroski, *U.S. Atomic Energy Commission Rep.* **GEAP-13539-15**, 1970.
73. J. Beaufrère and L. Valibus, *Mém. Sci. Rev. Mét.*, 1969, **66**, 433.
74. R. C. Andrews and L. H. Kirschler, *MSAR Rep.* (**66-174**), 1966.
75. R. C. Andrews, R. H. Hiltz, L. H. Kirschler and R. J. Udavcak, *ibid.*, (**67-216**), 1967.
76. N. P. Drozd and G. G. Maksimovich, *Fiz.-Khim. Mekhan. Mat.*, 1969, **5**, (4), 415.
77. R. C. Andrews and K. R. Barker, *MSAR Rep.* (**64-81**), 1964.
78. J. S. Coombs, R. E. Allen and F. H. Vitovec, *Trans. Amer. Soc. Mech. Eng.*, [D], 1965, **87**, 313.
79. F. H. Vitovec, *Proc. First Internat. Conf. Fracture, Sendai*, **1966**, (3), 1895.
80. E. N. da C. Andrade and R. F. Y. Randall, *Nature*, 1948, **162**, 890.
81. M. R. Pickus and E. R. Parker, *Trans. Amer. Inst. Mech. Eng.*, 1951, **191**, 792.
82. R. Hales, P. S. Dobson, and R. E. Smallman, *Acta Met.*, 1969, **17**, 1323.
83. E. D. Sweetland and E. R. Parker, *J. Appl. Mech.*, 1953, **20**, 30.
84. C. E. Price, *Acta Met.*, 1966, **14**, 1787.
85. H. J. Gough and D. G. Sopwith, *J. Inst. Metals*, 1932, **49**, 93.
86. H. J. Gough and D. G. Sopwith, *ibid.*, 1935, **56**, 55.
87. H. J. Gough and D. G. Sopwith, *ibid.*, 1946, **72**, 415.
88. N. Thomson, N. J. Wadsworth, and N. Louat, *Phil. Mag.*, 1956, **1**, 113.
89. N. J. Wadsworth and J. Hutchings, *ibid.*, 1958, **3**, 1154.
90. T. Broom and A. Nicholson, *J. Inst. Metals*, 1960–61, **89**, 183.
91. J. A. Bennett, 'Proc. Tenth Sangamore Army Mater. Res. Conf.', p. 209. **1964**: (Syracuse Univ. Press).

92. R. K. Ham, 'Thermal and High Strain Fatigue', p. 55. **1967**: London (Metals and Metallurgy Trust).
93. M. R. Achter, *Amer. Soc. Test. Mat. Spec. Tech. Publ.* **415**, p. 181, 1967.
94. M. Böhmer and D. Munz, *Metall. u. Technik.*, 1970, **24**, 446, 857.
95. F. S. Pettit and J. K. Tien, Ref. 2, p. 576.
96. M. Gell and G. R. Leverant, *Amer. Soc. Test. Mat. Spec. Tech. Publ.* **520**, p. 37, 1973.
97. J. R. Haigh, C.N.A.A. PhD Thesis, **1973**.
98. C. B. Harrison and G. N. Sandor, *Eng. Fracture Mech.*, 1971, **3**, 403.
99. L. A. James and E. B. Schwenk, *Metall. Trans.*, 1971, **2**, 491.
100. K. U. Snowden, *Acta Met.*, 1964, **12**, 295.
101. G. J. Danek, H. H. Smith and M. R. Achter, *Proc. Amer. Soc. Test. Mat.*, 1961, **61**, 775.
102. H. H. Smith and P. Shahinian, *Metall. Trans.*, 1970, **1**, 2007.
103. H. H. Smith and P. Shahinian, *J. Inst. Metals*, 1971, **99**, 243.
104. A. Coles, unpublished work, 1966.
105. B. J. Darlaston and T. R. Cook, *CEGB Rep.* **(RD/B/N 560)**, 1965.
106. P. E. Brookes, N. Kirby, and W. T. Burke, *J. Inst. Metals*, 1959–60, **88**, 500.
107. R. Doldon, *J. Nuclear Mat.*, 1963, **8**, 169.
108. L. E. Raraty and R. W. Suhr, *J. Inst. Metals*, 1966, **94**, 292.
109. R. P. Skelton, *Metal Sci. J.*, 1967, **1**, 140.
110. J. K. Beddow, *Metallurgia*, 1968, **77**, 185.
111. K. Endo, *Bull. Japan Soc. Mech. Eng.*, 1960, **3**, 76.
112. M. H. Weisman, *Amer. Soc. Mech. Eng. Paper* **(61-AV-21)**, 1961.
113. M. R. Achter, G. J. Danek, and H. H. Smith, *Trans. Amer. Inst. Mech. Eng.*, 1963, **227**, 1296.
114. R. L. Stegman and P. Shahinian, *Amer. Soc. Test. Mat. Spec. Tech. Publ.* **459**, p. 42, 1969.
115. H. H. Smith, P. Shahinian, and M. R. Achter, *Trans. Amer. Inst. Mech. Eng.*, 1969, **245**, 947.
116. K. U. Snowden, *Proc. First International Conference on Fracture, Sendai*, **1966**, (3), 1881.
117. L. F. Coffin, *Trans. Amer. Soc. Met.*, 1963, **56**, 339.
118. M. Gell and G. R. Leverant, 'Proc. Second Internat. Conference on Fracture, Brighton', p. 565. **1969**: London (Chapman and Hall).
119. J. E. Northwood, R. S. Smith. and N. Stephenson, *National Gas Turbine Establishment Memo.* **(M325)**, 1959.
120. H. D. Williams, *CEGB Rep.* **RD/L/N 72/66**, 1966.
121. H. H. Smith and P. Shahinian, *Trans. Amer. Soc. Metals*, 1969, **62**, 549.
122. G. J. Hill, Ref. 92, p. 312.
123. D. J. White, *Proc. Inst. Mech. Eng.*, 1969–70, **184**, 223.
124. H. G. Edmunds and D. J. White, *J. Mech. Eng. Sci.*, 1966, **8**, 310.
125. L. F. Coffin, Ref. 2, p. 590.
126. C. J. McMahon and L. F. Coffin, *Metall. Trans.*, 1970, **1**, 3443.
127. D. Walton and E. G. Ellison, *Internat. Metallurgical Rev.*, 1972, **17**, 100.
128. R. A. T. Dawson, *Proc. Inst. Mech. Eng.*, 1969–70, **184**, 234.
129. L. F. Coffin, Ref. 92, p. 171.
130. W. McFegan and J. A. Robinson, unpublished work, 1962.
131. K. B. Smith, unpublished work, 1967.
132. R. P. Skelton, *Metal Sci.*, 1974, **8**, 56.
133. G. H. Broomfield, J. Gravener, J. Mofford, and E. E. C. Hutchins, 'Proceedings of the Conference on Irradiation Embrittlement and Creep, Nov. 1972'. London (British Nuclear Energy Society).
134. H. W. Evans and G. H. Broomfield, unpublished work, 1970.
135. C. C. Addison, E. Iberson, and R. J. Pulham, 'Surface Phenomena of Metals', Monograph No. 28, p. 246. **1968**: London (Society of Chemical Industry).
136. K. Ingle and G. A. Trussler, unpublished work, 1966.
137. E. Glenny and T. A. Taylor, *J. Inst. Metals*, 1959–60, **88**, 449.
138. F. Malamand and G. Vidal, *Recherche Aeronautique*, 1957, (56), 47.
139. E. Glenny, *Metallurgical Rev.*, 1961, **6**, 387.
140. R. B. Evans quoted by G. P. Tilly, 'Thermal Stress and Thermal Fatigue' (ed. D. J. Littler), p. 47. **1969**: London (Butterworth).
141. L. Northcott and H. G. Baron, *J. Iron Steel Inst.*, 1956, **184**, 385.
142. H. B. Probst, 'Proc. Conf. Aerospace Structural Materials'. **(NASA-SP-227)**, p. 279. 1970.
143. S. D. Gertsriken, I. Ya. Dekhtyar, L. M. Kumok, V. V. Pihpenko and M. S. Khazanov, *Vop. Fiz. Met. i Metallov., Academy of Sciences, Kiev, Ukranian S.S.R.,* p. 132, 1963.
144. F. Bollenrath and R. Sonntag, *Cobalt*, 1962, **14**, 3.
145. D. A. Spera, 'Proc. Conf. Aerospace Structural Materials,' **(NASA-SP-227)**, p. 43, 1970.
146. A. F. Brown, *Appl. Mater. Research*, 1966, **5**, 67.
147. M. R. Achter, *Scr. Metall.*, 1968, **2**, 525.
148. K. U. Snowden, *J. Appl. Phys.*, 1963, **34**, 3150.
149. M. Knudsen, 'The Kinetic Theory of Gases: Some Modern Aspects'. 3rd Edition. **1950**: London (Methuen).
150. P. Clausing, *Ann. Physik.*, 1930, **7**, 489.
151. K. U. Snowden, *Phil. Mag.*, 1964, **10**, 435.
152. F. J. Bradshaw and C. Wheeler, *Internat. J. Fracture Mechanics*, 1969, **5**, 255.
153. S. Dushman, 'Scientific Foundations of Vacuum Technique'. **1960**: New York (John Wiley).
154. M. R. Achter, First Symposium 'Surface Effects on Spacecraft Materials' (edited by F. J. Clauss), p. 286. **1960**: New York (John Wiley).
155. L. D. Jaffe, *ibid.* (in discussion), p. 304.
156. F. A. Buttner, E. R. Funk, and H. Udin, *J. Phys. Chem.*, 1952, **56**, 657.
157. E. D. Hondros, *Acta Met.*, 1968, **16**, 1377.
158. J. A. Williams and I. G. Palmer, *Phil. Mag.*, 1971, **23**, 1155.
159. E. D. Hondros, 'Interfaces Conference', p. 77. **1969**: Melbourne (Butterworth).
160. H. R. Tipler and D. McLean, *Metal Sci. J.*, 1970, **4**, 103.
161. J. D. Noden, C. J. Knights, and M. W. Thomas, *Brit. Corrosion J.*, 1968, **3**, 47.
162. F. N. Rhines and J. S. Wolf, *Metall. Trans.*, 1970, **1**, 1701.
163. J. M. Francis, C. J. Lee, and J. H. Buddery, *J. Iron Steel Inst.*, 1968, **206**, 921.
164. P. R. Openshaw and L. L. Shreir, *Corrosion Sci.*, 1965, **5**, 665.
165. F. H. Fern and J. E. Antill, *ibid.*, 1970, **10**, 649.
166. D. Caplan and M. Cohen, *ibid.*, 1966, **6**, 321.
167. J. M. Francis and V. R. Howes, unpublished work, 1969.
168. D. E. Martin, *Trans. Amer. Soc. Mech. Eng.*, [D], 1965, **87**, 850.
169. C. L. Formby, unpublished work, 1972.
170. I. Le May, *Amer. Soc. Mech. Eng. Paper* **(67-WA/Met-14)**. 1967.
171. O. Kalvenes, K. Piene, and P. Kofstad, *Corrosion Sci.*, 1964, **4**, 211.
172. A. B. Knutsen, J. F. G. Condé, and K. Piene, *Brit. Corrosion J.,* 1969, **4**, 94.
173. C. E. Richards and T. C. Lindley, *Eng. Fracture Mech.*, 1972, **4**, 951.
174. R. S. Fidler and M. J. Collins, *Atomic Energy Rev.,* 1975, **13**, 1.

SECTION IV:
Super 12% Cr Steels

The Super 12% Cr Steels

J Z BRIGGS and T D PARKER

Applications

The commercial development of Super 12% Cr steels waited until the need for them arose. As far back as 1912, Krupp and Mannesmann are said to have made 12% Cr steels with 2 to 5% Mo, presumably for corrosion resistance. Applications, however, were lacking. The first impetus towards the development of the Super 12% Cr family came from the steam-turbine field where engineers wanted a little better hot strength than was available from the straight 12% Cr steel without sacrificing damping capacity, oxidation resistance or ease of fabrication. This led to the Class I steels, some of which still are used for turbine blades. The emergence of Class II steels with higher hot strength is closely tied to the jet plane and the desire of designers to have an oxidation-resistant ferritic steel for turbine blades that would meet strength requirements at room and elevated temperatures. Today the major field of Class II steels has shifted to steam power plants where they are gradually pushing out the Class I steels as well as replacing lower alloy chromium-molybdenum and chromium-molybdenum-vanadium steels. Meanwhile, Class III steels at the next level of hot strength have been introduced as contenders in the aerospace industry.

The following notes indicate some of the more interesting uses of the Super 12% Cr steels. The properties that explain their accelerating acceptance will be discussed in subsequent sections.

Steam Power Plants

Production of electricity is a highly competitive business. Even before the advent of nuclear power, makers of steam turbines progressively raised operating temperatures and pressures for more economical utilization of fuel and cheaper electricity to meet the growing demand, which has traditionally doubled every ten years. As temperatures, pressures and size have increased, it has been necessary to make more and more parts of alloy steel, so the steam-turbine field is one of the finest examples of careful choice of the alloy steel best suited to service.

Class I Super 12% Cr steels have long been an acknowledged material for turbine blades and related parts, such as locking pins, diaphragms, nozzle partitions, nozzle blocks, rings and shrouding. Class II steels have now not only replaced much of this Class I use but have also extended into other applications such as rotors, casings and bolts. Class III steels may offer promise for the future.

Turbine Blades. Steam turbine blades operate under a variety of conditions. The short inlet blades work at temperatures near to the maximum steam temperature so creep strength is important with the amount of permissible creep usually 0.1 or 0.2% deformation in 100,000 hours. The blades at the exhaust end, however, are equally important since they fix the amount of steam that can pass through the turbine efficiently, which determines the maximum efficient power rating. While the temperature of these blades is low and may be only slightly over room temperature, they must resist corrosion and erosion by the moist steam. Additional requirements in respect to room-temperature yield and fatigue strength result from the length of these blades—36 inches is normal and designers are talking about lengths up to 60 inches and more—and the high speed, which may be around 1500 feet per second. If the fir-tree-type serrated fastenings, considered by many as most efficient, are used, the turbine-blade steel must be resistant to the embrittling effect of notches to avoid premature failure. The steel must also be capable of being riveted to the tenon strips without cracking. Finally, but not least, good damping capacity is beneficial, especially in the longer blades, in view of the unavoidable fluctuations in steam conditions that could otherwise lead to destructive vibration especially when the natural frequency of the buckets is in resonance with the steam fluctuations.

The conventional 12% Cr steels have an ideal set of properties for turbine buckets as long as the temperature does not exceed about 900 F (480 C). Austenitic steels have been tried but in general have not been satisfactory except for the short blades at the inlet end where damping capacity is not as crucial as it is for the longer blades in the later stages. Use of Class I Super 12% Cr steels permits an increase in operating temperatures to perhaps 950 F (510 C) although the exact value varies from country to country and manufacturer to manufacturer. With Class II steels, the maximum temperature can be raised to as much as 1100 F (595 C), depending on design. In certain cases, however, lower maximums are prescribed as in naval steam-turbine blades where the US Navy specifies the maximum allowable temperature for a Class II steel as 1000 F (540 C).

The choice of a turbine-bucket material is to a large extent a function of design considerations. This becomes quite clear when some of the first turbines operating with 1050 F (565 C) steam or hotter are reviewed. Westinghouse Electric in the USA continued its long use of Class I steels and went to austenitic alloys for the inlet end when needed. An exception was the Eddystone 1 turbine where Class II buckets were used on the last two rows along with nonferrous K42B for the first four rotation rows. General Electric, on the other hand, has been most active in applying Class II steels to turbine blades as far back as Sewaren 2 in 1948. German practice for turbines operating with 1100 F (595 C) steam has been a combination of these choices: Class II buckets have been used for the later stages with austenitic alloys at the inlet end. This applies to some of the first high-temperature Siemens units starting with the 1955 Leverkusen 11 plant. But,

Siemens selected a Class II steel for all blades on the Arnsberg 1 turbine, which was put into operation in February 1957. In its Reutlingen turbine, Escher Wyss used only austenitic-steel buckets although the Swiss since have apparently made some use of Class II Super 12% Cr steels.

Turbine blades are generally machined from bar stock or forged, although some precision-cast stator blades have been used satisfactorily. Class II blades are always fully heat treated but Class I steels may be used in the hot-rolled or cold-drawn and tempered condition.

Casings (Turbine Shells). The major problem encountered with turbine casings is cracking due to transient thermal stresses. While procedures and devices can go far towards minimizing this, the casing itself must have maximum resistance to thermal shock.

Low-alloy chromium-molybdenum or chromium-molybdenum-vanadium steels are suitable for casings up to about 1020/1050 F (550/565 C), at which point their strength becomes too low and their scaling too high. Adoption of austenitic-steel castings for higher temperatures accentuates thermal stresses because of low thermal conductivity and high thermal expansion. Cast Super 12% Cr steels have appeared a logical intermediate choice because of their lower coefficient of thermal expansion and higher thermal conductivity as compared to austenitic steels and their higher hot strength and scale resistance as compared to the low-alloy steels. Terrell's theoretical approach has given an indication of the advantages of the Super 12% Cr steels (page 51). Laboratory tests have confirmed these calculations. At General Electric, Walker's tests based on strain fatigue at 950 F (510 C) with a hold in the tension part of the cycle showed the best results for cast 1% Cr—1% Mo steel in the annealed condition, followed very closely by a heat-treated Class I Super 12% Cr steel, which was much better

Source: *The Super 12% Cr Steels*, Climax Molybdenum Co., 1965

than the austenitic 316Cb and under some conditions had two- to five-times longer life. A different series of tests also intended to simulate casing service was conducted by Gysel, Werner and Gut at Georg Fischer in Switzerland. Of all the steels studied—which included carbon, low-alloy and precipitation-hardening austenitic compositions—Class II Super 12% Cr steel had by far the best resistance to thermal shock and showed practically no surface deterioration up to the highest temperature tested (1110 F—600 C). Stress-relieved welds in the Class II steel gave results comparable to those of the fully heat-treated parent metal.

While the above potential benefits of Super 12% Cr casings have been recognized everywhere, the major progress in their commercial utilization has been in Germany where high-temperature steam power plants since December 1955 (Leverkusen 11) have used Class II Super 12% Cr steel castings for either the inner or outer casing or both, depending on design.

An excellent example of Super 12% Cr steel casings is given by the Escher Wyss steam turbines for the 1110 F (600 C) Baudour central power station in Belgium. The entire high-pressure casing is made from cast Super 12% Cr steel with austenitic-steel fresh-steam inlets and stagnant-steam cooling between the inner and outer casings. Mannesmann connecting units (see page 183) are used between the ferritic casing and the austenitic pipes so no difficulties in welding were encountered. The junction is cooled by steam so its temperature is only a little over 930 F (500 C), where the stress-rupture strength of the 12% Cr steel is much greater than that of the austenitic composition at 1075 F (580 C).

Flatt has stated that the fuel savings by going to a 1110 F (600 C) unit at Baudour would have been cancelled out by added material costs if it had been necessary to go to austenitic steels for all high-temperature parts. The choice of Super 12% Cr

steel, however, for the casings and as many as possible of the other parts changed this picture entirely and made the high-temperature unit economical. As a matter of fact, the performance at Baudour has been so good that an identical turbine installation is in construction at Monceau.

Bolts and Nuts. Bolts for turbine cylinders and steam chests must have high enough stress-relaxation strength to maintain flange tightness against internal steam pressure throughout the life of the plant. In addition the bolting materials must have high room-temperature yield strength, good notch ductility and resistance to embrittlement by thermal aging and straining, sufficient creep ductility to accommodate cumulative plastic strains and no loss of stress-relaxation strength on retightening during the life of the plant.

High-temperature bolts in steam turbines were one of the first places where Class II Super 12% Cr steels won acceptance throughout the world. Two industry specifications have been established in this field. ASTM A 437—59 T covers a Class II steel with specified tensile and impact properties after a specific heat treatment. DIN 17 240 Sheet 2 lists two Class II compositions as suitable bolt materials for temperatures over 1000 F (540 C) to about 1200 F (650 C) with a Class I steel for nuts. The ASTM specification states that the Class II grade listed can also be used for nuts. In general, austenitizing temperatures are held on the low side of the range and tempering temperatures on the high side to ensure optimum ductility and thermal stability.

The temperature ranges where these bolts are selected vary considerably from country to country. As contrasted with the 1000/1200 F (540/650 C) in DIN 17 240, most other countries do not use these materials much over 1050 F (565 C) while some prefer them to the low-alloy bolting materials at temperatures as low as 900 F (480 C). Within this range, Class II steels have somewhat superior relaxation strengths to the conventional low-alloy steels as shown by Draper's results on bolts strained to 0.15% at 1155 F (625 C):

	Time for Stress to Fall to 9 ksi, hr
1% Cr-Mo-V	546
12% Cr-Mo-V	1095

The thermal stability and particularly the resistance to embrittlement at elevated temperatures under the influence of notches are much better for the Super 12% Cr steels than for either low-alloy or austenitic bolting materials. To take advantage of the Super 12% Cr steels in some cases it may be necessary to use steam cooling. At the Philo plant and other high-temperature General Electric turbines constructed from about 1950 on, steam cooling is used on five of the inner-casing flange bolts of Class II steel on either side of the turbine.

In use of Super 12% Cr bolts, attention must be given to differences in thermal expansion between the bolts and flanges, which may cause either excessive tightening or loosening of the joint on heating to operating temperatures. With a low-alloy-steel flange used at 1000 F (540 C), the stress at operating temperatures for a given tightening practice will be appreciably higher with Super 12% Cr bolts than with low-alloy bolts. In this particular case, the Super 12% Cr bolts should be tightened to lower initial stress values at room temperature to avoid overstressing during service. To prevent seizing and problems in dismounting, it is advisable to use Class II bolts with nuts of a different Class II grade, a Class I steel or, if the same grade is selected, a different hardness.

Rotors. Adapting Super 12% Cr compositions to heavy forgings, such as rotors, which may run up to 20 tons or more, has required some changes in analysis as indicated in general terms by Brennecke and Schinn and spelled out in more detail by Boyle and Newhouse.

The first commercial use of Super 12% Cr rotors dates back some ten years on the Continent. Class II rotors have been generally used on high-temperature steam turbines in Germany and Switzerland since that time. Even when the superheated steam goes up to 1255 F (680 C), design keeps the temperature of the rotor to about 1060 F (570 C) where the strength characteristics of the Class II steels are particularly outstanding.

In the USA a backup integral-disk rotor for the 1958 Eddystone plant was made of a Class II steel. Although it was not used since the first-choice austenitic superalloy (Discaloy) was satisfactory, the Class II rotor was produced without particular difficulty and tested. Mochel has stated that it "was a stable rotor and would give unlimited service at 1150 F (620 C) steam."

A new rotor steel, presumably the composition shown as "GE" on page 34, was announced by General Electric in 1964. It is said to have a 25% higher rupture strength above 1000 F (540 C) than conventional rotor steels. The result has been a decrease in the length of turbine-generators in the 550,000-kw range and above. Until now, rotors for tandem-compound generators with this rating have had to be made in at least four sections or casings. With the new alloy, only three are needed. The upshot of this is an increase in kilowatt capacity, a decrease in size and a "significant reduction" in power-plant costs. As of March 1965, nine of these rotors had been or were being manufactured.

The only major service problem encountered with rotors of Class II steels has been poor wear and friction properties at the bearing surfaces (see pages 86/87). The most common solution has been hard chromium plating of the rotor over the surfaces involved.

Boiler, Superheater and Reheater Tubes. These applications use large quantities of tubing. A Benson boiler, for example, may have close to 400,000 feet of tubing covering many grades from carbon to austenitic steel. In superheaters, reheaters and boilers, where each grade of tubing is chosen to meet a specific set of conditions, the large gap in design stresses between the ferritic and austenitic steels has long been evident. The interest in Class II tubing stems from the fact that this type of steel neatly fills the gap with design values roughly halfway between those of pearlitic and austenitic steels at the temperatures where the Super 12% Cr steel would be most likely to fit in. Moreover its resistance to scaling is considerably greater than that of other nonaustenitic steels.

In a 1959 paper, Class detailed the results of various test runs:

1. No particular irregularity in dissimilar welds between Class II high-pressure tubing and stabilized austenite tubing was observed after about 1½ years at 930 F (500 C).

2. Superheater tubing with dissimilar welds between a Class II steel and austenitic steel was removed from service after 12,000 hours and about 80 cooling cycles. The steam temperature was 1110 F (600 C) with a wall temperature of about 1220 F (660 C) for the Class II steel and 1255 F (680 C) for the austenitic steel. The austenitic steel had scaled considerably more than the Class II steel, probably because of carburization under operating conditions. Simple butt welds between the two types of tubing were in good shape so perhaps special junctions may not be necessary.

3. After 7082 hours, units were taken out for examination from an end superheater working with a steam temperature of 930/985 F (500/530 C), a combustion-gas temperature of 1470 F (800 C) and a pressure of 115/120 atmospheres. No damage was found on the cold-bent tubes of Class II tubing or at the welds. Butt joints with 2¼% Cr—1% Mo steel showed some decarburization and loss of hardness on the low-alloy side.

Bettzieche has indicated that the factors of safety in respect to life and stress are the same for Class II steel at 1110 F (600 C) as for 0.3% Mo steel at 985 F (530 C) and a 0.9% Cr—0.45% Mo steel at 1040 F (560 C). Class II tubing has been used to replace 2¼% Cr—1% Mo tubing at high temperatures and low pressures where the scale resistance of the latter was inadequate; to replace 9% Cr—1% Mo steel at 1020 F (550 C) where better stress-rupture characteristics were required and better weldability was desirable; and where better resistance to stress corrosion in presence of chloride ions was needed than could be obtained with the usual austenitic steels.

The final place of Class II steels in this application will depend on the cost of tubing relative to that of austenitic steel and low-alloy chromium-molybdenum steel. If the cost factor of the Class II tubing is in line, expanding use of this type of material in boilers, superheaters and reheaters can be expected, especially at temperatures up to 1075 F (580 C).

Other. Numerous other applications of Super 12% Cr steels in the turbine field have been satisfactory. Among these might be mentioned:

shaft labyrinth packings of 1000 F (540 C) turbine where a 6% Cr steel did not have sufficient strength but a Class I grade met the requirements

vacuum rings and nozzle boxes where Class II steel is serving without difficulties

valve stems and spindles are made of either Class I or Class II steel in the nitrided condition for increased wear resistance

valve casings of a Class II steel steam cooled are used in the high-pressure section of the Baudour turbine

rapid-closing regulation valves for steam turbines operating at 1040/1060 F (560/570 C) and 180/210 atmospheres have been made of a Class II steel welded to an austenitic steel, X 8 CrNiMoVNb 16 13, using a Mannesmann-type connecting junction

diaphragms to hold stator vanes and other parts where sliding motion is involved are made of a nonhardening Class I steel with 0.09% C max, 14/16% Cr, 0.05% Al max and 1.5% Mo min

seal strips and sections for dummies and glands of steam turbines have long been made of a Class I steel

valve seats of Class II steels

Gas Turbines

The desirable attributes of nonaustenitic gas-turbine wheels in jet aircraft led to the development of Class II Super 12% Cr steels. As compared to austenitic steels, the nonaustenitic steels are noteworthy for their high yield strength at room temperature, which is useful in start-ups, and their lower thermal stresses, resulting from the lower coefficient of thermal expansion and higher thermal conductivity. Among the nonaustenitic steels, the Class II and III compositions have the best resistance to corrosion and oxidation.

Gas-Turbine Wheels. Class II steels functioned satisfactorily in aircraft gas turbines and for a period were practically standard in British-built motors. The temperatures of the latest models, however, have increased to such an extent that Class II steels no longer meet requirements, although Class III may and is being used for at least one model in the UK.

In industrial gas turbines, however, Class II steels are widely chosen. Although some difficulties were encountered in the early stages of producing the large forgings needed, these have been overcome by improvements in composition, manufacture and testing, so discs of 70 inches in diameter and larger can be made with suitable strength and toughness.

Some of the characteristics of the Super 12% Cr steels have been shown in tests by Zonder, Rush and Freeman on turbine wheels:

	Stress, ksi, for 1% Deformation in 1000 hours	
	1000 F (540 C)	1100 F (595 C)
Class II (Crucible 422)		29
3% Cr-Mo-W-V (H-40)		31
1.25% Cr-Mo-Si-V ("17-22A" S)	51	17
4340	17	

While the Class II wheel did not have higher strength than the strongest of the low-alloy steels, it demonstrated by far the highest thermal stability as measured by elongation and reduction of area in rupture tests. Furthermore, unlike the other steels, the Class II steel had good resistance to oxidation so surface protection would not be needed even in long service.

Sakamoto of General Electric's Small Aircraft Engine Department also found that Class II wheels (approximately AISI 619) showed no signs of embrittlement even with improper heat treatment whereas the low-alloy "17-22A"V did. In this case, the Class II steel also had higher rupture strength at 1000 F (540 C). The lack of embrittlement of the Class II wheels will become more important in heavier wheels where the triaxial constraint becomes effective. A Class II grade has been selected for the rotors of the ML-1 gas turbine, where the maximum rotor temperature is expected to be 1000 F (540 C); this gas turbine is one component of the US Army Gas-Cooled Reactor-Systems Program.

Turbine Wheels for Cartridge Starter. The wheels in the starters of jet planes are subjected to temperatures of about 840 F (450 C) in the disc and 1200 F (650 C) in the blades with a speed of about 55,000 rpm. The service time, however, is very brief. Sulzer's precision-cast 2Cr120Mo8Co has proved satisfactory in integral wheels and blades and has withstood speeds as high as 70,000 rpm in acceptance tests.

Turbosuperchargers. In view of the relatively low temperatures prevailing in the wheel, cast Class II steels have given good service.

Source: *The Super 12% Cr Steels*, Climax Molybdenum Co., 1965

Gas-Turbine Buckets. Class II Super 12% Cr steels are used for buckets in some stages of industrial gas turbines. Here one of the advantages of the Super 12% Cr steel is the high damping capacity. In single-shaft machines in locomotive service, the early failure encountered with austenitic 16-25-6 buckets was attributed to their low damping capacity since Class II buckets were satisfactory. Hull feels that the buckets must be designed either to eliminate major resonant conditions or to use a material such as the Super 12% Cr steels with a very large amount of structural damping capacity. At 960 F (515 C) Class II buckets have lasted for 364 hours in a gas-fired turbine and 1190 in one with oil firing. Where higher temperatures and stresses are involved, as in the second stage of a 16,500-kw industrial gas turbine, the shrouds over the Class II buckets are supported by separate segmented rings of austenitic steel.

Compressor Wheels and Blades. The 12% Cr and Super 12% Cr steels offer significantly better corrosion and oxidation resistance than low-alloy steels. Originally stress-corrosion cracking caused some failures until the need for avoiding certain tempering temperatures (see pages 152/153) was realized. Class II steels are superior to the conventional 12% Cr steels in that they can be heat treated to the desired hardness and fatigue characteristics without tempering in the dangerous range. Class II steels have a wide use in certain stages of axial-flow compressors although other materials are used for the highest and lowest temperatures. In January 1965, Child stated that Class II steels were being used in compressor disks of Bristol Siddeley's Sapphire, Olympus, Orpheus and Pegasus engines as well as in Rolls Royce's Avon and Spey. Blades in some compressors are also made of Class I grades.

Other. Bolts for use at high temperatures are made of Class II Super 12% Cr steels for gas as well as steam turbines. A German aircraft-material specification sets a maximum service temperature of 1075 F (580 C) but usage varies in other countries. A low-carbon Class II steel goes into casings and structural-frame parts of aircraft gas turbines in the form of sheet and rings, which are welded to forgings.

Hydraulic Turbines

Class I Super 12% Cr steels either in the wrought or cast condition, have given excellent performance in hydraulic turbines. In the case of wrought material, one major factor is the better resistance of the molybdenum-containing steel to corrosion, particularly by acid river water. With castings, such as the 15,500-pound castings produced by Birdsboro for water-turbine impulse wheels having a bucket pitch diameter of 84 inches with disc diameters up to 8½ inches, Class I material was selected over straight 12% Cr because of its superior ductility and toughness characteristics. Some producers have also added vanadium to give a Class II type.

Propellers

Stainless-steel propellers on ships have certain distinct benefits including ability to serve in ice where bronze and cast iron are not suitable and resistance to corrosion in seawater. Hardenable chromium steels are preferred to austenitic propellers for most ships because of their high strength, good corrosion resistance and lower cost.

For some 25 years, many of the larger Swedish warships and ocean-going merchantmen have been equipped with cast Class I Super 12% Cr propellers. The corrosion resistance is sufficient for propellers continuously running in seawater but serious corrosion damage has occurred on steering propellers, which have long idle periods.

The US Navy ran into several failures with cast 12% Cr propellers fitted on icebreakers. The failures were diagnosed as due to excessively low toughness and ductility at low operating temperatures in the presence of small casting defects and deep hammer marks. To remedy this defect, a Class I cast steel with low silicon and additions of molybdenum and nickel was developed and is now included in Amendment-1 of MIL-S-16993A for "ship propellers and other applications requiring resistance to impact at low temperatures". Even at freezing temperatures, its plasticity and toughness are such that extremely large (and unlikely) flaws would be required to initiate failure under the expected conditions of plastic overload.

Aircraft and Missile Parts

Class II Super 12% Cr steels have found a number of applications for structural parts such as brackets and clamps in addition to fasteners and gas-turbine components discussed above. One of the noteworthy applications in the UK is a 39-foot-long keel boom and a midsection of the nacelles for the supersonic T 188. The high structural efficiency in torsion and tension may not be fully utilized until temperatures are higher than those corresponding to Mach 4.

At the time that the aircraft industry was examining high-strength steels for use in skin of supersonic planes, Class I and Class II of the Super 12% Cr steels were in existence but the necessary information on properties was not available for proper examination. Nevertheless, Class II steels did look promising despite their susceptibility to stress corrosion and the difficult weldability but development of other materials with superior properties in these respects put them out of competition. The vastly better properties of some of the Class III steels (see page 80) have brought the Super 12% Cr steels back into the running. Brazed honeycomb of Class II steel had excellent flexural strength but its relatively low corrosion resistance was considered inadequate without coatings. Here again, the emergence of the Class III materials may change the picture. The Class III grade, AFC-77, developed under sponsorship of the US Air Force, is considered to have particularly good prospects.

Martin concluded that a Class II steel would be a good material for small thin-wall pressure vessels but did not evaluate the possibility for larger vessels. Solar preferred pressure vessels of hot-work-die steels to those from a Class II Super 12% Cr steel because of the higher attainable strength even though the die steels gave brittle fractures and the Class II vessels ductile ones.

A cast Class I steel is used for liquid-oxygen valves in rocket engines. The molybdenum addition serves to increase the impact toughness at subzero temperatures for a given hardness as compared to the straight 12% Cr steel.

Pulp, Paper and Fiber Industries

Class I Super 12% Cr steels are widely used, especially in Scandanavia, for parts involving moderate corrosion and wear resistance, such as:

> filter and strainer plates in chemical and mechanical wood-pulp operations
>
> Hollander knives in paper mills
>
> parts of water-purification plant in cellulose mills
>
> evaporator and condenser tubes, and sieves in sulfate mills
>
> cast slashers
>
> beater and jordan bars (with carbon on the high side)
>
> knives for cutting artificial fibers

Petroleum Industry

Class I Super 12% Cr steels have been used for almost 30 years as linings for pressure vessels in the petroleum industry. The molybdenum addition serves to stabilize the carbides, slow down carbide precipitation, suppress embrittlement and prevent loss of corrosion resistance under conditions leading to embrittlement. The preferred composition is an

Source: *The Super 12% Cr Steels*, Climax Molybdenum Co., 1965

AISI 405 with chromium on the low side and 0.4/0.6% Mo. Hundreds of these lined vessels have been produced by A O Smith alone and have proved satisfactory with operating temperatures of 500/900 F (260/480 C).

In France, Class II steels are recommended for stressed parts operating in hot oil or oil vapor at temperatures of 930 F (500 C) and higher.

Nuclear Industry

The combination of good mechanical properties at temperatures up to 1050 F (565 C) along with good corrosion and erosion resistance to relatively moist steam encountered in nuclear-powered steam turbines makes 12% Cr steels desirable materials for certain components. Likewise, these steels have suitable properties for components exposed to the primary coolant, where good strength, good corrosion and erosion resistance to water, and suitable magnetic characteristics are required. For both applications it is necessary that the material be welded in restrained sections. Because of its experience in the steam-power industry, General Electric proposed substituting a Class I casting (with molybdenum and nickel) for Type 410. Tests showed these castings could be readily welded under conditions of high restraint with either the shielded-metal-arc or the submerged-arc welding process.

In the UK, a Class I grade is also being used for high-strength casings in boiling-water reactors and a Class I steel is included in a German series of steels specifically designed for nuclear applications.

The only Class II use that has been publicized is the turbine rotor in the ML-1 gas turbine included in the US Army Gas-Cooled Reactor-Systems Program.

High Pressures and High-Pressure Hydrogen

If the curve for design values of Class II steel (see page 123) is extended down to 930 F (500 C), it falls below the point for hot yield strength so even at this temperature stress-rupture values would form the basis for stress calculation. Class II steels would therefore have great value for all types of high-pressure equipment to be used at temperatures below 1110 F (600 C), a field of special interest to the chemical industry today. Austenitic steels are not suitable for this type of part because of their low yield strengths. As illustrated in the table on page 124, equipment operating at 930 F (500 C) under 300 atmospheres would require much thinner walls in Class II tubing than would be needed for any of the low-alloy high-temperature steels or for austenitic grades.

The resistance of Super 12% Cr steels to high-pressure hydrogen has long been realized as shown by Strauss's 1932 patent on Class I and II steels for this application. Their practical utilization, however, dates back to about 1943 in Germany. Prior to that time, high-chromium ferritic steels had been used for insert parts but did not have the hot strength needed for critical structural parts nor the resistance to hydrogen sulfide desirable in many cases. With the addition of molybdenum, tungsten and vanadium, the hot strength was increased sufficiently so the steel could be applied as pressure-bearing walls. Moreover, the Super 12% Cr steels had good resistance to hydrogen sulfide and, of advantage in some cases, lower expansion and higher thermal conductivity than austenitic grades. Stahl-Eisen-Werkstoffblatt 590-61 includes a Class II grade (X 20 CrMoV 12 1) for tubes and formed parts to be used in hydrogen under high pressures and temperatures but does not establish specific operating limits.

Tools, Dies and Molds

In view of their high hot strength, it is logical that the Super 12% Cr steels would be considered for hot-work applications. As a matter of fact, one Class

II and one Class III steel are marketed primarily for this purpose.

The main characteristic of the Class II grade is said to be high resistance to heat checking. Among the suggested applications for this steel are aluminum die-casting dies, core pins, extrusion press-forging dies, aluminum forging dies, copper forging dies and mandrels for extruding nonferrous materials.

The Class III steel was developed to provide optimum resistance to heat checking and to wear or washing during elevated-temperature service. Since 29 Rockwell C hardness is the lowest that can be obtained by tempering or annealing, this steel will not drop below this value during service. By contrast, most hot-work steels temper back to Rockwell C 15/22 on exposure to high service temperatures. Among the applications where this grade has proved superior to conventional tool steels has been brass die-casting dies. Another Class III steel, not primarily developed as a tool steel, has been proposed for die-casting dies and cores, extrusion dies and mandrels, and piercer points.

One place Class II steels have found a limited application is in plastic molds where the corrosion resistance of AISI 420 at Rockwell C 40/44 is not adequate.

A Class III steel is being considered for some types of glass molds where long life is expected because of its hot hardness and corrosion resistance.

Miscellaneous

Class I steels have been used for many types of miscellaneous applications, mainly where high strength or hardness is needed along with some degree of rust resistance. Among these applications are axles, spindles, bolts, chemical instruments, water pumps, separator balls, household articles that do not have to have exceptional edge properties, mirrors, leaf springs in electrical equipment, feeler gages, arch supports, shafts, propellor shafts for salt water and surgical equipment. Other uses for hardened and cold-rolled strip are measuring tapes, reels, gages, nail files, butcher saws, wire netting and reinforcing wire for heat-resisting concrete.

A Class II steel in the form of flat shims has been used in welded continuous rod mats. It gives high joint strength (through alloying) without use of high-strength rods, which do not have the required bend ductility.

A standard hex screw of a Class II Super 12% Cr steel fastens the manifold to the block on a diesel engine. The fastener must maintain tension over a wide range of temperatures. This fastener replaced a high-alloy high-tensile screw because of better corrosion resistance and lower cost. The fastener is hardened and tempered to a minimum hardness of Rockwell C 35.

Summary

From the above discussion it is clear that applications of Super 12% Cr steels are not restricted to steam and gas turbines, although these uses represent the greatest tonnage. On the contrary, the outstanding properties of these steels have found a variety of uses in a number of industries. As tubes and large forgings become more readily available and as Class III steels become better known, it is expected that the applications will grow even more rapidly than they have in the past.

Source: *The Super 12% Cr Steels*, Climax Molybdenum Co., 1965

Physical Properties

Physical properties are among the noteworthy advantages of the Super 12% Cr steels, especially in the case of high-temperature applications. Their lower thermal expansion as compared to that of austenitic steels is a significant factor. For instance, in a boiler with 100 feet of tubes heated to 1100 F (595 C), the expansion of an austenitic steel is almost a foot, clearly enough to cause problems. Moreover, this lower expansivity combined with the higher thermal conductivity of the Super 12% Cr steels means lower thermal stresses than would be set up in austenitic steels under the same conditions. Terrell and Class have used different means of calculation and presentation but both arrive at the same conclusion that a Super 12% Cr steel has lower thermal stresses and greater resistance to thermal distortion not only than the conventional types of austenitic steel that would be considered for the temperature range where the Super 12% Cr steels would be used but also than lower alloy and carbon steels. In turn, the considerably lower stresses in a turbine casing, for example, result in higher resistance to thermal-stress cracking on starting or changing loads. Since most turbine-shell cracks are due to transient thermal stresses, this characteristic alone can account for the gradual trend to Super 12% Cr steels for steam turbine casings.

Thermal Expansion

The coefficients of linear thermal expansion of all the Super 12% Cr steels fall within a relatively narrow band, which is well below that of low-alloy steels and much lower than that of austenitic steels. This difference in expansion must be taken into account in relation to both total expansion and tightening practices for bolts. As mentioned above, it is likewise a factor in the high resistance of Super 12% Cr steels to thermal stresses.

Thermal Conductivity

While no data are available on Class III steels, the thermal conductivity of Class I and II Super 12% Cr steels can be represented by a band, which is intermediate betwen those for lower alloy and austenitic steels. As can be seen, the thermal conductivity of Super 12% Cr steels varies little with composition and temperature. Differences in heat treatment may account for much of the width of the band, on the basis of Krzhizhanovskiy's data.

Thermal Diffusivity

The value of 0.27 square foot per hour reported for Crucible 422 can doubtless be accepted as representative for Class II steels.

Electrical Resistivity

While the electrical resistivity of metals generally changes inversely to the thermal conductivity, this is not precisely true with Super 12% Cr steels since heat treatment has more effect on the latter property than on the former and test temperature has less. Nevertheless, just as in the case of thermal conductivity, the electrical resistivity of Super 12% Cr steels is intermediate between that of lower alloy steels and that of austenitic steels. The room-temperature resistivity of Super 12% Cr steels falls within the range of 50/67 microhm-cm and increases regularly to about 110/120 at 1470 F (800 C).

Moduli of Elasticity

The moduli of elasticity of the Super 12% Cr steels in tension and shear decrease regularly with increasing temperature. The bands are based on a large number of tension tests but a relatively small number of shear tests. It is believed that the width

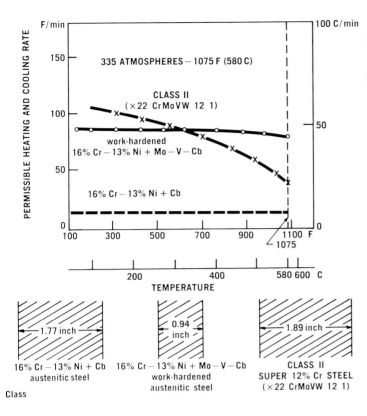

Calculated on basis of design stresses to give wall thickness needed to resist specified conditions in a steam chest

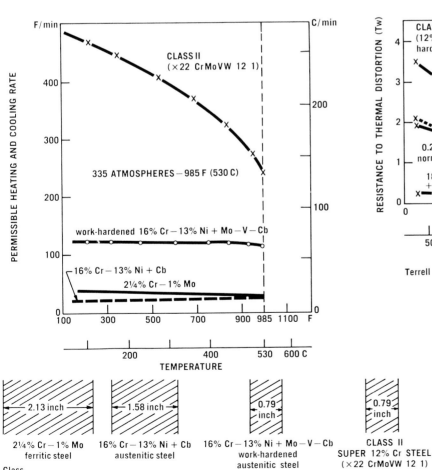

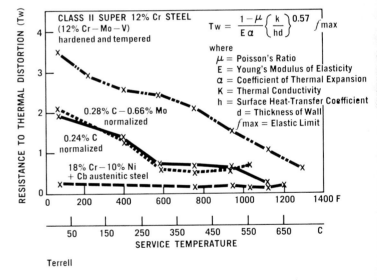

$$Tw = \frac{1-\mu}{E\alpha}\left\{\frac{k}{hd}\right\}^{0.57} f_{max}$$

where
μ = Poisson's Ratio
E = Young's Modulus of Elasticity
α = Coefficient of Thermal Expansion
K = Thermal Conductivity
h = Surface Heat-Transfer Coefficient
d = Thickness of Wall
f_{max} = Elastic Limit

Terrell

Source: *The Super 12% Cr Steels*, Climax Molybdenum Co., 1965

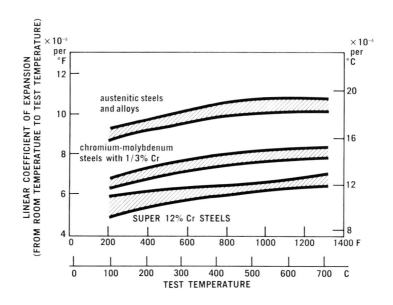

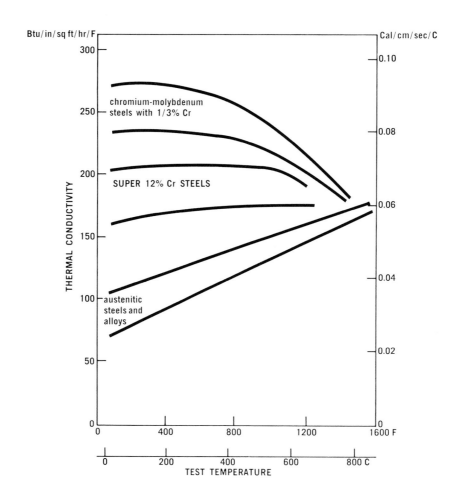

of the band in tension, which represents over 90% of the reported data, is less the result of composition than of test conditions. In tests on H46, Harris and Watkins observed a maximum difference between static and dynamic modulus of elasticity in tension of 625 ksi at 570 F (300 C).

Heat treatment and direction of test had such a marked effect on the static modulus-of-elasticity values in compression that it is impossible to draw a valid band. The only other data at hand, determined by the same method on Crucible 422 sheet, do not clarify the picture since they vary from 30,000/32,600 ksi at 72 F (21.7 C) to 30,500/31,700 ksi at 600 F (315 C) and 27,000/27,600 at 800 F (425 C).

Poisson's Ratio

The variations in reported values of Poisson's ratio are too great to allow presentation of a reliable band.

Specific Heat

No data are available on Class III steels. A value of 0.11 Btu/lb/F (0.11 cal/g/C) is most generally given for Class I and II grades at room temperature although individual figures go as high as 0.12 and as low as 0.10. These differences appear related to test procedures rather than composition. Ugine has indicated that the specific heat of Fluginox 62 and 65 increases from 0.11 at 68 F (20 C) to 0.135 at 570 F (300 C) and 0.19 at 1110 F (600 C).

Melting Point

Reported melting points for Class I steels range from 2640/2750 F (1450/1510 C) and for Class II steels from 2600/2750 F (1425/1510 C). It is doubtful if the difference in these figures is significant. No information has been found on Class III materials.

Density—Specific Gravity

Except for the oddly high value of 8.04 g/cc for Jethete M151, the reported values fall within the following limits:

	g/cc	lb/cu in.
Class I	7.70/7.80	0.277/0.281
Class II	7.75/7.85	0.279/0.283
Class III	7.80/7.90	0.281/0.284

Oliver and Harris found that the density of H46 decreased from 7.75 g/cc at 68 F (20 C) to 7.53 at 1470 F (800 C).

Magnetic Properties

Identical permeability values have been given for Crucible 422 and USS-12MoV:

Tempering Temperature			
F	700	800	900
C	370	425	480
Magnetic Permeability			
100 oersteds	85	75	93
maximum	85	92	100

Crucible Steel Company of America
USS Corp

The following magnetization and demagnetization properties apply to Jethete M152, presumably in the heat-treated condition:

Magnetization

Field Strength, H oersted	Induction, B gauss
5	360
10	770
25	5,350
50	11,500
100	14,500
250	16,950

Demagnetization from H = 250 oersteds	
Remanence, B_R, gauss	9770
Coercive Force, H_c, oersted	25.5

Samuel Fox & Company Ltd

Class reported a somewhat lower coercive force of 17 oersteds for a steel with 0.20% C, 11.3% Cr, 1.1% Mo, 0.3% V and 0.5% W after air cooling from 1920 F (1050 C) and tempering at 1380 F (750 C).

Source: *The Super 12% Cr Steels*, Climax Molybdenum Co., 1965

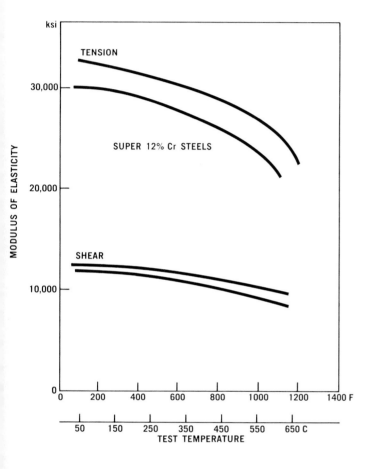

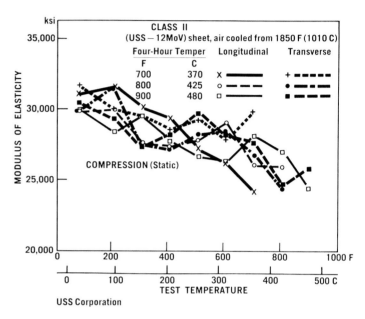

USS Corporation

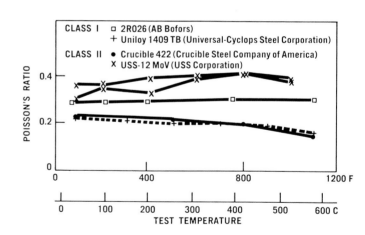

Emissivity

According to Goller, additions of molybdenum up to 0.44% had no appreciable effect on the emissivity of molten steels of this type.

Effect of Neutron Irradiation

Neutron irradiation at 390/930 F (200/500 C) decreased the ductility and increased the strength of 1Kh12MS with no significant difference in the response of annealed and heat-treated specimens.

	Irradiation						
			Neutrons/Cm²				
Condition	Temperature F	C	Overall	Neutrons with E >1 MW	0.2%-Offset Yield Strength ksi	Tensile Strength ksi	Elongation %*
A	...	...	...	...	63	92	29.5
	390/465	200/240	8×10^{20}	8×10^{19}	84	100	27.0
	610/680	320/360	1.2×10^{21}	3.6×10^{20}	88	106	25.0
	840/930	450/500	7×10^{21}	2.1×10^{21}	89	114	16.0
	held 16 days at 750 F (400 C) with no irradiation				62	94	28.5
B	...	...	...	...	97	126	21.5
	390/465	200/240	8×10^{20}	8×10^{19}	117	133	19.0
	610/680	320/360	1.2×10^{21}	3.6×10^{20}	115	136	20.0
	840/930	450/500	7×10^{21}	2.1×10^{21}	115	...	16.0
	held 16 days at 750 F (400 C) with no irradiation				95	126	21.0

* presumably in 50 mm

A annealed from 1650 F (900 C)

B oil quenched from 1830 F (1000 C) and tempered three hours at 1200 F (650 C)

Ibragimov, Voronin and Kruglov

Mechanical Properties

Most of the applications of Super 12% Cr steels depend on their mechanical properties, usually on more than one. For example, a combination of creep strength, fatigue strength and damping capacity is needed for steam-turbine blades.

Available information on mechanical properties will be discussed in the following order:

Tensile Properties	Friction and Wear
Compressive Properties	Stress-Rupture
Shear and Bearing Properties	Strength
Fatigue Strength	Creep Strength
Impact Toughness	Stress Relaxation
Fracture Toughness	Design Values
Damping Capacity	Hot Hardness

This summary documents various key features of the Super 12% Cr steels:

1. Some of the Super 12% Cr steels can be heat treated to very high room-temperature strength levels—over 200-ksi tensile strength.

2. In respect to hot strength, the Super 12% Cr steels are in general intermediate between the low-alloy chromium-molybdenum and chromium-molybdenum-vanadium steels, and the austenitic high-temperature alloys. The hot-strength characteristics increase from Class I to Class II to Class III of the Super 12% Cr steels although there can be a small overlap under some conditions.

3. Ductility and toughness characteristics are adequate for the intended applications, and are usually superior to those of the conventional 12% Cr steel.

4. Damping capacity is considerably better than that of the austenitic steels although lower than that of the straight 12% Cr steels.

Tensile Properties

Super 12% Cr steels can be heat treated to an extraordinary range of tensile strengths, going from less than 90 ksi for nonhardening Class I steels to over 300 for some Class III grades. Special thermomechanical working methods, such as ausforming and marstraining, offer the possibility of still greater strengths, as discussed on pages 166/169.

A number of tempering curves for Super 12% Cr steels are included in the section on Processing and Fabrication. The graphs applying only to hardness can be converted to tensile strength with Hagel and Becht's conversion graph. The lowest possible austenitizing temperatures are desirable from the standpoint of tensile ductility, measured by reduction of area.

Form. Tensile properties are influenced to a considerable extent by the form being tested. High strengths can be obtained in Class II **sheet and strip** as illustrated by the following longitudinal tests:

	Sheet Thickness in.	Yield Strength ksi	Tensile Strength ksi	Elonga-tion % in 2 in.	Tempering Temperature F	Tempering Temperature C
Crucible 422	0.025	194	268	6	400	205
		141	181	5	1100	595
Firth-Vickers 448	16-gage	155	171	10	1110	600
USS-12MoV	0.10	195	247	10.5	900	480

Crucible Steel Company of America, Firth-Vickers Stainless Steels Ltd, United States Steel Corporation

Castings generally have lower ductility than wrought products in the longitudinal direction but higher than transverse tests on wrought material. The ductility can also be affected by the method of casting. Usually the best ductility results in castings given the fastest solidification, as indicated by the following typical values on Firth-Vickers 507 in the hardened-and-tempered condition:

	Centri-Die	Centri-Sand
Yield Point, ksi	113	119
Tensile Strength, ksi	141	141
Elongation, % in 4 $\sqrt{A}$	16	12
Reduction of Area, %	40	25

Firth-Vickers Stainless Steels Ltd

The properties of **weldments** are covered in some detail on pages 181 and 183.

Since many of the applications of Super 12% Cr steels are in the form of **forgings,** their properties are of considerable interest. As can be seen from the compilation of tensile properties on various size forgings of a number of Super 12% Cr steels, segregation or presence of delta ferrite can have a marked influence in decreasing the tensile ductility. The results in this table on pages 59/60 also illustrate the effect of the direction of test. Directionality has little effect on strength but transverse tests have significantly lower ductility than longitudinal tests on wrought products.

Tensile Requirements

	Heat Treatment	Yield Strength (2)	Tensile Strength	Elongation % (3)	Reduction of Area %	Brinell Hardness
Class I						
AMS 5614	(1)	80 min	100 min	21 min	60 min	. . .
MIL-S-16993A(1)	(4)	65 min	90 min	18 min	30 min	. . .
Class 2 castings						
Stahl-Eisen-Werkstoffblatt 400-60						
X 15 CrMo 13 and X 20 CrMo 13						
	(5)	78 min	107/128	14 min	. . .	220/260
Class II						
AMS 5655	(6)	115 min	140 min	13 min	25 min	293/341
ASTM A 437-59 T						
Class B4B	(7)	100 min	145 min	13 min	30 min	(8)
Class B4C	(7)	80 min	115 min	18 min	50 min	(9)
Luftfahrt-werkstoff						
1.4934		92 min	114 min	14 min	. . .	235 min
MIL-S-861A						
Class 422	(10)	85 min	120 min	17 min	35 min	285 max
Stahl-Eisen-Werkstoffblatt 590-61						
X 20 CrMoV 12 1						
tubes and cylindrical hollow bodies						
	(11)	71 min	100/121	16 min	(12)	205/250
				12 min	(13)	
bars and small forgings	(11)	85 min	114/135	14 min	(12)	235/280
				10 min	(13)	
				12 min	(14)	

(1) 1750 F (955 C) air cool + one hour at 1100 F (595 C) or higher

(2) 0.2%-offset yield strength for USA specifications and the German Luftfahrtwerkstoff

(3) % in 4D for AMS specifications; % in 5D for German specifications; % in 2 inches for all USA specifications other than AMS

(4) recommended heat treatment 1825 F (995 C) for one hour per inch of thickness minimum, air cool to below 500 F (260 C) + 1300/1350 F (705/730 C)

(5) 1740/1830 F (950/1000 C) for X 15 CrMo 13 and 1785/1875 F (975/1025 C) for X 20 CrMo 13 oil quench + 1290/1380 F (700/750 C)

(6) 1925 F (1050 C) quench with double temper, first at 1100 F (595 C) or over and then at 1000 F (540 C) or over

(7) 1875/1925 F (1025/1050 C) for two hours minimum, oil quench + 1150 F (620 C) minimum or at least 100 F (55 C) higher than proposed operating temperature

(8) Brinell Hardness Number 293/341 for nuts and washers; 331 max for bolts and studs

(9) Brinell Hardness Number 229/277 for nuts and washers; 277 max for bolts and studs

(10) 1900 F (1040 C) oil quench + 1250/1325 F (675/720 C) for two hours minimum

(11) recommended heat treatment 1830/1960 F (1000/1070 C) oil quench or air cool + 1290/1435 F (700/780 C)

(12) longitudinal

(13) transverse

(14) tangential

Source: *The Super 12% Cr Steels*, Climax Molybdenum Co., 1965

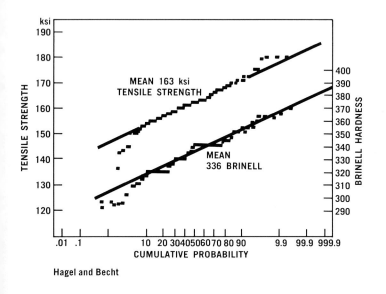

Hagel and Becht

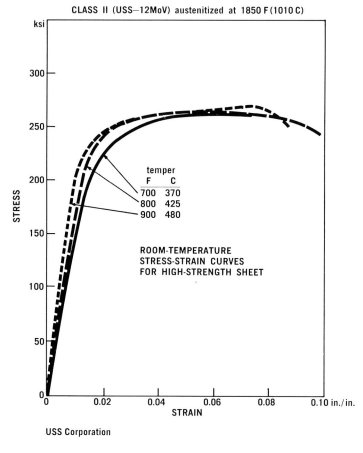

USS Corporation

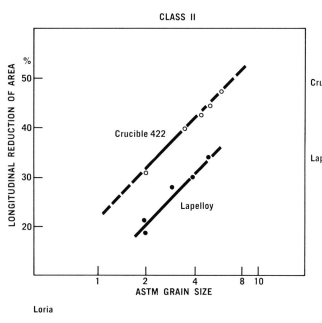

Loria

Crucible 422
2200 F (1205 C) oil
quench + 1200 F (650 C)
for two hours

Lapelloy
1900 F (1040 C) salt
quench + 1160 F (625 C)
for two hours and
1200 F (650 C) for
four hours

		Position of Test	Yield Strength % Offset	Yield Strength ksi	Tensile Strength ksi	Elonga-tion % (1)	Reduc-tion of Area %	
AISI 616	first-stage locomotive gas-turbine wheel. 2¾-inch bore. brittle failure in qualification tests. results given are GE acceptance tests	rim—radial	0.02	125	152	11.0	24.1	Heckman and Herbenar
		rim—tangential	0.02	122	151	12.5	28.2	
		core—axial	0.02	115	144	8.0	9.6	
	second-stage gas-turbine wheel. 74-inch diameter x 16½ inch. low-energy brittle failure during hot stressing. results given are GE acceptance tests	rim—radial	0.02	113	150	11.1	26.8	Heckman and Herbenar
		rim—tangential	0.02	113	142	13.0	27.8	
		core—axial	0.02	112	127	. . .	. . .	
	first-stage locomotive gas-turbine wheel. 49-inch diameter x 13 inch. recalled from service for reevaluation. results given are GE acceptance tests	rim—radial	0.02	126	155	11.8	23.2	Heckman and Herbenar
		rim—tangential	0.02	124	153	11.5	23.0	
		core—axial	0.02	122	150	8.5	6.6	
BVT 130 V	intermediate-pressure steam-turbine rotor. 200-inches long with four-inch bore	surface—tangential	0.2	78	110	22	55	Brennecke and Schinn
		center—axial	0.2	76	104	22	52	
Crucible 422	forgings processed from 36/55-inch-diameter ingots		0.02	100 min	140 min	12.0 min	25.0 min	specification Loria
			0.02	116	149	14.0	29.2	segregate-free
			0.02	112	125	1.0	1.6	massive carbides
			0.02	. . .	80	nil	nil	ferrite-carbide aggregates
	gas-turbine wheels. 19½-inches in diameter x 4⅝ inch	rim—surface plane radial	0.02	85					Zonder, Rush and Freeman
			0.2	118	144	16.0	38	
		rim—central plane radial	0.02	96					
			0.2	115	144	16.2	35.7	
		center—surface plane radial	0.02	88					
			0.2	109	144	9.2	20.0	
		center—central plane radial	0.02	90					
			0.2	107	132	8.4	15.9	
	integral-disc rotor forging			110	125	. . .	. . .	Trumpler, LeBreton, Fox and Williamson
Firth-Vickers 448	20-inch diameter x 3½-inch thick	rim—tangential	0.05	111				Firth-Vickers Stainless Steels Ltd
			0.5	129	141	18	51	
		center edge—diametral	0.5	123	135	11	21	
		center—diametral	0.5	132	139	17	47	
	23-inch diameter x 2¾-inch thick (stamping)	rim—tangential	0.5	125	138	20	48	Firth-Vickers Stainless Steels Ltd
		center	0.5	124	142	15	37	
	22-inch diameter x six-inch-thick hub; 1¾-inch-thick rim (stamping)	rim—tangential	0.05	117				Firth-Vickers Stainless Steels Ltd
			0.5	130	146	18	53	
		midway—tangential	0.05	116				
			0.5	130	140	18	50	
		center—diametral	0.05	115				
			0.5	128	140	16	40	
		radial	0.05	115				
			0.5	128	139	17	47	
Firth-Vickers 535	21-inch-diameter drop-stamped turbine disc with four-inch-thick hub	rim—tangential	0.05	118				Firth-Vickers Stainless Steels Ltd
			0.5	116	154	20	58	
		radial	0.05	117				
			0.5	143	154	18	57	
		center—transverse (mid-thickness)	0.05	119				
			0.5	145	155	17	50	

Source: *The Super 12% Cr Steels*, Climax Molybdenum Co., 1965

		Position of Test	Yield Strength % Offset	Yield Strength ksi	Tensile Strength ksi	Elongation % (1)	Reduction of Area %	
Firth-Vickers Molybdenum Vanadium Stainless Steel	forging 11-inch square x 24 inch	outside— longitudinal	0.05 / 0.2	90 / 94	114	26	62	Firth-Vickers Stainless Steels Ltd
		outside— transverse	0.05 / 0.2	90 / 95	115	20	38	
		center— longitudinal	0.05 / 0.2	86 / 91	110	29	65	
		center— transverse	0.05 / 0.2	89 / 93	112	20	38	
	14-inch-diameter valve forging x 10½-inch long	surface— tangential	0.5	98	115	23	63	Firth-Vickers Stainless Steels Ltd
GE	45-inch diameter x 40,000 pounds	midway—radial	0.02	106	143	16	41.6	Boyle and Newhouse
		surface—radial	0.02	100	138	11.5	27.7	
		midway—tangential	0.02	105	142	16	36	
HGT4	24-inch-diameter disc forging with three-inch-thick rim. boss two-inches thick x eight-inch diameter	periphery— tangential	0.1 / 0.2	117 / 122	141	15.5	42.9	Hadfields Ltd
		radial	0.1 / 0.2	116 / 120	139	15.0	37.9	
		center	0.1 / 0.2	112 / 116	134	15.0	41.3	
H59	four-inch-square billet	longitudinal	0.1	140	168	20.5	59.6	Jessop-Saville Ltd
Jessop-Saville H.46	approximate properties obtained on turbine-disc forgings		0.05 / 0.20	90 / 101	134	17	60	Oliver and Harris
Lapelloy (GE B5OR311— AISI 619)	range on seven wheels, 14-inch diameter x ¾ inch	tangential at rim	0.02	109 / 113	156 / 164	12.0 / 14.0	25.8 / 36.1	Sakamoto
Virgo 7	83,600-pound rotor forging, vacuum cast. 245-inches long with no bore	surface—longitudinal*	0.2	79	111	24	51	Société des Forges et Ateliers du Creusot (Usines Schneider)
		surface— tangential*	0.2	77	110	20	43	
		surface—radial*	0.2	76	109	18	36	
		4.3-inches in from surface—radial*	0.2	77	109	16	24	
		center—longitudinal	0.2	73	104	22	47	
X 20 CrMoWV 12 1	rotor for high-pressure steam turbine			78	110	22	55	Schinn

*all tests from "head" end of forging. Those from other end were similar

(1) in L = 5d on French and German tests; in L = 4 $\sqrt{A}$ on British tests; in L = 2 inches on USA except for Sakamoto where L = 1.5 inches on a 0.375-inch-diameter test specimen

Short-Time Elevated-Temperature Tensile Strength.

Bands representing over 90% of the short-time yield and tensile strengths are very broad. Part of the spread is doubtless due to testing method and especially holding time and strain rate. The effect of heating rate is illustrated by results obtained in rapid-heating constant-load tests, such as relate to the aerodynamic heating of missiles and reentry systems.

The width of the strength bands can be cut down by presentation of the strength at elevated temperatures as a percentage of the room-temperature value, but comparison with other materials is then more awkward. Such comparisons are generally based on "typical" values but are much more valuable and meaningful when based on minimum strengths. This is possible in countries such as Germany where minimum hot yield points are included in some of the specifications. Using such figures, the effect of changes in composition can be observed clearly and reliably.

The bands for tensile ductility (elongation and reduction of area) are even broader than those for strength and thus have little use. Any specific set of test data shows a slight drop in elongation and perhaps in reduction of area in the approximate range from 200/1000 F (95/540 C). The position at which the drop occurs is affected by strain rate.

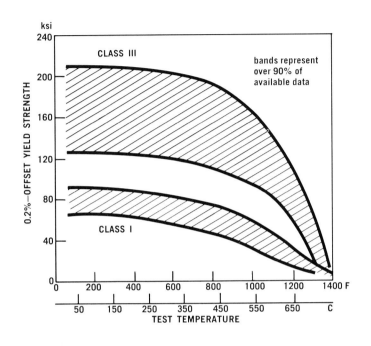

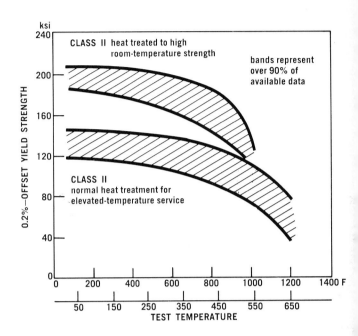

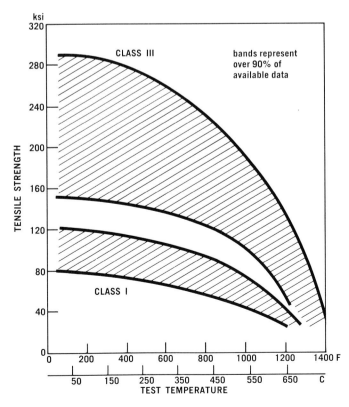

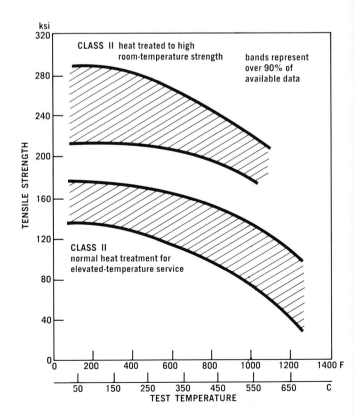

Source: *The Super 12% Cr Steels*, Climax Molybdenum Co., 1965

CLASS II (USS—12MoV)
1850 F (1010 C) air cool + 800 F (425 C)

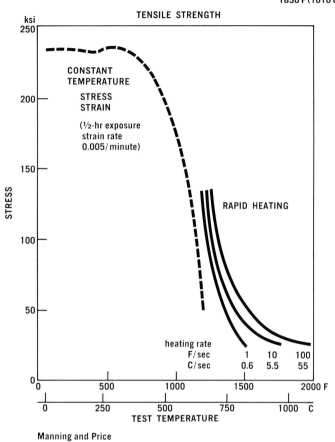

TENSILE STRENGTH

CONSTANT TEMPERATURE STRESS STRAIN

(½-hr exposure strain rate 0.005/minute)

RAPID HEATING

heating rate
F/sec   1   10   100
C/sec   0.6   5.5   55

Manning and Price

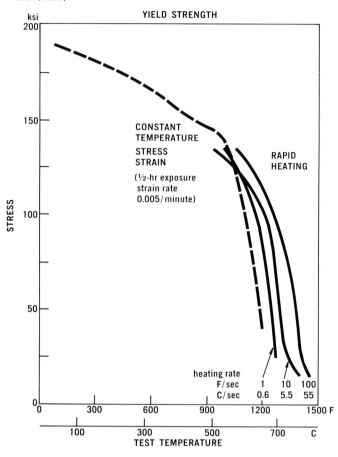

YIELD STRENGTH

CONSTANT TEMPERATURE STRESS STRAIN

(½-hr exposure strain rate 0.005/minute)

RAPID HEATING

heating rate
F/sec   1   10   100
C/sec   0.6   5.5   55

CLASS II (CRUCIBLE 422)
0.062-inch sheet   1900 F (1040 C) oil quench +
1000 F (540 C)

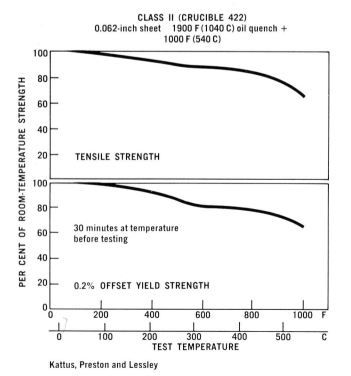

TENSILE STRENGTH

30 minutes at temperature before testing

0.2% OFFSET YIELD STRENGTH

Kattus, Preston and Lessley

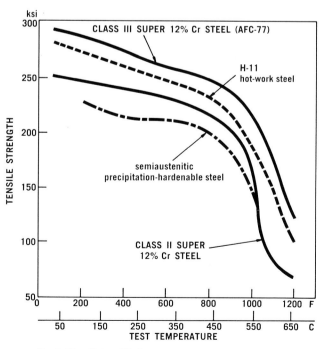

CLASS III SUPER 12% Cr STEEL (AFC-77)

H-11 hot-work steel

semiaustenitic precipitation-hardenable steel

CLASS II SUPER 12% Cr STEEL

Kasak, Chandhok and Dulis

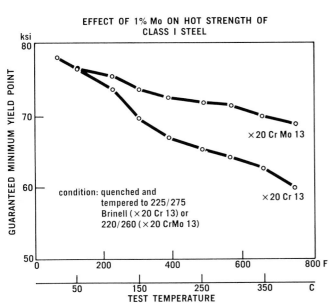

EFFECT OF 1% Mo ON HOT STRENGTH OF CLASS I STEEL

× 20 Cr Mo 13

× 20 Cr 13

condition: quenched and tempered to 225/275 Brinell (× 20 Cr 13) or 220/260 (× 20 CrMo 13)

GUARANTEED MINIMUM YIELD POINT

ksi

TEST TEMPERATURE

Stahl-Eisen Werkstoffblatt 400-60

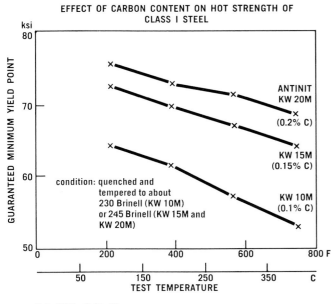

EFFECT OF CARBON CONTENT ON HOT STRENGTH OF CLASS I STEEL

ANTINIT KW 20M (0.2% C)

KW 15M (0.15% C)

KW 10M (0.1% C)

condition: quenched and tempered to about 230 Brinell (KW 10M) or 245 Brinell (KW 15M and KW 20M)

GUARANTEED MINIMUM YIELD POINT

ksi

TEST TEMPERATURE

Gebr Böhler & Co. AG

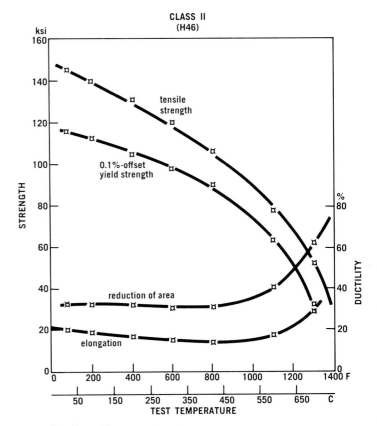

CLASS II (H46)

ksi

tensile strength

0.1%-offset yield strength

reduction of area

elongation

STRENGTH

DUCTILITY

%

TEST TEMPERATURE

Simmons and Cross

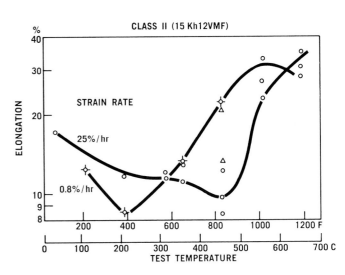

CLASS II (15 Kh12VMF)

%

STRAIN RATE

25%/hr

0.8%/hr

ELONGATION

TEST TEMPERATURE

Stanyukovich

Compressive Properties

Compressive characteristics are of prime interest in airframe design. While compressive strength is also a factor in some tool applications, the need for precise values is not as great as for airframes. Consequently, available data are confined to Class II grades that have been considered for airframes where much of the structure is loaded in compression. These data indicate the yield strength in compression is somewhat higher than that in tension, at least up to 1000 F (540 C). No data have yet been made available on Class III materials.

The ratio of compressive yield strength to density is often taken as a representative factor for depicting structural efficiency. This, of course, assumes a high degree of support of the compression sheet against structural instability. In the design range for a Mach 3 vehicle, the compressive yield strength/density ratio of Crucible 422 is comparable to that for PH 15-7 Mo and the 4% Al—3% Mo—1% V titanium alloy. This factor becomes important in the selection of honeycomb sections where one of the paramount reasons for a sandwich-type configuration is attainment of the desired structural stability at minimum weight. In tests on brazed honeycomb panels, the Class II Super 12% Cr steel detail had higher compressive strength than comparable panels of 17-7 PH and PH 15-7 Mo but Schwartz and Bandelin felt its relatively poor corrosion resistance would mitigate against its use in such thin sections.

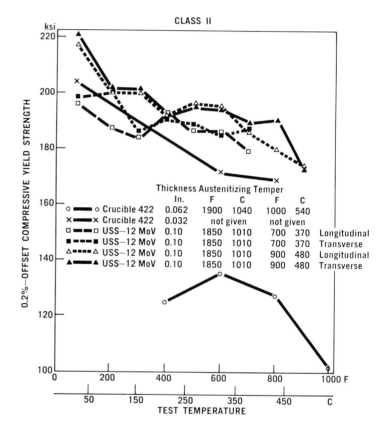

Crucible Steel Company of America;
Kattus, Preston and Lessley;
USS Corporation

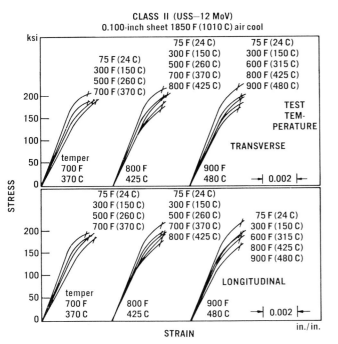

COMPRESSIVE STRESS-STRAIN CURVES

CLASS II (USS—12 MoV)
0.100-inch sheet 1850 F (1010 C) air cool

USS Corporation

CLASS II (CRUCIBLE 422)
0.062-inch thick 1900 F (1040 C) oil
quench + 1000 F (540 C)

Kattus, Preston and Lessley

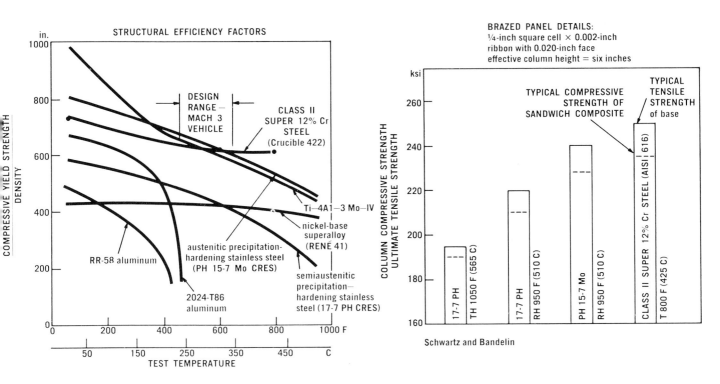

STRUCTURAL EFFICIENCY FACTORS

Crucible Steel Company of America (422);
Schwartz and Bandelin

BRAZED PANEL DETAILS:
¼-inch square cell × 0.002-inch
ribbon with 0.020-inch face
effective column height = six inches

Schwartz and Bandelin

Shear and Bearing Properties

Shear test data have been reported only on Class II steels. Increasing the test temperature from 400/1000 F (205/540 C) decreased the ultimate shear strength of sheet approximately proportional to the decrease that would be expected in tensile strength over this range. The picture is somewhat different with spot-welded sheet where the effect of welding was superimposed on the temperature effect.

With brazed honeycomb panels of AISI 616, static flexure tests gave considerably higher strengths than those of 17-7PH and PH 15-7 Mo. This is actually a measure of shear strength since failure occurs by shear in the core.

In his study of the suitability of various blade root forms for turbines, Siegfried used the ratio of nominal shear stress for fracture of the root to the nominal shear stress for fracture of smooth specimens as a measure of the strength of blade-root fixings. The ratio was 1.08 with fir-tree models of a Class II steel tested at 1020 F (550 C) for 1½ hours. This steel is considered outstanding because of its lack of sensitivity to notches.

The loss in bearing strength with increasing temperature of a Class II steel was comparable to the decrease for other types of strength.

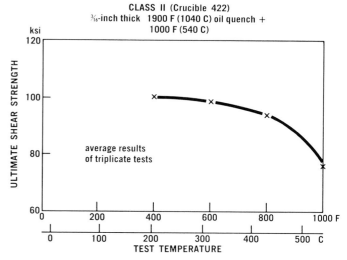

CLASS II (Crucible 422)
³⁄₁₆-inch thick 1900 F (1040 C) oil quench + 1000 F (540 C)

average results of triplicate tests

Kattus, Preston and Lessley

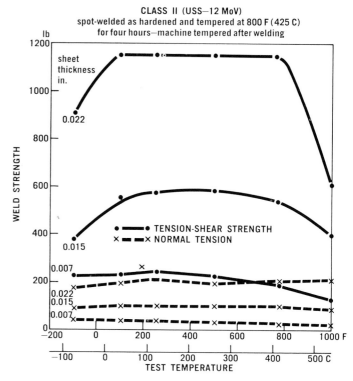

CLASS II (USS—12 MoV)
spot-welded as hardened and tempered at 800 F (425 C) for four hours—machine tempered after welding

● ────● TENSION-SHEAR STRENGTH
× ━ ━× NORMAL TENSION

Nippes, Savage, Ianniello and Owczarski

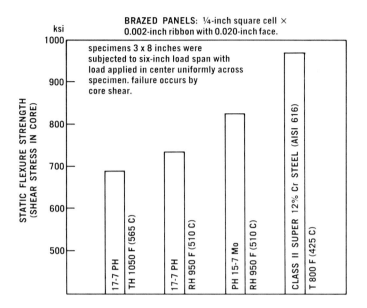

BRAZED PANELS: ¼-inch square cell × 0.002-inch ribbon with 0.020-inch face.

specimens 3 x 8 inches were subjected to six-inch load span with load applied in center uniformly across specimen. failure occurs by core shear.

STATIC FLEXURE STRENGTH (SHEAR STRESS IN CORE)

ksi

17-7 PH TH 1050 F (565 C)

17-7 PH RH 950 F (510 C)

PH 15-7 Mo RH 950 F (510 C)

CLASS II SUPER 12% Cr STEEL (AISI 616) T 800 F (425 C)

Schwartz and Bandelin

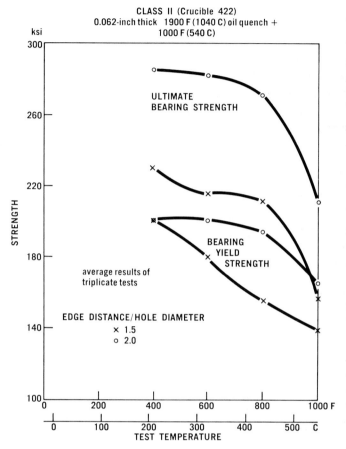

CLASS II (Crucible 422)
0.062-inch thick   1900 F (1040 C) oil quench +
1000 F (540 C)

ksi

STRENGTH

ULTIMATE BEARING STRENGTH

BEARING YIELD STRENGTH

average results of triplicate tests

EDGE DISTANCE/HOLE DIAMETER
× 1.5
○ 2.0

TEST TEMPERATURE

Kattus, Preston and Lessley

Fatigue Strength

Fatigue strength is important in many of the applications where Super 12% Cr steels are or might be used. Simply measuring the effect of a molybdenum addition—as Ono and Sasaki have done—is not enough because structure is significant as it is with most of the properties of these steels.

Role of Delta Ferrite. The influence of delta ferrite on fatigue characteristics has been the subject of much discussion but fortunately some reasonably clean-cut data are now available. Although a deleterious influence has often been attributed to delta ferrite, this statement is much too broad to stand without further qualification. At high hardnesses, delta ferrite undoubtedly lowers the fatigue resistance in both longitudinal and transverse tests. Massive segregates of ferrite and carbides at the surface can also decrease the fatigue strength. This has been demonstrated by Heckman and Herbenar as well as by Loria although Smeltzer in his discussion of Loria's paper stated that the room-temperature fatigue strength of Lapelloy turbine wheels and sheet was not decreased by the presence of such surface segregation.

Longitudinal room-temperature fatigue tests indicate no detrimental effect of about 5% ferrite if randomly distributed but the presence of 15/20% ferrite caused a significant lowering of the fatigue strength.

The most serious influence of ferrite, however, is evident in transverse tests where ferrite may be more damaging than nonmetallic inclusions. Practically this is of special importance in parts such as turbine wheels involving both radial and tangential stresses. The transverse fatigue strength of a high-quality Super 12% Cr steel with at the most small amounts of ferrite is about 80% of the longitudinal value; this ratio can drop to 65/70% in the presence of 15/20% delta ferrite. Little effect was observed with notched specimens since the weakening action of the ferrite is outweighed by that of the notch.

At temperatures around 500/700 F (260/370 C), 12% Cr steels with substantial amounts of delta ferrite show unusually high fatigue strengths, which may exceed the room-temperature values. Dulis, Chandhok and Hirth have tentatively attributed this enhanced strength to precipitation of alpha prime in the delta ferrite during fatigue testing. No data are available to show whether possible overaging might occur on long exposures in this temperature range.

Effect of Heat Treatment. Samuel Fox has reported that lowering the austenitizing temperature of Jethete M154 produced significantly higher fatigue strengths at the same tensile strength.

Austenitizing Temperature*		Tensile Strength ksi	Limiting Fatigue Strength ksi
F	C		
1650	900	161	91 and 99
1920	1050	155	83

*Specimens air cooled from austenitizing temperature and tempered at 1200 F (650 C)

Samuel Fox & Co Ltd

As with most materials, increased hardness or strength is generally accompanied by higher fatigue strength. Consequently, if high temperatures are not encountered in service, advantage may be taken of the higher strength obtained with low tempering temperatures. It should be pointed out, though, that while the lower tempering temperatures give higher fatigue strengths, the ratio of fatigue strength to tensile strength is lowered.

Nitriding might be expected to improve the fatigue properties of the Super 12% Cr steels just as it does with other steels. Dawes and Duce indicate this is so and further state that activated nitriding and salt-bath sulfidizing have essentially the same effect. A two-hour treatment of Firth-Vickers 448 at 1060 F (570 C) raised the endurance limit (for 10^7 cycles) from 75 to 81 ksi for unnotched specimens and from 42 to 66 for notched ones. Clearly, the chief merit of this treatment would be on parts containing notches or other stress concentrations.

CLASS II
(Crucible 422)
¾-inch-diameter bar stock or ¾-inch-thick plate
1900 F (1040 C) oil quench

% Delta Ferrite	Tensile Strength, ksi
0	155
5	160
15/20	~ 140

5% Ferrite, Tempered 1150 F (620 C)

0% Ferrite, Tempered 1150 F (620 C)

15/20% Ferrite, Tempered 1200 F (650 C)

LONGITUDINAL ELECTROMAGNETIC
CANTILEVER-BEAM TESTS

STRESS (ksi) vs CYCLES

Loria

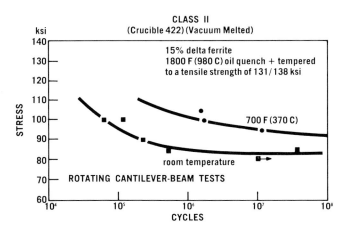

CLASS II
(Crucible 422) (Vacuum Melted)

15% delta ferrite
1800 F (980 C) oil quench + tempered
to a tensile strength of 131/138 ksi

700 F (370 C)

room temperature

ROTATING CANTILEVER-BEAM TESTS

STRESS (ksi) vs CYCLES

Dulis, Chandhok and Hirth

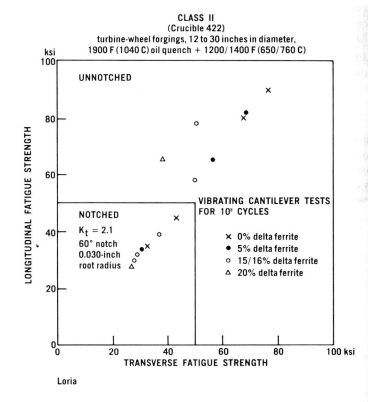

CLASS II
(Crucible 422)
turbine-wheel forgings, 12 to 30 inches in diameter,
1900 F (1040 C) oil quench + 1200/1400 F (650/760 C)

UNNOTCHED

NOTCHED

$K_t = 2.1$
60° notch
0.030-inch
root radius

VIBRATING CANTILEVER TESTS
FOR 10⁸ CYCLES

✕ 0% delta ferrite
● 5% delta ferrite
○ 15/16% delta ferrite
△ 20% delta ferrite

LONGITUDINAL FATIGUE STRENGTH vs TRANSVERSE FATIGUE STRENGTH

Loria

Source: *The Super 12% Cr Steels*, Climax Molybdenum Co., 1965

121

Cast vs Wrought. Too few comparative data are at hand to permit a general statement on the relative fatigue properties of cast and wrought parts of the same grade. Avesta gives a fatigue strength for annealed Avesta 739S castings 86% that of wrought steel of the same grade with the same heat treatment, while Jessop-Saville state that cast H46 has a fatigue strength about 60% that of the wrought material. This difference is presumably related to the fact that the one grade is a relatively simple Class I steel used in the annealed condition while the other is a more complex Class II steel used in the heat-treated state, but casting variables undoubtedly enter into the picture.

Effect of Temperature. Testing variables have a strong effect on high-temperature fatigue values. Room-temperature fatigue curves reach a constant value at perhaps 10^6 cycles but fatigue curves at elevated temperatures generally show a constant decrease. Data on Jessop-Saville's Vacumelt H.53 at 1110 F (600 C) show a steady decrease from 41 ksi at 20×10^6 cycles to 39 at 40×10^6 to 36 at 100×10^6.

Since testing variables have such an effect, it is difficult to compare elevated-temperature fatigue-strength values unless the tests were made with essentially the same set-up. When this is done, it is evident that the hot fatigue strengths of the three classes fall in the same order as is found with other types of test; namely, Class I is the weakest and Class III the strongest.

Only one set of data comparing longitudinal and tangential fatigue strengths at elevated temperatures has been found so it is impossible to state whether this is typical. In any case, as might be expected, the fatigue strength of transverse specimens is lower than that of longitudinal specimens.

In actual service, a constant stress is often superimposed on an alternating stress. The effect of this steady load becomes increasingly important as the temperature increases and failure may be due to rupture from the constant stress rather than a straight fatigue failure.

Thermal Fatigue. Thermal fatigue can be a significant factor in some types of service. The purpose of Walker's study was to evaluate the low-cycle thermal-fatigue-cracking characteristics of a number of cast steels, initially for steam-turbine shell materials. Shell cracks appear to be caused by the repeated presence of high residual tensile stresses, which arise due to temperature differences produced during turbine starts. The actual cracking seems to occur by creep-rupture mechanisms, primarily during long periods at normal operating temperatures. He used a reversed, uniaxial, strain-fatigue test at 950 F (510 C) with a hold in the tension part of the cycle. The best alloy was a 1% Cr—1% Mo steel with a Class I Super 12% Cr steel containing 0.5% Mo and 0.8% Ni tested in the heat-treated condition a close second. A 1% Mo steel and a columbium-stabilized Type 316 austenitic steel were poorer. The Super 12% Cr steel showed roughly a two-to-five-times longer life than the austenitic stainless steel under these conditions.

Clauss's investigation covered thermal fatigue and the effect of constrained and free thermal cycling on subsequent stress-rupture behavior. Specimens of Crucible 422 steel were alternately heated and cooled while constrained in a manner that prevented their free axial expansion and contraction, and the number of cycles to failure was determined. The maximum cycle temperature varied from 1000/1500 F (540/815 C) and the minimum cycle temperature was held constant at 200 F (95 C). Cycling to temperatures of 1200 F (650 C) and above caused failures in less than 2000 cycles while cycling to lower temperatures did not cause any failures in less than 20,000 cycles. Exposure to thermal fatigue or to thermal cycling alone in the absence of constraints significantly reduced the stress-rupture life. The reduction in rupture life undoubtedly resulted from a

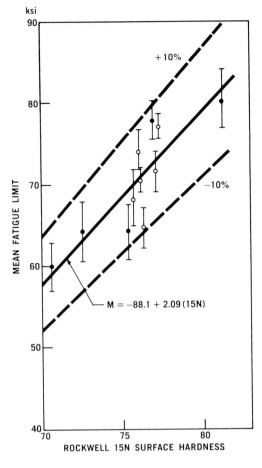

Morrow and Sinclair

CLASS II (Lapelloy)

2000 F (1095 C) salt quench,
martemper 650 F (345 C) +
temper 1150/1600 F (620/870 C)

CANTILEVER ROTATING-BEAM
TESTS

●   mechanically polished,
rms 2.5/4.0 microinches

○   other surface treatments,
including surface rolling,
rough grinding and rough
machining. rms 5.5/40
microinches

| 95% confidence limits

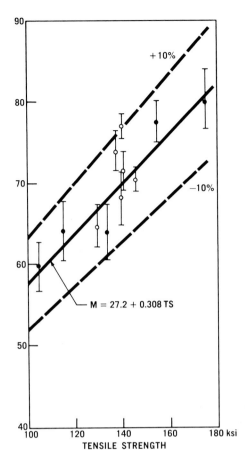

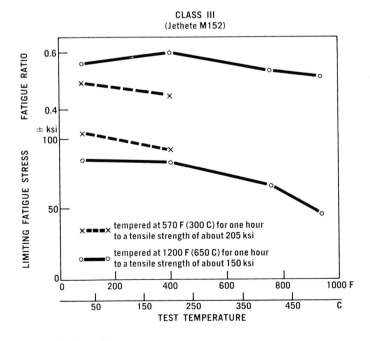

Irvine and Murray

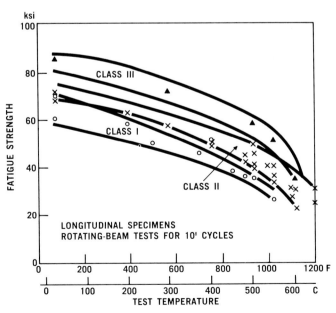

Firth-Vickers Stainless Steel Ltd
Jessop-Saville Ltd
Cyclops Corporation

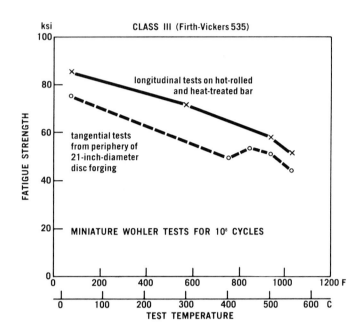

Firth-Vickers Stainless Steels Ltd

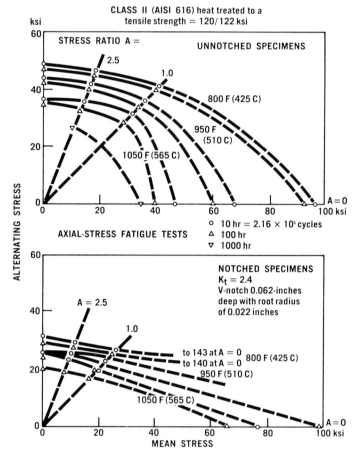

Matters and Blatherwick

change in structure caused primarily by temperature alone and not significantly affected by constraint during the cycling to temperatures of 1150 F (620 C) and higher. At lower cycle temperatures where the structure was more stable, cyclic stresses during constraint may be more important in accelerating structural changes.

Thus, in the absence of standard thermal-fatigue tests, little can be said about the resistance of Super 12% Cr steels to this type of stressing except to note that they appear to be considerably more resistant than austenitic materials in the temperature range where they would normally be used. This characteristic is undoubtedly related to the higher thermal-conductivity and lower thermal-expansion properties of the Super 12% Cr steels.

Thermal-shock tests, which can be considered as a form of thermal fatigue, have been made by Gysel, Werner and Gut as well as by Schinn and Ruttmann. Results of tests by the latter investigators are particularly interesting as they involved cycling between 1060 F (570 C) and 120/570 F (50/300 C) to determine whether dissimilar welds could withstand such conditions, which were far more severe than would be expected in service. No material separation or other joint damage was observed in joints between an austenitic steel and a Class II Super 12% Cr steel.

Corrosion Fatigue. According to Dorey, De Laval in Sweden undertook extensive experiments to assess the optimum fatigue properties of turbine-blading materials. The conclusion was that 0.09/0.14% C was most suitable in a 13% Cr—0.5% Mo steel for operating in moist steam and for welded designs. Moreover, acceptance is based on a maximum of 5% free ferrite. In addition, De Laval regard nickel as a detrimental addition as far as corrosion fatigue is concerned.

In a private communication, Avesta Jernverks AB has advised that the rotating-beam fatigue strength

in sea water of Avesta 739SG, long used for cast ship propellers, is about 26 ksi.

Impact Toughness

Notched-bar impact tests give an indication of the resistance of a material to brittle failure. In most applications of Super 12% Cr steels, impact toughness is of less significance than fracture toughness but industry specifications incorporating toughness requirements include impact- rather than fracture-toughness tests because of the existence of standard impact tests and the ease of making these tests. Typical room-temperature impact requirements are:

	Guaranteed Minimum Impact		Tensile Strength ksi
	type of test	ft lb	
Class I			
AMS 5614	Izod V-notch	50	100 min
Stahl-Eisen-Werkstoffblatt 400-60			
X 15 CrMo 13 and X 20 CrMo 13	DVM U-notch	52	107/128
Class II			
AMS 5655	Izod V-notch	8	140 min
ASTM A 437—59 T			
Class B4B	Charpy V-notch	10	145 min
Class B4C	Izod V-notch	25	115 min
Luftfahrtwerkstoff 1.4934	DVM U-notch	52	114 min
MIL-S-861A			
Class 422	Izod V-notch	25	120 min
Stahl-Eisen-Werkstoffblatt 590-61			
X 20 CrMoV 12 1			
tubes and cylindrical			
hollow bodies	DVM U-notch	62*	100/121
	DVM U-notch	41**	100/121
bars and small forgings	DVM U-notch	52*	114/135
	DVM U-notch	31**	114/135
	DVM U-notch	41***	114/135

*longitudinal   **transverse   ***tangential

Some of the many factors that can affect impact toughness are discussed briefly below. The effect of holding at elevated temperatures will be covered in the next section under Thermal Stability.

Heat Treatment. Holding the austenitizing temperature to the low side of the range has a favorable influence on impact toughness (see pages 155/156), while tempering at temperatures around 800/1000 F (425/540 C) has an embrittling effect on all 12% Cr steels. Specific data on Super 12% Cr steels are given on pages 162/164. The minimum impact, however, will probably be higher with Class I than with Class II steels.

Hardness and Strength. The effect of hardness and strength depends in part on the tempering temperature. If the hardness requires tempering in the embrittling range, the impact toughness will be less than it would be at higher or lower hardnesses involving tempering at lower or higher temperatures. If impact values corresponding to the trough of the impact—tempering-temperature curve are eliminated, tentative bands relating impact and strength can be established. The width of the band is probably attributable as much to heat-to-heat variations as to grade-to-grade variations as is indicated by the width of the impact band for Jethete M153.

Vacuum Melting. Vacuum melting has a favorable influence on impact characteristics, presumably largely as a result of minimizing nonmetallic inclusions and nitrogen content.

Structure. Although undesirable from other standpoints, free ferrite may improve the longitudinal impact properties of Super 12% Cr steels. This is generally observed when the tempering temperature is around 800/1000 F (425/540 C). In the transverse direction, however, it is always disadvantageous.

Section Size and Direction of Test. The effect of size and direction of test is illustrated in information from Samuel Fox on the impact and tensile properties of bars from 40 commercial heats of Jethete M152, air cooled from 1920 F (1050 C) and tempered at 1200/1245 F (650/675 C):

Size in.	Direction	Number of Results	Izod V-Notch Impact, ft lb	Tensile Strength ksi
1 1/8	longitudinal	62	44/76	137/161
2 1/2 / 3	longitudinal	10	51/72	139/159
4 / 4 1/2	longitudinal	12	44/67	139/157
	transverse	4	23/38	146/155
5 1/2 / 7	longitudinal	6	32/57	146/155
	transverse	6	6/31	146/155

These values show the lower impact in the transverse direction and also that both longitudinal and transverse impact drop off sharply over 4½-inches diameter for this particular grade. Presumably this would be tied in with the appearance of delta ferrite.

The histogram on page 165 includes data on the impact values of relatively small forgings, turbine blades, where the impact is comparable to expectations on small bars. With large forgings, however, the impact toughness is considerably less. Presence of ferrite and precipitated or segregated carbides would account for this. SFAC's data on the impact transition temperature of the Virgo 7 forging included in the table on page 60 illustrate the good toughness that can be obtained in heavy forgings with proper choice of composition and heat treatment.

Castings. Cast conventional 12% Cr steels have poor notch ductility when normalized and tempered to minimum tensile strengths of 90 ksi, even when the tensile ductility is satisfactory. Among the failures that have occurred because of this low toughness are several ship propellors.

In their determination of the toughness of cast 12% Cr steel, Brandt, Bishop and Pellini found that the higher silicon added by some foundries to increase fluidity had a deleterious effect on toughness. While this higher silicon led to the presence of free ferrite, the ferrite was not the sole cause of this brittleness. On the other hand, molybdenum has a favorable influence on both impact and fracture toughness. Supplemental additions of nickel were

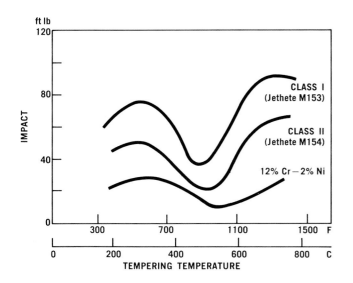

Irvine, Crowe and Pickering

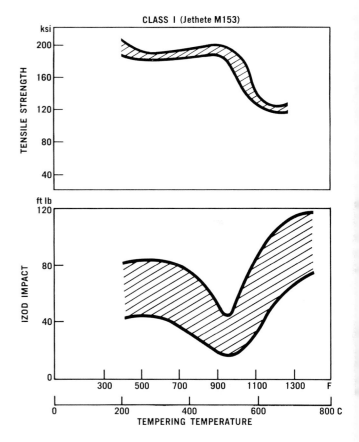

Samuel Fox & Co Ltd

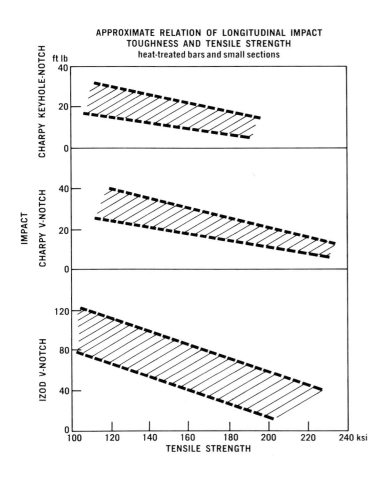

APPROXIMATE RELATION OF LONGITUDINAL IMPACT
TOUGHNESS AND TENSILE STRENGTH
heat-treated bars and small sections

Source: *The Super 12% Cr Steels*, Climax Molybdenum Co., 1965

EFFECT OF VACUUM MELTING ON CLASS II

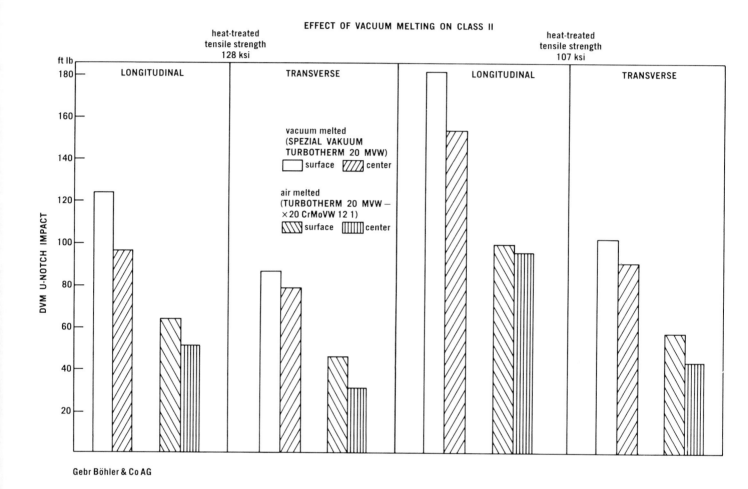

heat-treated
tensile strength
128 ksi

heat-treated
tensile strength
107 ksi

LONGITUDINAL TRANSVERSE LONGITUDINAL TRANSVERSE

vacuum melted
(SPEZIAL VAKUUM
TURBOTHERM 20 MVW)
☐ surface ▨ center

air melted
(TURBOTHERM 20 MVW –
✕20 CrMoVW 12 1)
▨ surface ⊞ center

DVM U-NOTCH IMPACT

ft lb

Gebr Böhler & Co AG

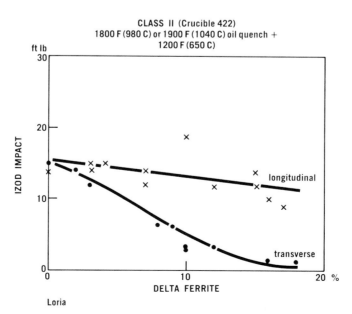

CLASS II (Crucible 422)
1800 F (980 C) or 1900 F (1040 C) oil quench +
1200 F (650 C)

ft lb

IZOD IMPACT

longitudinal

transverse

DELTA FERRITE

Loria

Class II		Impact			Tensile Strength ksi	
		position of test	type of test	ft lb		
AISI 616	first-stage locomotive gas-turbine wheel. 2¾-inch bore. brittle failure in qualification tests. results given are GE acceptance tests	rim—radial	Charpy V-notch	5.0, 6.0	155/157	Heckman
		rim—tangential	Charpy V-notch	4.0, 7.0	151/155	and
		core—axial	Charpy V-notch	2.0	150	Herbenar
	second-stage gas-turbine wheel. 74-inch diameter x 16½ inch. low-energy brittle failure during hot stressing. results given are GE acceptance tests	rim—radial	Charpy V-notch	8, 8.5	150.5	Heckman
		rim—tangential	Charpy V-notch	10, 11	142.5	and
		core—axial	Charpy V-notch	3.5	127	Herbenar
	first-stage locomotive gas-turbine wheel. 49-inch diameter x 13 inch. recalled from service for reevaluation. results given are GE acceptance tests	rim—radial	Charpy V-notch	6.5	152	Heckman
		rim—tangential	Charpy V-notch	6.5	151	and
		core—axial	Charpy V-notch	3.0	144	Herbenar
BVT 130 V	intermediate-pressure steam-turbine rotor. 200-inches long with four inch bore	surface—tangential	DVM U-notch	62, 63	110	Brennecke
		center—axial	DVM U-notch	98	104	and Schinn
Firth-Vickers Molybdenum Vanadium Stainless Steel	forging 11-inch square x 24 inch	outside—longitudinal	Izod V-notch	26	114	Firth-Vickers
		outside—transverse	Izod V-notch	12	112	Stainless
		center—longitudinal	Izod V-notch	36	110	Steels Ltd
		center—transverse	Izod V-notch	18	112	
	14-inch-diameter valve forging x 10.5-inches long	surface—tangential	Izod V-notch	33	115	Firth-Vickers Stainless Steels Inc
GE	45-inch diameter x 40,000 pounds	surface—radial	Charpy	12	138	Boyle and Newhouse
"prior art"	45-inch diameter x 40,000 pounds	surface—radial	Charpy	4	124	Boyle and Newhouse
GE B5OR311 (AISI 619)	gas-turbine wheel. 14-inch diameter x ¾ inch			Charpy V-notch	9.5, 11, 17, 19, 19	157/ 159.5	Sakamoto
X 20 CrMoWV 12 1	steam-turbine rotor			DVM U-notch	62, 65	104/110	Schinn

Source: *The Super 12% Cr Steels*, Climax Molybdenum Co., 1965

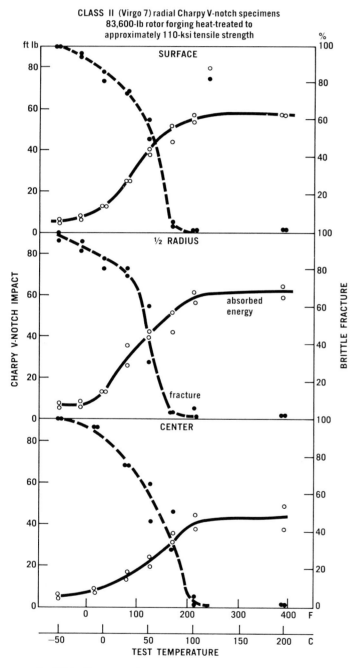

CLASS II (Virgo 7) radial Charpy V-notch specimens
83,600-lb rotor forging heat-treated to
approximately 110-ksi tensile strength

Société des Forges et Ateliers
du Creusot (Usines Schneider)

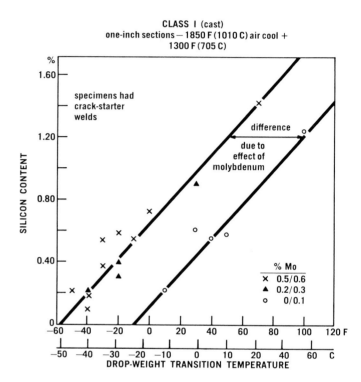

CLASS I (cast)
one-inch sections — 1850 F (1010 C) air cool +
1300 F (705 C)

Brandt, Bishop and Pellini

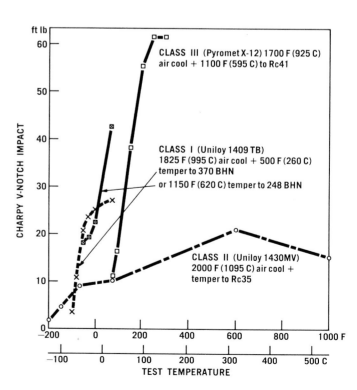

The Carpenter Steel Company
Cyclops Company

useful in larger sections but in the absence of molybdenum did not increase the toughness. Their recommendations were 0.5% Si max and 0.25% Mo in one-inch sections; and 0.5% Si max, 0.5/0.6% Mo and 0.6/0.7% Ni for four-inch sections to give the optimum toughness. These findings are now incorporated in a U.S. Department of Defense Specification (see page 37).

Gearhart of Birdsboro Corporation confirmed these results on large water-turbine impulse wheels. The following data apply to specimens taken from four-inch keel-block coupons, air cooled from 2000 F (1095 C) and tempered at 1200 F (650 C):

	Straight 12% Cr*	CLASS I SUPER 12% Cr**
Charpy V-notch Impact, ft lb		
32 F (0 C)		16
70 F (21 C)	11	
Yield Point, ksi	84	89
Tensile Strength, ksi	104	107
Elongation, % in 2 in.	21	22
Reduction of Area, %	49	54
Bend	120°	120°

*0.10% C, 0.78% Mn, 0.027% P, 0.011% S, 1.41% Si, 11.74% Cr, 0.03% Mo, 0.12% Ni

**0.11% C, 0.66% Mn, 0.029% P, 0.022% S, 0.55% Si, 12.40% Cr, 0.47% Mo, 0.80% Ni

In other words, at approximately the same strength level, the Class I Super 12% Cr steel showed a higher impact at a lower temperature than the conventional 12% Cr steel.

Class II Super 12% Cr steels have found a number of applications in the cast condition, one of the most important of which is steam-turbine casings. Brennecke and Schinn have given the toughness and strength of five casing castings of this type. Because of the critical nature of these castings, the properties are determined on specimens taken from the core of the casing or from trepanned specimens as indicated. As can be seen, the toughness is excellent. These casings operate at temperatures up to 1200 F (650 C) and up to 330 atmospheres.

Weldments. On weld deposits of a Class II steel with 0.12% C, 11.5% Cr, 0.72% Mo, 0.56% Ni, 0.29% V and 0.88% W, Kakhovskiy and Ponizovtsev obtained average impacts of 69 foot pounds on a U-notch Mesnager specimen at 109.5 ksi tensile strength after full heat treatment.

Effect of Temperature. The Super 12% Cr steels do not have high impact properties at cryogenic temperatures. On the other hand, their toughness at moderately elevated temperatures is considerably better than at room temperature. Results are shown for typical compositions of each of the three classes.

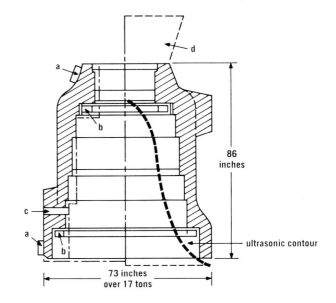

Properties of Casing Castings of GS X 22 CrMo(W)V 121 Weighing Over Three to 17.6 Tons

Casing Number	Grade	% C	% Mn	% Si	% Cr	% Mo	% Ni	% V	% W	Position of Specimen	DVM U-Notch Impact ft lb	0.2%-Offset Yield Strength ksi	Tensile Strength ksi	Elongation % in L = 5d	Reduction of Area %
1	VIS 11	0.23	0.7	0.33	10.7	1.1	0.6	0.4	0.4	a	91, 99 103, 113	83	108	23	58
										b	45, 77 88, 88	80	100	9	19
2	VIS 11	0.22	0.5	0.20	10.9	1	0.6	0.4	0.37	a	78, 78	81	105	21	57
										c	71	74	100	15	26
3	BVT 130V	0.20	0.6	0.30	11.7	1.1	0.7	0.34		a	62	87	114	21	52
										b	62, 65	80	111	18	37
4	BVT 130V	0.22	0.7	0.35	11.8	1.2	0.7	0.34		a	92	90	118	16	36
										b	80	85	114	16	36
5	CMW 11	0.21	0.7	0.23	12.5	1.1	0.6	0.27		a	74	94	112	20	50

Brennecke and Schinn

Fracture Toughness

Fracture toughness is a valuable guide to establishing allowable design stresses and permissible defects for parts made from various high-strength steels as it gives the relative capability of a material to sustain high stresses without failure in the presence of local stress raisers and the conditions under which rapid crack propagation may occur. With tensile strengths under 180 ksi, the possibility of low fracture toughness seldom arises with sheet although it may with heavy forgings. Fracture toughness is therefore of particular interest where Super 12% Cr steels are to be used at very high strengths as in some proposed aircraft, missile and pressure-vessel applications as well as in heavy forgings such as turbine wheels and rotors.

There is no standard test procedure for measuring fracture toughness but notched tensile tests are commonly used. With a steel having low fracture toughness and high susceptibility to stress concentrations, the notched strength will decrease sharply with increasing stress concentration factors.

The principal influence affecting fracture toughness —just as with impact toughness—is heat treatment and particularly tempering. Tempering at about 800/1000 F (425/540 C) leads to low fracture toughness (see pages 160, 162/164). This means that there is no single value of fracture toughness corresponding to a tensile strength because tempering temperature must also be taken into account. The graph from Espey, Jones and Brown illustrating this point also shows the superior fracture toughness of a Class II steel over a straight 12% Cr steel. Arrested quenching as in air cooling or marquenching narrows the tempering range where these low toughness values occur (page 162).

Effect of Composition. Too few data are available to clarify the effect of composition on fracture toughness except in the case of Class I castings, which have been discussed on pages 74, 79 and illustrated on page 78. Two of the Class II grades have a generally similar behavior while the only Class III steel on which information has been published displays unusually good fracture toughness within certain ranges of tempering temperature. On the whole, burst tests on pressure vessels and turbine wheels of Class II steels have been in agreement with these indications.

Fracture Toughness of AFC-77 (Class III)

Test Temperature F	C	Cold Reduction %	Tempering Temperature F	C	0.2%-Offset Yield Strength ksi	Tensile Strength ksi			Elongation % in 2 in.
						Edge Notched	Center Notched	Smooth	
−110	−80	none	700	370	208	195	149	260	11
72	22	none	700	370	182	213	183	237	11
650	345	none	700	370	146	160	152	234	12
−110	−80	10	700	370	284	87	48	294	6
72	22	10	700	370	255	259	172	264	5
650	345	10	700	370	218	175	166	251	9
−110	−80	none	750	400	219	190	59	264	10
72	22	none	750	400	187	238	172	237	11
650	345	none	750	400	152	180	159	235	11

longitudinal specimens, austenitized at 1900 F (1040 C) for ½ hour, oil quenched, refrigerated at −100 F (−75 C) for ½ hour or cold rolled 10%, and then tempered at indicated temperature for 2 + 2 hours

Kazak, Chandhok, Moll and Dulis

Presence of free ferrite appears to have no significant effect on longitudinal fracture toughness of Class II steels but segregation can have a pronounced influence. Consequently, any change in composition that would minimize segregation would also have a beneficial effect on fracture toughness.

CLASS II (USS — 12MoV)
1850 F (1010 C) oil quench + 900 F (480 C)

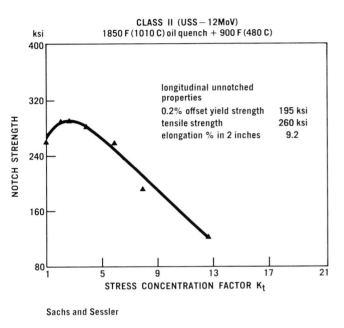

longitudinal unnotched
properties
0.2% offset yield strength   195 ksi
tensile strength   260 ksi
elongation % in 2 inches   9.2

Sachs and Sessler

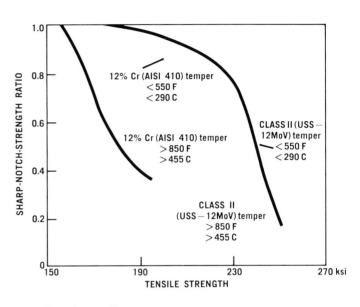

12% Cr (AISI 410) temper
< 550 F
< 290 C

12% Cr (AISI 410) temper
> 850 F
> 455 C

CLASS II (USS —
12MoV) temper
< 550 F
< 290 C

CLASS II
(USS — 12MoV) temper
> 850 F
> 455 C

Espey, Jones and Brown

CLASS II (USS — 12MoV)
0.041-inch-thick sheet air hardened +

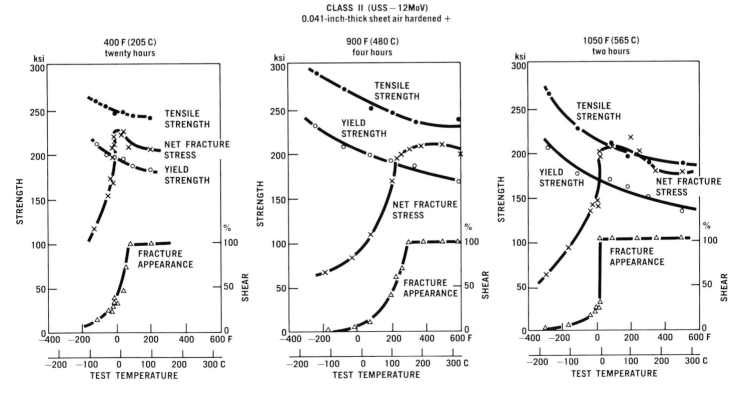

400 F (205 C)
twenty hours

900 F (480 C)
four hours

1050 F (565 C)
two hours

Srawley and Beachem

Source: *The Super 12% Cr Steels*, Climax Molybdenum Co., 1965

CLASS III (AFC-77)
2000 F (1095 C) oil quench + −100 F (−73 C)
+ two + two hours at temperature indicated

ksi

NOTCHED TENSILE
STRENGTH (K$_t$ = 3.9)

ROOM-TEMPERATURE
PROPERTIES

TENSILE STRENGTH

YIELD STRENGTH

STRENGTH

REDUCTION OF AREA

ELONGATION

DUCTILITY

%

TEMPERING TEMPERATURE

Kazak, Chandhok, Moll and Dulis

CLASS II (Crucible 422)
0.067-inch-thick specimens 1900 F (1040 C) oil quench +
−320 F (−195 C) + 900 F (480 C)

ksi

STRENGTH

TENSILE STRENGTH

0.2% YIELD STRENGTH

NET FRACTURE STRESS

SHEAR

%

FRACTURE TOUGHNESS BASED ON PERCENTAGE SHEAR

ksi √ in.

FRACTURE PROPAGATION RESISTANCE
(NET FRACTURE STRESS/YIELD STRENGTH)

▲ TRANSVERSE SPECIMENS
● LONGITUDINAL SPECIMENS

TEST TEMPERATURE

Banerjee and Hauser

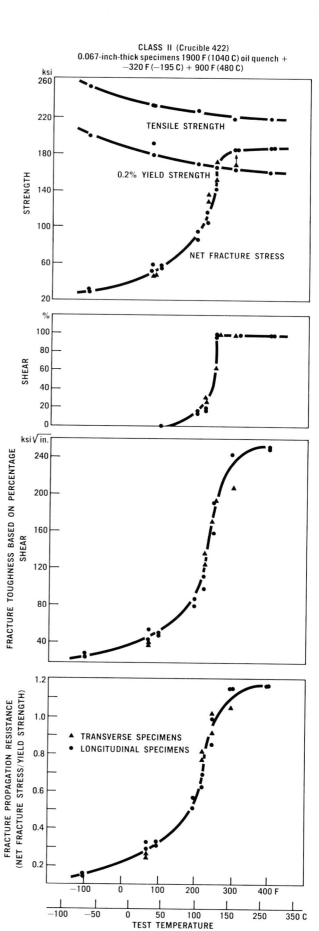

Effect of Test Temperature. Increasing test temperature also increases the fracture toughness. A steel that has inadequate fracture toughness at room temperature may well have satisfactory characteristics at 120 F (50 C). It is generally held that the fracture-appearance transition temperature (the lowest temperature at which a specimen would exhibit a 100% shear fracture) should not exceed the lowest operational temperature if the risk of brittle failure is to be minimized.

Effect of Sheet Thickness. The thinner the sheet, the lower is the fracture toughness transition temperature. The magnitude of this factor has been indicated for USS-12MoV hardened and tempered at 900 F (480 C).

Sheet Thickness in.	Fracture-Appearance Transition Temperature (FATT)		Net Fracture-Stress Transition Temperature	
	F	C	F	C
0.041	260/300	125/150	215	100
0.025	150/200	65/95	150	65
0.010	50/80	10/25	45	7

Srawley and Beachem

To carry the picture further, heavy forgings may have considerably higher transition temperatures than sheet.

Heavy Sections. Fracture toughness and the fracture transition temperature are important characteristics for turbine wheels and heavy rotors. Proper selection of composition is particularly significant for optimum results in order to avoid or minimize the deleterious effects of carbide segregates. With current practice it is feasible to obtain as good fracture toughness in large sizes as in small ones. Furthermore, because of their deep-hardening characteristics, the Super 12% Cr steels are not as sensitive to size as are low-alloy steels used for elevated-temperature applications.

Heavy forgings of Class II steels have at least as good fracture toughness as the low-alloy chromium-molybdenum-vanadium steels they are replacing for applications such as steam-turbine rotors. The SFAC data on page 78 show a fracture transition temperature (50% brittle fracture) of 105/140 F (40/60 C), comparable to that of heavy forgings of molybdenum-nickel-vanadium, chromium-molybdenum-nickel and chromium-molybdenum-nickel-vanadium steels and appreciably lower than that of high-temperature low-alloy chromium-molybdenum-vanadium grades, which in similar sections generally have a fracture transition temperature around 210/250 F (100/120 C).

Use of Low Tempering Temperatures. A logical inference to be drawn from these tempering-temperature—fracture-toughness diagrams would be use of a low-temperature temper to produce the optimum combination of strength and fracture toughness.

Notched Tensile Properties of AISI 619 Forgings

Test Temperature		Notched Strength Unnotched Strength	Unnotched Tensile Test				Notched Tensile Test		Charpy V-Notch Impact ft lb
F	C		0.02%-Offset Yield Strength ksi	Tensile Strength ksi	Elongation % in 1.5 in.	Reduction of Area %	Strength ksi	Reduction of Area %	
Normal Heat Treatment									
room		1.39	108.7	157.0	12.66	31.70	219.0	3.8	17, 19
room		1.37	110.5	159.5	12.00	28.69	218.0	3.8	9.5, 11, 19
1000	540	1.62	66.8	99.7	6.66*	51.99	161.5	6.0	30
1000	540	1.52	68.2	106.0	10.66	39.40	161.0	5.5	32
Abnormal Heat Treatment									
room		1.31	106.5	158.0	11.30	24.00	207.1	. .	15
room		. . .	105.1	158.5	12.00	24.00	. . .	. .	22
1000	540	1.52	72.4	100.1	6.70	29.40	152.3	. .	30
1000	540	. . .	71.6	99.6	8.00	38.60	. . .	. .	. .

*broke outside measured section

Forgings were 14 inches in diameter x ¾ inch. The "normal" heat treatment consisted of holding one hour at 2000 F (1095 C), oil quenching, tempering four hours at 1250 F (675 C) and air cooling. The "abnormal" treatment consisted of holding one hour at 2200 F (1205 C), oil quenching, tempering four hours at 1250 F (675 C) and air cooling.

Unnotched tensile tests were run on 0.375-inch-diameter proportional bars. Notched tensile tests were made on specimens having a large diameter of 0.505 inch; a notch diameter of 0.357 inch; a notch radius of 0.013 inch; a 60° V notch and a K_t of 3.4

Sakamoto

Source: *The Super 12% Cr Steels*, Climax Molybdenum Co., 1965

Banerjee, Hauser and Capenos have reported the very high fracture toughness values that can be obtained in laboratory tests on Crucible 422 hardened and tempered in this way (see Hucek's data on page 160).

Two problems arise, however, with utilization of this idea. The factors may not always be directly related to the operational strength of certain parts. For example, the Martin data on page 181 show that tempering at 600 F (315 C)—the lowest tempering temperature investigated—did not give as high bursting strengths as would have been expected and the best results were obtained by tempering at higher temperatures.

Secondly, use of a low tempering temperature limits the utility of the Super 12% Cr steels at elevated temperatures. A Class II steel tempered at perhaps 500 F (260 C) for good fracture toughness may show high tensile strengths in short-time tests at higher temperatures but the thermal stability will not be as high as that of a specimen tempered at higher temperatures.

Ausforming. Ausforming produces a considerable increase in hardness (see pages 166/168) but Brown, Thomas and Hardy have found that it may lower the resistance to crack propagation to an extent that represents an embrittlement.

Damping Capacity

Damping capacity, or internal friction, is a highly significant property in some of the present and potential applications of Super 12% Cr steels. A most notable example of this is steam-turbine buckets. The long, low-pressure buckets operating at high speeds will vibrate under some conditions because of vibrating forces set up by the steam itself. When the frequency with which steam passes coincides with the natural frequency of the buckets, the buckets will vibrate in resonance with great amplification of stresses. With present computers and calculating methods, it is possible to predict the vibratory characteristics so the bucket design can be selected or adjusted to meet operating conditions and keep the vibratory stresses within the capacity of the bucket material. Some bucket failures of austenitic steels in steam and commercial gas turbines have been attributed to the relatively low damping capacity of this material. On the other hand, the straight 12% Cr steel has a damping capacity peculiarly well adapted to steam-turbine blades. The Super 12% Cr steels may not be quite as good but still are far superior to austenitic steels.

On the basis of available data, Class I steels have roughly the same damping capacity as straight 12% Cr steel while the Class II grades fall somewhat lower. No figures on the damping capacity of Class III materials have been reported. This relative ranking is confirmed by other data. Williams, for instance, found the damping capacity of Crucible 422 was approximately half that of Type 403 at room temperature but was much higher than that of a low-alloy engineering steel, especially at torsional stresses of 20 ksi and lower.

	Specific Damping, %/cycle, at a torsional stress of		
	10	20	30 ksi
12% Cr steel (Type 403)	2.0	2.7	2.4
Class II steel (Crucible 422)	1.4	1.4	1.2
Low-alloy engineering steel (AISI 4140)	0.2	0.5	1.0

Williams

In steam-turbine work, the Super 12% Cr steels often come into consideration at temperatures where the only possible alternative would be an austenitic steel or alloy. Cochardt has demonstrated the pronounced superiority of a Class I material over an austenitic steel at these temperatures. In the light of the previously discussed tests, it can be assumed that the Class II Super 12% Cr steels would also be better than the austenitic material.

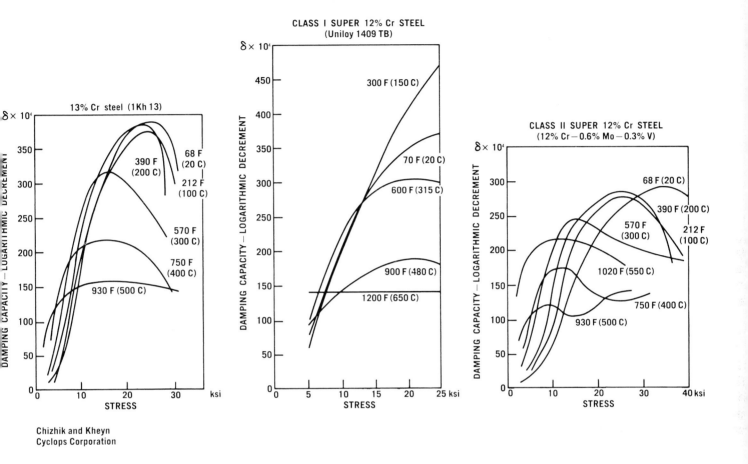

CLASS I SUPER 12% Cr STEEL
(Uniloy 1409 TB)

13% Cr steel (1Kh 13)

CLASS II SUPER 12% Cr STEEL
(12% Cr — 0.6% Mo — 0.3% V)

Chizhik and Kheyn
Cyclops Corporation

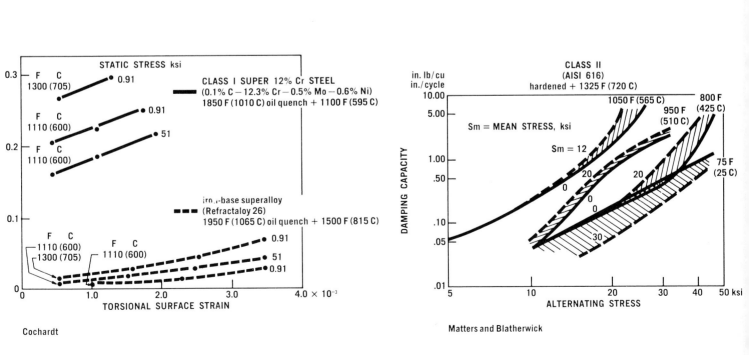

Cochardt

Matters and Blatherwick

Source: *The Super 12% Cr Steels*, Climax Molybdenum Co., 1965

Various factors affect the damping capacity. Super-imposed static stress had considerable effect on a Class I steel at 1110 F (600 C) but a smaller influence on a Class II steel at lower temperatures. Variations in damping characteristics will also result from differences in structure, which are not always spelled out clearly in test reports. Some evidence indicates that a steel with some delta ferrite present shows better damping characteristics at moderate temperatures than fully martensitic specimens of the same steel.

With plain 12% Cr steels, the damping characteristics show little change with increasing time that the specimen is exposed to fatigue conditions since the magnetomechanical component of damping is large. Pluhař found the same to be true for a Class II steel at 1110 F (600 C) but at room temperature the amount of plastic damping was appreciable and there was a significant change with time.

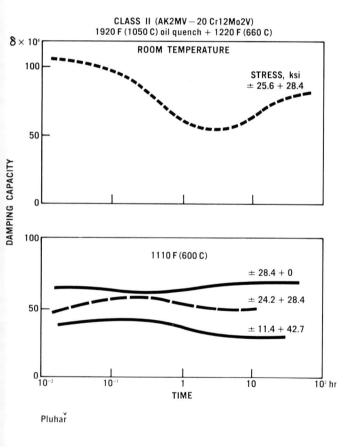

CLASS II (AK2MV — 20 Cr12Mo2V)
1920 F (1050 C) oil quench + 1220 F (660 C)

Pluhař

Friction and Wear

The fragmentary information available leads to the conclusion that the presence of free ferrite is the key factor in determining the frictional and wear properties of the Super 12% Cr steels.

Mochel of Westinghouse has repeatedly referred to the use of a Class I steel with 0.5% Mo for seal strips where one of the prime requirements is good rubbing characteristics against the rotating shaft or stationary cylinder without scoring or seizing. In addition the strips should wear rapidly if by chance they come in contact with the rotor, cylinder or blade shrouds during operation. The Class I steel has given excellent service for many years in contact with materials ranging from cast iron to austenitic steels. Since Mochel also mentions that the molybdenum leads to the formation of delta ferrite in addition to decreasing friction and producing nonseizing properties, it is reasonable to assume that the advantages of the Super 12% Cr steel in this application stem from the presence of free ferrite, which confers the wanted characteristics.

At least three journal failures have been reported on Class II steels in the heat-treated condition with, presumably, little if any free ferrite present. One of these was in a 1200 F (650 C) steam turbine at Leverkusen where premature failure occurred after only 600 hours, although—as is common with this type of bearing—the journals were running in white-metal bearings with pressurized-oil lubrication. Similar failures took place with a gas-turbine rotor during run-in tests and in a compressor wheel after 2½ years of service. Simple unlubricated wear and friction tests show that the straight 12% Cr steel as well as the Super 12% Cr steel have much greater wear and friction than carbon, low-alloy or austenitic steels. To avoid these failures, the general practice now is to use intermediate bushings of carbon or low-alloy steel between the Super 12% Cr journals and the white-metal bearings, or to hard-chromium plate the journals or to use sprayed coatings to ameliorate the bearing properties.

Erosion sometimes occurs on steam-turbine blading in the last stages of low-pressure turbines where the moisture content and the peripheral speeds are high. Cobalt-base wear-resistant inserts can be silver-soldered to the blade tips when this happens.

Where the wear resistance of Super 12% Cr steels as hardened is not adequate, parts such as valve stems and bushings are often nitrided. Newhouse, Seguin and Lape found that long holding at high temperatures decreased the room-temperature hardness of the nitrided steels but the nitrided Type 619 was still harder than a nitrided 12% Cr steel and had over Rockwell 15N 85 after the equivalent of 100,000 hours at 1050 F (565 C). In the parts in question, wear resistance would be expected to be related to hardness.

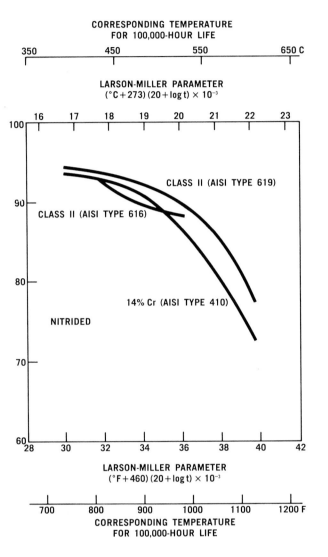

CORRESPONDING TEMPERATURE
FOR 100,000-HOUR LIFE

LARSON-MILLER PARAMETER
$(^\circ C + 273)(20 + \log t) \times 10^{-3}$

CLASS II (AISI TYPE 619)

CLASS II (AISI TYPE 616)

14% Cr (AISI TYPE 410)

NITRIDED

LARSON-MILLER PARAMETER
$(^\circ F + 460)(20 + \log t) \times 10^{-3}$

CORRESPONDING TEMPERATURE
FOR 100,000-HOUR LIFE

Newhouse, Seguin and Lape

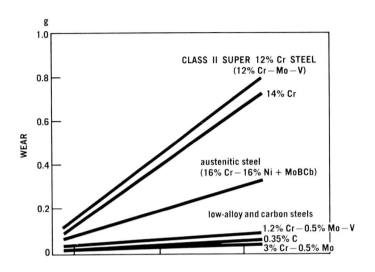

CLASS II SUPER 12% Cr STEEL
(12% Cr — Mo — V)

14% Cr

austenitic steel
(16% Cr — 16% Ni + MoBCb)

low-alloy and carbon steels

1.2% Cr — 0.5% Mo — V
0.35% C
3% Cr — 0.5% Mo

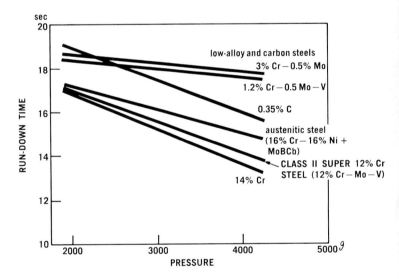

low-alloy and carbon steels
3% Cr — 0.5% Mo

1.2% Cr — 0.5 Mo — V

0.35% C

austenitic steel
(16% Cr — 16% Ni +
MoBCb)

CLASS II SUPER 12% Cr
STEEL (12% Cr — Mo — V)

14% Cr

PRESSURE

Unlubricated wear and friction test. Rods 0.2 inch
in diameter are pressed vertically against a steel
disk, which is rotated for ten seconds to give a rubbing
speed of about 30 ft/sec. Subsequently the disk
is disconnected and the run-down time of the disk
as well as the weight loss of the rods are determined.

Schinn

Source: *The Super 12% Cr Steels*, Climax Molybdenum Co., 1965

Stress-Rupture Strength

Stress-rupture strengths are used extensively in calculating design stresses, especially in the power field, since they show the stress needed to rupture the steel in a given time at a given temperature. The elongation at fracture and the stress-rupture strength determined on notched bars, which are useful indications of thermal stability and tendency towards embrittlement, will be discussed in the next chapter.

Because of the ease of making stress-rupture tests, considerable information is available on the Super 12% Cr steels. Much of the data in the following tables on the three classes are taken from catalogues and data sheets. They may therefore represent duplicate entries in cases where producers have relied on industry tests.

Long-Time Tests. Because of the straight-line log-log plots of short-time stress-rupture data, many have been tempted to extrapolate relatively short-time tests to 100,000 hours. A number of methods have been proposed for doing this more precisely. The fact remains, however, that there is no means of reliably predicting 100,000-hour strengths from short-time tests. Checking extrapolated data against actual long-time data seldom gives results more accurate than $\pm$ 25% and the variation may be as much as $\pm$ 65%, so great that the results are essentially useless.

One basic reason for the discrepancy between extrapolated and actual data is the occurrence of at least one point of inflection, believed related to precipitation phenomena. Stauffer and Keller have clearly illustrated the fallacious figures that can be obtained by extrapolation of short-time tests. Actual 100,000-hour test data on a Class I and a Class II steel show the differences in behavior that can be encountered.

Fortunately, some information is available on long-time tests extending up to about 100,000 hours (11 years) on Class I and II steels. In addition, SFAC's Larson-Miller curve on Virgo 7 includes test times up to 13,245 hours at 930 F (500 C) and 15,620 at 1020 F (550 C). Some of the data in catalogues and data sheets may belong in this category for much of this literature is vague as to the actual length of test.

Effect of Initial Strength. Various investigators have observed that room-temperature strength has a significant effect on the stress-rupture strength at lower temperatures and shorter times. This influence becomes negligible at 1110 F (600 C) and higher as well as with test times over about 50,000 hours. At considerably lower temperatures, such as those involved in supersonic airplanes, very high rupture strengths can be obtained by heat treating to high room-temperature strengths. The tempering temperatures needed to produce these high strengths, however, preclude utilization of these rupture values at higher temperatures over about 900 F (480 C).

Effect of Delta Ferrite. Data on the effect of delta ferrite are meager and conflicting. It appears clear from Loria's data that delta ferrite decreases the transverse stress-rupture strength. Loria considered that up to 18% ferrite had little or no apparent influence on longitudinal stress-rupture strength although the variation in values (from several heats) is so great as to leave this conclusion doubtful. In any case, ferrite is not necessarily deleterious as demonstrated by Thum and Richard's 100,000-hour tests on Class I steels where the molybdenum-containing steel was consistently stronger at long test times despite a much greater ferrite content. Presumably one factor would be whether the composition is such as to lead to strengthening precipitation in the ferrite during test.

REPRESENTATIVE STRESS-RUPTURE STRENGTHS OF CLASS I STEELS
100 and 1000 hours

Stress, ksi, for Rupture in

Material	Note	100 hours										1000 hours											
		930/500	975/525	1000/540	1020/550	1065/575	1100/595	1110/600	1200/650	1300/705	1400/760	900/480	930/500	975/525	1000/540	1020/550	1050/565	1065/575	1100/595	1110/600	1200/650	1290/700	1300/705
Firth-Vickers Molybdenum Stainless Steel																		26.9					
Fluginox 51													20.6					12.1					
410 + 0.5% Mo	(1)											43				14					5		
	(2)		39.9			25.1		20.2					34.9						18.8	11.6			
	(3)													26					13				2.9
410 + 1.2% Mo	(4)				15.0		9.0												11.0		5.8		
	(5)				20.0		10.0												15.0		6.8		
Jethete M153		53.8		36.0			22.6					41.4				25.3					15.0		
KP-1								21.1												14.2			
STM		46.9		32.7									42.7			27.0							
2 RO 26		55.5	41.2	30.5				19.2	10.0			45.5	32.7			24.2					12.8	6.0	
Uniloy 1409 TB	(6)			34.0		19.5		10.7		5.7	2.9					23.0			12.5		6.6		3.3
X 20 CrMo 13 1	(7)												54.0			39.1						3.7	
	(8)												56.9										
	(9)												61.2			43.4							
X 22 CrMo 14 1	(10)	46.5																					

10,000 and 100,000 hours

Stress, ksi, for Rupture in

Material	Note	10,000 hours										100,000 hours						
		900/480	930/500	975/525	1000/540	1020/550	1050/565	1065/575	1100/595	1110/600	1200/650	930/500	975/525	1000/540	1020/550	1100/595	1110/600	1200/650
Cromimphy A1007Mo												11.4 mean			5.7 mean			
D8512	(11)		39.8			25.6					10.0	29.9			17.1	5.7		
Firth-Vickers Molybdenum Stainless Steel						20.2									(14.6)			
Fluginox 51			15.6			9.2												
410 + 0.5% Mo	(1)	(42)						(11)			(3)							
	(2)			(30.5)				(12.1)		(6.7)								
410 + 1.2% Mo	(4)									(8.0)	(4.0)							
	(5)									(11.0)	4.3							
Jethete M153				(33.2)				(15.0)			(8.1)							
MTS 1	(11)		41.2			27.0					11.4	34.1			19.9	5.7		
RHM 7												28.5 min	19.9 min		14.2 min	4.3 min		
RHM 8												25.6 min			14.2 min			
Turbotherm 15 M			39.8			25.6					10.0	29.9			17.1	5.7		
Turbotherm 20 M			39.8			25.6					10.0	29.9			17.1	5.7		
2 RO 26			37.0	25.6		17.1				8.5	3.6							
Uniloy 1409 TB	(6)				16.0				8.3		4.0				10.7	5.2		2.4
Vaccutherm 5-32 and 5-32h	(11)		42.7			25.6					10.0	32.7			18.5	5.7		
X 20 CrMo 13 1	(7)		45.5			30.8												
	(8)		44.8															
	(9)		(49.0)			(32.0)												
X 22 CrMo 14 1	(10)		37.0									28.4						

( ) extrapolated
(1) 1450 F (790 C) air cool
(2) oil quench + temper
(3) 1750 F (955 C) quench + 1220 F (660 C)
(4) annealed
(5) 2280 F (1250 C) water quench + 1290 F (700 C)
(6) 1800 F (980 C) oil quench + 1000 F (540 C)

(7) for tests at 930 F (500 C) and 1020 F (550 C): 1870 F (1020 C) oil quench + 1275 F (690 C). for tests at 1290 F (700 C): 1870 F (1020 C) oil quench + 1290 F (700 C)
(8) 1800 F (980 C) oil quench + 1420/1200 F (770/650 C)
(9) 1920 F (1050 C) oil quench + 1165 F (630 C)
(10) 1785/1875 F (975/1025 C) oil quench + 1290/1380 F (700/750 C)
(11) heat treated

Source: *The Super 12% Cr Steels*, Climax Molybdenum Co., 1965

REPRESENTATIVE STRESS-RUPTURE STRENGTHS OF CLASS II STEELS
100 hours

		Stress, ksi, for Rupture in 100 hours											
		750 400	800 425	840 450	900 480	930 500	1000 540	1020 550	1100 595	1110 600	1200 650	1290 700	1300 F 705 C
AC-254	(1)	. . .	. . .	. . .	. . .	. . .	74.0	. . .	50.0	. . .	27.0	. . .	14.0
Carpenter 636 Alloy		. . .	. . .	. . .	. . .	. . .	63.0	. . .	46.0	. . .	25.0	. . .	. . .
Crucible 422	(2)	. . .	. . .	. . .	. . .	. . .	64/67	. . .	42/49	. . .	24/28	. . .	. . .
	(3)	. . .	235	. . .	. . .	. . .	. . .					. . .	. . .
	(4)	. . .	. . .	. . .	130	. . .						. . .	. . .
	(5)	. . .	210	. . .	. . .	. . .	68.0					. . .	. . .
	(6)	. . .	. . .	. . .	. . .	. . .					21.0	. . .	. . .
Firth-Vickers 448(B)		. . .	. . .	. . .	. . .	. . .		58/60		43/45	29/31		
Firth-Vickers 448(E)		. . .	. . .	. . .	. . .	. . .		67.2		42.6	31.4		
Fluginox 62		. . .	. . .	. . .	. . .	. . .		45.5		32.7			
Fluginox 65		. . .	. . .	. . .	. . .	. . .		58.3		42.7	22.8		
410 + 0.14% Cb + 0.7% Mo + 0.7% V	(7)	. . .	. . .	. . .	. . .	. . .	. . .	. . .	. . .	. . .	19.0	. . .	. . .
	(8)	. . .	. . .	. . .	. . .	. . .	. . .	. . .	. . .	. . .	21.0	. . .	. . .
410 + 0.6% Mo + 0.7% V	(7)	. . .	. . .	. . .	. . .	. . .	. . .	. . .	. . .	. . .	18.7	. . .	. . .
	(8)	. . .	. . .	. . .	. . .	. . .	. . .	. . .	. . .	. . .	21.0	. . .	. . .
410 + 1% Mo + 0.7% V	(7)	. . .	. . .	. . .	. . .	. . .	. . .	. . .	. . .	. . .	20.5	. . .	. . .
	(8)	. . .	. . .	. . .	. . .	. . .	. . .	. . .	. . .	. . .	23.0	. . .	. . .
422 M		. . .	. . .	. . .	. . .	. . .	86.0	. . .	71.0	. . .	44.0	. . .	. . .
G-X 20 CrMoVNb 12	(9)	. . .	. . .	. . .	. . .	. . .	. . .	. . .	. . .	34.1		. . .	. . .
Hecla HGT4		. . .	. . .	. . .	. . .	. . .					45.3	36.5	21.3
Jessop-Saville H46		105.3		. . .	. . .	77.3	. . .	60.5	. . .	45.9	30.9	19.7	
Jessop-Saville H59		. . .	. . .	(97.4)	. . .	77.3		53.8		33.6	. . .	. . .	
Jethete M151		98.1	. . .	. . .	. . .	58.2	. . .	. . .		22.4	. . .	. . .	
Jethete M152		. . .	. . .	. . .	. . .	81.8		53.8		37.0	. . .	. . .	
Jethete M154		. . .	. . .	. . .	. . .	80.6		. . .		33.6	. . .		
Jethete M160		. . .	. . .	. . .	. . .	78.4	. . .	61.6	. . .	44.8	31.4	. . .	
Lapelloy	(10)	. . .	. . .	. . .	. . .	. . .	65.0	. . .	45.0	. . .	25.0	. . .	
Lapelloy "C"		. . .	. . .	. . .	. . .	. . .	70.0	. . .	45.0	. . .	24.0	. . .	
Linco		. . .	. . .	. . .	. . .	. . .	69.0	. . .	49.0	. . .	27.0	. . .	
Mel-Trol H-46	(11)	. . .	. . .	. . .	. . .	. . .	75.0	. . .	55.0	. . .	35.0	. . .	
MTS 6	(12)	. . .	. . .	. . .	. . .	78.2	. . .	59.7	. . .	45.5	31.3		
Special Genco		. . .	. . .	. . .	. . .	. . .	63.0	. . .	46.0	. . .	25.0		
Uniloy 1420WM	(13)	. . .	. . .	. . .	. . .	. . .	63.0	. . .	46.0	. . .	25.0		
Uniloy 1430MV	(10)	. . .	. . .	. . .	80.0	. . .	65.0	. . .	49.5	. . .	26.5	. . .	
USS-12MoV	(14)	. . .	229.0	. . .	. . .	. . .	. . .				. . .	. . .	
	(15)	. . .	220.0	. . .	. . .	. . .					. . .	. . .	
	(16)	. . .	220.0	. . .	125.0	. . .	. . .				. . .	. . .	
Vaccutherm 5-40		. . .	. . .	. . .	. . .	. . .	. . .	59.7	. . .	45.5	29.9	. . .	
X 15 CrMoWV 12 1	(17)	. . .	. . .	. . .	. . .	. . .	. . .	42.7	. . .	. . .	. . .	. . .	. . .
	(18)	. . .	. . .	. . .	. . .	. . .	55.5		. . .	. . .	. . .	. . .	. . .

1000 hours

		Stress, ksi, for Rupture in 1000 hours													
		700 370	750 400	800 425	840 450	900 480	930 500	1000 540	1020 550	1040 560	1050 565	1065 575	1100 595	1110 600	1200 650	1290 F 700 C
AC-254	(1)	...	...	...	...	...	...	60.0					39.0	...	18.0	...
Carpenter 636 Alloy		...	...	...	...	...	...	57.0					37.0	...	15.0	...
Cromimphy A3200		...	...	...	...	...	...		39.8 min		33.4 min			27.0	...	...
Crucible 422	(2)	...	...	...	...	...	...	55/62					35/39		14/17	
	(3)	...	...	190	...	...	...									...
	(4)	...	...	...	89	...	...									...
	(5)	...	...	...	...	...	...	58.0								...
	(6)	...	...	...	...	...	...								14.0	...
EI 747	(19)	...					...		35.6	...	...		...			
EI 993	(20)	...								44.1	...			41.2		
	(21)	...									49.8					
56 TS		...	...						49.8		...	...		29.9	15.6	
Firth-Vickers Molybdenum Vanadium Stainless Steel		...	...	...	...		53.8	...	35.8		...	28.0	...	21.3	...	...
Firth-Vickers 448(B)		...	...	...					54/56					36/38	20.2	
Firth-Vickers 448(E)		...							58.2					35.8	22.4	
Fluginox 62		...							39.8					24.2	...	
Fluginox 65		...							51.2					31.3	15.6	
410 + 0.14% Cb + 0.7% Mo + 0.7% V	(7)	...													14.0	...
	(8)	...													14.5	...
410 + 0.6% Mo + 0.7% V	(7)	...													13.5	...
	(8)	...													15.0	...
422 M		...				89.0			75.0					49.0	...	29.0	
G-X 20 CrMoVNb 12	(9)	...												27.1		
G-X 20 CrMoWV 12 1	(22)	...							39.1					22.5	...	...
Hecla HGT4														37.0	24.0	10.1
Jessop-Saville H46		...	96.3				71.7		51.5					38.1	22.0	11.2
Jessop-Saville H59					92.0		72.8		...						...	
Jethete M151		...	89.6				50.4		34.0					14.6		
Jethete M152							76.2		33.6					24.6		
Jethete M154							78.4		...					24.6	...	
Jethete M160							72.8		54.9					37.0	21.3	
Lapelloy	(10)							41/68					22/42	...	12/20	
Lapelloy "C"	(23)							55.0					35.0	...	13.0	
	(24)							60.0					32.0	...	10.0	
Linco								60.0					35.0	...	17.0	
Mel-Trol H-46	(11)							65.0					43.0	...	25.0	
MTS 6	(12)						71.1		51.2					38.4	22.8	
1Kh12MF									38.4					...		
1Kh12VMF									40/41					...		
1Kh12V1MFT														26.3		
Soleil T1	(25)								49.8					29.9	15.6	
Soleil T2	(8)								39.8					27.0	...	
Special Genco								55.0					34.0	...	15.0	
Uniloy 1420WM	(13)							57.0					37.0	...	17.0	
Uniloy 1430MV	(10)					72.0		57.5					38.0	...	15.0	
USS-12MoV	(14)	...		180.0									...	...	...	...
	(26)	225.0+	...										...	...	...	...
	(15)	235.0+	...	170.0		...							...	...	...	...
	(16)		...	190.0	...	82.0							...	...	...	...
Vaccutherm 5-40									51.2					37.0	21.3	...
Virgo 8									49.8					29.9	15.6	
X 15 CrMoWV 12 1	(17)	...							32.7					17.8	...	
	(18)	...							42.7					24.2	...	
X 20 CrMoWV 12 1	(27)								38.4					25.6	...	
X 20 CrMo(W)V 12 1	(28)						46/57		34/38					21/31		
Z07CDWVB														28.4	11.4	
Z15CDNbV 12									51.2					29.9	...	
Z15CDWV 12		...	...	...	...				51.2					29.9	...	

Source: *The Super 12% Cr Steels*, Climax Molybdenum Co., 1965

		Stress, ksi, for Rupture in 10,000 hours													
		750 400	800 425	900 480	930 500	950 510	1000 540	1020 550	1040 560	1050 565	1065 575	1100 595	1110 600	1200 650	1300 F 705 C
AC-254	(1)	...	...	...	...	...	(48)	...	...	...	...	(30)	...	(10.5)	...
Carpenter 636 Alloy		...	...	...	...	...	49.0	...	...	...	...	23.0	...	...	...
Cromimphy A3200		...	...	...	...	...	...	33.4	...	...	...	...	15.6	...	...
Crucible 422	(2)	...	...	58/75	...	...	34/50	...	...	...	...	17/25	...	...	...
	(29)	...	84.0	...	...	46.0	...	...	...	31.0	...	...	...	...	...
D8514	(30)	...	...	...	39.8	...	...	28.4	...	...	...	...	11.4	...	...
	(31)	...	...	...	52.6	...	...	35.6	...	...	...	...	17.1	...	...
D8518	(32)	...	...	...	44.1	...	...	31.3	...	...	...	...	12.8	...	...
D8518 W	(32)	...	...	...	44.1	...	...	31.3	...	...	...	...	12.8	...	...
EI 747	(19)	...	...	...	...	...	...	28.4	...	...	...	...	...	...	...
EI 993	(20)	...	...	...	...	...	...	...	35.6	...	...	...	32.0	...	...
	(21)	...	...	...	...	...	...	...	...	...	38.4	...	...	...	...
56 TS		...	...	...	...	...	...	39.8	...	...	...	...	21.3	11.4	...
Firth-Vickers Molybdenum Vanadium Stainless Steel		...	...	...	47.0	...	...	29.1	...	...	17.9	...	13.4	...	...
Firth-Vickers 448(B)		...	...	...	...	...	...	33.6	...	...	...	...	14.6	7.8	...
Firth-Vickers 448(E)		...	...	...	...	...	...	42.6	...	...	...	...	20.2	12.3	...
Fluginox 62		...	...	...	...	...	...	31.3	...	...	...	...	15.6	...	...
Fluginox 65		...	...	...	...	...	...	39.8	...	...	...	...	21.3	11.4	...
G-X 20 CrMoVNb 12	(9)	...	...	...	...	...	...	...	...	...	...	...	12.8	...	...
G-X 20 CrMoWV 12 1	(22)	...	...	...	...	...	...	30.5	...	...	...	...	13.2	...	...
Hecla HGT4		...	...	...	...	...	...	...	...	...	...	...	30.7	15.7	4.9
HT9		...	...	...	40.0	...	...	28.0	...	...	...	...	14.0	...	...
Inoxesco 13		...	...	...	...	...	...	...	...	...	...	...	12.7	6.7	3.6
Jessop-Saville H46		...	...	...	...	...	...	44.8	...	...	...	...	17.2	7.8	...
Jethete M151		85.1	...	...	47.0	...	...	24.2	...	...	...	...	8.5	...	...
Jethete M152		...	...	...	69.4	...	...	21.7	...	...	...	...	16.1	...	...
Jethete M154		...	...	...	...	...	...	...	...	...	...	...	15.7	...	...
Jethete M160		...	...	...	67.2	...	...	37.0	...	...	...	...	16.8	8.5	...
Lapelloy	(10)	...	...	...	...	...	42.0	...	...	...	...	22.0	...	9.0	...
Lapelloy "C"		...	...	...	...	...	(42)	...	...	...	...	(22)	...	...	...
Mel-Trol H-46	(11)	...	...	...	...	...	49.0	...	...	...	...	22.0	...	...	...
MTS 5	(12)	...	...	...	44.1	...	...	29.9	...	...	...	...	14.2	...	...
1Kh12MF		...	...	...	...	...	...	32.0	...	...	...	...	...	...	...
1Kh12VMF		...	...	...	...	...	...	35.6	...	...	...	...	...	...	...
1Kh12V1MFT		...	...	...	...	...	...	...	...	...	...	...	22.8	...	...
Soleil T1	(25)	...	...	...	...	...	...	39.8	...	...	...	...	21.3	11.4	...
Soleil T2	(8)	...	...	...	...	...	...	33.4	...	...	...	...	15.6	...	...
Turbotherm 20MV		...	...	...	49.8	...	...	32.0	...	...	...	...	17.1	...	...
Turbotherm 20MVNb		...	...	...	55.5	...	...	38.4	...	...	...	...	21.3	...	...
Turbotherm 20MVW	(33)	...	...	...	49.8	...	...	32.0	...	...	...	...	17.1	...	...
	(34)	...	...	...	42.7	...	...	28.4	...	...	...	...	15.6	...	...
Uniloy 1420WM	(13)	...	...	...	...	...	49.0	...	...	...	...	23.0	...	...	...
Uniloy 1430MV	(10)	...	...	65.0	...	...	48.0	...	...	...	...	23.3	...	...	...
Vaccutherm 5-34 and 5-36	(12)	...	...	...	45.5	...	...	29.9	...	...	...	...	14.2	...	...
Virgo 8		...	...	...	...	...	...	39.8	...	...	...	...	21.3	11.4	...
X 08 CrMoWV 14 1		...	...	...	...	...	...	29.9	...	...	...	...	...	...	...
X 15 CrMoVNb 12	(35)	...	...	...	...	...	...	...	...	...	...	...	14.2	...	...
	(36)	...	...	...	...	...	...	...	...	...	...	...	18.5	...	...
X 15 CrMoWV 12 1	(17)	...	...	...	...	...	...	24.2	...	...	...	...	10.0	...	...
	(18)	...	...	...	...	...	...	31.3	...	...	...	...	10.0	...	...
	(37)	...	...	...	...	...	...	29.9	...	...	...	...	(10.0)	...	...
X 20 CrMoWV 12 1	(33)	...	...	...	...	...	...	...	...	...	...	...	8.5	...	...
	(27)	...	...	...	...	...	...	29.9	...	...	...	...	15.6	...	...
	(38)	...	...	...	46.2	...	...	32.7	...	...	...	...	20.6	10.7	...
X 20 CrMo(W)V 12 1	(28)	...	...	...	39/46	...	...	27/34	...	...	...	...	14/18	...	...
X 22 CrMoWV 12 1		...	...	...	44.1	...	...	28.5	...	...	...	...	...	...	...
X 23 CrNiMoVNbW 12 1		...	...	...	...	...	...	21.0	...	...	...	...	...	...	...
Z07CDWVB		...	...	...	54.0	...	...	37.0	...	...	...	...	17.1	7.1	...
Z15CDNbV 12		...	...	...	...	...	...	39.8	...	...	...	...	19.9	...	...
Z15CDWV 12		...	...	...	...	...	...	39.8	...	...	...	...	19.9	...	...
Z20CDNbV 11	(39)	...	...	...	...	...	...	42.7	...	...	...	...	...	10.7	...

		Stress, ksi, for Rupture in 100,000 hours										
		900 480	930 500	975 525	1000 540	1020 550	1050 565	1065 575	1100 595	1110 600	1140 615	1200 F 650 C
Cromimphy A2200		...	...	...	...	27.0	...	20.6	...	15.6	13.2	...
Cromimphy A3100		...	...	...	...	14.2	...	...	...	...	...	...
Cromimphy A3200		...	...	...	...	24.2 min	20.6 min	...	...	...	...	...
D8514	(30)	...	35.6	...	...	19.9	...	...	...	6.4	...	...
	(31)	...	42.7	...	...	25.6	...	...	...	8.5	...	...
D8518	(32)	...	38.4	...	...	24.2	...	...	...	7.1	...	...
D8518 W	(32)	...	38.4	...	...	24.2	...	...	...	7.1	...	...
Firth-Vickers Molybdenum Vanadium Stainless Steel		...	...	...	...	(21)	...	(11)	...	(8)	...	...
Fluginox 62		...	...	...	...	24.2	...	...	...	10.0	...	...
Fluginox 65		...	...	...	...	31.3	...	...	...	15.6	...	8.5
GE	(40)	56.0	...	...	37.0	...	28.0	...	18.0	...	...	...
	(41)	57.0	...	...	36.0	...	25.0	...	...	...	...	...
HT9		...	33.0	...	...	21.0			...	9.0	...	...
Inoxesco 13		...	...	...	...	...	...	...	...	8.8	...	4.5
Jessop-Saville H46		...	...	...	...	...	...	...	...	(9.0)	...	...
MTS 5	(12)	...	37.0	...	...	22.8	...	...	...	7.1	...	...
1Kh12MF		...	...	...	...	27.0	...	...	...	...	...	...
1Kh12VMF		...	...	...	...	31.3	...	...	...	...	...	...
1Kh12V1MFT		...	...	...	...	...	...	...	...	19.9	...	...
RHMV4		...	29.9 min	23.5 min	...	17.8 min	...	...	...	6.4 min	...	...
Soleil T1	(25)	...	...	...	...	27.0	...	18.5	...	12.8	...	...
	(42)	...	...	...	...	31.3	...	21.3	...	14.7	...	...
Soleil T2	(8)	...	...	...	...	18.5	...	12.1	...	8.5	...	...
T58		...	...	...	...	28.4	...	...	...	12.2	...	...
Turbotherm 20MV		...	37.0	...	...	22.0	...	...	...	7.8	...	...
Turbotherm 20MVNb		...	45.5	...	...	28.4	...	...	...	11.4	...	...
Turbotherm 20MVW	(33)	...	37.0	...	...	22.0	...	...	...	7.8	...	...
	(34)	...	34.1	...	...	20.6	...	...	...	7.8	...	...
Vaccutherm 5-34 and 5-36	(12)	...	37.0	...	...	22.8	...	...	...	7.1	...	...
Virgo 7		...	34.1 min	25.6 min	...	18.2 min	...	11.7 min	...	6.4 min	...	...
Virgo 8		...	39.8 min	31.8 min	...	24.3 min	...	17.6 min	...	11.9 min	...	...
X 15 CrMoWV 12 1	(17)	...	...	...	...	(21.3)	...	...	...	(7.1)	...	...
	(37)	...	...	...	...	21.3	...	...	...	15.6	...	...
X 20 CrMoV 11	(43)	...	...	...	...	24.2	...	...	...	...	...	...
X 20 CrMoV(W) 12	(38)	...	38.4	...	...	25.0	...	...	...	12.9	...	7.1
	(44)	...	...	...	...	24.2	...	...	...	...	...	...
X 20 CrMo(W)V 12 1	(28)	...	31/37	...	...	21/23	...	...	...	6.8	...	...
X 22 CrMoWV 12 1		...	31.3	...	...	17.1	...	...	...	...	...	...
Z15CDNbV 12		...	...	...	...	27.0	...	...	...	12.8	...	...
Z15CDWV 12		...	...	...	...	27.0	...	...	...	12.8	...	...

( ) extrapolated
(1) 2100 F (1150 C) quench + 1350 F (730 C)
(2) bar stock quenched + 1200 F (650 C)
(3) sheet, 0.025-in. thick, quenched + 800 F (425 C)
(4) sheet, 0.025-in. thick, quenched + 900 F (480 C)
(5) sheet, 0.025-in. thick, quenched + 1000 F (540 C)
(6) sheet, 0.025-in. thick, quenched + 1200 F (650 C)
(7) as cast
(8) cast + heat treated
(9) cast + heat treated to a minimum tensile strength of 151 ksi
(10) 2000 F (1095 C) oil quench + 1300 F (705 C)
(11) 2100 F (1150 C) air cool + 1200 F (650 C)
(12) heat treated
(13) 1900 F (1040 C) oil quench + 1200 F (650 C)
(14) sheet, 0.050-in. thick, 1850 F (1010 C) quench + 900 F (480 C)
(15) sheet, 0.10-in. thick, 1850 F (1010 C) quench + 800 F (425 C)
(16) sheet, 0.10-in. thick, 1850 F (1010 C) quench + 900 F (480 C)
(17) 1920 F (1050 C) oil quench + 1290 F (700 C) + 1380/1435 F (750/780 C)
(18) 1920 F (1050 C) oil quench + 1290 F (700 C)
(19) 1920 F (1050 C) quench + 1335 F (725 C)
(20) 2100 F (1150 C) quench + 1200 F (650 C)
(21) 2100 F (1150 C) quench + 1290 F (700 C)
(22) cast + 1920 F (1050 C) air cool + 1365 F (740 C)
(23) 1900 F (1040 C) + 1200 F (650 C)
(24) 1900 F (1040 C) + 1050 F (565 C)

(25) heat treated to a mean tensile strength of 121 ksi
(26) sheet, 0.10-in. thick, 1850 F (1010 C) quench + 700 F (370 C)
(27) 1920 F (1050 C) oil quench + 1365 F (740 C)
(28) range applies to material falling within three tensile-strength ranges: 92/121, 121/142 and over 142 ksi
(29) bar stock quenched + 1325 F (720 C)
(30) 1870/1920 F (1020/1050 C) oil quench + 1290/1380 F (700/750 C)
(31) 2010/2100 F (1100/1150 C) oil quench + 1200/1290 F (650/700 C)
(32) 1870/1940 F (1020/1060 C) oil quench + 1240/1330 F (670/720 C)
(33) heat treated to a tensile strength of 114/135 ksi
(34) heat treated to a tensile strength of 100/121 ksi
(35) 2100 F (1150 C) oil quench + 1265 F (685 C)
(36) 2100 F (1150 C) oil quench + 1265 F (685 C) + 1380 F (750 C)
(37) 2100 F (1150 C) oil quench + 1290 F (700 C)
(38) 2050 F (1120 C) oil quench + 1290 F (700 C)
(39) 2010/2100 F (1100/1150 C) oil quench + 1265/1290 F (675/700 C)
(40) surface radial properties of 45-in.-OD forging, 40,000 pounds, 1925 F (1050 C) oil quench + double tempered at 1050 F (565 C) + 1150 F (620 C)
(41) transverse bore properties of 45-in.-OD forging, 40,000 pounds, 1925 F (1050 C) oil quench + double tempered at 1050 F (565 C) + 1150 F (620 C)
(42) heat treated to a mean tensile strength of 142 ksi
(43) heat treated to a martensitic structure and a tensile strength of about 128 ksi
(44) heat treated to a tensile strength of about 111 ksi

Source: *The Super 12% Cr Steels*, Climax Molybdenum Co., 1965

REPRESENTATIVE STRESS-RUPTURE STRENGTHS OF CLASS III STEELS

Stress, ksi, for Rupture in

		100 hours								1000 hours								10,000 hours				100,000 hours
		930 500	1000 540	1020 550	1050 565	1075 580	1100 595	1110 600	1200 650	930 500	1000 540	1020 550	1050 565	1075 580	1100 595	1110 600	1200 650	1000 540	1020 550	1075 580	1110 600	1075 F 580 C
AFC-77 (1)		...	160	...	...	...	90	...	33	...	...	...	...	...	...	...	...	...	...	...	...	... ...
(2)		...	102	...	76	...	...	...	...	...	80	...	56	...	...	...	...	...	...	...	...	... ...
Firth-Vickers 535		...	...	...	...	...	...	...	...	...	...	71.7	...	...	...	38.1	...	...	(47)	...	...	... ...
H58		...	...	...	...	...	80.6	...	...	...	...	...	...	...	...	68.3	...	...	...	...	...	... ...
Jethete M210		107.6	...	71.7	...	...	47.0	...	...	105.3	...	61.6	...	...	...	34.7	...	...	...	...	...	... ...
Pyromet X-12		...	75	...	...	...	45	...	30	...	65	...	...	...	(35)	...	(20)	(46)	...	...	...	... ...
2Cr120Mo8Co (3)		...	...	...	51/58	...	...	...	...	...	...	...	46/57	...	...	...	...	...	...	43/50	...	(36/51)
(4)		...	83	...	...	...	...	...	...	...	57	...	...	...	...	...	...	...	37	...	...	... ...
(5)		...	46	...	...	...	...	...	...	...	39	...	...	...	...	...	...	...	30	...	...	... ...
(6)		...	43	...	...	...	...	...	...	...	28	...	...	...	...	...	...	...	...	...	...	... ...
Vacumelt H53 (7)		...	...	82.4	...	...	...	53.8	31.4	...	...	68.3	...	...	...	38.1	24.6	...	...	...	...	... ...

(1) quench + 1100 F (595 C)
(2) 1900 F (1040 C) oil quench + −100 F (−73 C) + 1400 F (760 C)
(3) cast
(4) cast + 2010 F (1100 C) quench + 1330 F (720 C)
(5) cast + 2010 F (1100 C) quench + 1150 F (620 C)
(6) cast + 2190 F (1200 C) quench + tempered to a tensile strength of 128 ksi
(7) 2140 F (1170 C) oil quench + 1130 F (610 C) + 1185 F (640 C)

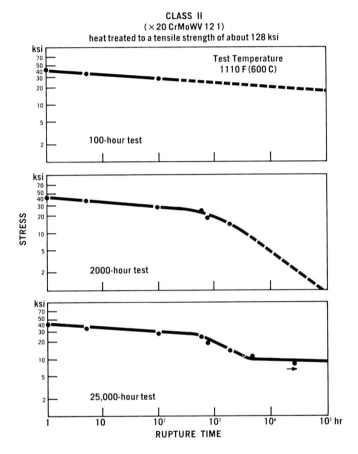

CLASS II
(×20 CrMoWV 12 1)
heat treated to a tensile strength of about 128 ksi

Test Temperature
1110 F (600 C)

100-hour test

2000-hour test

25,000-hour test

Stauffer and Keller

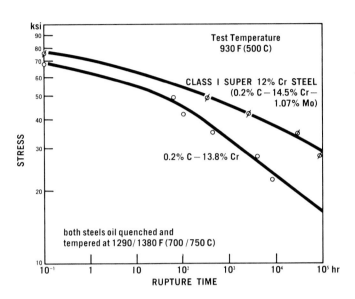

Test Temperature
930 F (500 C)

CLASS I SUPER 12% Cr STEEL
(0.2% C — 14.5% Cr —
1.07% Mo)

0.2% C — 13.8% Cr

both steels oil quenched and
tempered at 1290/1380 F (700/750 C)

Thum and Richard

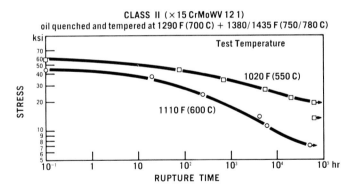

CLASS II (×15 CrMoWV 12 1)
oil quenched and tempered at 1290 F (700 C) + 1380/1435 F (750/780 C)

Test Temperature

1020 F (550 C)

1110 F (600 C)

Bennewitz

Source: *The Super 12% Cr Steels*, Climax Molybdenum Co., 1965

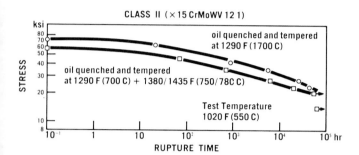

CLASS II (×15 CrMoWV 12 1)

oil quenched and tempered
at 1290 F (1700 C)

oil quenched and tempered
at 1290 F (700 C) + 1380/1435 F (750/78C C)

Test Temperature
1020 F (550 C)

STRESS

RUPTURE TIME

Bennewitz

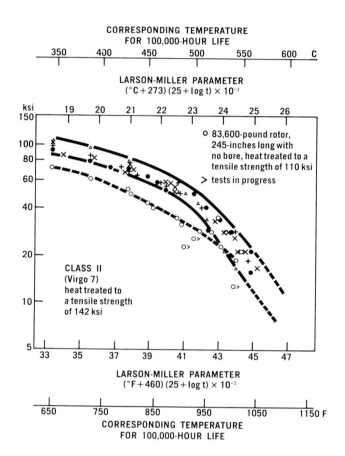

CORRESPONDING TEMPERATURE
FOR 100,000-HOUR LIFE

LARSON-MILLER PARAMETER
(°C + 273) (25 + log t) × 10⁻³

○ 83,600-pound rotor,
245-inches long with
no bore, heat treated to a
tensile strength of 110 ksi
> tests in progress

CLASS II
(Virgo 7)
heat treated to
a tensile strength
of 142 ksi

LARSON-MILLER PARAMETER
(°F + 460) (25 + log t) × 10⁻³

CORRESPONDING TEMPERATURE
FOR 100,000-HOUR LIFE

Société des Forges et
Ateliers du Creusot
(Usines Schneider)

148

Stress-Rupture Data on Some Class I Steels Based on Test Times over 10,000 Hours

		Stress, ksi, for Rupture in						
		10,000			25,000	30,000	50,000	100,000 hr
		930 / 500	1020 / 550	1200 / 650	930 / 500	1020 / 550	930 / 500	930 F / 500 C
Firth-Vickers Molybdenum Stainless Steel 410 + 1.2% Mo	(1)	...	20.2	4.3	...	16.8	...	...
X 20 CrMo 13 1	(2)	45.5	30.8	...	42.7	...	...	...
	(3)	44.8	...	...	40.6	...	37.7	...
X 22 CrMo 14 1	(4)	37.0	...	...	...	...	...	28.4

(1) 2280 F (1250 C) water quench + 1290 F (700 C)
(2) 1870 F (1020 C) oil quench + 1275 F (690 C)
(3) 1800 F (980 C) oil quench + 1420/1200 F (770/650 C)
(4) 1785/1875 F (975/1025 C) oil quench + 1290/1380 F (700/750 C)

Fabritius
Firth-Vickers Stainless Steels Ltd
General Electric Company as quoted in
Metals and Alloys 18 (1943) 61
Thum and Richard

Stress-Rupture Data on a Class III Steel Based on Test Times over 10,000 Hours

		Stress, ksi, for Rupture in 10,000 hours at 1020 F (550 C)
2Cr120Mo8Co	(1)	37.0
	(2)	30.0

(1) cast + 2010 F (1100 C) quench + 1330 F (720 C)
(2) cast + 2010 F (1100 C) quench + 1150 F (620 C)

Schneider

Stress-Rupture Data on Some Class II Steels Based on Test Times over 10,000 Hours

		Stress, ksi, for Rupture in										
		10,000						25,000		30,000		50,000 hours
		800 / 425	930 / 500	1020 / 550	1065 / 575	1110 / 600	1200 / 650	1020 / 550	1110 / 600	1020 / 550	1110 / 600	1020 F / 550 C
Crucible 422	(1)	84.0	...	...	...	...	...	...	...	...	...	...
E! 747	(2)	...	...	28.4	...	...	...	...	...	...	...	...
Firth-Vickers Molybdenum Vanadium Stainless Steel		...	47.0	29.1	17.9	13.4	...	...	...	25.8	10.1	...
Firth-Vickers 448 (B)		...	...	33.6	...	14.6	7.8	...	...	...	...	...
Firth-Vickers 448 (E)		...	...	42.6	...	20.2	12.3	...	...	...	...	...
G-X 20 CrMoVNb 12	(3)	...	...	...	...	12.8	...	...	11.4	...	...	...
G-X 20 CrMoWV 12 1	(4)	...	...	30.5	...	13.2	...	24.9	10.5	...	...	...
Jessop-Saville H46		...	...	44.8	...	17.2	7.8	...	...	...	12.3	...
Jethete M151		...	47.0	24.2	...	8.5	...	...	...	21.1	6.9	...
X 15 CrMoVNb 12	(5)	...	...	...	...	14.2	...	...	...	...	...	...
	(6)	...	...	...	...	18.5	...	...	...	...	...	...
X 15 CrMoWV 12 1	(7)	...	...	31.3	...	10.0	...	...	...	...	...	22.7
	(8)	...	...	24.2	...	10.0	...	...	...	...	...	17.1
X 20 CrMoNiV (W,Nb) 12	(9)	...	...	27.0 (mean)	...	12.8 (mean)	...	...	...	...	...	...
	(10)	...	...	28.4 (mean)	...	13.5 (mean)	...	...	...	...	...	...
	(11)	...	...	35.6 (mean)	...	15.6 (mean)	...	...	...	...	...	...
X 20 CrMoWV 12 1	(10)	...	...	...	...	8.5	...	...	7.8	...	...	...
	(12)	...	...	29.9	...	15.6	...	...	12.2	...	...	...
	(13)	...	...	32.7	...	20.6	10.7	...	...	...	...	...
Z20CDNbV 11	(14)	...	...	42.7	...	10.7	...	...	...	...	...	...

(1) quench + 1325 F (720 C)
(2) 1920 F (1050 C) quench + 1335 F (725 C)
(3) cast + heat treated to a tensile strength of 151 ksi
(4) cast + 1920 F (1050 C) air cool + 1365 F (740 C)
(5) 2100 F (1150 C) oil quench + 1265 F (685 C)
(6) 2100 F (1150 C) oil quench + 1265 F (685 C) + 1380 F (750 C)
(7) 1920 F (1050 C) quench + 1290 F (700 C)
(8) 1920 F (1050 C) quench + 1290 F (700 C) + 1380/1435 F (750/780 C)
(9) heat treated to a tensile strength of 92/114 ksi
(10) heat treated to a tensile strength of 114/135 ksi
(11) heat treated to a tensile strength of 135/164 ksi
(12) 1920 F (1050 C) oil quench + 1365 F (740 C)
(13) 2050 F (1120 C) oil quench + 1290 F (700 C)
(14) 2010/2100 F (1100/1150 C) oil quench + 1265/1290 F (675/700 C)

Bennewitz
Chizhik and Kheyn
Constant
Fabritius
Firth-Vickers Stainless Steels Ltd
Samuel Fox & Co Ltd
Jessop-Saville Ltd
Matters and Blatherwick
Schinn
Simmons and Cross
Stauffer and Keller
Zschokke, Stauffer and Felix

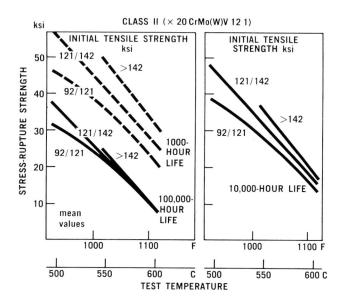

Schinn as quoted by Fabritius

CLASS II STEELS (for composition range, see page 99)
INITIAL ROOM-TEMPERATURE TENSILE STRENGTH, ksi

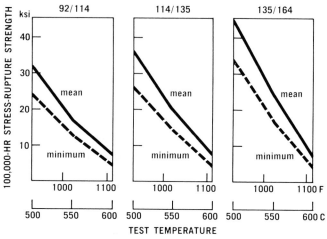

Zschokke, Stauffer and Felix

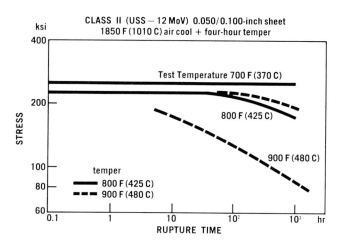

USS Corporation

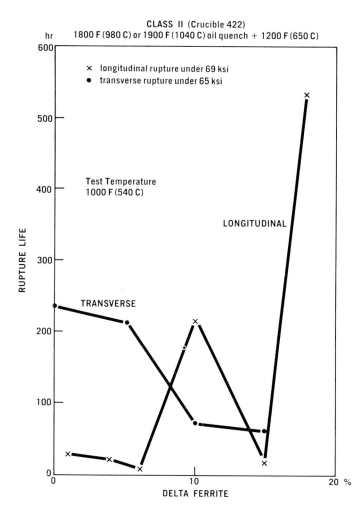

150

Loria

Reproducibility. Available information indicates good reproducibility of stress-rupture strength from heat to heat. Crucible's rupture results on more than 100 heats of Crucible 422 give a reasonably narrow band and Samuel Fox's figures on the stress-rupture strength of three heats of Jethete M151 are in close agreement.

Of even greater interest is a Swiss statistical evaluation of 380 tests carried out at 1020 F (550 C) and 1110 F (600 C) by Brown Boveri, Escher Wyss and Sulzer Brothers on a number of Class II steels within the following chemical range:

% C	0.16/0.26	% Mo	0.6/1.5
% Cb	0/0.7	% Ni	0.2/1.2
% Cr	10.3/13.2	% V	0.2/0.7
		% W	0/0.8

The results were evaluated on the basis of minimum and mean stress-rupture strengths; particular weight is placed on the former as ideally it would be the best base for design stresses. Initial strength had a pronounced effect on the lower temperature values but once this was taken into account, the data show amazingly good reproducibility, especially in view of the varying composition and the fact that the tests were made on turbine blades, bars and forgings from Austrian, British, French, German and Swedish steelworks.

Comparative Strengths. With the information in the preceding tables, it is possible to compare the three classes of Super 12% Cr steels. Despite a little overlap, strength in the temperature range of 900/1100 F (480/595 C) increases from Class I to Class II to Class III if the values on the very-low-cobalt cast Class III steel, 2 Cr120Mo8Co, are ignored. At lower and higher temperatures, the differences among the classes are not as clean cut. As expected, however, these steels are superior to the straight 12% Cr steel.

In respect to stress-rupture strength, Class II steels fall almost half way between the low-alloy chromium-molybdenum and chromium-molybdenum-vanadium steels and the high-temperature austenitic steels. This is illustrated by data on high-temperature bolting materials and on steels for use in high-temperature high-pressure hydrogen.

A rating of steels coming into consideration for application in sheet form for aircraft use demonstrates that Crucible 422 is comparable to AM 350 and considerably stronger than 17-7 PH.

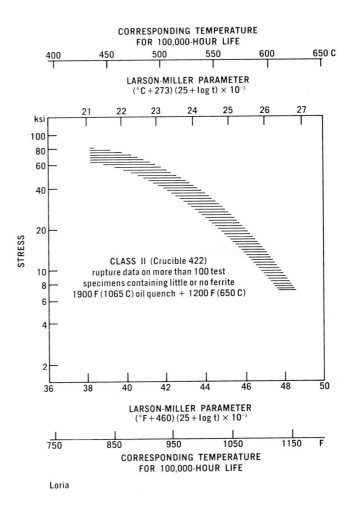

Loria

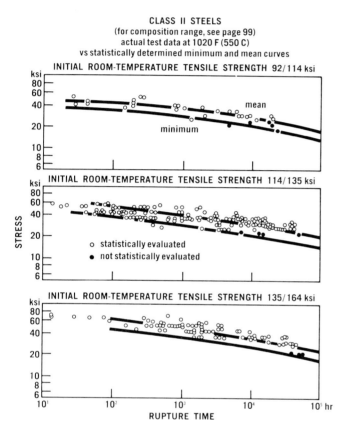

CLASS II STEELS
(for composition range, see page 99)
actual test data at 1020 F (550 C)
vs statistically determined minimum and mean curves

INITIAL ROOM-TEMPERATURE TENSILE STRENGTH 92/114 ksi

INITIAL ROOM-TEMPERATURE TENSILE STRENGTH 114/135 ksi

INITIAL ROOM-TEMPERATURE TENSILE STRENGTH 135/164 ksi

o statistically evaluated
• not statistically evaluated

mean

minimum

STRESS

RUPTURE TIME

Zschokke, Stauffer and Felix

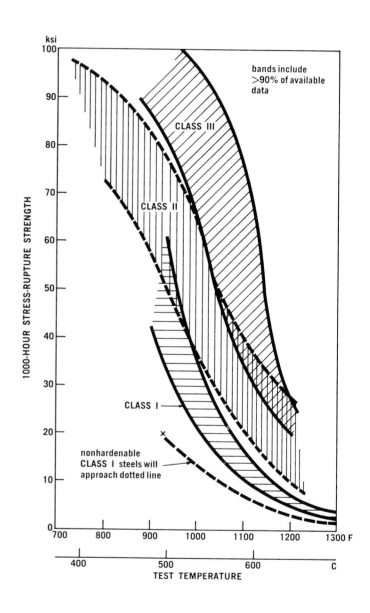

bands include
>90% of available
data

CLASS III

CLASS II

CLASS I

nonhardenable
CLASS I steels will
approach dotted line

1000-HOUR STRESS-RUPTURE STRENGTH

TEST TEMPERATURE

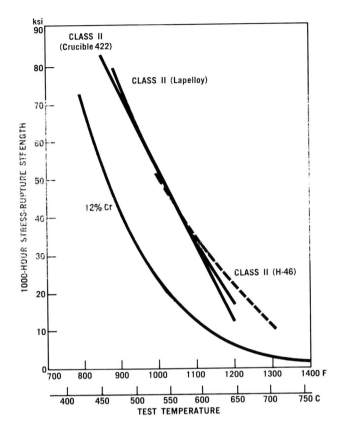

Simmons and Cross

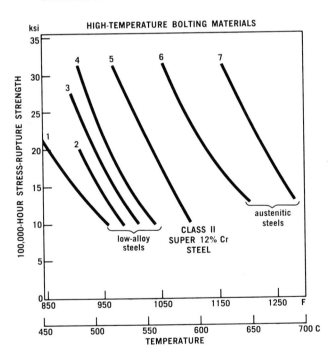

Schinn and Ruttman

	% C	% Cr	%Mo	% Ni	% V
1	0.12	0.7	1.0	1.5	—
2	0.31	2.7	0.3	—	0.3
3	0.24	1.3	0.6	—	0.2
4	0.21	1.3	1.1	—	0.3
5	0.22	12.0	1.0	0.5	0.3
6	0.09	16.0	1.8	15.0	— 1% Cb/Ta + 0.1% N
7	0.07	16.5	1.8	16.5	— 0.7% Cb/Ta

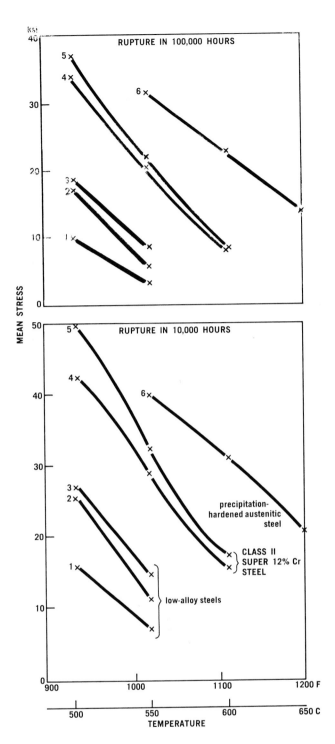

Stahl-Eisen-Werkstoffblatt 590-61

STEELS RESISTANT TO HYDROGEN UNDER PRESSURE

		%C	%Cb	%Cr	%Mo	%Ni	%V
1	10 CrMo 11	0.1	—	2.8	0.25	—	—
2	17 CrMoV 10	0.2	—	2.8	0.25	—	0.15
3	20 CrMoV 135	0.2	—	3.2	0.55	—	0.50
4 } 5 }	× 20 CrMoV 12 1	0.2	—	11.8	1.0	0.5	0.30
6	× 8 CrNiMoVNb 16 13	0.1 max	0.1	16.5	1.3	13.5	0.70

4 pertains to material heat treated to a minimum room-temperature yield point of 71 ksi; 5 to steel heat treated to a minimum room-temperature yield point of 85 ksi

Source: *The Super 12% Cr Steels,* Climax Molybdenum Co., 1965

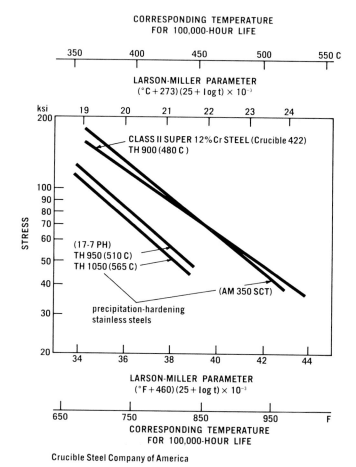

CORRESPONDING TEMPERATURE
FOR 100,000-HOUR LIFE

LARSON-MILLER PARAMETER
$(°C + 273)(25 + \log t) \times 10^{-3}$

CLASS II SUPER 12% Cr STEEL (Crucible 422)
TH 900 (480 C)

(17-7 PH)
TH 950 (510 C)
TH 1050 (565 C)

(AM 350 SCT)

precipitation-hardening
stainless steels

STRESS

LARSON-MILLER PARAMETER
$(°F + 460)(25 + \log t) \times 10^{-3}$

CORRESPONDING TEMPERATURE
FOR 100,000-HOUR LIFE

Crucible Steel Company of America

Castings. Data from a number of sources, some of which are included in the previous tables, indicate that the stress-rupture strength of castings is equivalent to that of wrought material of comparable composition and treatment. The most convincing proof of this comes from Shahinian and Lane, who reported on the stress-rupture strength of the same steels as cast and in the wrought condition.

	Alloying Additions*			Stress, ksi, for Rupture in 1000 Hours							
				1100 F (595 C)				1200 F (650 C)			
						Forgings				Forgings	
% Cb	% Mo	% V		Castings as cast	homog- enized	nor- mal- ized	nor- malized and tem- pered	Castings as cast	homog- enized	nor- mal- ized	nor- malized and tem- pered
0.14	0.5	0.7		25.0	28.0	25.5	23.0	14.5	14.5	13.5	12.0
. . .	0.6	0.7		22.5	26.0	28.5	24.5	12.5	14.5	14.5	14.0
. . .	1.0	0.7		28.0	32.0	32.5	31.0	14.5	15.5	16.5	16.0

*base steel contained about 0.18% C and 11.5% Cr. castings homogenized at 2100 F (1150 C). forgings tempered four hours at 1250 F (675 C)

Shahinian and Lane

Heavy Sections. Since many of the present and potential applications of Super 12% Cr steels are in the form of heavy sections, their strength at elevated temperatures is of concern. The limited published data indicate stress-rupture strengths comparable to those of bars, with no pronounced directional effects. A graphic representation of the stress-rupture strength of X 20 CrMoWV 12 1 shows the points for specimens from the axial bore of a turbine shaft fall exactly on the curve representing minimum values based on an industry evaluation of all German test data on this grade. SFAC data on Virgo 7 (page 96) indicate a somewhat lower strength at lower values of the Larson-Miller parameter in a heavy forging than actually found in small sections heat treated to a somewhat higher strength but at higher parameter values the difference disappears.

In a preliminary evaluation of integral-disc rotors for the Eddystone superpressure turbine, Crucible 422 and a low-alloy chromium-molybdenum-vanadium steel had equivalent stress-rupture strengths, which were appreciably lower than those of an iron-base superalloy.

Representative Stress-Rupture Strengths of Large Forgings of Class II steels

				Test Temperature		Stress, ksi, for Rupture in				
				F	C	100	1000	10,000	100,000 hr	
Crucible 422	jet-turbine wheels	19½-in. dia x 4⅝-in. thick	1900 F (1040 C) OQ + 1200 F (650 C)	1000	540	60	56	..	..	Zonder, Rush and Freeman
				1100	595	43	36	..	..	
				1200	650	22		..	..	
			1900 F (1040 C) AC + 1200 F (650 C)	1100	595	39	32	..	..	Rush and Freeman
Firth-Vickers 448	jet-turbine wheels		2150 F (1175 C) AC + 1200 F (650 C)	1100	595	53	41	..	..	Timken
			2150 F (1175 C) OQ + 1200 F (650 C)	1100	595	48	40	..	..	
GE	rotor	45-in. dia x 40,000 lb	1925 F (1050 C) to oil at 400 F (205 C), equalize at 300 F (150 C) + 1050 F (565 C), cool to 150 F (65 C), hold 24 hr + 1150 F (620 C)	900	480	..	..	..	56.5	Boyle and Newhouse
				1000	540	..	..	..	36.5	
				1050	565	..	..	..	26.5	
				1100	595	..	..	..	18	
Lapelloy	gas-turbine wheel	14-in. dia x ¾-in. thick	2000 F (1095 C) OQ + 1250 F (675 C)	1000	540	80	(72)	..	..	Sakamoto
12Cr-Mo-V	rotor			1020	550	..	..	..	17.2 (center) 19.0 (surface)	Buckley, Caplan, Johnson and Kent
				1110	600	..	..	..	7.2 (center) 7.8 (surface)	
X 20 CrMoWV 12 1	axial core of turbine shaft	forged from 44-ton billet	1905 F (1040 C) AC + 1255 F (680 C)	1020	550	31	28	22	17	Bandel as quoted by Schinn
X 22 CrMoV 12 1	large steam-turbine rotor forging		heat treated to a tensile strength of 111 ksi	1110	600	..	16	13.5	7.8	Stauffer and Keller

Source: *The Super 12% Cr Steels*, Climax Molybdenum Co., 1965

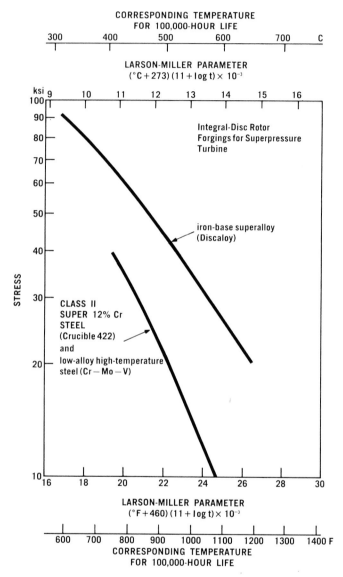

CORRESPONDING TEMPERATURE
FOR 100,000-HOUR LIFE

LARSON-MILLER PARAMETER
$(°C + 273)(11 + \log t) \times 10^{-3}$

Integral-Disc Rotor
Forgings for Superpressure
Turbine

iron-base superalloy
(Discaloy)

CLASS II
SUPER 12% Cr
STEEL
(Crucible 422)
and
low-alloy high-temperature
steel (Cr — Mo — V)

LARSON-MILLER PARAMETER
$(°F + 460)(11 + \log t) \times 10^{-3}$

CORRESPONDING TEMPERATURE
FOR 100,000-HOUR LIFE

Trumpler, LeBreton,
Fox and Williamson

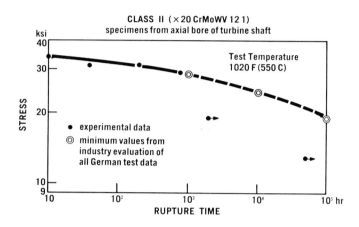

CLASS II ($\times 20$ CrMoWV 12 1)
specimens from axial bore of turbine shaft

Test Temperature
1020 F (550 C)

• experimental data
◎ minimum values from
industry evaluation of
all German test data

STRESS

RUPTURE TIME

Bandel as quoted by Schinn

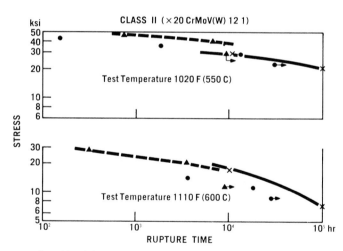

CLASS II ($\times 20$ CrMoV(W) 12 1)

Test Temperature 1020 F (550 C)

Test Temperature 1110 F (600 C)

STRESS

RUPTURE TIME

▲ weld metal
• weldment
× parent metal

Müller and Kautz

156

Weldments. Kakhovskiy and Ponizovtsev considered that the absence of free ferrite and of vanadium in solution was the main prerequisite for high stress-rupture strength in weldments of Class II steels. They developed a 15Kh12NMVFB filler rod for optimum results in arc welding with carbon-dioxide shielding; its composition was:

% C	0.13/0.18	% Cr	11.0/12.5
% Mn	0.9/1.3	% Mo	0.7/1.1
% Si	0.3/0.6	% Ni	0.9/1.3
% Cb	0.15/0.20	% V	0.2/0.4
		% W	0.8/1.2

Weld deposits had a 100,000-hour stress-rupture strength of 14.2/17.1 ksi at 1110 F (600 C). Müller and Kautz, who ran tests on weld metal as well as on weldments, concluded that the stress-rupture characteristics were in line with those of the matching parent metal.

Crucible Steel Company has reported a 100-hour stress-rupture strength of about 72 ksi at 1000 F (540 C) for 0.078-inch-thick Crucible 422 sheet welded with a similar filler metal and subsequently heat treated.

Transverse specimens removed from MIG and submerged-arc welds in a Class I casting with 0.67% Mo and 1.71% Ni were stress-rupture tested at temperatures from 1000 F (540 C) to 1250 F (675 C). There was no significant difference among the two types of welds and the two post-heat treatments. But, contrary to the results of Kakhovskiy and Ponizovtsev on Class II welds, Kreischer, Cothren and Near found that the stress-relieved MIG weld, which had the greatest ferrite content, also had the highest rupture strength at parameter values beyond 100,000 hours at 1050 F (565 C) and the highest ductilities over the full range of test conditions.

Schinn and Ruttmann tested dissimilar welds at 1110 F (600 C). Transition joints of an austenitic steel and a Class II Super 12% Cr steel had a stress-rupture strength similar to or higher than that of the Super 12% Cr steel. Zemzin also studied dissimilar welds between a Class II steel and an austenitic steel. EI 802 was joined to an austenitic steel (Kh15N35V3T) with an austenitic filler metal (Kh15N25M6). If the EI 802 was heat treated to a yield strength of 78 ksi before welding, the stress-rupture strength at 1075 F (580 C) was high, about 18.5 ksi for rupture in 100,000 hours, and the welds showed high ductility. When the EI 802 was heat treated to a higher yield strength of 97 ksi before welding, failure in rupture tests lasting over 1733 hours occurred in the heat-affected-zone on the EI 802 side. The ductility was not high but the 100,000-hour stress-rupture strength was about 23 ksi.

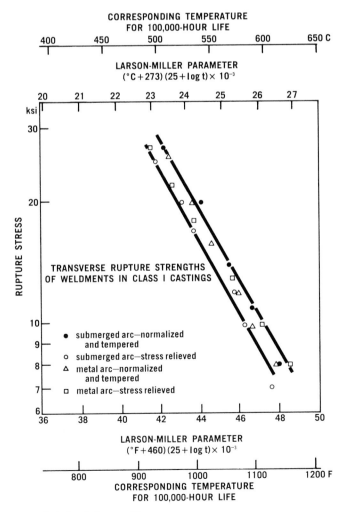

Kreischer, Cothren and Near

Source: *The Super 12% Cr Steels*, Climax Molybdenum Co., 1965

Creep Strength

Creep strength is significant for many of the applications where the Super 12% Cr steels come into consideration although the governing criteria vary. In steam power plants, a field where these steels are assuming increasing importance, the usual criterion is the stress that will give 1 or 0.1% creep in 100,000 hours. The key property in modern aerospace systems, on the other hand, is often the strength that will produce 1% creep in 1000 hours.

Less information is available on creep than on rupture strength because creep tests are more complicated to make. Despite this, sufficient data have been reported to illustrate the creep characteristics of the three classes of Super 12% Cr steels.

Long-Time Tests. Use of conventional terms such as "1% elongation in 100,000 hours" is misleading since no creep test on a Super 12% Cr steel has been run this long. While it is usually deemed acceptable to extrapolate for one cycle, many of the figures in the tables have been extrapolated to a greater degree. Liberman's data on EI 802 show clearly the effect of the length of testing on the reported creep rate. In light of this and the growing demand for reliable long-time data in the steam- and industrial gas-turbine fields, the small amount of information at hand on tests running over 10,000 hours has been compiled in the two tables on Class I and II steels on page 111. No tests in this category have been found for Class III steels.

Isochronous Curves. Isochronous or constant-time stress-strain curves constructed from creep data are used in the aircraft field to simplify the evaluation of airframe design. The deviation of these curves from the basic tensile-test stress-strain curve shows the stress at which creep becomes an important factor under the applicable conditions of time and temperature.

Comparative Strengths. A comparison of the relative creep strengths of the three classes shows that Class III steels are clearly the strongest up to about 1050 F (565 C) and Class I the weakest. At higher temperatures, overaging makes the Class III steels weaker than Class II materials although the exact temperature limit where this crossover occurs may shift when more complete data are available on Class III grades. The superiority of Class I steels to the straight 12% Cr steel is not great but still enough to warrant use of the former steels. For instance, the straight chromium steel would be limited to about 840 F (450 C) in parts requiring a maximum creep deformation of 0.1% in 100,000 hours under a load of 11.2 ksi but this temperature can be raised to 915/930 F (490/500 C) for a Class I steel with 0.5% Mo. The width of the band for Class II steels may result in part from the effect of chromium content and prior treatment. Holdt has concluded that steels with a chromium content of about 12% or higher have better creep strengths than steels with the chromium near the lower limit of 10%. Moreover, at the temperatures under consideration here, heat treatment to a higher room-temperature strength generally increases the creep strength, at least up to about 1050 F (565 C).

It is now possible to see how the Super 12% Cr steels compare with other elevated-temperature steels. On the basis of 0.2% elongation in 100,000 hours, a Class II steel (X 20 CrMoV 12 1) lies only slightly higher than the conventional low-alloy high-temperature bolting materials with about 1/1.5% Cr. The superiority of the Super 12% Cr steel is therefore not as great as would have been surmised from rupture-strength values. Thus attempts to calculate creep strength from rupture strength appear to have only restricted validity.

When compared to lower carbon steels of the types used for tubing, both Class I and II steels appear to have a distinct creep-strength advantage. At temperatures below about 1110 F (600 C), Class II steels even show a useful margin over austenitic steels

Representative Creep Strengths of Class I Steels

		Creep Strength, ksi											
		840 450	900 480	930 500	975 525	1000 540	1020 550	1050 565	1100 595	1110 600	1200 650	1250 675	1300 F 705 C
		1% elongation in 10,000 hours											
D8512				31.3			18.5			7.1			
Firth-Vickers Molybdenum Stainless Steel	(1)				10.1*		6.7*						
Jethete M153				11.9*			4.9*			2.7*			
MTS 1	(1)			34.1			19.9			7.1			
Turbotherm 15M				31.3			18.5			6.4			
Turbotherm 20M				31.3			18.5			6.4			
Vaccutherm 5-32 and 5-32h	(1)			32.7			18.5			7.1			
X 20 CrMo 13 1	(2)			37.0			23.5						
	(3)			38.4			21.9						
		minimum creep rate 1% in 10,000 hours											
405 + 0.3% Al + 0.5% Mo	(4)		18.0			9.0							
410 + 0.4% Mo	(5)										1.8		
410 + 0.5% Mo	(6)		26.0			12.8		7.8	4.4		2.3	1.5	1.1
410 + 0.6% Mo	(7)			15.6									
2 RO 26		35.6		21.3	15.6		12.8				4.3	1.8	
Uniloy 1409 TB	(8)		32.5			11.0			5.0		3.0		

Representative Creep Strengths of Class I Steels (continued)

		Creep Strength, ksi										
		800 425	840 450	885 475	900 480	930 500	975 525	1000 540	1020 550	1100 595	1110 600	1200 F 650 C
		1% elongation in 100,000 hours										
D8512						24.2			14.2		4.3	
Firth-Vickers Molybdenum Stainless Steel	(1)					12.3**	5.6/ 6.7**		(3.4)**			
MTS 1	(1)					24.2			13.5		3.6	
Turbotherm 15M						24.2			14.2		4.3	
Turbotherm 20M						24.2			14.2		4.3	
Vaccutherm 5-32 and 5-32h	(1)					25.6			14.2		4.3	
Virgo 3 and 3S			29.9	22.8		15.6	8.5					
Virgo 6			29.9	22.8		15.6	8.5					
		minimum creep rate 1% in 100,000 hours										
405 + 0.3% Al + 0.5% Mo	(4)	18.0			11.8				5.5			
410 + 0.4% Mo	(5)											0.8
Uniloy 1409 TB	(8)					16.0			6.0		3.0	1.5

*0.1% elongation in 1000 hours   **0.1% elongation in 10,000 hours

( ) extrapolated
(1) hardened and tempered
(2) 1870 F (1020 C) oil quenched + 1275 F (690 C)
(3) 1920 F (1050 C) oil quenched + 1165 F (630 C)
(4) normalized and tempered

(5) annealed
(6) oil quenched + 1220 F (660 C)
(7) oil quenched + 1290 F (700 C)
(8) 1800 F (980 C) oil quenched + 1150 F (620 C)

Source: *The Super 12% Cr Steels*, Climax Molybdenum Co., 1965

Representative Creep Strengths of Class II Steels

		\#\#\# Creep Strength, ksi									
		750 / 400	800 / 425	840 / 450	900 / 480	930 / 500	1000 / 540	1020 / 550	1065 / 575	1100 / 595	1110 / 600	1200 F / 650 C
					1% elongation in 10,000 hours						
D8514	(1)					31.3		19.9			7.8	
D8514	(2)					39.8		25.6			11.4	
D8518 and 8518 W	(3)					35.6		21.3			8.5	
Firth-Vickers Molybdenum Vanadium Stainless Steel						29.1*		16.8*	11.2*		7.8*	
Firth-Vickers 448						38.1*		25.8*			14.6*	6.7*
56 TS								31.3			17.1	
Fluginox 65						31.3		17.1			8.5	
G-X 20 CrMoWV 12 1	(4)							19.9			9.2	
Hecla HGT4								43.2			28.9	13.4
Jethete M151		52.6*				23.6*					4.8*	
Jethete M152				51.5*		33.2*		18.8*			5.6*	
Jethete M154	(5)	60.5*									9.0*	
Mel-Trol H-46							23.8*			16.2*		7.5*
MTS 5	(6)					38.4		24.2			9.2	
Soleil T1								31.3			17.1	
Turbotherm 20MV	(6)					37.0		22.8			10.0	
Turbotherm 20MVNb	(6)					39.8		28.4			17.1	
Turbotherm MVW	(7)					39.8		25.6			11.4	
Vaccutherm 5-34 and 5-36	(6)					37.0		24.2			9.2	
X 15 CrMoWV 12 1	(8)							24.9			5.7	
	(9)							(11.9)			(5.7)	
X 20 CrMoV 12 1	(10)							18.5/ 22.8				
X 20 CrMoV(W) 12 1	(11)							17.1/ 21.3				
X 20 CrMoWV 12 1	(12)					35.6		19.9			8.5	
	(12)							22.0			10.8	
	(13)					37.0		26.5			17.1	8.5
X 22 CrMoV 12 1						35.6		21.3				
X 22 CrMoWV 12 1						35.6						
Z07CDWV 13						44.1		27.0			12.8	
Z15CDNbV 12						34.1						
Z15CDWV 12						34.1						
					minimum creep rate 1% in 10,000 hours						
Carpenter 636	(14)						39.0			19.0		
Crucible 422	(14)						32.0/ 40.0			13.0/ 17.0		
Jessop-Saville H46	(15)			69.4		58.2		40.3			23.5	12.1
Lapelloy	(16)						39.0			19.0		5.0
Modified 422							(50.0)					(20.5)
Uniloy 1420 WM	(14)						39.0			20.0		
USS-12MoV	(17)		160.0									
	(18)		140.0		48.0							

Representative Creep Strengths of Class II Steels (continued)

		Creep Strength, ksi									
		700 370	750 400	840 450	930 500	975 525	1000 540	1020 550	1065 575	1075 580	1100 595	1110 F 600 C
					1% elongation in 100,000 hours							
Cromimphy A2200	(19)	. . .	. . .	. . .	. . .	. . .	. . .	24.2	18.5	. . .	. . .	13.9
D8514	(1)	. . .	. . .	. . .	27.0	. . .	. . .	17.1	. . .	. . .	. . .	5.0
D8514	(2)	. . .	. . .	. . .	32.7	. . .	. . .	21.3	. . .	. . .	. . .	6.4
D8518 and 8518 W	(3)	. . .	. . .	. . .	29.9	. . .	. . .	18.5	. . .	. . .	. . .	5.7
Firth-Vickers Molybdenum Vanadium Stainless Steel		. . .	. . .	. . .	24.6**	. . .	. . .	7.8/9.0**	(5.6)**	. . .	. . .	4.5**
Firth-Vickers 448		. . .	. . .	. . .	24.6/ 26.9**	. . .	. . .	16.8**	. . .	. . .	. . .	6.7**
Fluginox 65		. . .	. . .	. . .	24.2	. . .	. . .	10.0	. . .	. . .	. . .	5.7
Hecla HGT 4		. . .	. . .	. . .	. . .	. . .	. . .	17.9**	. . .	. . .	. . .	. . .
Jethete M151		. . .	(50.4)**	. . .	15.7**	. . .	. . .	7.4**	. . .	. . .	. . .	(1.1)**
Mel-Trol H-46		. . .	. . .	. . .	. . .	. . .	14.8**	. . .	. . .	. . .	8.3**	. . .
MTS 5	(6)	. . .	. . .	. . .	29.9	. . .	. . .	17.1	. . .	. . .	. . .	5.0
T56		. . .	. . .	. . .	24.2	. . .	. . .	12.8	. . .	. . .	. . .	. . .
Turbotherm 20MV	(6)	. . .	. . .	. . .	31.3	. . .	. . .	17.1	. . .	. . .	. . .	5.7
Turbotherm 20MVNb	(6)	. . .	. . .	. . .	31.3	. . .	. . .	17.1	. . .	. . .	. . .	7.1
Turbotherm MVW	(7)	. . .	. . .	. . .	31.3	. . .	. . .	17.1	. . .	. . .	. . .	5.7
Vaccutherm 5-34 and 5-36	(6)	. . .	. . .	. . .	30.0	. . .	. . .	17.1	. . .	. . .	. . .	5.0
Virgo 7	(20)	. . .	. . .	. . .	34.1	25.6	. . .	18.5	12.1	. . .	. . .	6.4
Virgo 8	(20)	. . .	. . .	. . .	39.8	31.2	. . .	24.1	16.3	. . .	. . .	10.7
X 20 CrMoV 12 1	(10)	. . .	. . .	. . .	. . .	. . .	. . .	12.8/ 18.5	. . .	. . .	. . .	. . .
X 20 CrMoV(W) 12 1	(11)	. . .	. . .	. . .	. . .	. . .	. . .	11.4/ 14.2	. . .	. . .	. . .	. . .
X 20 CrMoWV 12 1	(12)	. . .	. . .	. . .	27.0	. . .	. . .	15.6	. . .	. . .	. . .	4.3
X 22 CrMoV 12 1		. . .	. . .	. . .	25.6	. . .	. . .	. . .	. . .	. . .	. . .	. . .
X 22 CrMoWV 12 1		. . .	. . .	. . .	25.6	. . .	. . .	12.8	. . .	. . .	. . .	. . .
Z15CDNbV 12		. . .	. . .	. . .	. . .	. . .	. . .	24.2	. . .	. . .	. . .	. . .
Z15CDWV 12		. . .	. . .	. . .	. . .	. . .	. . .	24.2	. . .	. . .	. . .	. . .
					minimum creep rate 1% in 100,000 hours							
Crucible 422	(14)	. . .	. . .	. . .	. . .	. . .	16.0/ 28.0	. . .	. . .	. . .	. . .	. . .
EI 747	(21)	. . .	. . .	. . .	. . .	. . .	. . .	13.5	. . .	. . .	. . .	. . .
EI 802		. . .	. . .	. . .	. . .	. . .	. . .	14.2/ 17.1	. . .	10.0/ 12.5	. . .	. . .
Jessop-Saville H46	(15)	. . .	. . .	53.8	45.9	. . .	. . .	31.8	. . .	. . .	11.2	. . .
Uniloy 1420 WM	(14)	. . .	. . .	. . .	. . .	. . .	26.0	. . .	. . .	. . .	. . .	. . .
USS-12MoV	(17)	100	. . .	. . .	. . .	. . .	. . .	. . .	. . .	. . .	. . .	. . .

*0.1% elongation in 1000 hours   **0.1% elongation in 10,000 hours

( ) extrapolated

(1) 1870/1920 F (1020/1050 C) oil quench + 1290/1380 F (700/750 C)
(2) 2010/2100 F (1100/1150 C) oil quench + 1200/1250 F (650/700 C)
(3) 1870/1940 F (1020/1060 C) oil quench + 1240/1330 F (670/720 C)
(4) cast + 1920 F (1050 C) air cool + 1365 F (740 C)
(5) 1920 F (1050 C) air cool + 1200 F (650 C)
(6) heat treated
(7) heat treated to a tensile strength of 114/135 ksi
(8) 1920 F (1050 C) oil quench + 1290 F (700 C)
(9) 1920 F (1050 C) oil quench + 1290 F (700 C) + 1380/1435 F (750/780 C)
(10) heat treated to a tensile strength of 128 ksi
(11) heat treated to a tensile strength of 111 ksi

(12) 1920 F (1050 C) oil quench + 1365 F (740 C)
(13) 2050 F (1120 C) oil quench + 1290 F (700 C)
(14) 1900 F (1040 C) quench + 1200 F (650 C)
(15) average values for bars and gas-turbine discs tempered at 1240/1275 F (670/690 C)
(16) oil quench + 1300 F (705 C)
(17) 0.10-in. sheet 1850 F (1010 C) oil quench + 800 F (425 C)
(18) 0.10-in. sheet 1850 F (1010 C) oil quench + 900 F (480 C)
(19) 2100 F (1150 C) oil quench + 1290 F (700 C)
(20) 1920 F (1050 C) oil quench + 1290/1380 F (650/750 C)
(21) 1920 F (1050 C) air cool + 1335 F (725 C)

Source: *The Super 12% Cr Steels*, Climax Molybdenum Co., 1965

Representative Creep Strengths of Class III Steels

	Creep Strength, ksi								
	0.1% Elongation in 1000 hr					0.1% Elongation in 10,000 hr			
	750 400	840 450	930 500	1020 550	1110 600	840 450	930 500	1020 550	1110 F 600 C
Firth-Vickers 535	85.1	80.6	65.0	38.1	. . .	(74)	(56)	(20)	. . .
Jethete M210	. . .	. . .	57.1	. . .	8.7	. . .	. . .	. . .	3.6

( ) extrapolated

CLASS II (EI 802)

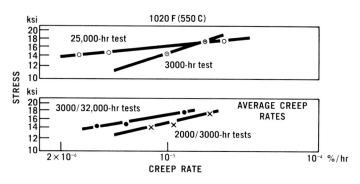

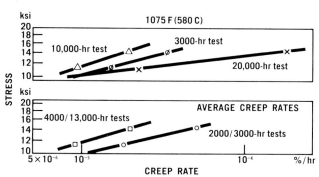

Liberman

Creep-Strength Data on Class I Steels Based on Test Times over 10,000 Hours

	Creep Strain %	Time hr	Creep Strength, ksi			
			930 500	975 525	1020 550	1110 F 600 C
Firth-Vickers Molybdenum Stainless Steel	0.1	10,000	12.3	5.6/6.7	...	...
	0.2	10,000	...	10.1	6.7/7.8	...
	0.2	30,000	...	...	5.6	...
Jethete M153	0.2	10,000	...	...	...	2.2
12.3% Cr—1% Mo	1.0	10,000	19.9	...	...	...
X 20 CrMo 13	0.2	10,000	20.3	...	...	...
	0.2	25,000	17.4	...	...	...
	0.2	50,000	15.6	...	...	...
	1.0	10,000	37.0	...	23.5	...
			38.4*	...	21.9*	...
	1.0	25,000	34.1	...	19.9	...
	1.0	50,000	32.0	...	...	...

*1920 F (1050 C) oil quenched + 1165 F (630 C); the other data on X 20 CrMo 13 were obtained with specimens oil quenched from 1870 F (1020 C) and tempered at 1275 F (690 C)

Fabritius
Firth-Vickers Stainless Steels Ltd
Samuel Fox & Co Ltd
Stauffer and Keller

Creep-Strength Data on Class II Steels Based on Test Times over 10,000 Hours

		Creep Strain %	Time hr	Creep Strength, ksi						
				930 500	1020 550	1050 565	1065 575	1075 580	1110 600	1200 F 650 C
EI 747	(1)	1 x 10⁻⁵%/hr*		...	13.5	...	...	...	...	...
EI 802	(2)	0.65	26,000	...	...	...	...	11.4	...	...
	(2)	0.6	25,000	...	17.1	...	...	...	...	...
	(2)	0.45	28,000	...	...	14.2	...	...	...	...
	(3)	0.45	40,000	...	...	14.2	...	...	...	...
	(4)	0.35	10,000	...	...	...	...	8.5	...	...
	(3)	0.3	38,000	...	14.2	...	...	...	...	...
Firth-Vickers Molybdenum Vanadium Stainless Steel		0.1	10,000	24.6	7.8/9.0	...	...	...	4.5	...
		0.2	10,000	...	14.6/15.7	...	10.1	...	5.6/6.7	...
		0.2	30,000	...	10.1	...	6.7	...	...	...
Firth-Vickers 448		0.1	10,000	24.6/26.9	16.8	...	...	...	6.7	...
		0.2	10,000	35.8	21.3	...	...	...	10.1	4.5
		0.5	10,000	...	31.4	...	...	...	9.0	...
G-X 20 CrMoWV 12 1	(5)	1.0	10,000	...	19.9	...	...	...	9.2	...
	(5)	1.0	25,000	...	17.8	...	...	...	7.5	...
Hecla HGT4	(6)	0.293	10,000	...	...	...	...	...	17.9	...
Jessop-Saville H46		0.3	11,423	49.3	...	...	...	...	...	...
		0.38	46,195	...	26.9	...	...	...	...	...
		0.5	10,000	...	...	...	...	...	15.7	...
		1.0	17,000	...	...	...	...	...	17.9	...
Jethete M151		0.1	10,000	15.7	...	...	...	...	...	...
X 20 CrMoWV 12 1	(7)	0.5	10,000	...	16.1	...	...	...	7.7	...
	(7)	0.5	25,000	...	14.2	...	...	...	...	...
	(7)	1.0	10,000	...	22.0	...	...	...	10.8	...
	(8)			...	27.7	...	...	...	...	...
	(7)	1.0	25,000	...	19.9	...	...	...	9.2	...
	(9)			...	...	...	...	...	8.5	...

*longest test time 10,454 hours

(1) 1920 F (1050 C) quench + 1335 F (725 C)
(2) 1830 F (1000 C) oil quench + 1255 F (680 C)
(3) 1830 F (1000 C) oil quench + 1255 F (680 C) + 1290 F (700 C)
(4) cast + double normalize (2010/1930 F—1100/1050 C) + 1255/1290 F (680/700 C)

(5) cast + 1920 F (1050 C) air cool + 1365 F (740 C)
(6) center of disc forging
(7) 1920 F (1050 C) oil quench + 1365 F (740 C)
(8) 2050 F (1120 C) oil quench + 1290 F (700 C)
(9) heat treated to a tensile strength of about 128 ksi

Chizhik and Kheyn
Fabritius
Firth-Vickers
Stainless Steels Ltd
Samuel Fox & Co Ltd
Hadfields Ltd
Jessop-Saville Ltd
Liberman
Stauffer and Keller
Theis

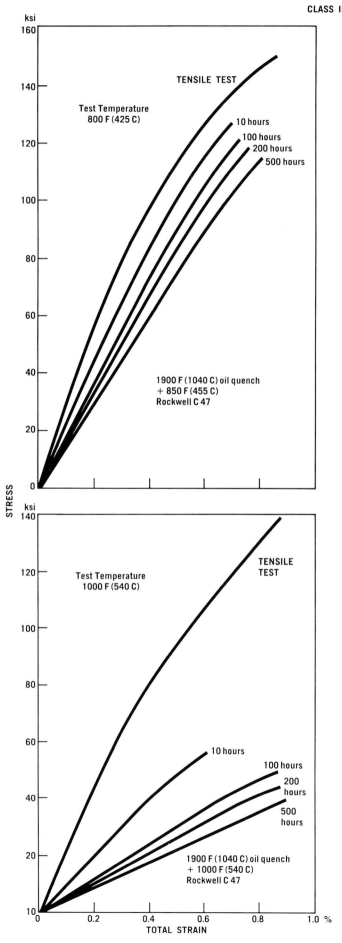

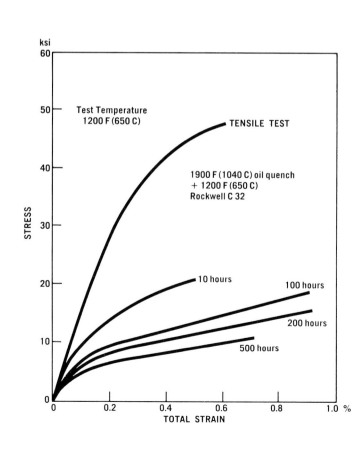

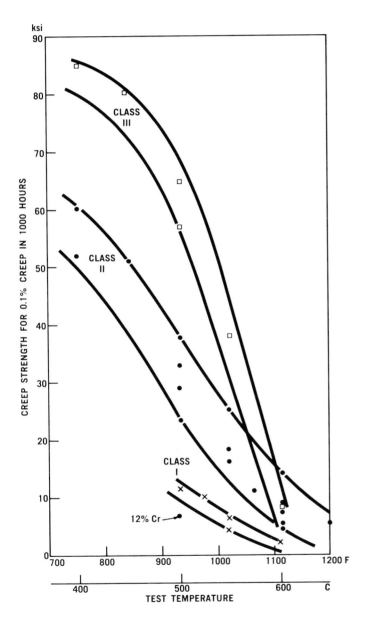

CLASS III

CLASS II

CLASS I

12% Cr

ksi

CREEP STRENGTH FOR 0.1% CREEP IN 1000 HOURS

TEST TEMPERATURE

Firth-Vickers Stainless Steels Ltd
Jessop-Saville Ltd

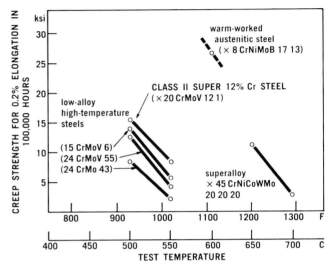

ksi

CREEP STRENGTH FOR 0.2% ELONGATION IN 100,000 HOURS

warm-worked
austenitic steel
(× 8 CrNiMoB 17 13)

CLASS II SUPER 12% Cr STEEL
(× 20 CrMoV 12 1)

low-alloy
high-temperature
steels

(15 CrMoV 6)
(24 CrMoV 55)
(24 CrMo 43)

superalloy
× 45 CrNiCoWMo
20 20 20

TEST TEMPERATURE

Holdt

Source: *The Super 12% Cr Steels*, Climax Molybdenum Co., 1965

containing various combinations of columbium, molybdenum and vanadium. This margin is greater for an elongation of 1% in 10,000 hours than for 1% in 100,000 hours.

Temperature Fluctuations. The effect of cycling temperatures ($\pm$ 50 F—28 C) on the creep strength of a Class I steel, quenched and tempered, was evaluated by Smith and Houston. With a mean temperature of 1000 F (540 C), cycling resulted in a minimum creep rate slightly but probably not significantly greater than the computed rate. Cycling at 1300 F (705 C), however, gave an observed minimum creep rate some 15-times greater than the rate computed from constant-temperature tests.

Load Fluctuations. Bungardt, Bauser and Sychrovsky investigated the reverse creep, or contraction that occurs when the load is removed from a part at elevated temperatures, of a Class I steel with 0.15% C, 12.6% Cr and 1.05% Mo. The major factor affecting the reverse creep was the prior stressing. Annealed steel behaved about the same as heat-treated steel. Subsequent creep behavior was not significantly affected by the length of the unloading period. The investigators proposed that precipitation and recrystallization influenced the creep and reverse-creep results.

Castings. The creep strength of castings is of interest since Super 12% Cr steels, at least Classes I and II, are frequently used in this condition. The German tests included in the creep tables indicate that cast steels have a slightly lower creep resistance than wrought steels of the same composition and treatment. Similar data have been obtained by Liberman, Silyayev, Fedortsov-Lutikov and Sheshenev as well as by the Naval Research Laboratory.

Heavy Sections. Many of the applications of Super 12% Cr steels involve heavy sections so knowledge of their creep strength is needed. Data appear in producers' catalogues for Firth-Vickers Molybdenum

Vanadium Stainless Steel, Firth-Vickers 448, Firth-Vickers 535 and Hecla HGT 4. Creep data have also been obtained on large integral-disc rotor forgings produced of Crucible 422 for a superpressure turbine (Eddystone Unit No. 1). These data are in good agreement with the properties obtained on small sections. In this case, the creep characteristics of the Class II steel (Crucible 422) rotor forgings were identical with those of the low-alloy chromium-molybdenum-vanadium steel and inferior to those of an austenitic superalloy.

The most direct comparison is given by Rush and Freeman's tests on jet turbine wheels of Crucible 422 since the variables of composition and heat treatment were eliminated. In tests lasting about 1000 hours at 1100 F (595 C), the maximum difference between bar stock rough machined to 0.8-inch diameter and wheels approximately 19.5-inch OD x 4⅝-inch thick was 1 ksi or under 5%.

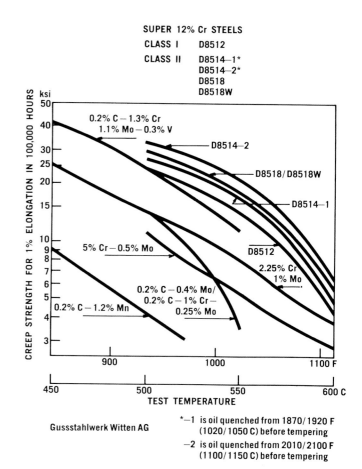

SUPER 12% Cr STEELS	
CLASS I	D8512
CLASS II	D8514—1*
	D8514—2*
	D8518
	D8518W

Gussstahlwerk Witten AG

*—1 is oil quenched from 1870/1920 F (1020/1050 C) before tempering

—2 is oil quenched from 2010/2100 F (1100/1150 C) before tempering

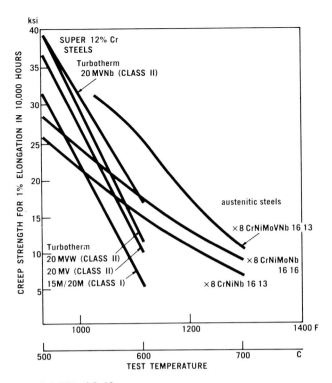

Gebr Böhler & Co AG

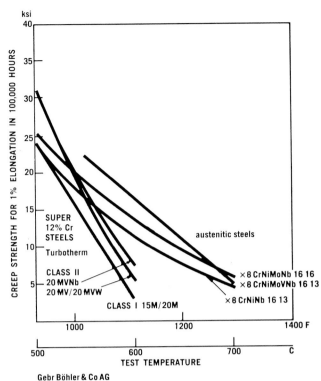

Gebr Böhler & Co AG

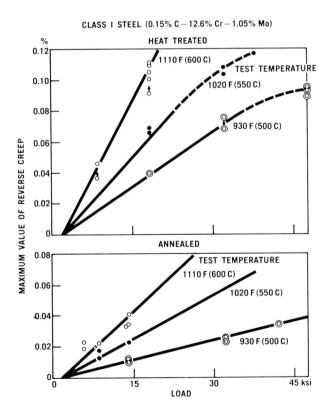

Bungardt, Bauser and Sychrovsky

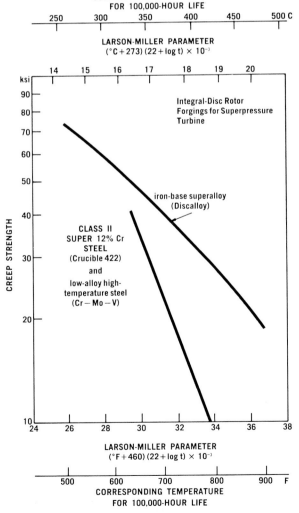

Trumpler, LeBreton, Fox and Williamson

Source: *The Super 12% Cr Steels*, Climax Molybdenum Co., 1965

Stress Relaxation

Stress relaxation is a particularly important property for bolts, which are initially tightened so some known strain is imposed. During service at high temperatures, this stress and therefore the load in the bolt are reduced by creep relaxation. The residual stress remaining in the bolt must not fall below a minimum permissible value derived from design considerations regarding the pressure between the cylinder flanges needed to keep the joint tight and the maximum area of bolt cross section. Laboratory tests give useful comparative relaxation properties although they do not exactly simulate service conditions in which elastic follow-up from the flange assembly may maintain a high residual stress in the bolt despite creep.

Turbine cylinder bolts are usually tightened to a strain of 0.15% although some turbine builders use higher strains. With ferritic bolt materials, the joints are usually designed so the minimum bolt stress to maintain the joint is about 7/11 ksi.

As would be expected from their creep properties, the Super 12% Cr steels lie between the low-alloy ferritic and the high-alloy austenitic steels. The Class I steels are superior to the straight 12% Cr steel and the Class II steels to those in Class I. No data on Class III steels have been found but presumably they would show higher residual stresses than Class II steels.

Effect of Heat Treatment. Conflicting data have been published on the effect of tempering temperature. Soviet investigations on EI 993 indicated that the residual stress increased with the tempering temperature but USA data on Crucible 422 with a similar composition showed the opposite result.

Retightening. In practice bolts are usually refitted after each turbine overhaul and therefore are retightened several times during service. For a power-station turbine designed for a 20-year life, which would normally be overhauled at 30,000-hour intervals, this would mean the bolts would be retightened six times. Creep-relaxation behavior is not necessarily the same in each operating period between tightening so this factor must be studied for any material to be considered for high-temperature bolting materials in the power industry.

Draper of Associated Electrical Industries and Jessop-Saville have used different methods to study the effect of retightening but both have concluded that the Class II Super 12% Cr steels are not damaged by retightening and some improvement may even be observed in subsequent periods.

Draper further gave the time for the stress to fall to 8.96 ksi in the first loading period; these data can serve as indicators of the relative stress-relaxation strength of these two alloys:

	Time, hours, for Stress to Fall to 8.96 ksi	
	1155 F (625 C)	1200 F (650 C)
1% Cr-Mo-V	546	157
12% Cr-Mo-V	1095	178

Draper

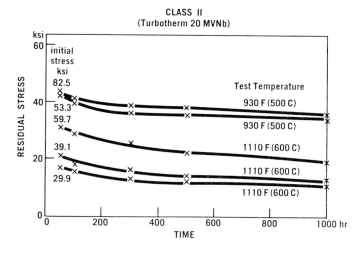

CLASS II
(Turbotherm 20 MVNb)

Gebr Böhler & Co AG

APPROXIMATELY COMPARABLE RELAXATION STRENGTHS

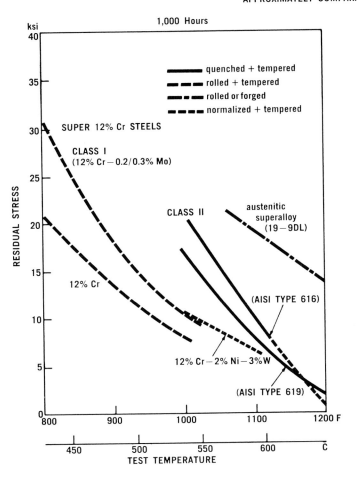

Freeman and Voorhees

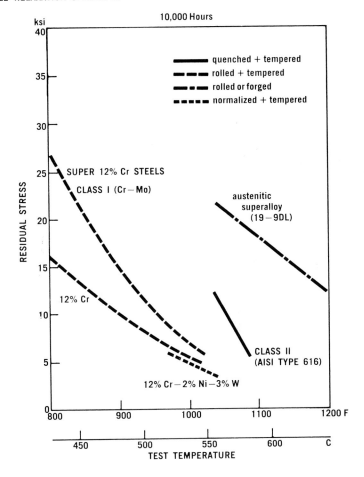

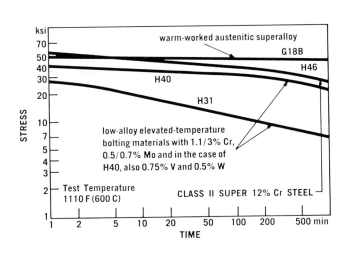

Oliver and Harris

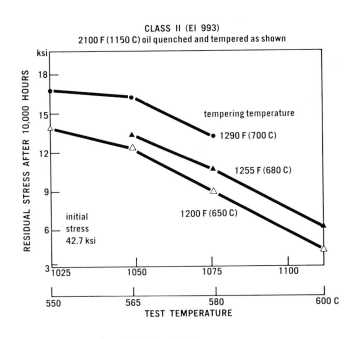

Petropaviovskaya, Borzdyka and Merlina

Source: *The Super 12% Cr Steels*, Climax Molybdenum Co., 1965

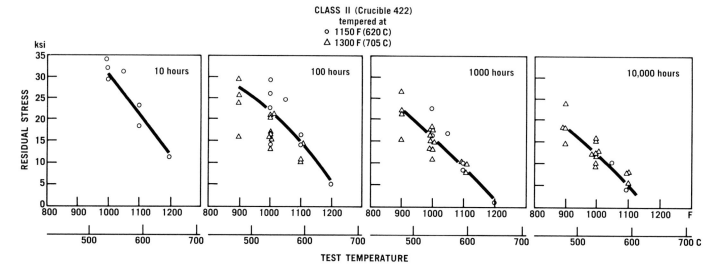

CLASS II (Crucible 422)
tempered at
○ 1150 F (620 C)
△ 1300 F (705 C)

TEST TEMPERATURE

Metals Handbook

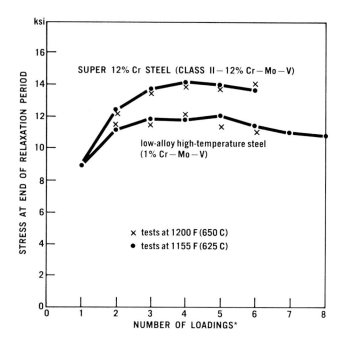

Draper

*Initial strain 0.15%. Relaxation allowed to take place until residual stress has fallen to 8.96 ksi. With the specimen still at test temperature, the strain is increased to the original value of 0.15%.

Effect of Restraining on Stress-Relaxation Characteristics of Jessop-Saville H46*

Time hr	Initial Straining	Residual Stress, ksi		
		Restraining		
		First	Second	Third
1	24.8	27.3	24.4	26.5
10	21.5	25.1	22.2	24.1
50	18.2	22.7	19.8	22.3
100	16.7	21.3	18.5	21.2
300	14.1	19.0	16.4	19.3
1000	12.5	16.8	13.8	16.1

*Initial straining 0.15% total constant strain at 1110 F (600 C). After each 1000-hour run, the residual stress was totally relaxed and the specimen restressed to 29.8 ksi, the nominal stress required to produce the 0.15% total plastic strain for the first run

Jessop-Saville Limited

Design Values

Design values vary greatly depending on the application. Expected life, temperature and allowable deformation are among the key items that must be taken into consideration in most cases.

To give the designer an overall view of the characteristics of a specific grade, producers publish various types of charts that summarize the creep and rupture strengths and sometimes additional information. Such charts are merely a starting point, as will be illustrated briefly for two types of potential uses of Super 12% Cr steels.

Airframe. Short-time tensile strength is a prime requirement in designing wings and fuselage of planes such as the supersonic transport, since the time and temperature conditions prevailing will probably not be sufficient to cause creep or rupture. In many evaluations, the ratio of tensile strength to density is selected as a measure of structural efficiency. Bringing density into the picture can make a pronounced change in the relative ratings of a series of steels. Raring has indicated that fatigue strength and the ratio of compressive strength to density should also be considered.

Because of the critical nature of the application and the serious results of any failure, the designer must also look at the toughness and resistance to crack propagation of the materials he is studying. Loria has proposed tear strength for this purpose, while Raring prefers notched tensile strength.

Steam Turbines. Creep and rupture strengths are the main factors that enter into designs for steam turbines. The part determines the permissible deformation. For example, it may be as much as 1% for short turbine buckets but under 0.1% for turbine wheels.

Long-time test data are needed for accurate evaluation because of the relative unreliability of extrapolations and other prediction methods. In view of the greater amounts of long-time tests that have been run, the bases for calculation have become more precise. It is now often feasible to estimate the minimum 100,000-hour stress-rupture strength as well as the average. This means that the factor of safety can be adjusted to actual requirements and must not also cover ignorance of the behavior of the material.

The maximum allowable stresses and the maximum operating conditions in superheaters show clearly the place of the Class II Super 12% Cr steel in filling the gap between low-alloy high-temperature steels and austenitic steels. In certain ranges, the Super 12% Cr steel even allows thinner walls than can be used with austenitic steels, especially if the stresses for the latter are based on the yield strength rather than on rupture. The thinner walls are accompanied by a number of advantages, most notable of which is lower thermal stresses. Above a certain temperature limit, however, the strength margin of the Super 12% Cr steel over austenitic steels vanishes. Bettzieche in 1962 has discussed at some length the effect of fluctuations in temperature on the factor of safety from this angle.

Designers must take into account scaling resistance and tube wastage. The greater resistance of the Super 12% Cr steels to oxidation and corrosion comprises one of their significant benefits over the low-alloy steels.

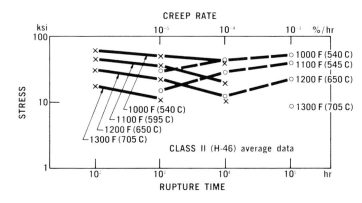

CLASS II (H-46) average data

Simmons and Cross

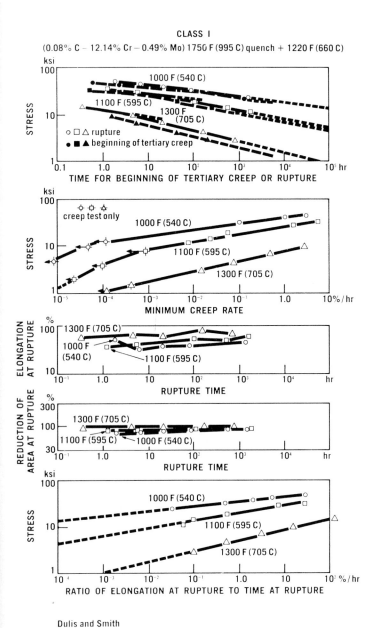

Dulis and Smith

CLASS II (Crucible 422)
1900 F (1040 C) oil quench + 1200 F (650 C)

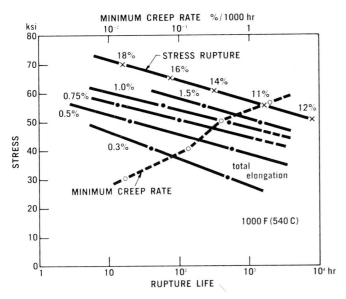

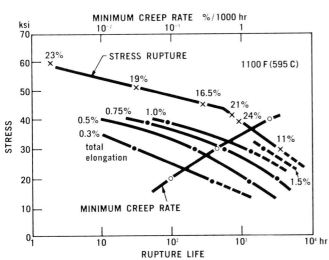

Crucible Steel Company of America

Source: *The Super 12% Cr Steels*, Climax Molybdenum Co., 1965

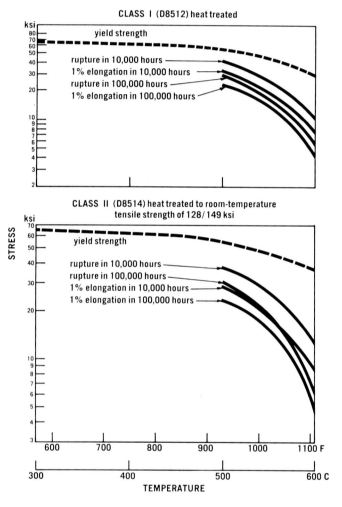

CLASS I (D8512) heat treated

CLASS II (D8514) heat treated to room-temperature
tensile strength of 128/149 ksi

Gussstahlwerk Witten AG

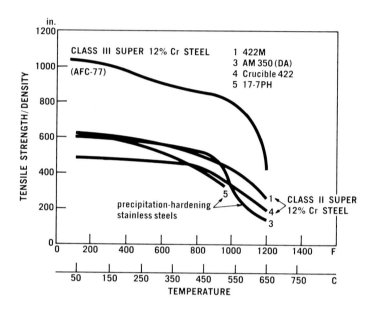

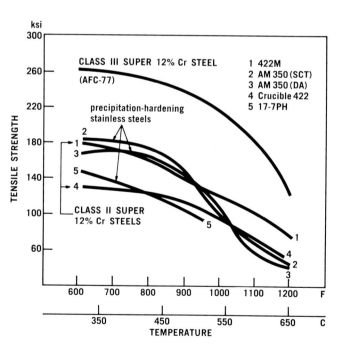

Crucible Steel Company of America;
Loria

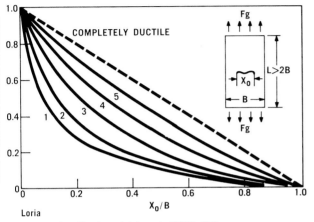

COMPLETELY DUCTILE

Loria

1 cutlery-type stainless steel (AISI 420)
2 precipitation-hardening stainless steels (17-7PH)
3 CLASS II SUPER 12% Cr STEEL (Crucible 422)
4 hard-cold-rolled stainless steel (AISI 301)
5 precipitation-hardening stainless steel (AM 350)

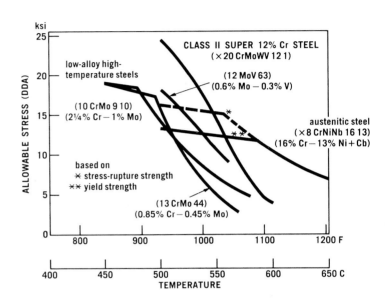

Bettzieche

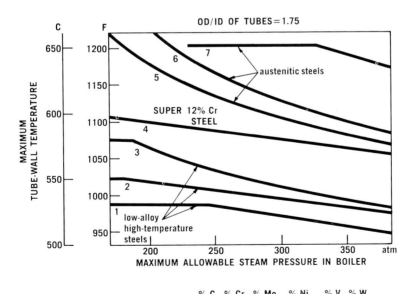

OD/ID OF TUBES = 1.75

Tweer as given by Brühl

		% C	% Cr	% Mo	% Ni	% V	% W	
low-alloy high-temperature steels	1 15 Mo 3	0.16	.	0.30	. . .	. . .	. .	
	2 13 CrMo 44	0.14	0.85	0.45	. . .	. . .	. .	
	3 10 CrMo 9 10	≦0.15	2.25	1.0	. . .	. . .	. . .	
CLASS II SUPER 12% Cr STEEL	4 ×20 CrMoWV 12 1	0.20	12.0	1.0	0.45	0.30	0.50	
austenitic steels	5 ×8 CrNiNb 16 13	0.10	16.0	. . .	13.0	. . .		+ 1.2% Cb/Ta max
	6 ×8 CrNiMoNb 16 16	0.10	16.5	1.8	16.5	. . .		+ 1.2% Cb/Ta max
	7 ×8 CrNiMoVNb 16 13	0.10	16.5	1.3	13.5	0.75	. . .	+ 1.2% Cb/Ta max + 0.10% N

Source: *The Super 12% Cr Steels*, Climax Molybdenum Co., 1965

Relative Wall Thickness of Superheater Tubing with 0.79-inch ID Operating at 1060 F (570 C)

Low-Alloy Steels			CLASS II SUPER 12% Cr STEEL (X 20 CrMoWV 12 1)	Austenitic Steel (X 8 CrNiNb 16 13)
13 CrMo 44	10 CrMo 9 10	12 MoV 63		
1	0.49	0.36	0.31	0.23
2.04	1	0.74	0.63	0.47
2.78	1.36	1	0.86	0.64
3.22	1.58	1.16	1	0.74
4.4	2.13	1.57	1.35	1

Bettzieche

Required Wall Thickness in inches for Superheater Tubes with an Inside Diameter of 1.2 inches Operating under a Pressure of 300 atmospheres

Temperature		Low-Alloy Steels			CLASS II SUPER 12% Cr STEEL (X 20 CrMoWV 12 1)	Austenitic Steel (X 8 CrNiNb 16 13)	
F	C	13 CrMo 44	10 CrMo 9 10	12 MoV 63			
930	500	0.18	0.29	0.16	0.11	0.23*	(0.18)**
1020	550	0.98	0.51	0.31	0.21	0.25*	(0.19)**

*calculated on basis of 0.2%-offset yield strength
**calculated on basis of stress-rupture strength

Bettzieche

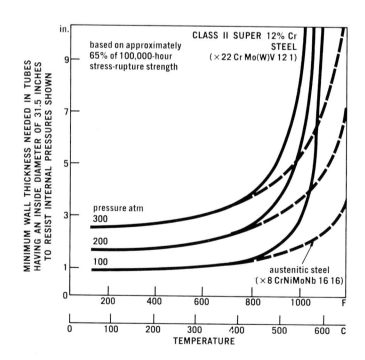

Brennecke and Schinn

Hot Hardness

Although hot hardness increases with hot strength, specific hot-hardness figures are of interest in the case of hot-work dies where hardness rather than strength has conventionally been a measure of the resistance to wear and erosion at operating temperatures. The limited available data indicate that the hardness retention of Class I and II Super 12% Cr steels at temperatures up to 1200 F (650 C) is comparable to that of H-13, a standard hot-work steel and all display a sharp drop-off at temperatures over about 1000 F (540 C). It should be noted, how-ever, that the attainable hardness of the Super 12% Cr steels is less than that of hot-work steels such as H-13 because of a lower carbon content. Consequently, a tempering temperature of only 980 F (525 C) gives a room-temperature hardness of Rockwell C 51 in Crucible 422, while a temper at about 1125 F (605 C) will produce Rockwell C 46.5 in H-13. Under these conditions, the Crucible 422 would be less stable at 1000 F (540 C) and more likely to soften on longer exposure than H-13. Uniloys 1420 WM and 1430 MV tempered at higher temperatures would be expected to be as thermally stable as H-13.

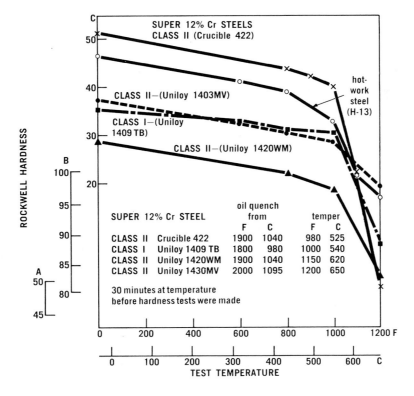

Crucible Steel Company of America
Cyclops Corporation

Source: *The Super 12% Cr Steels*, Climax Molybdenum Co., 1965

SECTION V:
Stainless Steels

Strength of Stainless Steels at Elevated Temperature

T. D. Parker

WHILE STAINLESS steels were originally developed for their resistance to corrosion, some of the important industrial applications of these steels involve elevated temperatures. One of the reasons for the selection of stainless steels for these uses is their

The author is Manager, Metallurgical Development, Climax Molybdenum Co., New York, N.Y. This paper was presented at the AISI Stainless Steel Symposium held 16 October 1967 at the Materials Engineering Congress, Cleveland.

resistance to corrosion and oxidation at high temperatures, but strength is also significant in the majority of uses and in some cases may be the key factor governing the choice of a stainless steel.

Industry has been accumulating and publishing data on the high-temperature strength of these steels for almost 40 years, with Norton's 1929 book one of the first (1). Much data have been compiled since that

time, specially on the 18-8 type and its modifications, which have accounted for the bulk of the stainless steel tubing and piping used in the power and process industries over this period. It is not feasible to present all this information in this paper, therefore attention will be directed to some of the factors involved in selecting a stainless steel for elevated-temperature use from the standpoint of strength and toughness.

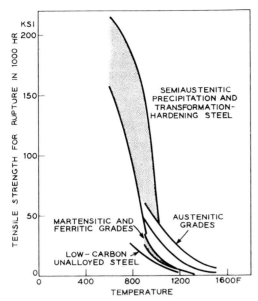

Fig. 1. General comparison of the hot-strength characteristics of austenitic, martensitic and ferritic stainless steels with those of low-carbon unalloyed steel and semi-austenitic precipitation- and transformation-hardening steels.

The first factor to be considered is hot strength, as this is decisive in determining the deformation over the expected life. Thermal stability is the second factor, since this may set limits to a particular grade from the standpoint of softening or, more commonly, embrittlement. Physical properties are also of significance in certain cases.

The potential service life can vary from a matter of seconds in some aerospace components to 25 years or so for parts such as boiler tubes; the allowable deformation can cover at least as wide a range. Consequently, the choice of the most suitable and

economical steel for a particular application depends on an accurate knowledge of the service conditions to which the part will be subjected. Often this information can be obtained only through service testing, which should be done whenever possible unless adequate guides in the form of industry codes are available.

Elevated-Temperature Strength Characteristics

Figure 1 gives a broad concept of the strength ranges we are considering. As shown, the stainless steels have higher hot strength (characterized here by 1000-hr stress-rupture

strength) than low-carbon unalloyed steel, with the austenitic grades of the 300 Series displaying considerably higher strengths than the martensitic and ferritic grades of the 400 Series. Precipitation-hardening stainless steels have higher hot strength than the standard stainless steels at lower temperatures but their strength advantage disappears when they begin to overage. To give an indication of this, Fig. 1 includes a band representing the approximate values for one class of these steels, namely the semi-austenitic precipitation- and transformation-hardening steels.

The standard stainless steels do not display useful degrees of precipitation hardening, but other strengthening mechanisms such as cold working and heat treatment can be beneficial in the temperature range where these steels still behave in an elastic manner. This would mean up to about 800-1000 F for moderate life requirements but might extend to higher temperatures for extremely short periods. Figure 2 illustrates the strength advantage that can be gained by cold working Type 301. Above about 800 F, design would normally be based on creep or rupture strength, where there is no essential difference between the strength of Type 301 and that of the other 18-8 types. Figure 3 shows the higher short-time tensile and yield strength that can be gained in the lower temperature range by using Type 410 in the quenched-and-tempered condition. In most high-temperature applications, how-

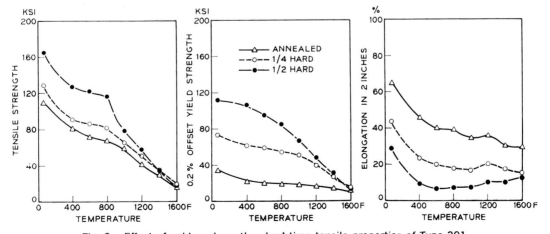

Fig. 2. Effect of cold work on the short-time tensile properties of Type 301.

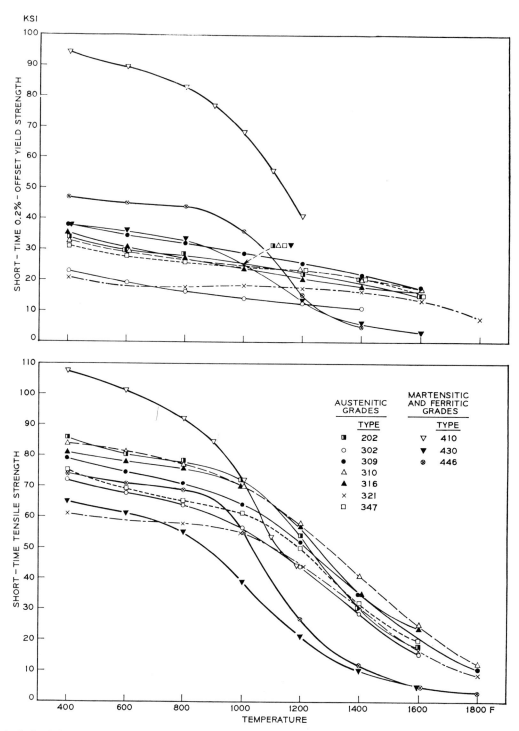

Fig. 3. Typical short-time tensile strengths of various standard stainless steels. All steels were tested in the annealed condition except for the martensitic Type 410, which was heat treated by oil quenching from 1800 F and tempering at 1200 F.

ever, the only strengthening mechanism that applies to the standard stainless steels is solid-solution strengthening.

As indicated in Table 1, the creep and rupture strengths of annealed steels of the 400 Series are closely comparable, with Type 410 having a slight superiority under most conditions. Figures 4 and 5 show the same relationship for the most commonly used austenitic steels, Types 304, 316, 321 and 347, with Types 316 and 347 displaying a modest strength advantage. As can be seen, the variation among test data is often greater than the variation from one grade to another. It is for this reason that increasing reliance is being placed on statistical analysis of high-temperature data, one example of

Table 1. Rupture and Creep Characteristics of Chromium Stainless Steels in Annealed Condition

| | Stress, ksi, for rupture in | | | | | | | | | | | | | | | |
| | 1000 hr | | | | | | | | 10,000 hr | | | | | | | |
Type	800	900	1000	1100	1200	1300	1400	1500 F	800	900	1000	1100	1200	1300	1400	1500 F
405	...	25.0	16.0	6.8	3.8	2.2	1.2	0.8	...	22.0	12.0	4.7	2.5	1.4	0.7	0.4
410	54.0	34.0	19.0	10.0	4.9	2.5	1.2	...	42.5	26.0	13.0	6.9	3.5	1.5	0.6	...
430	...	30.0	17.5	9.1	5.0	2.8	1.7	0.9	...	24.0	13.5	6.5	3.4	2.2	0.7	0.5
446	...	...	17.9	5.6	4.0	2.7	1.8	1.2	...	...	13.5	3.0	2.2	1.6	1.1	0.8

| | Stress, ksi, for a creep rate of | | | | | | | | | | | | | | | |
| | 0.0001%/hr | | | | | | | | 0.00001%/hr | | | | | | | |
Type	800	900	1000	1100	1200	1300	1400	1500 F	800	900	1000	1100	1200	1300	1400	1500 F
405	...	43.0	8.0	2.0	...	...	...	...	...	14.0	4.5	0.5	...	...	...	...
410	43.0	29.0	9.2	4.2	2.0	1.0	0.8	...	19.5	13.8	7.2	3.4	1.2	0.6	0.4	...
430	23.0	15.4	8.6	4.3	1.2	1.4	0.9	0.6	17.5	12.0	6.7	3.4	1.5	0.9	0.6	0.3
446	31.0	16.4	6.1	2.8	1.4	0.7	0.3	0.1	27.0	13.0	4.5	1.8	0.8	0.3	0.1	0.05

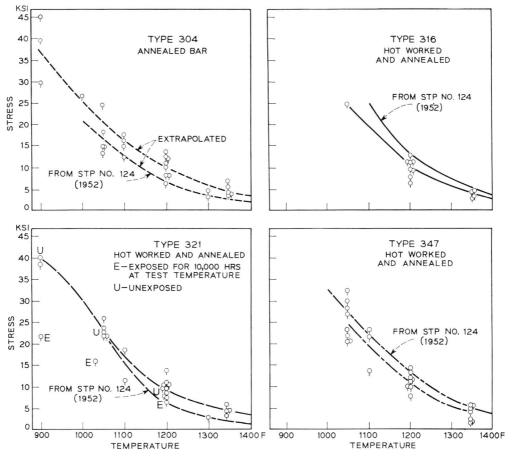

Fig. 4. Comparative 100,000-hr stress-rupture data for Types 316 and 347 tube and pipe as presented in the 1952 (2) and 1965 (3) ASTM publications.

which is shown in Fig. 6 for Type 304H.* While this approach is most satisfactory, it is feasible only where

---

* ASTM has established "H" modifications of the AISI stainless steels for use at elevated temperatures.

---

sufficient test data are at hand. This is not the case with the 400 Series (Table 1) nor with any of the steels at temperatures over 1500 F (Table 2).

Test data such as have just been

presented generally apply to commercial material as normally furnished. Processing variables may have an effect. The establishment of the "H" modifications for tubular products has led to narrower bands. This has been

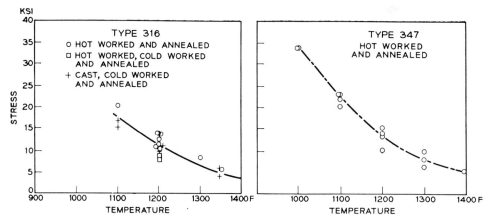

Fig. 5. Comparative 100,000-hr stress-rupture data for Types 316 and 347 tube and pipe as presented in the 1965 ASTM (3) publication.

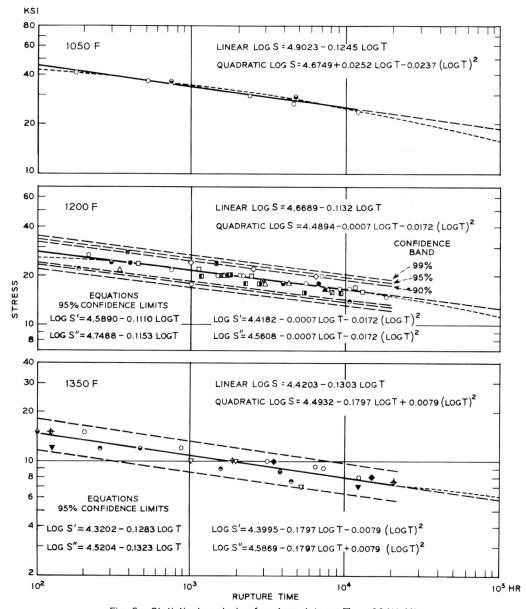

Fig. 6. Statistical analysis of rupture data on Type 304H (4).

Table 2. Rupture and Creep Characteristics of Six Standard Grades at 1600-2000 F*

AISI Type	Testing Temperature, F	Stress, psi				Extrapolated elongation at rupture in 10,000 hr, %
		Rupture time, hr			Creep rate 0.01%/hr	
		100	1000	10,000*		
302	1600	4700	2800	1750	2500	150
	1800	2450	1550	960	1300	30
	2000	1300	760	460	620	18
309S	1600	5800	3200	**	3500	**
	1800	2600	1650	1000	1000	105
	2000	1400	830	480	760	42
310S	1600	6600	4000	2500	4000	30
	1800	3200	2100	1350	1750	60
	2000	1500	1100	760	800	60
314	1600	4700	3000	1950	2300	110
	1800	2600	1700	1100	1000	120
	2000	1500	1120	850	900	82
316	1600	5000	2700	1400	2600	30
	1800	2650	1250	600	1200	35
	2000	1120	360	**	400	**
446	1600	1100	720	**	800	**
	1800	700	500	**	420	**

* Data from Brickner, Ratz and Domagala (5).
† Extrapolated.
‡ Testing time not long enough to determine accurate values.

particularly significant with Type 321 for which final heat treatment is most important and has eliminated the possibility of a fine grain size and the resultant low hot strength, which had led to creep swelling and rupture in Type 321 steam superheater tubes heat treated to a fine grain size to facilitate fabrication. On the other hand, hot finishing or cold reducing tubing prior to solution treating has little effect on hot strength properties as shown by the rupture data in Fig. 7 for Type TP 304H tubing.

Large forgings of stainless steel are another product that is coming increasingly into consideration, for instance in the nuclear field for temperatures up to about 800 F. In Pryle and Wessel's study (6) of large forgings of Types 304 and 347, the variation in properties correlated with variations in hardness as furnished and was attributed to the use of different solution treatments. At equivalent hardness levels, the strength over a range of section sizes was comparable. As indicated in Fig. 8, the yield and ultimate strengths up to 800 F were unaffected by either specimen orientation, or radial or axial position in a heavy Type 347 forging for a nuclear reactor. The fracture strength, however, was dependent on both orientation and axial location. While the transverse ductility was considerably less than the longitudinal ductility, particularly at the center, Pryle and Wessel (6) concluded that this did not lead to notch sensitivity or embrittlement problems.

Many, if not most, high-temperature components are welded; thus, the effect of welding must also be taken into account. Welding decreases the rupture and creep strengths at higher temperatures, but with good welding practices, reliable values can be obtained. The uniform creep and rupture strengths of weldments are illustrated (Fig. 9) in tests made by Rowe and Stewart (7) in connection with power-generation systems for aerospace applications. The weld efficiency, however, decreased with increasing temperature and time:

Temperature, F	Rupture Time, hr	Weld Efficiency, %
1350	100	100
	1000	80
	5000	70
1500	100	80
	1000	60
	5000	50
1650	100	80
	1000	70
	5000	60

They associated the reduction in long-time rupture strength of weldments with the lack of long-time ductility in the weld deposit, since all but one failure occurred in the weld bead.

Because many elevated-temperature components are subjected to bi-

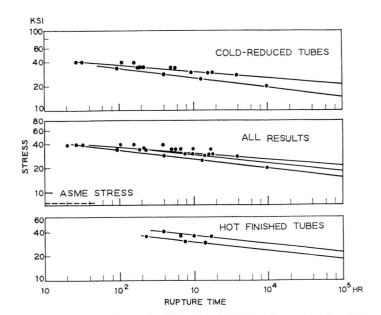

Fig. 7. Effect of production method on the 1100 F rupture strength of Type TP 304H tubing. Tubes, which ranged in size from 1.900 × 0.400 in. to 7¾ × 1¾ and 8.766 × 1.030 in., were water quenched from 1950 F after hot finishing or cold reducing. Specimens were taken from tube wall.

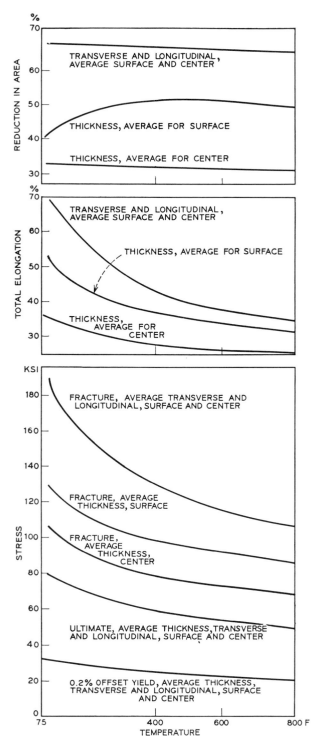

Fig. 8. Effect of specimen position and orientation on the hot strength and ductility characteristics of a Type 347 forging. The forging was a disk, 83 in. diam × 18 in. thick, machined from a forged slab, 92 × 85 × 21 in., formed from a 33-ton ingot. All specimens were taken from cores. The specimen axis of the longitudinal tests was in the 92-in. direction, that of the transverse tests in the 85-in. direction and that of the thickness tests in the 21-in. direction (6).

axial tension or pressure, some questions might be raised as to the validity

of uniaxial tension data such as are generally reported. Davis's (8) tests

on Type 316 specimens cut from large sleeve forgings indicate (Fig. 10) that the rupture points for all biaxial and triaxial tests fall on the safe side of the curve for uniaxial tension. Thus, uniaxial-tension creep and rupture data constitute a safe basis for design if the design stresses are calculated in terms of the average effective stress.

Where strength requirements are vital, as in pressure vessels, most elevated-temperature equipment is designed in line with the *Boiler and Pressure Vessel Code* of the American Society of Mechanical Engineers. This code, originally issued strictly as a safety code, has kept pace with advances in technology. Today, it is said to represent an excellent compendium on minimum requirements for design, fabrication, inspection and construction of safe equipment capable of a long period of usefulness (9). Separate codes are issued for various types of equipment, such as unfired pressure vessels, power boilers and nuclear vessels. Since specified design stresses may change, the designer should refer to the latest applicable revision.

The current provisions for establishing allowable stress values for unfired pressure vessels and power boilers are the following:

Unfired Pressure Vessels

Elastic range — Lowest value of:
25% of specified minimum tensile strength at room temperature, or
25% of estimated minimum tensile strength at temperature, or
62.5% of specified minimum 0.2% yield strength at room temperature, or
62.5% of estimated minimum 0.2% yield strength at temperature.

Plastic range — 100% of stress for a creep rate of 0.01% in 1000 hr as based on a conservative average of the reported tests as evaluated by the committee. The resulting stress, with a few exceptions, does not exceed 60% of the average stress

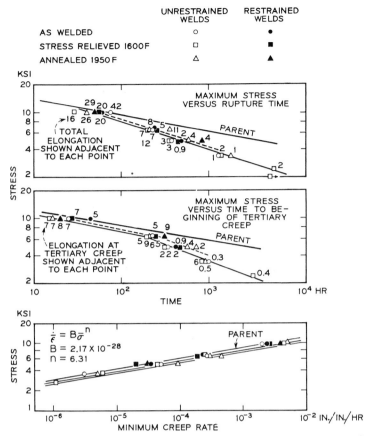

Fig. 9. Effect of welding and postweld treatments on the rupture and creep characteristics at 1500 F of TIG-welded Type 316 (7).

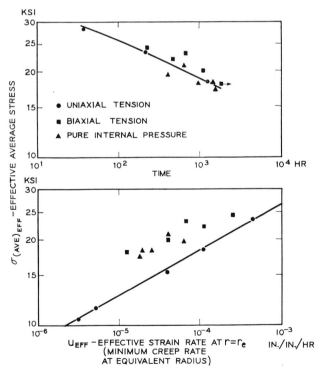

Fig. 10. Rupture and creep characteristics of Type 316 at 1200 F under uniaxial tension, biaxial tension and pure internal pressure. The effective average stress is based on the average values of the three principal stress components (8).

to produce rupture at the end of 100,000 hr determined from available extrapolated data. With successful service experience as a guide, some values have been established up to a maximum limit of 100% of the estimated minimum stress to produce rupture at the end of 100,000 hr.

Power Boilers

Elastic range	As for unfired pressure vessels.
Plastic range	Lowest value of: 100% of average stress for a creep rate of 0.01% in 1000 hr, or 60% of average 100,000-hr rupture strength, or 80% of minimum 100,000-hr rupture strength.

Figure 11 illustrates how these criteria for power boilers are applied to Type 304. The established stresses are based on weighted test data as evaluated by a group of competent analysts, who rely heavily on service experience with evidence of satisfactory performance being preferred to all forms of test data.

Design philosophy accounts for part of the difference between the requirements for power boilers and unfired pressure vessels; power boilers are expected to have a useful life of at least 25 years, whereas many chemical and oil-refining plants become obsolete in ten years or less. While the criteria now in use make no mention of design life, Brister and Leyda (9) and Bolton (10) have pointed out that the evaluation of rupture life for power boilers is expected to assure a material life greater than 25 years, but the basis for setting unfired pressure-vessel stresses implies that the vessel need last no longer than about 11 years. Consequently, Bolton (10) has suggested that design criteria be developed that will allow components to be designed for the intended use life with consideration of the limiting total creep deformation. Since creep and stress-rupture strengths are time-dependent functions, vessels that

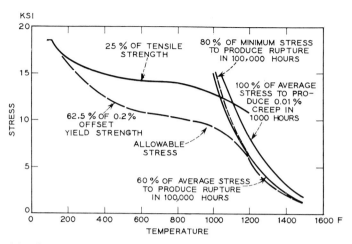

Fig. 11.   Criteria in ASME Code for Power Boilers applied to Type 304 (9).

operate at design temperatures for only a short period of time would have higher allowable stresses than vessels operating at design temperatures for longer periods of time.   One example of a design criterion based on a specific life in a specific environment is given in Fig. 12, which shows the developed design stresses for a 30-year life of Type 316 nuclear reactor components in a sodium environment.

High-temperature compressive and bearing properties are of interest for certain elevated-temperature parts.

Data for Type 302 are given in Table 3.   Fatigue strength, which is closely related to thermal stresses, will be discussed later under physical properties.

Thermal Stability

Thermal stability is important for any material to be used at elevated temperatures.   With time, some changes in structure and mechanical properties will occur with almost any steel or alloy; the significant factor is their relevance to service perform-

ance.   Formation of sigma phase, for instance, is undesirable because it leads to embrittlement.   Its presence does not necessarily destroy the utility of the part; many stainless steels containing sigma phase have given excellent service.

The structural changes that occur in stainless steels are coagulation or precipitation of carbides, and formation of sigma, chi and related phases. Spheroidization or coagulation of carbides and consequent softening are found largely in the martensitic steels when exposed to temperatures approaching or exceeding the original tempering temperature.   Precipitation of carbides in the 300 Series in the temperature range of 800 to 1600 F causes a loss of toughness and—except for the low-carbon "L" modifications, and Types 321 and 347—may make the steel subject to intergranular corrosion in certain mediums at or near room temperature.   This indicates that special attention must then be paid to the possibility of aqueous corrosion, for instance in sulfuric, sulfurous and polythionic acids formed during shut-down periods.

Formation of sigma and chi occurs at about 1200-1400 F in any of the steels with over about 12% Cr and generally takes place more rapidly in the straight chromium steels than in the austenitic steels.   Chromium steels of the 400 Series may also become embrittled after exposure at around 900 F due to the formation of what has been described as a "high-chromium ferrite phase," which is related to but not the same as sigma.

Provided the steel is sound and uncracked after being subjected to these conditions, any structural changes can be overcome by re-heat treatment of the steel.   The only form of structural change that cannot be corrected by subsequent heat treatment is the grain coarsening that may occur in Type 446 at 1800 F and higher.

Lack of thermal stability is evidenced after service by softening or loss of room-temperature toughness or, in the stressed condition, by embrittlement at elevated as well as lower temperatures.   While softening is a fact of life that clearly impairs

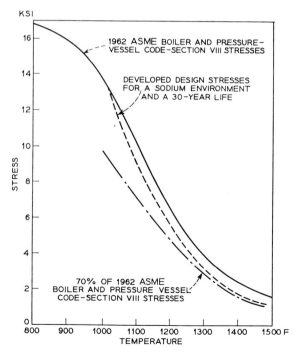

Fig. 12.   Developed design stresses for a 30-yr life of Type 316 in a sodium environment (11).

Table 3. Effect of Temperature on the Tensile, Compressive and Bearing Properties of Annealed Type 302*

| Temperature, F | Exposure time before test, hr | 0.2% Offset yield strength, ksi | | | Ultimate strength, ksi | |
		Tensile	Compressive	Bearing	Tensile	Bearing
78	..	45	46	73	89	168
400	0.5	33	36	62	69	111
	2	31	36	56	67	110
	10	32	37	57	68	111
	100	32	37	55	67	110
600	0.5	31	35	57	68	109
	2	31	34	54	68	106
	10	31	32	54	66	109
	100	30	33	58	67	109
800	0.5	29	31	55	65	104
	2	28	31	53	63	105
	10	28	30	53	66	113
	100	29	31	50	65	100
1000	0.5	25	29	46	60	93
	2	24	29	50	61	94
	10	25	28	48	63	93
	100	24	29	45	62	94
1200	0.5	23	..	44	54	81
	2	23	..	44	54	81
	10	24	..	43	54	80
	100	22	..	42	52	78

Data from Doerr (12).

load-carrying ability, embrittlement is a far more complex phenomenon and its implications must be considered in the light of actual service requirements.

Softening generally is significant only with martensitic or cold worked steels. It sets a definite limit to the allowable conditions of exposure that can be tolerated without softening to a point where the part would be unusable. One example is given in Fig. 13, which indicates that Type 440C can be held at 900 F for only relatively short periods if the high hardness needed for bearings and similar parts is not to be sacrificed.

Even when not accompanied by any sensitivity to embrittlement at operating temperatures, loss of room-temperature toughness is important because it means that the equipment must be handled carefully to avoid impact loading when it is cooled down for maintenance work. Table 4 shows the effect of prolonged holding at temperatures of 900 to 1200 F on the room-temperature toughness of various standard stainless steels, while Fig. 14 puts this embrittlement in bet-

ter perspective by comparing the effect of holding for 10,000 hr in the same temperature range on the impact—temperature curves of a typical martensitic, a ferritic and an austenitic steel. The transition temperature for the martensitic Type 410 is close to room temperature. However, moderate variations in residual elements, presence of small amounts of strain or slight modifications in operating conditions can shift this transition so the brittleness would be evident at room temperature as it is with the ferritic Type 430. While there is some loss of toughness in the austenitic Type 304, the remaining impact strength is adequate for parts such as superheaters. (Weld metals can become more embrittled than the parent metal, as indicated below.)

There is no marked alteration in the room-temperature tensile properties of the most commonly used austenitic grades, even after prolonged holding at high temperatures (Fig. 15). Moreover, in most instances, the effect on properties is less at operating temperatures than at room temperature.

Poor thermal stability under stress can lead to brittle failures under load, especially in the presence of notches. It is difficult to evaluate both because of the diversity in operating conditions (where few parts are exposed to constant temperature and load for their entire design life) and because of the complexity in interpreting data. This type of embrittlement is largely a problem at temperatures of 1000-

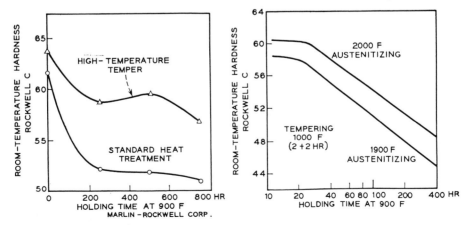

Fig. 13. Softening of Type 440C on extended holding at 900 F.

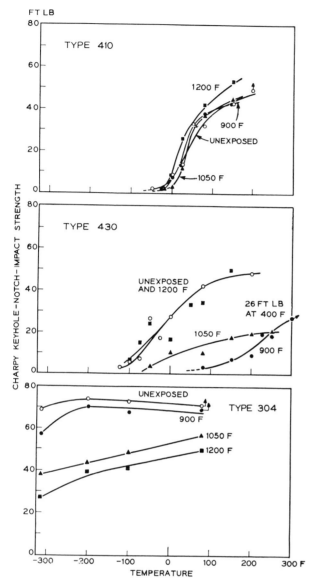

Fig. 14. Effect of holding 10,000 hr at 900, 1050 and 1200 F on the impact characteristics of Types 410, 430 and 304. Hardness values were as follows (13):

	DPN Hardness			
		After exposure for 10,000 hr at		
Type	Unexposed	900	1050	1200 F
410	125	125	124	123
430	185	274	198	169
304	138	140	147	141

Type 347 in attempts to determine the reason for cracking observed in welded heavy sections of Type 347 and to find alternative materials not subject to this embrittlement. A notch-brittle condition in the heat-affected zone as-welded, combined with high residual welding stresses, is considered to be a major factor contributing to the premature service cracking. The behavior of Type 347 is demonstrated in Fig. 16 on the basis of notched and welded stress-rupture tests where a value of notched strength lower than the unnotched strength is evidence of embrittlement. These tests show a definite embrittlement of Type 347 welded with Type 347, and only slightly better results were reported by Christoffel (18) with 16Cr-8Ni-2Mo filler metal. While this embrittlement could be overcome by a post-weld solution treatment, which would be the safest procedure for removing the residual welding stresses, this is not feasible or desirable with large heavy installations. Similar tests on Type 316 (also presented in Fig. 16) illustrate that Type 316 weldments are capable of absorbing the strains produced by relief of residual stresses during service when installed in the as-welded condition. This basic difference in behavior has been confirmed in service.

The difficulty in evaluating various types of tests is shown in Table 6, which presents other data developed in connection with the properties of welded heavy-walled main steam piping of Types 316 and 347 under consideration for the Eddystone Unit No. 1 of Philadelphia Electric. While there is some difficulty in a precise correlation of the results, it was concluded that Type 316 with a 16Cr-8Ni-2Mo weld metal was resistant to embrittlement under load. As will be discussed later in more detail, industry's experience with stainless steel steam piping has pointed out the need for service tests in evaluating thermal stability.

Physical Properties

Certain physical properties are of interest to the designer: linear expansion, thermal conductivity and elastic

1500 F, since at higher temperatures the limiting factor is strength, not ductility (see Table 2).

Various methods have been utilized in detecting susceptibility to embrittlement under stress. The most common of these are: Elongation at rupture; tensile or impact tests on specimens that have been under stress at elevated temperatures; and notched rupture tests.

Rupture elongation as determined on plain unnotched bars (Table 5) is not as informative as tests on weldments. Extensive investigations along this line have been conducted on

Table 4. Effect of Prolonged Holding at 900-1200 F on Room-Temperature Toughness and Hardness

AISI Type	Room-temperature Charpy keyhole impact strength, fl=lb							Room-temperature Brinell hardness						
	Unexposed	after 1000 hr at			after 10,000 hr at			Unexposed	after 1000 hr at			after 10,000 hr at		
		900	1050	1200 F	900	1050	1200 F		900	1050	1200 F	900	1050	1200 F
304	91	87	75	60	79	62	47	141	145	142	143	143	132	143
304L	89	93	76	72	85	71	63	137	140	134	134	143	143	143
309	95	120	85	43	120	51	44	109	114	109	130	140	153	159
310	75	...	48	29	62	29	2	124	119	119	130	152	174	269
316	80	86	72	44	87	49	21	143	151	148	170	145	163	177
321	107	101	90	69	88	72	62	168	143	149	166	156	151	148
347	56	60	55	49	63	51	32	169	156	167	169	156	169	123
405	35	...	36	26	...	39	34	165	...	143	137	...	143	143
410	33	...	41	27	39	3	21	143	...	114	154	124	143	128
430	46	...	32	34	1	3	4	184	...	186	182	277	178	156
446	1	...	1	1	1	1	1	201	...	211	199	369	255	239

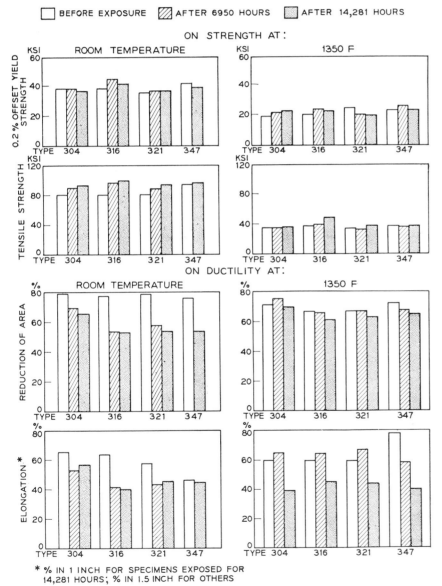

□ BEFORE EXPOSURE   ▨ AFTER 6950 HOURS   ▥ AFTER 14,281 HOURS

ON STRENGTH AT:

ROOM TEMPERATURE — 1350 F

0.2 % OFFSET YIELD STRENGTH, KSI

TENSILE STRENGTH, KSI

ON DUCTILITY AT:

ROOM TEMPERATURE — 1350 F

REDUCTION OF AREA, %

ELONGATION*, %

* % IN 1 INCH FOR SPECIMENS EXPOSED FOR 14,281 HOURS; % IN 1.5 INCH FOR OTHERS

Fig. 15. Tensile-strength data on tubes of Types 304, 316, 321 and 347 after exposure to 2000-psi, 1250 F steam for 6950 and 14,281 hr (14).

constants. The fact that the austenitic steels of the 300 Series have considerably greater thermal expansion than the martensitic and ferritic grades of the 400 Series (Fig. 17) is of special concern when dissimilar materials are joined. Moreover, the high expansion of the austenitic grades must be taken into consideration when providing holding straps or other connections for high-temperature equipment.

The effect of temperature on thermal conductivity is different for the three classes of stainless steel (Fig. 18). Thermal conductivity, however, is only one factor in the over-all heat-transfer coefficient, where film and scale resistance on both sides of the metal are also to be considered. Film resistance is dependent on the type of fluid involved, as well as its velocity. Scale resistance is caused by corrosion of the metal or fouling by built-up deposits. Consequently, in the design of heat exchangers and similar equipment, where the over-all heat-transfer coefficient is used in calculations, the differences among the various types of stainless steel are not as great as would be concluded from the thermal conductivity alone.

The table on elastic constants (Table 7) includes both dynamic constants as well as the conventional static constants. The dynamic constants were determined by an ultrasonic pulse technique that, unlike the usual static tensile test, is not influenced by creep or deformation. Use of the dynamic constants is therefore preferable at higher temperatures and under rapidly changing temperature conditions. The above physical properties are further of vital importance in determining the thermal stresses that occur when thermal expansion or contraction is partially or completely restrained. The following simplified factor can be used to calculate thermal stress:

$$\frac{\text{modulus of elasticity} \times \text{coefficient of thermal expansion} \times \text{temperature}}{1 - \text{Poisson's ratio}}$$

Thermal conductivity is not directly included in this factor but is decisive in establishing the uniformity of temperature distribution. In complicated fabricated structures subjected to alternate heating and cooling, high thermal stresses are a particularly significant problem to designers in the petroleum, power plant and nuclear fields. The importance of local deformation and relaxation during temperature cycling of piping has been recognized for some time.

The fluctuating thermal stresses resulting from periodic changes in temperature can lead to thermal fatigue. In most cases, the number of temperature reversals is relatively limited as compared to the number of stress reversals in room-temperature fatigue. For a 20-year life, a figure of 7000 cycles—corresponding to one cycle a day—has been used in piping design while the number of major temperature swings for process equipment over the same period has been placed at 40,000.

As a result of their low thermal conductivity and high thermal expansion, the austenitic stainless steels of the 300 Series are more sensitive to thermal fatigue than steels of the 400 Series. Figure 19 shows the marked difference between full and partial constraint, while Fig. 20 indicates that in constrained thermal cycling there is no need to differentiate between mechanical and thermally induced strains nor between compressive and tensile strains.

Damping capacity, which is hardly a physical nor a mechanical property, should not be overlooked. Martensitic stainless steels such as Type 410

Table 5. Rupture Ductility of Some Standard Stainless Steels*

AISI Type	Prior treatment	Extrapolated percent elongation at rupture in 10,000 hr					
		900	1050	1100	1200	1300	1500 F
304	1900 F quench	20.0	16.0	10.0	12.0	23.0	16.0
304L	1900 F quench	...	...	7.0	...	9.5	6.0
316	2000 F quench	...	...	2.5	...	30.0	20.0
316L	2000 F quench	...	...	44.0	...	28.0	14.0
321	1900 F air cool	...	...	4.0	...	7.0	6.0
	1900 F quench + 1550 F quench	40.0	28.0	...	20.0	...	...
347	1950 F quench	...	...	24.0	...	28.0	9.5
	1900 F quench + 1550 F quench	...	<1.0	...	29.0	...	...
410	1750 F quench + 1150 F air cool	...	66.0*	...	72.0	...	...
430	1425 F air cool	110.0	45.0	...	115.0	80.0**	...

* Data from Dulis, Smith and Houston (15) and Dulis and Smith (16).
† 1000 F
‡ 1350 F

have appreciably higher damping capacity than the austenitic grades. This is important for parts such as steam-turbine buckets operating at high speeds, which will vibrate in resonance with great amplification of stresses when the frequency with which steam passes coincides with the natural frequency of the buckets.

Applications

Stainless steels have been used at elevated temperatures up to about 2000 F for some 40 years. In these

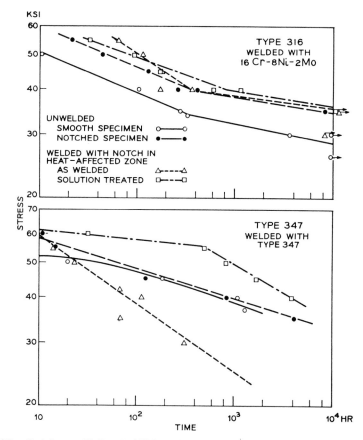

Fig. 16. Notch sensitivity at 1100 F of the heat-affected zone of welded Types 316 and 347 (17, 18).

Table 6. Comparison of Several Methods of Evaluating Thermal Stability of Heavy Welded Pipe, 11.5 in. OD × 5.5-in. ID × 3 in. Wall

Effect of thermal aging on notch toughness of each weld joint
Charpy keyhole impact, ft-lb

Weld joint Base metal	Weld Metal	Aging time at 1300 F, hr	As-welded Weld metal	HAZ	Base metal	1950 F air cool Weld metal	HAZ	Base metal
Type 347	Type 347	none				23	42	52
Type 347	16-8-2	none	33	40	52	56	48	53
Type 316	Type 316	none	27	60	83	42	83	83
Type 316	16-8-2	none	34	61	88	57	86	89
Type 316	16-8-2	none	33	67	81	49	83	87
Type 347	Type 347	1000				9	38	43
Type 347	16-8-2	1000	25	31	43	31	39	45
Type 316	Type 316	1000	21	41	55	23	46	49
Type 316	16-8-2	1000	28	41	56	45	64	65
Type 316	16-8-2	1000				32	43	45
Type 347	Type 347	10,000				4	36	42
Type 347	16-8-2	10,000	20	30	36			
Type 316	Type 316	10,000	4	25	32			
Type 316	16-8-2	10,000	22	33	41			
Type 316	16-8-2	10,000				25	27	26

Effect of Thermal Aging on Notch Toughness of Component Materials

Charpy keyhole impact, ft-lb

Aging time at 1300 F, hr	As-welded Base metal Type 316	Type 347	Weld metal Type 316	16-8-2	HAZ Type 316	Type 347	1950 F air cool Base metal Type 316	Type 347	Weld metal Type 316	16-8-2	Type 347	HAZ Type 316	Type 347
None	84	52	27	33	63	40	86	53	42	51	23	84	45
1,000	55	43	21	27	41	31	53	44	23	36	9	51	38
10,000	36	36	4	21	29	30	26	42		26	4	27	36

Summary of Rupture Test Results

Weld metal	Postweld heat treatment	Aging time at 1300 F, hr	100,000-hr stress-rupture strength, ksi 1100 F	1150 F	1200 F	Long-time test Time, hr	Elongation, % in 3 in.	Predominant break
			Type 316 Base Metal					
ASME allowable stress			10.4	8.5	6.8			
Type 316	none	none	18.0	14.0	11.0	3,858	4.1	Weld metal
Type 316	none	1000	17.0	13.5	10.3	3,664	4.5	Weld metal
Type 316	1950 F	none	18.5	15.0	11.7	5,338	10.1	Weld metal
Type 316	1950 F	1000	18.0	14.5	11.0	9,114	10.5	Weld metal
16-8-2	none	none	20.5	16.5	12.5	10,535	12.0	Base metal
16-8-2	none	1000	20.0	16.0	12.5	11,808	13.9	Base metal
16-8-2	1950 F	none	20.0	15.5	11.7	2,595	20.8†	Base metal
16-8-2	1950 F	1000	19.0	15.3	12.0	4,829	17.3	Weld-base
			Type 347 Base Metal					
ASME allowable stress			12.5	8.0	5.0			
Type 347	1950 F	none	22.0	18.5	14.7	1,894	6.8	Base metal
Type 347	1950 F	1000	19.5	16.0	12.5	4,943	9.0	Base metal
16-8-2	none	none	22.0	17.5	14.0	6,030	1.4	Weld-base
16-8-2	none	1000	21.5	17.0	13.5	4,490		Weld-base
16-8-2	1950 F	none	21.0	16.5	12.7	6,821	7.7	Weld metal
16-8-2	1950 F	1000	20.5	17.0	13.3	9,680	19.6	Weld-base

* Data from Caughey and Benz (19) and Garofalo, Malenock and Smith (21).
† 1250 F

largely industrial uses, it is picked not for its attractive appearance, but because it permits an operation to be carried out better or more economically. Strength is significant in many of these applications, but not to the extent that it can be divorced from corrosion and oxidation resistance in discussing the reasons for selection of stainless steel. In fields such as steam power generation, for example, it is the higher allowable design stresses for stainless steel that have permitted use of smaller parts and operation at higher temperatures and pressures than would have been feasible with lower alloy steels—although resistance to scaling and hot corrosion is also important in meeting the desired life requirements.

In certain applications, as in many furnace parts, precipitation-hardening stainless steels are able to provide greater strength at higher temperatures. One example in the latter category would be steam-turbine buckets where Types 405 and 410 have been to a large degree replaced by the super 12% Cr steels with higher contents of strengthening elements. Another would be the use of precipitation-hardening stainless steel in high-performance aircraft and space vehicles. About 68% by weight of the XB-70 aircraft, for instance, is made of a precipitation-hardening austenitic stainless steel, largely in the form of a honeycomb sandwich with a corrugated core spot-welded to face sheets. This product was selected for most of the wing and the upper and lower fuselage on the basis of its superior strength in the design temperature range of 450-630 F. Experience with the XB-70 design has permitted use of a similar construction for parts of the Apollo spacecraft. The standard stainless steels are used for a wide variety of parts, ranging from uses where the requirements for hot strength are not great to those where it is crucial, as in pressure applications.

Automotive mufflers, exhaust collectors of industrial gas turbines and chimneys are typical applications, where the imposed stresses are low and the support adequate. In these cases, the hot strength must only be sufficient to support the weight of the part without excessive distortion at service temperatures.

Retention of hardness and hot hardness rather than hot strength are deciding factors in some instances. With cyclone dust collectors, for example, the imposed mechanical stresses are low, yet the steel must retain sufficient hardness to resist wear by the abrasive dust particles. Retention of hardness is also important with parts such as bearings where high hardness is needed for proper operation, including frictional properties, wear and resistance to upsetting.

Stainless-clad equipment is utilized at high temperatures as well as at room and lower temperatures. The importance of the hot strength of the stainless cladding depends on the method of cladding, type of vessel and design procedure. The ASME Boiler and Unfired Pressure Vessel Code permits allowance for the strength of the cladding in the design of unfired pressure vessels under some conditions where the clad plate conforms to certain specifications and the joints are completed by depositing stainless steel weld metal over the base-plate weld to restore the cladding. A minimum shear strength of 20,000 psi is also specified for acceptance testing of the clad plate. In the case of nuclear vessels, no structural strength may be attributed to the cladding under the Code except where bearing stress is involved.

Stainless steel has given good performance in heat exchangers, Fig. 21. The role of hot strength here depends largely on design. When the exchanger is used under pressure, strength at temperature is again a factor that must enter into the design calculations.

Strength at temperature is very definitely involved in high-temperature rupture discs. During normal-pressure-system operation, rupture discs remain inactive. When overpressure builds up to a predetermined pressure setting, the disc ruptures instantaneously and the pressure is reduced to safe operating limits. Therefore,

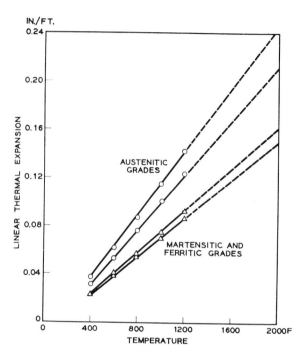

Fig. 17. Linear thermal expansion of the three main classes of stainless steel.

it is essential that a reliable strength is maintained over perhaps years at temperature. Figure 22 shows a typical rupture disc, before and after rupture.

As nuclear power assumes a more vital role in civilian and military installations, there are an increasing number of potential high-temperature applications for stainless steel, al-

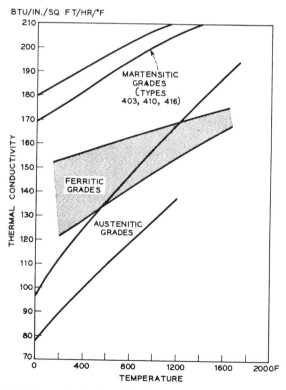

Fig. 18. Thermal conductivity of the three main classes of stainless steel.

though the large number of reactor types makes specific predictions hazardous. Certainly wider adoption of sodium- or organic-cooled reactors would be accompanied by a marked growth in the use of stainless steel. As previously indicated in Fig. 12, the design stresses for a 30-year life of Type 316 in a sodium environment are practically as high as for more conventional mediums. But, while some uses such as fuel-element cladding and reactor pressure vessels will depend on reactor type, other applications such as heat exchangers and fuel-element reprocessing equipment will probably be made of stainless steel regardless of the reactor type.

Most of the pressure vessels for nuclear-reactor containment have been clad or lined with stainless steel. The use of solid stainless steel for these shells has in general been restricted to cases where both high temperatures and corrosion are encountered as in liquid-metal-cooled reactors. This is not universally true, as some solid stainless-steel reactor pressure vessels are used even with pressurized-water reactors. One example is the 45-ton stainless-steel pressure vessel used in the floating nuclear-power generating station under construction for the US Army Corp of Engineers. This pressure vessel will operate with pressures to 1600 psi and coolant temperatures over 600 F, Fig. 23.

Types 410 and 420 have long been used for spindles, seats and other parts of high-temperature valves at temperatures up to about 1000 F. Retention of hardness, strength, good seating characteristics and dimensional stability are important.

Cracking-still tubes for the petroleum industry represent one of the earliest high-temperature applications of stainless steel, as 18-8 tubes were serving this purpose as early as 1928. Stainless-steel cracking tubes are still used, generally at temperatures of 1150 to 1300 F, both because of hot strength and corrosion resistance. Tube producers have reported on the properties of specimens cut from tubes that had seen up to almost 100,000 hr of service at 1200 to 1250 F. The

strengths obtained on these specimens at room and elevated temperatures have been within the range expected from unused material. Many of these could have been used longer but some refineries have the practice of removing these tubes after a certain period of time, such as 60,000 hr, regardless of how good their condition is.

The best-documented and one of the most exciting fields from the standpoint of hot strength is the generation of steam in central-station power plants, where there is a long history of service for the austenitic stainless steels in superheaters, reheaters, feedwater heaters, valves and main steam and reheat lines. Recent detailed reviews of this subject have been given by Baker and Soldan (25), Blumberg (24), Lien (25) and Rohrig (26), so only some of the broad developments will be mentioned here, with particular attention to superheaters and main steam lines.

To a large extent, the use of stainless steel for superheaters and main

Table 7. Effect of Temperature on Elastic Constants of Standard Stainless Steels*

	Temperature, F							
	75	300	500	700	900	1100	1300	1500 F
Young's Modulus (Tension), 10^3 ksi								
Martensitic grades								
Type 410 Static	29.8	27.8	26.5	25.4	23.8	18.9	...	...
Dynamic	31.8	30.7	29.6	28.5	26.8	23.8	20.4	...
Ferritic grades								
Type 430 Static	29.8	28.6	27.2	25.6	24.4	19.4	...	...
Dynamic	33.6	32.4	31.0	29.6	28.0	26.2	23.6	21.0
446 Static	30.2	29.1	27.8	26.8	26.0	22.0	...	...
Dynamic	30.4	29.4	28.3	26.9	...	...	...	...
Austenitic grades								
Type 304 Static	28.4	27.4	26.6	25.4	23.0	22.2	20.5	18.0
Dynamic	29.0	27.3	26.0	24.8	23.6	22.4	21.2	20.0
309 Static	28.1	26.8	25.6	24.4	23.2	21.4	20.4	16.9
310 Static	28.4	27.4	26.2	25.4	24.2	22.8	21.2	18.8
Dynamic	29.0	27.5	26.2	24.9	23.6	22.4	21.2	19.8
316 Static	28.4	27.2	26.4	25.6	23.8	22.6	21.0	18.6
321 Static	27.8	26.3	25.3	24.2	23.2	21.7	20.2	17.6
Dynamic	28.9	27.3	25.8	24.5	23.2	21.9	20.4	19.1
347 Static	28.6	27.2	26.1	24.6	23.2	21.6	20.6	17.3
Dynamic	28.9	27.5	26.1	24.8	23.4	22.0	20.7	19.4
Modulus of Rigidity (Shear) 10^3 ksi								
Martensitic grades								
Type 410 Static	12.0	11.4	11.0	10.5	10.0	8.2	...	...
Dynamic	12.5	12.1	11.7	11.2	10.5	9.1	7.9	...
Ferritic grades								
Type 430 Static	13.0	12.4	12.0	11.7	10.9	8.8	...	...
Dynamic	13.5	12.7	12.0	11.5	10.8	10.0	8.8	7.8
446 Static	12.9	12.5	12.0	11.4	11.0	9.6	...	...
Dynamic	11.9	11.4	10.9	10.4	...	...	...	...
Austenitic grades								
Type 304 Static	11.4	10.6	10.2	9.6	8.0	8.6	8.0	7.2
Dynamic	11.2	10.4	9.8	9.3	8.8	8.4	7.9	7.5
309 Static	11.0	10.4	9.8	9.4	9.0	8.4	7.7	6.8
310 Static	10.6	10.4	10.0	9.7	9.2	8.5	7.9	7.3
Dynamic	11.2	10.6	10.0	9.4	8.8	8.2	7.6	6.9
316 Static	11.3	10.8	10.2	9.6	9.2	8.6	8.0	7.5
321 Static	11.0	10.6	10.1	9.5	8.9	8.4	8.2	7.5
Dynamic	11.2	10.6	9.9	9.4	8.8	8.2	7.7	7.1
347 Static	11.0	10.5	10.0	9.5	8.8	8.2	7.6	6.8
Dynamic	11.4	10.7	10.1	9.5	8.9	8.3	7.8	7.2

Table 7—Continued

Poisson's Ratio

Martensitic grades									
Type 410	Static	0.24	0.21	0.20	0.20	0.20	0.16	...	...
	Dynamic	0.27	0.27	0.27	0.27	0.28	0.29	0.32	...
Ferritic grades									
Type 430	Static	0.14	0.15	0.13	0.10	0.12	0.10	...	...
	Dynamic	0.28	0.29	0.29	0.30	0.30	0.31	0.32	0.33
446	Static	0.17	0.16	0.16	0.18	0.18	0.15	...	...
	Dynamic	0.28	0.28	0.29	0.29	...	...	...	...
Austenitic grades									
Type 304	Static	0.24	0.28	0.30	0.32	0.28	0.30	0.28	0.25
	Dynamic	0.30	0.31	0.31	0.32	0.32	0.32	0.33	0.34
309	Static	0.27	0.29	0.30	0.30	0.30	0.28	0.32	0.25
310	Static	0.32	0.32	0.32	0.31	0.32	0.34	0.34	0.29
	Dynamic	0.29	0.30	0.30	0.31	0.32	0.32	0.33	0.34
316	Static	0.26	0.26	0.30	0.34	0.30	0.32	0.31	0.24
321	Static	0.26	0.24	0.26	0.27	0.30	0.30	0.23	0.18
	Dynamic	0.28	0.29	0.30	0.31	0.32	0.33	0.34	0.35
347	Static	0.30	0.30	0.31	0.30	0.33	0.31	0.35	0.28
	Dynamic	0.28	0.29	0.30	0.31	0.32	0.24	0.33	0.34

* Data from Fredericks (20) and Garofalo, Malenock and Smith (21).

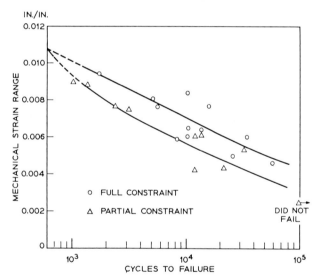

Fig. 19. Effect of constraint on resistance of solution-treated Type 347 to thermal cycling between 210 and 1110 F (22).

This potential offered by the stainless steels in raising steam temperatures and improving efficiency has been an important factor in the advancement of utility power plants over the past 15 to 20 years. The increase in steam temperature reached a peak at Eddystone Unit No. 1 of the Philadelphia Electric Co., where 1200 F with 5000 psi was used. For a while it looked as if there would be a steady increase in steam temperatures beyond the capability of the standard stainless steels. However, there has since been a downward trend in the newest designs to 1000-1050 F, which may represent only a temporary plateau from which new ideas and designs will develop. In part, it has been due to the cost of constructing plants for temperatures of the order of 1100 to 1200 F and in part to the fact that studies have indicated a decrease in operating availability as pressure, temperature and size of the units have increased.

This current change in philosophy, therefore, brings stainless steel into an area where it is competitively not supreme. Many of the earlier troubles encountered with the first uses of stainless steel in the power industry were unquestionably growing pains, and have been analyzed and understood. These difficulties were two-fold. As already mentioned, one was the low strength of some Type 321 superheater tubing, which was overcome by the adoption of the Type 321H modification. The second was cracking of some welded Type 347 heavy-wall steam piping; this was countered by use of a 16Cr-8Ni-2Mo filler metal in welding the already existing Type 347 pipe and by adoption of Type 316 (with some Type 304) in the newer installations. Thus, today, the major factors governing the selection of stainless steel are design and over-all economy.

Superheaters operate at the highest steam lines is influenced by the maximum temperature of the steam. With steam temperatures over 1050 or 1060 F, the austenitic stainless steels are usually selected because of their higher allowable stresses, which mean significantly smaller and lighter-walled tubes and pipe. This can be illustrated by the allowable stresses for Type 304 seamless pipe and tubes in unfired pressure vessels, as compared with those for the most popular of the low-alloy chromium-molybdenum steels (see next column):

	Allowable Stress, psi at maximum metal temperature (°F) of						
	900	950	1000	1050	1100	1150	1200
TP 304	10,000	9,750	9,450	9,000	8,250	6,900	5,500
2¼Cr-1Mo	13,100	11,000	7,800	5,800	4,200	3,000	2,000

Thus, it is clear that entirely apart from the question of scaling and corrosion, there is no question of the economics of using stainless steel at the higher temperatures. At 950 F and below, there may be no advantage to using stainless steel, while at 1000 F the strength advantage might be outweighed by cost factors.

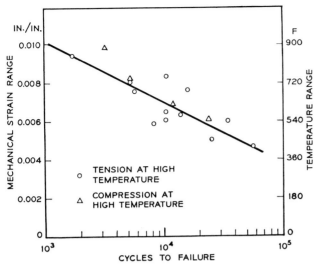

Fig. 20. Thermal-stress fatigue of solution-treated Type 347 as a function of type of high-temperature stress (22).

Fig. 22. Stainless-steel rupture discs are precision pieces of foil designed to rupture at a predetermined pressure and temperature setting. Types 316 and 347 are furnished in diameters from ⅛ through 44 in., bursting pressures from 2 to 100,000 psi and with a maximum temperature of 900 F for Type 316 and 750 F for Type 347.

steam pressures in the system and at metal temperatures approximately 100 F higher than the steam they carry. Major amounts of stainless steel, today primarily Types 304, 321, and 347, have been used successfully in superheaters and reheaters in power boiler service since about 1949. It is common practice to use austenitic steel in the higher temperature areas of superheaters and reheaters having

outlet steam temperatures of 1000 F and higher. Stainless steel is used exclusively for temperatures over 1100 F and can be economical for lower temperatures, especially for pressures above 1200 psi. Part of the boiler tubes for the steam generator for the Eddystone Unit No. 1 are of Types 316 and 321; Fig. 24 illustrates the size of a single Type 304 superheater section. Obviously, from the

size of these units, hundreds of thousands of feet of stainless steel superheater and reheater tubing are in use in the USA today.

Main steam piping (turbine leads) of stainless steel first went into service in November 1948 at the Sewaren 2 unit, operating at 1050 F and 1500 psi. Since that time, stainless steel has been used for main steam piping in a number of plants, largely those owned by Public Service Electric and Gas Co. of New Jersey (15 units) and American Electric Power Corp. (8 units). The early installations were all Type 347, which was later switched to Type 316 to overcome the cracking problems encountered with welded heavy-wall piping of Type 347.

Austenitic stainless steels are essential for steam conditions of 1060 F and higher; some power-plant engineers have felt its use at temperatures as low as 1000 F is desirable. Between 1000 and 1050 F, for example, stainless steel is economical if the plant layout is so arranged that short stainless steel pipes produce permissible forces and moments at boiler and turbine terminal connections, while longer loops of heavier wall ferritic chromium-molybdenum pipes would be required for comparable flexibility.

In this brief summary of high-temperature applications of stainless steel, some representative parts where stainless steel has proved essential or economical, have been discussed. It is the author's belief that increasing use

Fig. 21. Spiral heat exchangers consist of one spiral wrapped around another so each medium has the same flow characteristics. They are selected when conditions call for a compact design having the ability to handle sludges and slurries with low fouling incidence. This unit, fabricated of 18-18-2, is used to combat a problem of stress-corrosion cracking in a condenser employed in a tar plant.

Fig. 23. Type 316 stainless steel forging weighing 22,655 lb serves as head for reactor pressure vessel.

Fig. 24. Type 304 is used for boiler superheaters such as in this section being hoisted into place.

will be made of the hot strength of stainless steel with the growing trend of industry towards higher temperatures for greater efficiency.

REFERENCES

1. F. H. Norton, The Creep of Steel at High Temperatures McGraw-Hill, New York (1929).
2. Report on the Elevated Temperature Properties of Stainless Steel, issued under the auspices of The Data and Publications Panel, ASTM-ASME Joint Committee on The Effect of Temperature on the Properties of Metals, ASTM STP 124 (1952).
3. Report on the Elevated Temperature Properties of Stainless Steel, issued under the auspices of The Data and Publications Panel, ASTM–ASME Joint Committee on The Effect of Temperature on the Properties of Metals, ASTM Data Series Publications DS 5-S1 (1965).
4. T. M. Krebs and N. Soltys, A Comparison of the Creep-Rupture Strength of Austenitic Steels of the 18-8 Series, Joint International Conference on Creep 1963; papers published by Institution of Mechanical Engineers (1963) 6-101.
5. K. G. Brickner, G. A. Ratz and R. F. Domagala, Creep-Rupture Properties of Stainless Steels at 1600, 1800, and 2000 F, Advances in the Technology of Stainless Steels, ASTM STP 369 (1965) 99.
6. W. H. Pryle and E. T. Wessel, Tensile Properties of AISI Type 304 and 347 Stainless Steels at Moderate Temperatures for Section Sizes Ranging from Bars to Extremely Large Forgings, ASME paper 60-WA-9.
7. G. H. Rowe and J. R. Stewart, Creep-Rupture Behavior of Type 316 Stainless Steel Weldments Prepared With and Without Restraint, Welding Research Council, 27 (1962) 534-s.
8. E. A. Davis, Creep Rupture Tests for Design of High-Pressure Steam Equipment, ASME paper 59-MET-14.
9. P. M. Brister and W. E. Leyda, Establishing Allowable Design Stresses for Boilers and Pressure Vessels at Elevated Temperatures, Joint International Conference on Creep 1963; papers published by Institution of Mechanical Engineers (1963) 4-65.
10. B. E. Bolton, New Stress Criteria for Designing Economical Higher Steam Temperature Boilers, Joint International Conference on Creep 1963; papers published by Institution of Mechanical Engineers (1963) 4-55.
11. R. C. Anstine, The Combined Effects of a Sodium Environment and Extended Life on Type 316 Stainless Steel and Croloy 2-1/4 Alloy Steel Design Stresses; Report No. BW-67-3 The Babcock & Wilcox Company, Barberton, Ohio USAEC Contract AT(11-1)-1280 (Sept 1964).
12. D. D. Doerr, Determination of Physical Properties of Ferrous and Non-ferrous Structural Sheet Materials at Elevated Temperatures, WADC TR 6517, Part 2 (1954).
13. G. V. Smith, W. B. Seens, H. S. Link and P. R. Malenock, Microstructural Instability of Steels for Elevated Temperature Service, Proc ASTM 51 (1951) 895.
14. J. Hoke and F. Eberle, Experimental Superheater for Steam at 2000 Psi and 1250 F, Report After 14281 Hours of Operation, ASME Paper 55-A-102.
15. E. J. Dulis, G. V. Smith and E. G. Houston, Creep and Rupture of Chromium-Nickel Austenitic Stainless Steels, Trans ASM, 45 (1953) 42.
16. E. J. Dulis and G. V. Smith, Creep and Creep-Rupture of Some Ferritic Steels Containing 5 per cent to 17 per cent Chromium, Proc ASTM 53 (1953) 627.
17. R. J. Christoffel, Notch-Rupture Strength of Type 347 Heat-Affected Zone, Welding Res Council 25 (1960) 315-s.
18. R. J. Christoffel, Notch Sensitivity of the Heat-Affected Zone in Type 316 Material, Welding Res Council 28 (1963) 25-s.
19. R. H. Caughey and W. G. Benz, Jr., Material Selection and Fabrication, Main Steam Piping for Eddystone Trans ASME 82 (1960) Series D, 293.
20. J. R. Fredericks, A Study of the Elastic Properties of Various Solids by Means of Ultrasonic Pulse Techniques. Doctorate, University of Michigan (1947). Elastic Constants of Steels at Elevated Temperatures. Resume of High Temperature Investigations Conducted During 1948-50, Timken Roller Bearing Co. (1950).
No. 1 1200-F and 5000-Psi Service,
21. F. Garofalo, P. R. Malenock and G. V. Smith, The Influence of Temperature on the Elastic Constants of Some Commercial Steels, symposium on Determination of Elastic Constants, ASTM STP 129 (1952) 10.
22. L. F. Coffin, Jr, An Investigation of Thermal-Stress Fatigue as Related to High-Temperature Piping Flexibility, Trans ASME, 79 (1957) 1637.
23. R. A. Baker and H. M. Soldan, Service Experiences at 1050° F and 1100° F of Piping of Austenitic Steels, Joint International Conference on Creep 1963; papers published by Institution of Mechanical Engineers (1963) 4-85.
24. H. S. Blumberg, Type 316 for Elevated-Temperature Power-Plant Applications, Climax Molybdenum Co. (1965).
25. G. E. Lien, Experience with Stainless Steels in Utility Power Plants, Advances in the Technology of Stainless Steels and Related Alloys, ASTM STP 369 (1965) 136.
26. I. A. Rohrig, Residual Elements and Their Effect on the Applications of Austenitic Stainless Steels in the Power Industry, ASTM Symposium on Effects of Residual Elements on the Properties of Austenitic Stainless Steels (1966) paper No. 19.

Corrosion Resistance of Stainless Steels at Elevated Temperatures

L. A. Morris

STAINLESS STEELS are among the most popular construction materials for elevated-temperature process systems. Proper selection of the optimum grade for a particular set of environmental conditions presents numerous problems to the materials engineer. Consideration must be given to the high-temperature strength and structural stability of these steels; the economics of the selection must be feasible on a cost-to-performance basis. Adequate corrosion resistance is necessary for efficient operation of process equipment. Although it is recognized that all facets must be considered, the present discussion will be limited to the subject of corrosion resistance at elevated-service temperatures (greater than approximately 1000 F). Furthermore, since the selection of a material for a particular application is made from a series of available products, specific reference will be made to the standard AISI grades of stainless steel.

Much attention has been given to the compatibility of stainless steels with air or oxygen. However, recent trends in reactor design and steam generation equipment have resulted in renewed interest in steel oxidation in carbon monoxide, carbon dioxide and water vapor. There is evidence that the mechanism of attack is similar

The author is associated with the Metallurgical Laboratories of Falconbridge Nickel Mines Ltd., Thornhill, Ontario, Canada. This paper was presented at the AISI Stainless Steel Symposium held 16 October 1967 at the Materials Engineering Congress, Cleveland.

in these gases although the reaction rates are different. Exposure to mild conditions leads to the formation of a protective oxide film; when conditions are severe, film breakdown occurs and thick stratified scales are produced. The onset of this transition is unpredictible and is sensitive to alloy composition and atmospheric conditions.

Other gases, such as sulphur dioxide, hydrogen sulphide, hydrocarbons and the halogens, and contaminants which arise from the combustion of residual fuels (vanadium pentoxide and alkali salts) may be present in variable proportions and strongly affect corrosive conditions. Information concerning the practical limits of operation are available; nevertheless, considerable disagreement exists regarding the mechanisms of attack. Much the same is true for high-temperature corrosion by liquid metals and salts, although the former has received some attention by way of nuclear reactor operation.

This paper is intended to be a summary of the practical information with respect to the compatibility of stainless steels with the environments mentioned above; wherever possible, the important details of the reaction mechanisms are presented. In order to clarify the information related to scaling mechanisms, a general review of metal oxidation precedes the discussion of stainless steel corrosion. A detailed description of oxidation theory and mechanisms may be had by consulting one of several excellent monographs recently published (1-5).

Oxidation of Metals and Alloys

Oxide Structures

Practically all oxides are semiconductors, and electrical conduction may occur either by electron holes, (p-type semiconduction), or by electrons, (n-type semiconduction). Oxide semiconductors are not of exact stoichiometric composition, but may contain an excess of either cations or anions. This excess is accomplished by having cation or anion vacancies or ions in interstitial positions in the lattice. Thus, the model for a p-type semiconducting oxide is one in which the cation lattice contains vacant sites, and electrical neutrality is maintained by the formation of cations of higher valency or electron holes. Electrical conductivity occurs by the movement of electron holes, and ions via cation vacancies. The model for a metal excess or n-type oxide is one in which there are metal ions and electrons in interstitial positions or anion vacancies in the lattice.

The oxides formed on alloys are ternary semiconducting layers and the dissolution of solute metal ions into the oxide layer of the solvent element affects the concentration of defects. The rate of growth of these compounds depends on the defect concentration; therefore, reactions kinetics may be increased or decreased by the addition of solute ions.

The variable composition range of oxides is one of the most important factors when considering oxidation reactions. The nature of the defect structure of an oxide, which thickens by a diffusion mechanism, determines the oxidation rate. For an oxide layer growing on a metal, an oxygen pressure gradient and a defect gradient exist across the oxide. A concentration gradient within a single solid phase causes diffusion of ions via interstitial or vacant lattice sites. Therefore, a knowledge of the defect structure and nature of the diffusion species is necessary in order to understand the oxidation mechanism.

Oxidation Rates and Mechanisms

Although compact oxides grow by a diffusion mechanism, other processes may be rate controlling, giving rise to a number of empirical rate laws. A major consideration of gas-metal reaction studies is the determination of an empirical rate equation and, if possible, theoretical rate equations based on physical theory to explain the kinetics of the reaction.

The simplest rate expression is the linear equation, where the thickness of the oxide film has no influence on the rate of uptake of oxygen, and is directly proportional to time. If an oxide is non-protective, offering no barrier between the gas phase and metal surface, the linear law is expected to hold. Porous or cracked oxides are formed on a number of metals which show a strong tendency to oxidize at a linear rate. These metals generally have a low or high volume ratio of oxide to metal consumed. A low volume ratio suggests that the metal is always exposed to the gas phase.

Under special conditions, the growth of a compact pore-free scale may follow a linear rate law. For example, owing to the exceptionally high concentration of vacant lattice sites in some oxides, ions can diffuse rapidly to the oxide-gas interface, and the surface reaction becomes rate controlling when the oxidizing potential of the gas phase is low.

The parabolic rate has been observed in numerous cases and is based on rate control by either cation or anion diffusion across a barrier film. The oxidation rate is inversely proportional to the film thickness, and there-

fore, oxide growth decreases with increasing time. Wagner has described the mechanism for scale growth on a pure metal according to this law (6). The rate of oxidation is controlled by the diffusion of reactants on a concentration gradient existing across the scale. Wagner derived an expression for the rate of oxidation in terms of the specific conductivity of the oxide, the transport numbers of ions and electrons and the free energy decrease of the oxidation reaction. Experimental verification was obtained for several metals.

Kinetic data for thin film formation has been found to obey cubic and logarithmic rate laws. It has been suggested that thin films ($<10^{-4}$ cm) form by ion migration under the influence of a strong electric field across the oxide, whereas thick films are formed by thermally activated diffusion in an electrically neutral oxide.

The oxidation rates of alloys rarely agree with a given rate law over a range of conditions. Deviations often occur and a combination of rate equations is required to describe the reaction kinetics. Therefore, it is not possible to predict the effects to be expected from the addition of an alloying element to a pure metal by a unified theory. Numerous factors must be considered; for example, the affinity of the alloy constituents for the components of the reactant gas, the solubility limits of the phases, the diffusion rates of atoms in alloys and in the oxides, the formation of ternary compounds, the relative volumes of the various phases, all of which may be functions of temperature and pressure. Several limiting cases of alloy oxidation have been theoretically treated, and it is hoped that advancements in knowledge will permit the derivation of a more universal theory.

Oxide Morphologies

Moreau and Bènard (7) have presented a general classification of the different scale morphologies which may occur on an alloy A-B based on experimental observations. The classes are illustrated diagrammatically in Fig. 1. Class I refers to the case of selective oxidation (preferen-

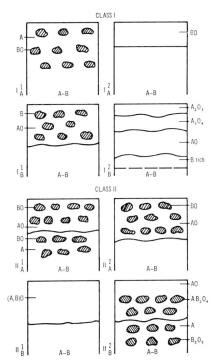

Fig. 1. Oxidation modes of alloys. (After Moreau and Benard, Ref. 7.)

tial oxidation of a single element). This classification can be divided into two subgroups, I_A and I_B, where the former concerns the selective oxidation of the minor addition and the latter the major component. In each case, there appears two possibilities. The oxide of the minor more reactive element may be nucleated and precipitate within the matrix of the major element. The formation process is called internal oxidation, (I_A^1). Another possibility is the formation of an oxide film of the minor element, (I_A^2).

Again two possibilities exist in the case of the selective oxidation of the major element. A thick oxide layer may be formed on the alloy surface, which contains entrapped globules of the addition element, (I_B^1), or the more noble element may concentrate at the metal-oxide interface, and subsequently diffuse back into the metal phase (I_B^2).

Class II deals with the simultaneous oxidation of both elements. Again, it is possible to subgroup the various modes of oxidation. In Class II_A, the oxides are insoluble, and either the minor or the major element has a greater affinity for oxygen. If the minor element has a higher affinity, its oxide may nucleate on the surface or precipitate as internal oxide. Since the oxide of the major element can form, it will entrap these nuclei or precipitates as the metal-oxide interface recedes. The end result is a dispersion of one oxide in the other and internal precipitates of the former in the metal phase, (II_A^1). If the major element has a greater affinity for oxygen, internal oxidation does not occur as above. Therefore, the reaction product consists of a dispersion of oxide particles in the oxide of the major element, (II_A^2).

The second subgroup refers to mixed oxides, that is oxides in which the two alloying elements are associated. These oxides occur most frequently at high temperatures and oxygen potentials. The simplest case to describe is the formation of a solid solution of one element in the oxide of the other, (II_B^1). The other case concerns the formation of oxides which approach stoichiometric composition, (II_B^2). This reaction often gives rise to a spinel oxide, AB_2O_4.

The above classification is quite general, and introduces some clarity to the complex problem of alloy oxidation. It is unfortunate that variations from these simple cases may occur as a result of mechanical stresses, coalescence of oxides, and other factors often characteristic of a particular system.

In the above classification, Moreau and Bènard also considered the possible formation of multilayered scales. If a metal exhibits several stable oxides, they may appear in various proportions, depending on conditions of temperature and pressure. The oxide richest in metal will be located adjacent to the metal phase, whereas that richest in oxygen will be nearest the gas phase. Even under conditions of high oxidation potentials, where only the appearance of the most stable oxide is expected, the potential at the metal-oxide interface favors the formation of the lower oxides, and these will grow. Examples illustrating these characteristics are copper, which forms cuprous and cupric oxides, and iron, which forms wüstite, magnetite and hematite.

High Temperature Corrosion of Stainless Steels

REACTION WITH OXIDIZING GASES

Oxidation reactions are the more frequent causes of high-temperature corrosion of stainless steels. This is not surprising because most environments are either air, oxygen, carbon dioxide and steam, or complex atmospheres containing one or more of these gases; (sulphur dioxide may also be included but has been deferred to the following section). Most studies have been associated with air or oxygen (dry and damp), and a considerable amount of fundamental and practical data has been recorded. Rapid accumulation is presently occurring with respect to corrosion by carbon dioxide and steam at temperatures up to 2200 F. Nevertheless, the oxidation behavior of stainless steels still defies description by a unified mechanism partly due to the complex nature of the problem, and partly due to disagreement and lack of reproducibility concerning scale structure and kinetic data. The large number of combinations that exist in gaseous and alloy compositions adds further complications. However, with the information at hand, it is usually possible to specify the most suitable grade of stainless steel for mildly corrosive conditions; when operating conditions become severe, it is often necessary to test coupons in actual or carefully simulated service conditions (8).

Reaction Kinetics

Most of the empirical rate expressions have been observed with chromium and chromium-nickel stainless steels and, therefore, it is difficult to discuss any particular rate equation. A general formulation of rate data has been presented by Wood (9) and is illustrated in Fig. 2. Under mild conditions, a protective film grows according to curve OAD and the rate of oxidation decreases with time (the parabolic rate may be closely approximated in this range). For severe environments, an initial (induction) period OA is followed, then a sudden increase in rate AB occurs (film

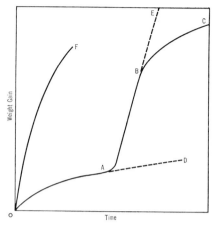

Fig. 2. Typical rate data for the oxidation of chromium and nickel-chromium stainless steels at elevated temperatures. (After Wood, Ref. 9.)

breakthrough). Depending on alloy composition and oxidizing conitions, either self-healing of the oxide will occur and curve BC is followed for some time after which breaks of the type AB are repeated, or a non-protective oxide is maintained and BE is followed. Under extreme conditions, OF is followed directly and the oxide is non-protective. The actual geometry of the curves and the occurrence of "breakthrough" are dependent on time, alloy composition and environment. Reproducibility with standard conditions is often poor. For these reasons, and the fact that corrosion data are reported in a variety of ways, (weight gain per unit area, weight loss per unit area, penetration rate or scale thickness), it is difficult to precisely compare the performance of the different grades under the same conditions. Nevertheless, general trends can be indicated.

Oxidation Mechanisms

When iron is heated in highly oxidizing gases at temperatures greater than 1000 F, a multilayered scale is formed consisting of wüstite at the metal-oxide interface, an intermediate layer of magnetite, and hematite at the oxide-gas interface. Oxidation proceeds at a rapid rate, and this has been attributed to the occurrence of the wüstite phase. This oxide contains an inordinate number of cation vacancies and a large composition range, which are conducive to rapid diffusion of iron ions across this compound. Thus, the oxide thickens rapidly at the expense of the metallic phase. It has been established that additions of chromium and nickel to iron impart oxidation resistance by removal of wüstite as a stable phase. Examination of the Fe-Cr-O (10) and Fe-Ni-O (11) phase diagrams illustrates that chromium additions are much more effective than nickel in this respect, and for equilibrium conditions (at least at temperatures > 1800 F) quantities greater than 13% Cr promote the formation of oxides rich in chromium. The excellent resistance of the ferritic chromium and austenitic chromium-nickel stainless steels to mildly oxidizing conditions is attributed to the protective nature of these compounds.

When chromium and chromium-nickel stainless steels are heated to elevated temperatures in an oxidizing gas, chromium is selectively oxidized producing a thin film of primarily chromic oxide, "Cr_2O_3" (the quotation marks indicate that some iron is in solid solution). Oxidation occurs at a markedly slow rate according to OA in Fig. 2, since cations diffuse through this film at extremely slow rates. This film is referred to as a Type A scale (12, 13-16) and is characteristic of low oxidation rates at low temperatures and/or high chromium contents. After a period of time, depending on alloy composition and environment, film "breakdown" occurs and a rapid period of oxidation, typified by AB in Fig. 2, is experienced. This results in the formation of a Type B scale, which is a duplex structure consisting primarily of an inner layer of spinel oxide, $Fe\,Fe_{(2-x)}Cr_xO_4$ where $0 < x < 2$, which may contain nickel, manganese and silicon in solution, and an outer layer of ferric oxide, "Fe_2O_3", containing chromium and other elements in solution. The spinel oxide can range in composition from $FeCr_2O_4$ at the metal oxide interface to Fe_3O_4 at the spinel-ferric oxide interface.

The breakdown of the Type A scale to the Type B scale is classified as a Type I break (12, 13-16). In atmospheres of low oxidizing potential (CO/CO_2, H_2/H_2O, etc.), wüstite containing small quantities of chromium and nickel, (Fe, Cr, Ni)O, may occur in the scale on alloys of marginal chromium content ($< 15\%$). In this case, a complex stratified scale may be formed consisting of "Cr_2O_3" (as internal oxide or a thin continuous surface layer), the spinel phase $Fe\,Fe_{(2-x)}Cr_xO_4$ with $x \sim 2$, (Fe, Cr, Ni)O, Fe_3O_4 and Fe_2O_3 where Fe_2O_3 is nearest the gas phase. If the oxygen pressure is reduced below the dissociation pressure of Fe_2O_3, then Fe_3O_4 will appear as the outermost phase. A further reduction would establish (Fe, Cr, Ni)O as the outer phase.

The mechanism and morphological developments of breakthrough are not clearly understood. It appears that breakthrough occurs at highly localized sites randomly distributed over the specimen surface; this results in the production of warts or nodules of stratified Type B scale (14, 15). Depending on alloy composition and oxidizing conditions, the nodules may expand laterally until the entire surface is covered with stratified scale (9). In some cases, breakthrough remains localized, and the protective oxide rich in "Cr_2O_3" persists for long periods of time (17).

Two theories have been presented to account for the Type A to Type B scale change. It has been suggested that either a chemical mechanism is operative, whereby the protective scale is penetrated by iron ions causing transformation to the spinel oxide and producing iron (ferric) oxide at the outer surface (14-16), or scale cracking occurs and the underlying alloy (depleted in chromium) reacts directly with the atmosphere producing the spinel oxide and outer iron oxides. In either case, stratified scales are generated over whole or part of the surface; the scale may again become protective, establishing kinetics according to curve BC, or remain non-protective and follow BE. Only one Type I break will occur per sample and indeed may never occur for alloys of high chromium content at low temperatures (14).

If kinetics are established according to curve BC, subsequent breaks of similar geometry to AB may occur. These have been classified as Type II breaks and are characterized by cracks in scales which are thicker than about 10 mu (14). After the break, the scale becomes protective and rates are re-established according to the BC geometry. Oxidation continues, and this pattern may be repeated. At temperatures greater than 1600 F, the occurrence of Type II breaks appears to be the general rule rather than the exception. It has been suggested that accumulation of voids and silica at the metal/oxide interface contributes to scale cracking (19).

Internal oxidation of the alloy phase also occurs when the metal is in contact with the spinel oxide. If localized breakdown occurs, sub-scale precipitation is only noted below the nodular growths. The oxide particles are "Cr_2O_3" or spinel of higher chromium content than the spinel phase in the scale. The chrome-rich subscale can develop below a spinel rich in iron since the oxygen potential at the metal/oxide interface is sufficient to form these oxides. Subscale formation is not observed beneath "Cr_2O_3" films in pure alloys; however, a silica subscale has been reported for commercial steels (17).

The surface reactions which occur on stainless steels in oxidizing gases are extremely complex. In addition, impurities in commercial materials generate further complications and many investigators are confining their research efforts to pure binary alloys in an attempt to clarify some of the discrepancies of stainless steel oxidation.

Practical Considerations

General. It is apparent from the above discussion that when stainless steels are exposed to an oxidizing atmosphere, chromium is selectively oxidized to form "Cr_2O_3." Cation diffusion rates in the compound are extremely slow, and therefore, this oxide is protective. If conditions are severe, film breakdown occurs leading to the formation of stratified scales, consisting of an inner spinel oxide and an outer ferric oxide containing some chromium in solution ("Fe_2O_3"). These oxides offer some protection to the metallic phase. Extreme conditions give rise to stratified scales in very short time periods (< 1 hr) (14). These conditions are encountered primarily at high temperatures and low chromium and nickel contents. Reaction mechanisms, although desirable, are still not reliable as a basis for alloy selection.

Experience has shown that as the chromium and nickel contents in iron are increased, the oxidation rate is reduced. The most pronounced reduction occurs when chromium is present in excess of 12 to 13% and scales rich in "Cr_2O_3" are produced. Since reaction depletes chromium in the alloy phase, greater contents are required at higher temperatures in order to maintain this film. Table 1 indicates the maximum recommended service temperatures in air as a function of chromium and nickel contents (standard AISI grades) to prevent excessive scaling. For isothermal operation (continuous service), the beneficial effect of chromium is apparent. For an approximately constant chromium content, increasing nickel compositions permit higher operating temperatures. It has been suggested that even a small amount of nickel ions dissolved in "Cr_2O_3" reduces the rate of cation diffusion and the rate of oxidation (9). Furthermore, it appears that nickel retards the transformation of Type A ("Cr_2O_3") scales to Type B scales (spinel and iron oxides) (19).

In many processes, isothermal conditions are not maintained and operating temperatures must be lowered. The temperature limits in this case are shown in the column, "Intermittent Service" in Table 1. Expansion and contraction differences between the alloys and scales during heating and cooling cause cracking or spalling of the protective scales. This allows the oxidizing media to attack the exposed metal surface. (Cyclic operation is rarely considered in stud-

Table 1. Recommended Maximum Service Temperatures in Air

AISI type*	Intermittent Service °C	°F	Continuous Service °C	°F
201	815	1500	845	1550
202	815	1500	845	1550
301	840	1550	900	1650
302	870	1600	925	1700
304	870	1600	925	1700
308	925	1700	980	1800
309	980	1800	1095	2000
310	1035	1900	1150	2100
316	870	1600	925	1700
317	870	1600	925	1700
321	870	1600	925	1700
"330"	1035	1900	1150	2100
347	870	1600	925	1700
410	815	1500	705	1300
416	760	1400	675	1250
420	735	1350	620	1150
440	815	1500	760	1400
405	815	1500	705	1300
430	870	1600	815	1500
442	1035	1900	980	1800
446	1175	2150	1095	2000

* Type 330 not an AISI type.

ies related to oxidation mechanism). The expansion characteristics of the straight chromium grades are more compatible than the austenitic chromium-nickel steels and this is reflected in the temperature limits in Table 1. Howes (20), however, has reported that for oxide thicknesses of approximately 1 mu on an Fe-28% Cr alloy, oxide reaction to cyclic conditions is sensitive to temperature of oxidation, rate of cooling and final temperature. Experiments demonstrated that even for slow cooling, (20 C/hr), oxides thicker than 1 mu formed above 1292 F spalled on cooling below 932 F. Oxides greater than 50 mu did not spall on slow cooling.

The spalling resistance of the chromium-nickel steels is greatly improved at higher nickel levels (21) and is illustrated in Fig. 3. Nickel reduces the differential thermal expansion between alloy and oxide and thereby reduces stresses at the alloy/oxide interface during cooling. The austenitic grades are often specified at intermediate and elevated temperature since they are more resistant to brittle sigma phase formation and have better load-bearing capacity than the ferritic steels.

Effect of Atmosphere. Although

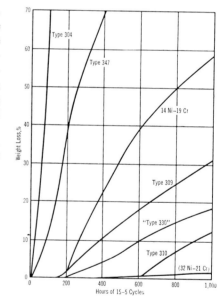

Fig. 3. Resistance of austenitic stainless steels to cyclic conditions at 1800 F (cycle consists of 15 min at temperature and 5 min in air). (After Eiselstein and Skinner, Ref. 21.)

reaction mechanisms are probably similar in air, oxygen, water vapor and carbon dioxide (formation and breakdown of protective oxides), reaction rates may vary considerably. For example, similar scaling behavior has been observed for chromium and chromium-nickel steels in air and oxygen; however, the reaction rates are greater in the initial stages in pure oxygen and scale breakdown occurs more rapidly as the oxygen potential (in oxygen-nitrogen atmospheres) is increased (22). Results obtained in air should be applied with care when considering service in pure oxygen. Nitrogen in air may substitute for oxygen ions in the protective "Cr_2O_3" scale and assist in reducing oxidation rates (9).

In almost all cases considered, an increase in corrosion rate is observed for chromium and chromium-nickel steels in the presence of large or small amounts of water vapor (14, 23-25). Exceptional behavior was noted for Type 446; corrosion was less in moist air than in dry air (14).

Figure 4 illustrates the effect of moist air on the oxidation of AISI Type 302 and a 15Cr-35Ni alloy generally referred to as "Type 330." Type 302 undergoes rapid corrosion in wet air at 2000 F, whereas a protective film is formed in dry air. The higher nickel "Type 330" is less sensitive to the effects of moisture. In a recent literature survey (26), it was illustrated that increasing chromium and nickel contents permit higher operating temperatures in moist air. Type 446 is usable at temperatures approaching 2000 F; however, Types 310 and 309 are superior at temperatures greater than 1800 F. For the other grades listed in Table 1, the temperature limits for service in moist air should be adjusted downwards. Ruther and Greenberg have shown that the addition of moisture to oxygen significantly increases the corrosion rates of Types 304, 321, 316 and 347, see Table 2 (27).

Incorporation of nuclear steam superheaters before turbogenerators has stimulated investigations on the corrosive effects of steam (27-30). Ruther and Greenberg (27) studied the effects of steam temperature, pressure, oxygen and hydrogen content and velocity on the general corrosion of AISI Type 304 and an alloy (essentially AISI Type 405 containing higher aluminum) generally referred to as "Type 406". Several tests were carried out on other 300 and 400 Series steels. The corrosion of steels is extremely sensitive to the method of surface preparation (27, 30). Treatments which introduced surface cold work resulted in a substantial decrease in corrosion rate for times up to 2000 hr. The effect of cold work is reflected in the data reported in Table 2 (compare corrosion of samples prepared by electropolishing and wet grinding). The total corrosion of mechanically treated (MT) and electropolished (EP) Type 304 with increasing temperature is shown in Fig. 5. Caplan (31) has reported that cold work improves the resistance of Fe-24.3 Cr and Fe-26.2 Cr alloys to oxidation in damp argon at 1112 F by suppressing internal oxidation, probably as a result of increased diffusivity of chromium in the cold worked metal phase. Once a barrier oxide film is formed in the early stages of reaction, chromium depletion is not so severe and internal

Table 2. Defilmed Metal Weight Losses of Stainless Steel Tested at 1200 F for Seven Days

Sample and Surface Preparation	Defilmed Metal Weight Loss, mg/cm²		
	dry O_2	wet O_2	Steam (600 psig) (30 ppm O_2)
304 wet ground	0.19	3.6	6.6
electropolished	0.16	5.9	7.5
321 wet ground	0.27	1.2	1.8
electropolished	0.16	6.5	9.2
316 wet ground	0.15	1.8	0.34
electropolished	—	6.6	9.0
347 wet ground	0.24	2.3	3.9
electropolished	0.36	6.4	11.2
403 electropolished	—	—	13.5
405 electropolished	—	—	18.9
"406" electropolished	—	—	2.0
410 electropolished	—	—	11.4
430 electropolished	—	—	5.0
446 electropolished	—	—	0.34

(After Ruther and Greenberg (27)).

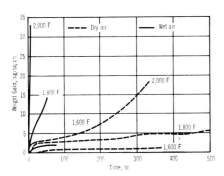

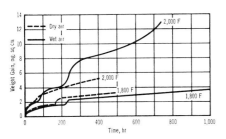

Fig. 4. Oxidation of AISI Type 302 and "Type 330" in wet and dry air. Upper curves, 302; lower curves, "330". (After Caplan and Cohen, Ref. 14.)

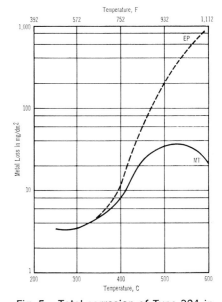

Fig. 5. Total corrosion of Type 304 in superheated steam after 1000 hr; MT = mechanically treated; EP = electropolished. (After Warzee et al, Ref. 30.)

oxidation is suppressed; the beneficial effect of surface cold work is still evident after long exposure times even though the metal is annealed at temperature.

At 1112 F, flowing steam (30 to 91 m/sec) was noted to be approximately 1.5 times more corrosive to Type 304 than static gas; oxygen contents up to 30 ppm have no effect (27). Flaking and loss of the outer corrosion products were severe in the dynamic tests. The 400 Series exhibited improved corrosion resistance as the chromium content was increased; and a dramatic decrease was observed at levels > 16% Cr (Types 430 and 446), see Table 2. Type 406 had high resistance to steam and it was assumed that the aluminum content was responsible for reduced corrosion rates.

It is difficult to indicate maximum service temperatures for steam service, especially in view of the sensitivity of corrosion rate to surface condition. It has been stated that the austenitic grades may be used up to 1600 F and Types 309, 310 and 446 at higher temperatures (28-32). Types 304, 321 and 347 are being used to produce low-pressure steam at temperatures approaching 1400 F (25). Scales on Types 304, 347 and 316 tend to exfoliate at higher temperatures (1500 F) (33).

The oxides and oxide morphologies produced on stainless steels in steam and simulated steam, (water vapor-argon atmospheres) have not been clearly established. Several investigators have reported scale structures similar to the Type B scales found on alloys oxidized in air, namely, an inner spinel layer and an outer ferric oxide layer (28, 29, 31). This type of structure was noted for AISI Type 304 and the so-called Type 406 in superheated steam and on an Fe–24 Cr alloy in simulated steam (Ar saturated at 77 F). Only the spinel layer was observed on Type 347 in one atmosphere steam in the temperature range 1110-1290 F (24). Large amounts of wüstite occur on iron-chromium alloys in simulated steam (O.1 H$_2$O, 0.9 Ar) at 1290 to 2010 F for alloys containing less than 15%

Cr (34). A characteristic scale consisting of an outer layer of wüstite and a porous inner layer of wüstite and iron-chromium spinel is produced. At temperatures below 1650 F a layer of higher oxides of iron is formed on the outer wüstite scale. At 1470 F, the layer is magnetite (Fe$_3$O$_4$) and at 1290 F hematite (Fe$_2$O$_3$) is produced on magnetite. The authors postulated that the outer wüstite scale grows by cation migration and the porous inner scale by a dissociative mechanism (vapor transport mechanism). This mechanism may be important in cases where porous scales are formed on steels.

It appears that the duplex spinel-ferric oxide scale structure is formed on alloys rich in chromium in high-pressure steam. Wüstite may occur as a major reaction product on steels of marginal chromium content (13 to 15%) at high temperatures (> 1600 F) in water vapor-argon atmospheres (simulated steam) or atmospheres of low oxidizing potential. Layers of magnetite and hematite atop the wüstite scale are favored at lower temperatures.

Interest in the oxidation of stainless steels in carbon dioxide and carbon dioxide/carbon monoxide atmospheres in the temperature range 1100 to 1800 F has been revived since austenitic steels are scheduled for use in gas-cooled nuclear reactors (35-38). McCoy (36) conducted a detailed study on the oxidation of Type 304 in carbon dioxide over the temperature range 1200 to 1800 F. The reaction kinetics were very complex, demonstrating essentially protective behavior although several transitions occurred which were approximately linear. The kinetic data for one atmosphere carbon dioxide are shown in Fig. 6. At 1200 F the scale was primarily ferric oxide, "Fe$_2$O$_3$". At 1300 to 1800 F, the major oxides were spinel and "Fe$_2$O$_3$". These scales are similar to the Type B scales observed in air oxidation (see above). Oxidation rates were less for cold worked material, however, the rates were similar to annealed material after 100 hr at temperature. Draycott and Smith (39) noted the formation of chromic

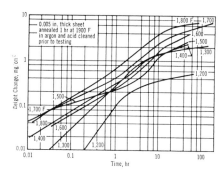

Fig. 6. Oxidation of Type 304 in carbon dioxide at 1 atm pressure. (After McCoy, Ref. 36.)

oxide "Cr$_2$O$_3$" on cold worked surfaces of 18 Cr-8 Ni-Ti steel oxidized at 1020 to 1290 F in carbon dioxide; cold worked surfaces were more resistant than etched surfaces. A multilayered scale was formed in the latter case.

Figure 7 illustrates the effect of carbon dioxide pressure on the total corrosion of Type 304 at 1500 F. A rapid increase in scaling rate occurs at approximately 10 torr; rates are insensitive to pressure above 100 torr. The steel is carburized in carbon dioxide at these temperatures and the effect of pressure is noted in Fig. 7 (carburization will be discussed below). McCoy concluded that oxidation is greater in carbon dioxide than in air.

Type 406 is highly resistant to carbon dioxide at 1700 F; a protective Al$_2$O$_3$ film was the only oxide observed (36). Protective scales of "Cr$_2$O$_3$" and spinel oxides are formed on 20 Cr-25 Ni steels in carbon dioxide over the range 1380-1560 F (35-37). Weight gains are very low (0.1 mg/cm^2 and 0.2 mg/cm^2 after

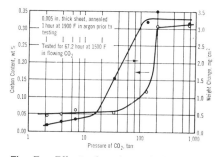

Fig. 7. Effect of carbon dioxide pressure on the carburization and oxidation of 304 at 1500 F. (After McCoy, Ref. 36.)

100 hr at 1380 and 1560 F, respectively) and the better resistance (compared to Type 304 stainless steel) is primarily due to the "Cr_2O_3" film. It appears that higher nickel contents promote the formation of protective chromic oxide scales.

Oxidation tests in dry static carbon dioxide at 500 psi indicated that scaling of Types 314 and 321 become critical at 1300 F; Type 347 showed acceptable corrosion up to 1500 F.

Effect of Impurities. In most studies on the oxidation properties of commercial stainless steels, the effects of trace elements are often overlooked. These elements may affect the rate controlling reactions producing important changes in corrosion behavior. Solution of trace elements in protective scales may influence film breakdown.

Silicon has been studied under a variety of conditions; the protective nature and subsequent breakdown of scales formed on a Type 446 steel containing 0.44% Si has been attributed to the formation of silica films at the metal/oxide interface (12, 14, 15). In a recent note, Caplan and Cohen illustrated that 0.5% Si in pure Fe-26 Cr alloys does not improve oxidation resistance, Fig. 8. Francis (42), however, found a sharp increase in oxidation rate in carbon dioxide when silicon was removed from a 20Cr-25Ni steel; this investigator postulated that silicon assisted the growth of a chromium-rich oxide film.

Manganese has been found to be a deleterious element in Type 446 (41) and in a 20Cr-25Ni steel (42). A pure Fe-26Cr-1Mn alloy oxidized

at the same rate as a Type 446 steel containing 0.75% Mn, both of which were greater than the pure binary (Fe-Cr) as shown in Fig. 8. A spinel oxide ($MnO \cdot Cr_2O_3$) occurs at the expense of the protective "Cr_2O_3" film.

The additions of small quantities of rare earth elements to an Fe-26Cr alloy improved the scale adherence (43). Internal oxide precipitates of the rare earth elements mechanically key the outer "Cr_2O_3" scale. Additions of thorium are reported to improve the oxidation resistance of austenitic chromium-nickel steels (44).

Small quantities of molybdenum are not deleterious; however, larger additions can lead to catastrophic failure of the oxide (45, 46). Extreme conditions depend on the molybdenum and chromium contents and temperature. Aluminum (as in AISI Type 405 and in the nonstandard Type 406) improves oxidation resistance when present in sufficient amounts to form aluminum-rich films (Al_2O_3 or Al-bearing spinels) (28, 36). However, "Cr_2O_3" scales are preferred when cyclic conditions are encountered (44).

REACTION WITH SULPHIDIZING GASES

Reaction with sulphur-bearing gases is another common form of high-temperature corrosion. Alloys react with sulphur in the form of hydrogen sulphide, sulphur dioxide or sulphur vapor to produce sulphides, oxides or sulphur-oxygen compounds with one or more of the alloying elements. Complex reaction products may form on a steel during high-temperature corrosion in sulphur dioxide since two competing reactions may occur:

$$SO_2 + M \rightarrow MS + O_2$$
$$SO_2 + 2M \rightarrow 2MO + S$$

In hydrogen sulphide atmospheres, hydrogen is evolved, and reaction takes place in a reducing environment. Fundamentally, the same mechanisms described above for alloy oxidation apply to the formation of sulphide scales.

Reaction Kinetics

It is sufficient to note that the empirical rate equations noted above apply equally as well to the formation of sulphide scales. Often the linear rate equation holds due to the porous nature of many metal sulphide scales. In some cases, the parabolic law is obeyed when compact scales are formed.

Sulphidation Mechanisms

The mechanisms of sulphidation of metals and alloys have yet to be explored in detail. Most studies have been confined to pure metals (1, 2). Fundamental investigations have been hampered because the number of stable sulphides is much greater than for oxides; the low melting points of metal sulphides and metal-sulphide eutectics have complicated matters. In sulphur dioxide, metal sulphates, sulphides and oxides may be formed. Sulphides formed by reaction with sulphur or hydrogen sulphide are usually more voluminous than corresponding oxides. and porous or cracked scales are generated.

The sulphide scale formed in reducing atmospheres on stainless steels containing less than 20% Cr (Types 304, 410, 430) is primarily iron sulphide (FeS) (47, 48, 104). At low chromium contents the scale may be duplex, consisting of an inner layer of iron sulphide containing appreciable amounts of dissolved chromium and an outer layer of iron sulphide. Higher chromium contents (Types 430 and 304) promote the formation of a single-layered scale consisting of iron sulphide containing chromium in solution (104). The occurrence of chromium sulphides is promoted on alloys containing more than 20% Cr. These sulphides may give rise to complex stratified scales containing layers of CrS, Cr_2S_3, FeS and FeS_2 (49, 50). It is generally accepted that the transport of iron ions through the sulphide layers is the predominant step. The low melting point nickel-nickel sulphide eutectic may be formed on the austenitic steels containing greater than 25% Ni even in the presence of high chromium. The

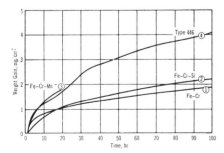

Fig. 8. Oxidation of Fe-26Cr, Fe-26Cr-0.5Si, Fe-26Cr-Mn alloys and Type 446 in dry oxygen at 1 atm, 1994 F. (After Caplan and Cohen, Ref. 41.)

occurrence of molten phases during high-temperature service leads to catastrophic destruction of the alloy. It is impossible to present a section on the modes of sulphide scale formation on stainless steels since the details of sulphidation mechanisms are few. Most data refer to practical situations and are discussed in the following sections.

Practical Considerations

Reaction with Sulphur Dioxide. Stainless steels containing more than 18 to 20% Cr are resistant to dry sulphur dioxide (51-54). In 24-hr tests over the temperature range 1100 to 1600 F, only a heavy tarnish was formed on Type 316 in atmospheres varying from 100% oxygen to 100% sulphur dioxide (55). The corrosion rate of Type 316 in sulphur dioxide-oxygen-nitrogen atmospheres was 4.9 mpy at 1185 to 1210 F. Fe–15 Cr (Types 430, 440), Fe–30 Cr (~ 446) and Fe–18Cr–8Ni (304) show increasing resistance to sulphur dioxide in the order indicated (2).

Reaction with Hydrogen Sulphide. Most of the recent data concerning high-temperature corrosion of stainless steels by hydrogen sulphide have stemmed from the use of these steels in catalytic reformers and desulphurizers, therefore, published information is given for hydrogen-hydrogen sulphide atmospheres. It was shown in an earlier study (56) that chromium and chromium-nickel stainless steels are rapidly corroded at high temperatures in 100% hydrogen sulphide, as reported in Table 3. The

Table 4. Corrosion Rates of 300 and 400 Series Stainless Steels in Hydrogen-Hydrogen Sulphide Atmospheres

Test Conditions							
Temperature (°F)	950	960	970	985	1120	1390	1390
H₂ Pressure (psig)	175	485	485	485	485	185	185
H₂S Conc (vol %)	0.10	0.80	0.20	0.75	0.05	0.05	0.15
Time (hr)	598	461	234	222	415	458	468
Materials	Corrosion Rate (mpy)						
Type 405 (12 Cr)	75	—	—	—	—	13	180
Type 410 (12 Cr)	—	220	190	300	100	—	160
Type 430 (16 Cr)	—	39	30	60	—	16	240
Type 446 (26 Cr)	10	15	26	42	41	76	230
Cr-Ni Steels	7	23	27	40	12	5	65

(After Backensto et al (57)).

data, however, indicate the beneficial effect of increasing the chromium content of steels to values greater than 12 to 15%. Incremental additions of chromium up to approximately 25% impart progressively greater resistance. Nickel contents up to 20% do not appear to be detrimental in the presence of high chromium contents (20 to 25% Cr); experience has shown that higher nickel contents are deleterious.

Backensto et al (57) and Sorell and Hoyt (47) have reported data for a wide range of operating conditions which might be encountered in catalytic reformers and desulphurizers. Table 4 compares the performance of the 400 and 300 Series stainless steels over the temperature range 950 to 1390 F and hydrogen sulphide contents of 0.05 to 0.80 vol % based on laboratory data. At 1390 F, none of the steels has sufficient corrosion resistance for practical use except at very low hydrogen sulphide contents. At approximately 1000 F, the 16% Cr, 26% Cr and the austenitic chromium-nickel steels corrode at approximately the same rate. At higher temperatures, the straight chromium steels are corroded at considerably higher rates.

Tests on manganese-modified steels (Type 202) revealed that hydrogen sulphide corrosion was equivalent to the conventional austenitic steels up to 900 F, but may exhibit higher rates above this temperature (57). Merrick and Mantell (58) inferred that Types 201 and 202 would not be

Table 3. Corrosion Rates of Chromium and Chromium-Nickel Steels in 100% H₂S at Atmospheric Pressure (120 hour test)

Material	Corrosion Rate (mpy)*	
	752 F	932 F
5 Cr	240	1000
9 Cr	200	700
12 Cr	130	400
17 Cr	90	200
25 Cr	—	100
18Cr-9Ni	80	200
26Cr-20Ni	60	100

* mpy = mils per year (After Naumann (56)).

more resistant than the conventional austenitic steels.

The 18Cr–8Ni stainless steels have been used extensively in industry to combat hydrogen-hydrogen sulphide atmospheres (47). Figure 9 demonstrates the effect of hydrogen sulphide content and temperature on

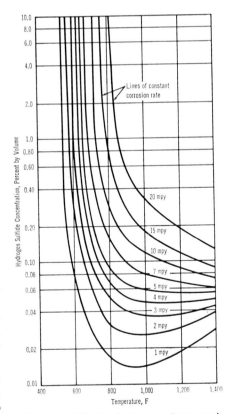

Fig. 9. Effect of temperature and hydrogen sulphide content on the corrosion rate of austenitic stainless steels (pressure range is 175 to 500 psig: exposure times greater than 150 hr.) (After Backensto and Sjoberg, Ref. 58.)

the corrosion rate of austenitic chromium-nickel steels in general.

Reaction with Sulphur Vapor. The austenitic steels are readily attacked by sulphur vapor at temperatures approaching 1000 F as shown in Table 5. Increased resistance is obtained by increasing the chromium content to greater than 20%. West (59) has indicated that Type 310 is the best selection of the austenitic steels for sulphur vapor service up to 1300 F. Krebs (25) has reported that Type 310 tubes have been used for 4½ years in a chemical plant for heating sulphur to about 1100 F.

Effect of Sulphur in Flue and Process Gases. It is extremely difficult to generalize corrosion rates in flue and process gases. The gas composition and temperature may vary considerably in the same process unit. White (8) and Collins (60) have expressed the necessity for field tests.

Combustion gases normally contain sulphur compounds; sulphur dioxide is present in an oxidizing gas along with carbon dioxide, nitrogen, carbon monoxide and excess oxygen. Protective oxides are generally formed, and depending on exact conditions, the corrosion rate may be approximately the same as in air (61) or slightly greater. This refers to clean oxidizing combustion atmospheres in the absence of corrosive fuel-ash products. (The effect of ash compounds will be discussed in another section). The resistance of stainless steels to normal combustion gases is increased by successive increments in chromium content, as shown in Fig.

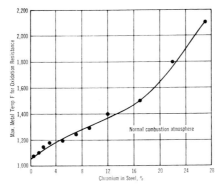

Fig. 10. Effect of chromium on the oxidation resistance of steel in a normal combustion atmosphere. (After Krebs, Ref. 25.)

10. Table 6 indicates the beneficial effect of chromium and the influence of fuel source. Corrosion rates of 1 to 2 mpy have been reported for Types 304, 321, 347 and 316 in the temperature range 1200 to 1400 F. High nickel-chromium steels are often used in more aggressive atmospheres.

Reducing flue gases contain various amounts of hydrogen sulphide, hydrogen, carbon monoxide, carbon dioxide and nitrogen. The corrosion rates encountered in these environ-ments are sensitive to hydrogen sulphide content and temperature, and satisfactory material selection often necessitates service tests. For example, Guthrie and Merrick (62) reported that commercial steam reforming tubes of Type 310 could accept feed stocks containing about 2.5 wt % S even though hydrogen sulphide was present in the effluent gas. These investigators concluded that either the steam in the feed gas maintained oxidizing conditions in the high-temperature zone or a protective coke film formed a barrier between the tube and feed. Table 7 illustrates the effect of sulphur content on the corrosion of AISI Types 309 and 310 and the non-standard Type 330 in oxidizing and reducing flue gases. The deleterious effect of high nickel content is apparent (see Type 330).

REACTION WITH CARBURIZING GASES

Stainless steels will react with hydrocarbon or carbonaceous gases at elevated temperature. Carbon is absorbed, and subsequently diffuses into the alloy interior. Carbides of the

Table 6. Corrosion Rates of Stainless Steels in Flue Gases (Exposure 3 months)

Material AISI type	Coke Oven gas (1500 F)	Coke Oven gas (1800 F)	Natural gas (1500 F)
430	91	236†	12
446(26 Cr)	30	40	4
446(28 Cr)	27	14	3
302B	104	225†	—
309S	37*	45	3
310S	38*	25	3
314	23*	94	3

* Pitted specimens—average pit depth.
† Specimens destroyed.
(After White (8)).

Table 5. Corrosion of Stainless Steels in Sulphur Vapor at 1060 F

Material AISI type	Corrosion Rate* (mpy)
314	16.9
310	18.9
309	22.3
304	27.0
302B	29.8
316	31.1
321	54.8

* Corrosion rates based on 1295 hr tests, International Nickel Co. data.

Table 7. Corrosion Rates of Stainless Steels Air, Oxidizing, Reducing Flue Gases*

Material	Air 2000 F	Air 2200 F	Oxidizing Flue Gas (mpy) 2000 F 5gS†	Oxidizing Flue Gas (mpy) 2000 F 100gS	Reducing Flue Gas (mpy) 2000 F 5gS	Reducing Flue Gas (mpy) 2000 F 100gS
AISI Type 309	40-90	60	50-70	40-100	20-50	30
AISI Type 310	40	50-80	50	40	20-50	30
"Type 330"	50	100-1000	60-300	100-500	50-200	300-800

* Data for cast stainless steels; H. S. Avery, Materials Technology in Steam Reforming Processes, C. Edeleanu (Ed.) Pergamon Press, N.Y. (1966) 73.
† Grains of sulphur per 100 cu ft.

most reactive alloying additions, (Cr, Cb, Ti) will be precipitated in the steel matrix and grain boundaries. The reaction is broadly termed carburization and the usual effects are reductions in ductility and impact properties, depending on the degree and distribution of absorbed carbon. The extent of carburization is a function of the alloy content, temperature, service time and the chemistry of the environment. Carburization has been observed to occur in hydrocarbon, carbon monoxide–carbon dioxide atmospheres, pure carbon dioxide and molten metals containing dissolved carbon. Spalling of heavily carburized surfaces may result in metal loss. (Carburization is often not referred to as a corrosion process: however, it is, in the sense that internal oxidation is a corrosion process; the mechanism of internal oxidation is analogous to carburization.)

Of the elements present in stainless steels, field experience and laboratory analyses have demonstrated that carburization resistance is strongly dependent on chromium content. Carburization is reduced as chromium is progressively increased. Table 8 indicates the increase in carbon content of chromium and chromium-nickel steels obtained in pack carburizing tests at 1800 F. Comparison of Types 430 and 446 illustrates

Table 8. Pack Carburization Tests*

Material† AISI Type	Composition	Si Content (%)	Increase in bulk carbon Content (%)
"330"	15 Cr-35 Ni	0.47	0.23
"330"	15 Cr-35 Ni + Si	1.00	0.08
310	25 Cr-20 Ni	0.38	0.02
314	25 Cr-20 Ni + Si	2.25	0.03
309	25 Cr-12 Ni	0.25	0.12
347	18 Cr-8 Ni + Cb	0.74	0.57
321	18 Cr-8 Ni + Ti	0.49	0.59
304	18 Cr-8 Ni	0.39	1.40
302B	18 Cr-8 Ni + Si	2.54	0.22
446	28 Cr	0.34	0.07
430	16 Cr	0.36	1.03

* 40 Cycles of 25 Hours at 1800 F
† Type 330 not an AISI Type.
(After Mason, Moran, Skinner (61)).

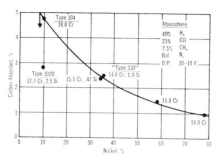

Fig. 11. Effect of nickel on gas carburization of 15% Cr steels at 1785 F, 1500 hr. (International Nickel Co.)

the effectiveness of chromium. Minor alloying additions also influence the degree of carbon absorption. Silicon, titanium and columbium retard carburization, as shown in Table 8. Increasing quantities of nickel are advantageous in carburizing systems. The effect of nickel is illustrated in Table 8 (see "Type 330") and in Fig. 11 and 12. Common selections for

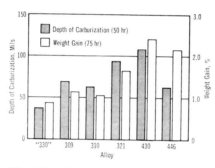

Fig. 12. Carbon absorption in pack carburization tests at 2000 F. (After Collins, Ref. 60.)

high temperature service in carburizing atmospheres are AISI Types 309 and 310 and the non-standard Type 330.

Carburization of stainless steels oxidized in carbon monoxide-carbon dioxide and pure carbon dioxide has been reported (35, 36, 38, 64). The gases are normally considered noncarburizing. Two mechanisms have been suggested to account for the occurrence of carburization. Jepson et al (64) suggest that carbon is deposited on compact oxide surfaces and then is transferred through the oxide to the alloy by a diffusion mechanism. There is some doubt as to the origin of carbon deposition. On the other hand, McCoy (36) has pos-

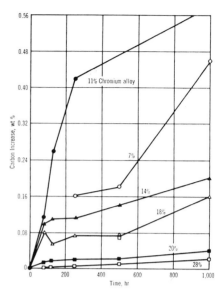

Fig. 13. Effect of chromium content on the carburization of a 9% Ni steel in carbon dioxide at 1428 F. (After Martin and Weir, Ref. 65.)

tulated that porous oxide in the outer scale provides pockets for accumulation of carbon monoxide. Conditions near the metal surface become reducing as well as oxidizing and carburization occurs. Both mechanism may be valid, since the oxide is compact in one case and porous in the other.

The nature of the oxide formed definitely influences the tendency for the steel to carburize. Martin and Weir (65) have shown that the formation of oxides rich in Cr_2O_3 impedes carburization. Figure 13 illustrates the effect of chromium content on the carburization of a 9% Ni steel in carbon dioxide. When the chromium content exceeds approximately 11%, Cr_2O_3 is established as the primary oxide and carburization decreases. McCoy (36) noted that no detectable carburization occurred in the nonstandard Type 406 due to the formation of a very protective Al_2O_3 film. This author concluded that greater than 25% Cr would be required in the absence of aluminum to suppress carburization, in agreement with Fig. 13.

Another form of metal deterioration encountered in carburizing gases (with or without oxygen-bearing components) at elevated temperature occurs by the metal dusting phenomena (66, 67). Corrosion is generally in the temperature range 800 to 1500 F and alloy attack occurs by localized pitting or over-all surface wastage. The corrosion products consist of dust or powder containing metal carbides, metal oxides and graphite. Carburization of the surfaces from which the metal is lost is usually observed. Of a number of stainless steels exposed to atmospheres which cause metal dusting, none were found to be immune to this type of attack over long exposure times (67).

Several mechanisms have been proposed to explain this form of attack. Hoyt and Caughey proposed that carbon, produced by catalytic decomposition of carbon monoxide on iron- or nickel-rich sites on the surface, enters the steel, forming chromium carbide, initially at grain boundary sites and eventually in the matrix (66). An appreciable density change occurs so that the grain is either disintegrated or detached from the steel at grain boundaries. Mechanisms based on volatile carbonyl formation (68) and alternate cycles of carburization and oxidation (69) have been proposed.

REACTION WITH OTHER GASES

Hydrogen. Attack by hydrogen may be encountered in ammonia synthesis, hydrodesulphurization, hydrogenation and oil refining equipment. Carbon steels are inadequate for high-pressure hydrogen systems at temperatures exceeding approximately 450 F. Although the mechanism is not fully understood, various authorities believe that hydrogen diffuses into the steel and reacts with iron carbide at grain boundary sites or pearlite colonies, producing methane gas. The methane gas, which cannot diffuse out of the steel, collects to form blisters and/or cracks in the material (70, 71). To prevent methane formation, cementite (Fe_3C) must be replaced by more stable carbides. Elements such as chromium, vanadium, titanium or columbium are added to steels to form stable carbides. Nelson (72), in a comprehensive survey of all available service

data demonstrated that increasing chromium contents in steel permit higher operating temperatures and hydrogen partial pressures. Chromium carbide forms in these steels and is stable in the presence of hydrogen. Under severe conditions of operation, (temperatures greater than 1100 F), chromium steels, containing greater than 12% Cr, and the austenitic stainless steels are resistant to attack in all known applications.

Nitrogen, Ammonia. Most metals and alloys are inert towards molecular nitrogen at elevated temperatures. However, atomic nitrogen will react with and penetrate many steels, producing hard brittle nitride surface layers. Iron, aluminum, titanium, chromium and other alloying elements may take part in these reactions. One of the main sources of atomic nitrogen is the dissociation of ammonia in ammonia converters, plant lines, heaters and nitriding furnaces which operate in the temperature range of 700 to 1100 F at pressures varying from atmospheric to 15,000 psi. In these atmospheres (atomic nitrogen and hydrogen), the chromium carbide present in low chromium steels, Cr_7C_3, may be attacked by atomic nitrogen to produce chromium nitride (more stable than Cr_7C_3) and release carbon to react with hydrogen, forming methane gas (70). Either blistering or cracking or both can occur as described above. However, with chromium contents in excess of 12%, it appears that the carbide in these steels, $Cr_{23}C_6$, is more stable than chromium nitride and this reaction does not occur. Hence, the stainless steels are employed for high-temperature service in hot ammonia atmospheres.

The behavior of stainless steels in ammonia depends on temperature, pressure, gas concentration and the chromium and nickel contents of the steel. Results from field tests have demonstrated that the corrosion rates (depth of altered metal or case) for chromium stainless steels are greater than those for the austenitic grades. In the latter case, alloys with higher nickel contents had better resistance to corrosion (70, 73). Higher corrosion rates are experienced with in-

Table 9. Corrosion Rates of Stainless Steels in Ammonia Converter and Plant Line

Material* AISI Type	Nominal Ni Content %	Corrosion Rate (mpy)	
		Ammonia† Converter	Ammonia‡ Plant Line
446	—	1.12	164.5
430	—	0.9	—
302B	10	0.73	—
304	9	0.59	99.5
316	13	0.47	>520
321	11	0.47	—
309	14	0.23	95
314	20	0.1	—
"330"	34	0.06	—
"330"	36	0.02	—

* Type 330 not an AISI Type.
† 5-6% NH_3: 29,164 hr at 915-1024 F
‡ 99.1% NH_3: 1540 hr at 935 F.
(After Moran et al (73)).

creased ammonia contents. Corrosion rates for several stainless steels exposed in an ammonia converter and plant line are given in Table 9.

Cihal (70) has also reported that the austenitic stainless steels give better performance than the chromium steels. Corrosion rates of approximately 10 mpy in synthesis gases containing up to 16% NH_3 were observed for several austenitic steels in the temperature range 840 to 970 F.

Halogens. Austenitic stainless steels are severely attacked by halogen gases at elevated temperatures. Fluorine is more corrosive than chlorine, and the upper temperature limits for dry gases are approximately 480 and 600 F, respectively, for the high chromium-nickel grades (32).

REACTION WITH FUEL-ASH DEPOSITS

Catastrophic Oxidation

When stainless steels (and other metals and alloys) are heated in the presence of certain low melting point metallic oxides, corrosion occurs at an extremely rapid rate; this phenomenon is referred to as catastrophic oxidation (74). Most investigations have been confined to the corrosive action of MoO_3 (74, 75), PbO (76) and V_2O_5 (74, 77) in the temperature range 1100 to 1700 F although other oxides are known to cause the effect.

Catastrophic oxidation has received much attention since consideration has been given to the use of low-grade fuels (residual and crude oils) in super-heater and gas turbine units.

The presence of vanadium in the fuel results in the formation of vanadium pentoxide on combustion. This oxide, which is retained on metallic surfaces as fuel-ash deposit, promotes this form of oxidation. In this area, the pentoxide is not the only deleterious compound in fuel-ash deposits. Sodium, (present in the atmosphere as a chloride or in the fuel), and sulphur react to form corrosive sulphates and vanadates, the latter by reaction between sodium sulphate and vanadium pentoxide. Sodium chloride, by itself or as an impurity in these compounds, greatly influences the rates of corrosion. Sodium sulphate by itself is not particularly corrosive, at least at temperatures below the melting point (~1625 F); Danek (78) and Bergman (79) have indicated that this compound causes sulphidation and catastrophic oxidation of gas turbine components operating at temperatures above the sulphate melting point. Molten vanadium pentoxide (melting point ~1200 F) is very corrosive, and mixtures of sulphate and pentoxide are extremely reactive. A particular corrosive mixture contains approximately 80 mol % V_2O_5 (80).

Oxidation Mechanisms

Catastrophic oxidation takes place at the metal-oxide interface and is generally associated with the formation of voluminous corrosion products. Marker experiments and the occurrence of linear reaction rates support this premise. Most investigators purport that the deposition of or formation of liquid or semi-liquid phases are the essential steps in catastrophic oxidation (75, 81). The liquid destroys the normally protective oxide and oxidation proceeds at a rapid rate. Numerous studies have demonstrated that oxidation occurs at normal rates up to the melting point of the deposited compounds or the eutectic temperature of binary or ternary mixtures. Above a definite temperature, corresponding to the melting point, oxidation is catastrophic.

Although the actual details of the corrosion mechanism have not been developed, some additional postulates have been given. Fitzer and Schwab (82) have suggested that vanadates, produced by reaction of sulphates and vanadium pentoxide, may: (a) act as an oxygen carrier, and since these mixtures have a high absorbing capacity for oxygen, there exists a plentiful supply of oxygen; (b) be molten and dissolve or destroy the oxide barrier film and expose the metal surface (accepted mechanism); (c) distort the lattice structure of the oxide, allowing reactants to permeate the film at uninhibited rates. As mentioned above, most experimental data supports (b).

Danek (78) has suggested that in fuel-ash deposits consisting mainly of molten sodium sulphate, sulphidation of the underlying metal occurs. The molten slag penetrates the oxide film and reaction between sulphur in the slag and alloying additions forms metal sulphides, depleting the alloy matrix of elements which form the protective oxide. As a result, oxidation of the depleted matrix occurs at high rates. Furthermore, the sulphides initially formed are oxidized and sulphur is released. This sulphur is available to form new sulphide and the process becomes autocatalytic. A similar mechanism has been suggested to account for sulphide reaction products in stainless steels in chloride-contaminated sulphates (83).

Other experimenters have offered a different mechanism to account for the rapid corrosion and intergranular attack of stainless steels in dry sodium chloride-oxygen atmospheres (84) and the catastrophic oxidation of stainless steels in lead oxide (76) at temperatures below the melting points of deposits or corrosion product-deposit mixtures. It is suggested that the compounds (NaCl or PbO) react with or prevent the formation of the protective chromic oxide film, "Cr_2O_3," forming porous, flaky, nonprotective chromates. Rapid oxidation occurs in the absence of the protective film.

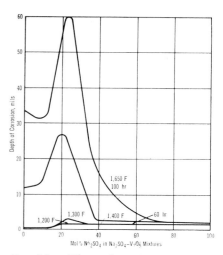

Fig. 14. Effect of sodium sulphate-vanadium pentoxide mixtures on the oxidation of Type 304. (After Cunningham and Brasunas, Ref. 86).

The reactions suggested by Beck et al. are (84):

(a) $2NaCl + Cr + 2O_2 \rightarrow Na_2CrO_4 + Cl_2$
(b) $8NaCl + 2Cr_2O_3 + 5O_2 \rightarrow 4Na_2CrO_4 + 4Cl_2$
(c) $46NaCl + Cr_{23}C_6 + 52O_2 \rightarrow 23Na_2CrO_4 + 6CO_2 + 23Cl_2$

Reaction (b) destroys the protective oxide film, and (c) results in rapid intergranular penetration via grain boundary carbide.

Although the details of the mechanism are not entirely understood, catastrophic oxidation has stimulated research with the object of producing alloys resistant to this type of attack or slag additives which prevent the formation of liquid phases. Some success has been experienced in both areas (85-87), however, much more study is required for economical operation of equipment.

Practical Considerations

All the stainless steels (and other popular engineering alloys) are severely attacked by low melting point oxides and fuel-ash containing vanadium pentoxide and/or chlorides with or without sodium sulphate. Depending on prevailing conditions, rapid corrosion may be experienced at temperatures greater than 1100 F.

Figures 14, 15 and 16 illustrate the effect of various contaminants on the corrosion of stainless steels; all

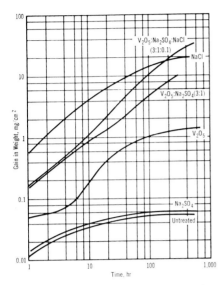

Fig. 15. Effect of various salts on the oxidation of Type 347 at 1202 F. (After Alexander and Marsden, Ref. 81.)

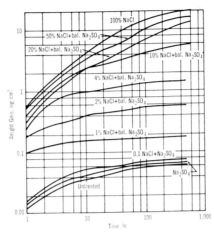

Fig. 16. Effect of sodium-chloride-sodium sulphate mixtures on the oxidation of Type 347 at 1202 F. (After Lucas, Ref. 88.)

indicate that sodium sulphate alone is not harmful, at least up to 1650 F. Figures 14 and 15 show that molten vanadium pentoxide is corrosive and that sulphate-pentoxide mixtures are extremely corrosive, especially in the concentration range 20 to 30 mol % Na_2SO_4. Corrosion is not severe at temperatures below the melting point of the slag, Fig. 14. Figures 15 and 16 demonstrate the deleterious effect of sodium chloride in sulphate or sulphate-pentoxide deposits. Sulphide precipitation in the steel below the oxide scale is apparent in the presence of chloride-contaminated slags (83). Either the sulphidation-oxidation or chromate scaling mechanism described above may be operative. Figures 15 and 16 demonstrate the high rates of attack in 100% NaCl. In this respect, Pickering et al. (84) found sodium fluoride to be more aggressive and sodium bromide less aggressive than the chloride. Alexander (89) reported increasing corrosion in the order, calcium, potassium, sodium and lithium chlorides.

Lewis (105), on the basis of a series of tests on austenitic and Type 446 stainless steels in Na_2SO_4-NaCl and Na_2SO_4-V_2O_5 mixtures at 1290 to 1830 F, reported that Type 446 offered the best over-all resistance to corrosion. Nevertheless, several aus-

tenitic grades gave better performance at specific salt compositions and temperatures. Corrosion was most severe in mixtures containing 80% V_2O_5-20% Na_2SO_4, in agreement with Fig. 14. The significance of nickel in the austenitic steels (AISI Types 304, 309, 310, and "Type 330") could not be generalized since the effect of nickel on corrosive attack varied with test temperature.

REACTION WITH LIQUID METALS AND MOLTEN SALTS

Liquid Metals

Liquid metals possess high thermal conductivity and specific heat and for these reasons are used as heat transfer mediums in nuclear reactors. Molten metals are also used for heat treating baths, and in hot-dip zinc and tin plate production. The high temperatures involved may offset any beneficial effects of using liquid metal heat transfer systems due to their corrosive nature towards metal containers. Since sodium has a low cross-section for neutron capture and a low melting point, interest has mainly been centered on this element as a reactor coolant.

Liquid metal corrosion depends primarily on the rate of solution of the solid container and the solubility limit of the dissolved element in the molten metal. However, a number of factors complicate this simple situation by affecting the rates and extent of solu-

bility in the liquid phase. The occurrence of temperature and concentration gradients and impurities strongly influence the corrosive nature of a particular system. The several types of corrosive attack between liquid and solid metals that have been observed are: (a) simple solution and alloying, (b) temperature gradient mass transfer, (c) concentration gradient mass transfer, and (d) impurity reaction. These types of corrosion are discussed in the following section.

Liquid Metal Corrosion Mechanisms

In simple solution attack, the solid metal or alloying elements in an alloy dissolve in the liquid phase. For a static system, in which there are no temperature gradients or dissimilar metals, the extent of damage depends primarily on the solubility limit of the dissolving species in the liquid, and the ratio of the surface area of the solid to the volume of liquid. Small surface-to-volume ratios will increase the extent of damage. The rate at which simple solution occurs may be influenced by other factors, such as impurities in the solid and liquid. Alternately, the liquid metal may alloy with the solid, producing solid solutions and undesirable brittle intermetallic phases. Uniform attack or deep intergranular penetration may result in simple solution or alloying attack. Intergranular penetration of liquid, which results in the formation of intermetallic phases, and the preferential leaching of alloying elements from grain boundary sites have been observed.

Temperature gradient mass transfer occurs in dynamic solid-liquid systems when one portion of the system is at a higher temperature than another. The driving force for mass transfer is the difference in the solubility of the dissolved element at the temperature limits of the liquid phase in the system. Figure 17 illustrates the basic steps in temperature gradient mass transfer. Atoms go into solution at the hot metal surface and diffuse through a boundary layer (lamellar layer) into the bulk liquid. In the case of selective dissolution of an alloying element, solid state diffusion

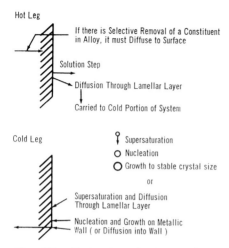

Hot Leg

If there is Selective Removal of a Constituent in Alloy, it must Diffuse to Surface

Solution Step

Diffusion Through Lamellar Layer

Carried to Cold Portion of System

Cold Leg

♀ Supersaturation
○ Nucleation
◎ Growth to stable crystal size

or

Supersaturation and Diffusion Through Lamellar Layer

Nucleation and Growth on Metallic Wall (or Diffusion into Wall)

Fig. 17. Basic steps in temperature gradient mass transfer. (After Manly, Ref. 90.)

of this element to the alloy surface precedes the dissolution step. Metal atoms are carried by the liquid stream to the cooler portion of the system where supersaturation and precipitation occur either in the liquid or on the container walls. Reaction rates are influenced by impurities, temperature and temperature difference. Although the rate-determining step has not been determined, the rate of diffusion of reaction products through the lamellar layer at the hot surface has been suggested (91). This is based on experiments which demonstrated direct influence of fluid velocity on reaction rates. Fluid flow affects the lamellar layer thickness and turbulence which affect flux rates through the film. This form of attack can be quite damaging due to the accumulation of deposits in small diameter sections in addition to thinning of the corroded material.

Concentration gradient mass transfer or dissimilar metal mass transfer occurs in systems consisting of two or more different metals or alloys. In this process, atoms are removed from one metal or alloy and reach the surface of a dissimilar metal or alloy either by diffusion and/or movement of the liquid. Alloying takes place and the dissolved atoms are removed from the liquid phase, which allows the reaction to continue to destruction. The driving force for dissimilar metal transfer is the activity gradient of the

corroded species in the metal-liquid metal system. Carburization and decarburization of metals and alloys in contact with a liquid metal may be rationalized on the same basis. When carbon-bearing alloys of differing carbon activity are in contact with the same liquid, carbon migrates from the material of higher activity to the material of lower activity. The carbon activity tends to equilibrium with all materials in contact with the liquid phase. The strength of the decarburized alloy may be greatly decreased whereas the carburized material may be embrittled.

Impurities, such as carbon, nitrogen and oxygen, may influence reaction rates in liquid metal systems. Although the mechanisms by which impurities affect corrosion rate are not completely understood, several investigations have demonstrated that impurities: (a) may change solubility limits, (b) may accelerate dissolution or participate directly in the corrosion reaction. For example, it has been established that sodium is not corrosive until contaminated by oxygen. To explain this effect, it has been suggested that oxygen in sodium promotes the formation of complexes of the type $Na_xM_yO_z$. Iron is believed to be transferred to the liquid by a hot zone reaction:

$$Fe + 3\,Na_2O \rightarrow 2\,Na + (Na_2O)_2 \cdot FeO$$

In the cold zone, the reverse reaction occurs and iron is precipitated. Also, impurities may: (c) affect the surface tension of the solid-liquid interface, and (d) form barrier films inhibiting the reaction. Oxygen contamination of lead may result in the formation of protective oxide films on metals and alloys since many elements in corrosion resistant materials are able to reduce lead oxide.

Practical Considerations

Details of the corrosivity of all liquid metals in contact with stainless steels would require an inordinate amount of space, therefore, the following section will be limited to the more common low melting-point metals and alloys. Numerous systems are reported in the corrosion section

of the Liquid Metals Handbook (92).

Sodium and Sodium-Potassium Alloys. A considerable amount of data has been published on the compatibility of sodium and sodium-potassium with stainless steel (91-94). The common chromium and austenitic stainless steels (Types 304, 316 and 347) can be used in sodium (and sodium-potassium) flow systems up to 1000 F with virtually unlimited life, provided oxygen and carbon contents of the liquids are low. The austenitic grades are preferred at high temperatures for strength considerations. Static test data are many and varied; typical results for the austenitic grades are: (a) negligible weight change at 930 F, < -0.1 mg/cm^2/month, (b) slight weight gain at 1100 F, $+0.1$ mg/cm^2/month, and (c) substantial weight gain at 1830 F, $+40$ mg/cm^2/month. Maximum service temperatures in static systems approach 1500-1600 F. In dynamic systems, austenitic steels are good construction materials, providing hot zone temperatures do not exceed 1200 F and temperature differences of 300 F.

The presence of oxygen and carbon seriously affects service performance. High oxygen levels greatly increase the corrosion rate. These contents are controlled to low levels by hot or cold traps. Sodium may be contaminated by carbon, and carburize contacting metals. Zirconium getters are used to control carbon contents. At 1200 F, the equilibrium surface carbon concentration of Type 304 is 0.04, 1.0 and 3.0 wt % at sodium carbon contents of 15, 25 and 60 ppm, respectively (91). Nitriding of Types 304, 347 and 410 in sodium under a nitrogen blanket has been reported. Nitride penetration in Type 410 was greater by a factor of two (95).

Potassium. Tests on the corrosion resistance of Type 316 in potassium illustrated that this liquid is not as corrosive as sodium (at least in the case of this alloy). Figure 18 shows the corrosion rates of Type 316 in identical dynamic sodium and potassium test loops at 1575 F, temperature gradient 475 F. At 5000 hr, the corrosion rate in potassium corre-

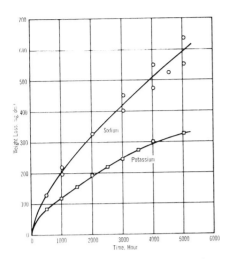

Fig. 18. Sodium-potassium corrosion of Type 316 in dynamic test loop at 1575 F. (After Ref. 96.)

Table 10. Penetration of Stainless Steel by Molten Lithium at 1500-1300 F

Material AISI Type	Time, hr	Max. Depth of Attack (mils) Intergranular	Solution*	Deposit Thickness (mils)
316	52-155	1.0-4.5	3.9-8.5	34.2-41.0
304	105-138	1.0-5.0	3.8-6.0	18.5-20.0
310	64-96	3.0-4.0	2.2-4.7	19.5-24.0
321	69-200	nil-2.0	6.2-6.4	25.0-32.0
347	82-160	0.5-0.6	4.3-4.9	33.0-40.0

* Decrease in wall thickness.
(After Freed and Kelly (97)).

sponded to 0.12 mpy; the rate in sodium was approximately three times as great.

Lithium. In general, the chromium and chromium-nickel stainless steels may be used up to 1500 F for static containments and up to 1180 F in dynamic systems. Table 10 gives the maximum depth of intergranular penetration and surface thinning of several austenitic steels in 1500-1300 F lithium forced convection loops for various exposure times.

Lead. Lead has a high solubility for several metals, and therefore, simple solution corrosion may be encountered. In the temperature range 1210 to 1740 F, lead diffuses into grain boundaries of Type 304 (98). Fissures are formed at the grain boundaries. At the higher temperature range, chromium diffuses to the alloy surface, forming a brittle compound. Certain high-nickel steels are satisfactory for molten lead containment up to 1600 F. Large pots are regularly made of the "Type 330" composition (32).

Zinc. All austenitic grades are readily attacked by molten zinc at temperatures greater than 900 F. From studies on the compatibility of stainless steels with zinc vapor at 1650 F, it appears that the 400 Series (at least Types 405 and 440) are more stable than the 300 Series, although large weight losses were observed in both cases (99).

Mercury. In static tests, the corrosion rates of Types 304 and 310 are 20 and 47 mpy, respectively, at 1210 F. In dynamic tests in liquid mercury at 1095 F, Type 304 is rapidly attacked by transgranular leaching of nickel. Type 410 was much more resistant. High nickel contents should be avoided in dynamic systems. Nejedlik and Vargo (100) have shown that liquid mercury corrosion of iron-base alloys increases as the nickel + chromium + manganese content rises. Hence, Type 410 was found to be more resistant than the higher chromium-bearing Type 446.

Molten Salts

Molten salts may corrode metals according to the reaction processes described for liquid metal corrosion, namely, simple solution, temperature and concentration gradient mass transfer and impurity reactions. Corrosion may also occur by direct chemical reaction. Edeleanu and Gibson (101) have postulated that simple solution corrosion is not possible since a number of metals have no true solubility in salts, and oxidation of the metal to metal ions is required to initiate transfer.

Most information refers to the corrosive nature of heat treating salts or salt mixtures. Jackson and LaChance (102) have conducted a comprehensive investigation concerning the corrosive nature of molten salts to (cast) austenitic stainless steels. The resistance of various compositions to specific salts is described below.

Practical Considerations

Chloride Salts. Barium, potassium and sodium chloride are common heat treating salts for use at elevated temperatures. They may be used alone or in combination. The salts or salt mixtures become increasingly corrosive as they are contaminated by impurities (oxygen). Results on cast stainless steels exposed to ternary salt mixtures (55% $BaCl_2$, 25% KCl, 20% NaCl) at 1600 F demonstrated that corrosion rates increased as the chromium content increased. Nickel lowered the corrosion rate for a given chromium content. Intergranular attack was severe for all compositions. Steels high in nickel and low in chromium, ("Type 330"), are usually selected for chloride salt containments.

Nitrate and Nitrite Salts. These salts may be used at temperatures up to 1100 F (potassium nitrate, sodium nitrite). The austenitic stainless steels show excellent resistance to potassium nitrate at the upper service temperatures. Only "Type 330" is attacked intergranularly to any extent. (32).

Cyanide Salts. Molten cyanides

Table 11. Corrosion of Stainless Steels by Cyanide Salts*

Material† AISI Type	Nitrogen Penetration (mils)‡	Depth of Carburization (mils)‡
"330"	8	33
"330 + Si"	5.5	33
310	10.5	33
314	20	30
309	14	39
316	21.5	62
321	19	59
304	18	52
302B	42.5	56

* Salt composition: 23-30% NaCN, 2-7% NaOCN, 0.6-1.5% NaCl, balance Na_2CO_3.
† Type 330 not an AISI Type.
‡ Exposure conditions—672 hr at 1300-1500 F.
(International Nickel Co. data).

are used as carburizing, nitriding, and carbonitriding media. It was illustrated above that chromium inhibits carburization but promotes nitridation. Nickel inhibits both reactions, therefore, steels of low chromium and high nickel contents are used in service ("Type 330"). The nitrogen and carbon penetration for several grades are given in Table 11.

Carbonate Salts. Janz and Conte have shown that Type 347 is highly resistant to sodium-lithium carbonate mixtures over the temperature range 1110-1290 F (103). These investigators suggested that the oxide film ($LiFeO_2$) is responsible for protection.

Summary

A considerable amount of fundamental and practical data concerning the corrosion resistance of stainless steels at elevated temperatures has been reported. However, from the present discussion, it is apparent that much remains to be done in both areas. Operating experience has to a great extent documented the uses and limitations of stainless steels in many situations; nevertheless, due to the urgency of practical problems, compatibility tests in numerous industrial environments are often required. Future advances in fundamental research may eliminate the need for many field tests and permit higher operating limits for the present and/or slightly modified grades through a knowledge of the detailed corrosion mechanisms. Comprehensive reports of practical operating experience can contribute to basic studies. The effects of thermal cycling, high gas pressures, surface preparation (industrial finishes), surface geometry and stress on service life may provide important data. Fundamental studies on the defect structure of oxides, the effects of impurities on corrosion rate and the plastic and adherence properties of oxides are some areas that would provide beneficial data to the industrial field.

It is hoped that the present discussion will assist the engineer in understanding the nature of corrosion problems and in the selection of stainless steels for particular corrosive environments.

REFERENCES

1. K. Hauffe, Oxidation of Metals, Plenum Press, N.Y. (1965).
2. O. Kubaschewski and B. E. Hopkins, Oxidation of Metals and Alloys, Butterworths, London (1962).
3. J. Benard, Oxidation des Metaux, Gauthier-Villars, Paris (1962-64) 2 volumes.
4. P. Kofstad, High Temperature Oxidation of Metals, Wiley and Sons, N.Y. (1966).
5. R. A. Rapp, Corros, 21 (1965) 382.
6. C. Wagner, Atom Movements, ASM, Metals Park, Ohio (1953) 153.
7. J. Moreau and J. Bènard, Rev Met, 59 (1962) 161.
8. W. F. White, Mater Protect, 2 (1963) 47.
9. G. C. Wood, Corros Sci, 2 (1962) 173.
10. A. U. Seybolt, J Electrochem Soc, 107 (1960) 147.
11. M. J. Brabers and C. E. Birchenall, Corros, 14 (1958) 179t.
12. H. J. Yearian, E. C. Randell and T. A. Longo, Corros, 12 (1956) 515t.
13. H. J. Yearian, W. Derbyshire and J. F. Radavich, Corros, 13 (1957) 597.
14. D. Caplan and M. Cohen, Corros, 15 (1959) 141t.
15. D. Caplan and M. Cohen, J Metals, 4 (1952) 1057.
16. J. O. Edström, J Iron Steel Inst, 185 (1957) 450.
17. G. C. Wood and M. G. Hobby, J Iron Steel Inst, 203 (1965) 54.
18. G. C. Wood, T. Hodgkiess and D. P. Whittle, Corros Sci, 6 (1966) 129.
19. H. J. Yearian, H. E. Boren, Jr., and R. E. Warr, Corros, 12 (1956) 561t.
20. V. R. Howes, Corros Sci, 6 (1966) 549.
21. H. L. Eiselstein and E. N. Skinner, ASTM STP No. 165 (1964).
22. H. M. McCullough, M. G. Fontana and F. H. Beck, Trans ASM, 43 (1951) 404.
23. F. Eberle, F. G. Ely and J. A. Dillon, Trans ASME, 76 (1954) 665.
24. I. LeMay, Paper presented at Conference of Metallurgists, Montreal, (1964).
25. T. M. Krebs, Process Industry Corrosion Notes, Ohio State University, September, (1960).
26. D. A. Jones, AEC Research and Development Report, HW-76952, (1963).
27. W. E. Ruther and S. Greenberg, J Electrochem Soc, 111 (1964) 1116.
28. W. E. Ruther, R. R. Schlueter, R. H. Lee and R. K. Hart, Corros, 12 (1966) 147.
29. G. P. Wozaldo and W. L. Pearl, Corros, 21 (1965) 355.
30. M. Warzee, J. Hannaut, M. Maurice, C. Sonnen, J. Waty and Ph. Berge, J Electrochem Soc, 112 (1965) 670.
31. D. Caplan, Corros Sci, 6 (1966) 509.
32. Corrosion Resistance of the Austenitic Chromium–Nickel Stainless Steels in High Temperature Environments, Inco Bull, (1963).
33. F. Eberle and C. H. Anderson, ASME, Paper No. 61-Pur-3 (1961).
34. C. T. Fujii and R. A. Meussner, J Electrochem Soc, 111 (1964) 1215; 110 (1963) 1195.
35. H. T. Daniel, J. E. Antill and K. A. Peakall, J Iron Steel Inst, 201 (1963) 154.
36. H. E. McCoy, Corros, 21 (1965) 84.
37. J. M. Francis and W. H. Whitlow, J Iron Steel Inst, 203 (1965) 468.
38. C. T. Fujii and R. A. Meussner, J Electrochem Soc, 114 (1967) 435.
39. A. Draycott and R. Smith, AAEC/-E-52 (1960).
40. W. A. Maxwell, TID-7597, Book 2, from US/UK Meeting on Compatibility Problems of Gas-Cooled Reactors (1960).
41. D. Caplan and M. Cohen, Nature, 205 (1965) 690.
42. J. M. Francis, J Iron Steel Inst, 204 (1966) 910.
43. E. J. Felton, J. Electrochem Soc, 108 (1961) 490.
44. H. Krainer, L. Wetternik and C. Carius, Arch Eisenhuttenw, 22 (1951) 103.
45. A. DeS. Brasunas and N. J. Grant, Trans ASM, 44 (1952) 1117.
46. S. S. Brenner, J Electrochem Soc, 102 (1955) 16.
47. G. Sorell and W. B. Hoyt, NACE Technical Committee Reports, Publication 56-7.
48. D-X Sunray Oil Company, Inspection of Catalytic Reforming Units at Tulsa, Okla. (1955).
49. A. Davin and D. Coutsouradis, Corrosion et Anticorrosion, 11 (1963) 347.
50. A. Davin, D. Coutsouradis, M. Urbain and L. Habraken, Belgische Chem Ind, 30 (1965) 340.
51. M. Farber and D. M. Ehrenberg, J Electrochem Soc, 99 (1952) 427.
52. E. Wellman, Z Electrochem 37 (1931) 142.
53. M. M. Hallett, JISI 170 (1952) 321.
54. H. Ipavic, Her Vakuumschmelze (1923-33) 290.
55. J. H. Nicholson and E. J. Kwasney, Trans Electrochem Soc, 91 (1947) 681.
56. F. K. Naumann, Chemische Fabrik, 11 (1938) 365.
57. E. B. Backensto, R. D. Drew, J. E. Prior, and J. W. Sjoberg, NACE Technical Committee Report, Publication 58-3.
58. E. B. Backensto, J. W. Sjoberg, Corrosion, 15 (1959) 125t.
59. J. R. West, Chem Eng, 58 (1951) 276.
60. W. J. Collins, Prod Eng, November 25 (1963) 82.
61. J. F. Mason, J. J. Moran and E. N. Skinner, Corrosion, 16 (1960) 593t.
62. J. E. Guthrie and R. D. Merrick, Mater Protect, 3 (1964) 82.
63. H. Inouye, Proc Conf Corrosion of Reactor Materials, June 4-8, (1962) (IAEA, Vienna 1962).

64. W. B. Jepson, J. E. Antill and J. B. Warburton, Brit Corrosion J, 1, (1965) 15.

65. W. R. Martin and J. R. Weir, J Nuclear Matr, 16 (1965) 19.

66. W. B. Hoyt and R. H. Caughey, Corrosion, 15 (1959) 627t.

67. P. A. Lefrancois and W. B. Hoyt, Corrosion, 19 (1963) 360t.

68. C. J. Slunder, NACE Technical Committee Report, Publication 60-6.

69. F. Eberle and R. D. Wylie, Corrosion, 15 (1959) 622t.

70. V. Cihal, Proc 1st Int Cong Met Corrosion, London 591 (1961).

71. H. G. Geerlings and J. C. Jongebreur, Proc 1st Int Cong Met Corrosion, London (1961) 573.

72. G. A. Nelson and R. T. Effinger, Welding J Res Supl, 34 (1955) 125: Shell Development Co. Emeryville, Corrosion Data Survey, 1960.

73. J. J. Moran, J. R. Mihalisin and E. N. Skinner, Corrosion, 17 (1961) 191t.

74. W. C. Leslie and M. G. Fontana, Trans ASM 41 (1949) 1213.

75. G. W. Rathenau and J. L. Meijering, Metallurgia, 42 (1950) 167.

76. J. C. Sawyer, Trans AIME, 227 (1963) 346.

77. C. Sykes and H. T. Shirley, ISI Special Report No. 43 (1951) 153.

78. G. J. Danek, Jr., Naval Eng J (1965) 859.

79. P. A. Bergman, Corrosion, 23, (1967) 72.

80. R. T. Foley, NACE Technical Committee Report, Publication 58-11.

81. P. A. Alexander and R. A. Marsden, The Mechanism of Corrosion by Fuel Impurities, Butterworths, London (1963) 542.

82. E. Fitzer and J. Schwab, J. Berg-und-Hullen Monats, 98 (1953) 1.

83. H. T. Shirley, The Mechanism of Corrosion by Fuel Impurities, Butterworths, London (1963) 617.

84. H. W. Pickering, F. H. Beck and M. G. Fontana, Trans ASM, 53 (1961) 793.

85. W. A. Mueller and J. J. O. Gravel, Paper presented at the Chemical Institute of Canada Conference, Hamilton (1964).

86. G. W. Cunningham and A. DeS. Brasunas, Corrosion, 12 (1956) 389t.

87. A. Rahmel, The Mechanism of Corrosion by Fuel Impurities, Butterworths, London (1963) 556.

88. D. H. Lucas, J Inst Fuel, 36 (1963) 206.

89. P. Alexander, The Mechanism of Corrosion by Fuel Impurities, Butterworths, London (1963) 556.

90. W. D. Manly, Corrosion, 12 (1956) 336t.

91. J. H. Stang, E. M. Simons, J. A. DeMastry and J. M. Genco, Compatibility of Liquid and Vapour Alkali Metals with Constructional Materials, DMIC Report 227 (1966).

92. Liquid Metals Handbook, Dept of the Navy, Washington, D.C. Navexos, P-733 (1952) Chapter 4.

93. L. R. Kelman, W. D. Wilkinson and F. L. Yaggee, Argonne National Lab, Report No. ANL-4417 (1950).

94. E. G. Brush, Corrosion, 11 (1955) 229t: General Electric Co., Report No. GEAP-4832 (1965).

95. E. G. Brush and C. R. Todd, Knolls Atomic Power Lab, Report No. KAPL-M-EGB-21 (1955).

96. Proc. NASA-AEC Liquid Metals Corrosion Meeting, Volume I, Lewis Research Center, Cleveland, Ohio, Report NASA-SP-41 (1964) 213.

97. M. S. Freed and K. J. Kelly (Eds.) Corrosion of Columbium Base and Other Structural Alloys in High Temperature Lithium, Pratt and Whitney Aircraft, Report No. PWAC-355 (1961).

98. M. C. Naik and R. P. Agarwala, JISI, 203 (1965) 1024.

99. A. H. Fleitman, A. J. Romano and C. J. Klamut, Corrosion, 22 (1966) 137.

100. J. F. Nejedlik and E. J. Vargo, Electrochem Tech, 3 (1965) 250.

101. C. Edeleanu and J. G. Gibson, J Inst Metals, 88 (1960) 321.

102. J. H. Jackson and M. H. LaChance, Trans ASM, 46 (1954) 157.

103. G. J. Janz and A. Conte, 20 (1964) 237t.

104. F. J. Bruns, NACE Technical Committee Reports, Publication 57-2.

105. H. Lewis, J of Inst Fuel, 39 (1966) 8.

SECTION VI:
Heat-Resistant Alloy Castings

Heat-Resistant Alloy Castings

*By the ASM Committee on Heat-Resistant Castings**

CASTINGS are classed as heat resistant if they are capable of sustained operation when exposed, either continuously or intermittently, to operating temperatures that create metal temperatures in excess of 1200 F. Alloys used in castings for such service include: iron-chromium ("straight chromium" alloys), iron-chromium-nickel, iron-nickel-chromium, nickel-base, and cobalt-base.

Many alloys of the same general type are used also for their resistance to corrosive mediums at temperatures below 1200 F, and castings intended for such service are classed as corrosion resistant. Although there is usually a distinction between the alloys of the two classes, based on carbon content, the line of demarcation is vague, particularly for alloys used in the range from 900 to 1200 F.

Heat-resistant castings, as defined, are used in metallurgical furnaces, oil-refinery furnaces, cement-mill equipment, petrochemical furnaces, power-plant equipment, steel-mill equipment, turbo-superchargers, gas turbines, and in the manufacture of glass and synthetic rubber.

Although special heat-resistant alloys have been developed and produced in large quantities for use in military equipment, the major commercial application is in furnaces for the treatment of metals. Alloys of the iron-chromium and iron-chromium-nickel groups are of the greatest commercial importance.

The percentage of total production of these alloys accounted for by each group is shown in Fig. 1.

General Properties

The general characteristics of the five types of heat-resistant alloys are as follows:

1. Iron-chromium alloy castings that contain from 10 to 30% Cr and little or no nickel. These alloys are useful chiefly for resistance to oxidation. They have low strength at elevated temperatures. The use of these alloys is dictated by conditions, either oxidizing or reducing, that involve low static loads and uniform heating. The chromium content depends on the temperature to which the alloys are to be exposed in service.

2. Fe-Cr-Ni Alloys — Austenitic alloys that contain more than 18% Cr and more than 7% Ni — always more chromium

*N. J. GRANT, *Chairman,* Professor of Metallurgy, Massachusetts Institute of Technology; J. KEVERIAN, *Vice Chairman,* Manager, Applied Research and Development, Foundry Department, General Electric Co.; W. M. BOAM, Assistant Dept. Head, Research & Materials, Aerojet-General Corp.; A. G. BUCKLIN, Thomson Engineering Laboratory, Small Aircraft Engine Dept., General Electric Co.; WILLIAM J. COLLINS, Manager, Metallurgical Dept., Corning Glass Works; G. A. FRITZLEN, Technical Director of Development, Haynes Stellite Co., Union Carbide Corp.

ROGER P. KING, Metallurgist, Libbey-Owens-Ford Glass Co.; R. A. MILLER, Division Metallurgist, Electro-Alloys Div., American Brake Shoe Co.; F. H. PENNELL, Chief Metallurgist, De Laval Steam Turbine Co.; ROBERT M. PLATZ, Assistant Chief, Materials Laboratory, Fairchild Engine Div., Fairchild Engine & Airplane Corp.; C. D. PREUSCH, Materials & Process Engineer, Crucible Steel Co. of America; E. A. SCHOEFER, Executive Vice President, Alloy Casting Institute; C. W. SCHWARTZ, Technical Director, Misco Precision Casting Co.

F. A. SETTINO, Senior Group Leader, Metallurgy Group, Pittsburgh Plate Glass Co.; S. A. SHERIDAN, Ford Aircraft Engine Div., Ford Motor Co.; W. H. STRAUTMAN, Superintendent, Southwestern Portland Cement Co.; E. T. VITCHA, Manager, Chemical and Metallurgical Laboratories, Jet Div., Thompson Ramo Wooldridge, Inc.; R. M. WOODWARD, Metallurgical Research Laboratory, Owens-Corning Fiberglas Corp.; R. D. WYLIE, Chief Metallurgist, Boiler Div., Babcock & Wilcox Co.

than nickel. These alloys are used ordinarily under oxidizing or reducing conditions similar to those withstood by the iron-chromium alloys, but in service they have greater strength and ductility than the straight chromium alloys. They are used, therefore, to withstand greater loads and moderate changes of temperature. These alloys are also used in the presence of oxidizing and reducing gases that contain sulfur.

3. Fe-Ni-Cr Alloys — Austenitic alloys that contain more than 10% Cr and more than 25% Ni — always more nickel than chromium. These alloys are used for withstanding a reducing as well as an ox-

idizing atmosphere, except where sulfur content is appreciable (for example, 0.05% H_2S or more). In such atmospheres the iron-chromium-nickel alloys of type 2 are recommended. The type 3 alloys do not carburize rapidly or become brittle and do not take up nitrogen in a nitriding atmosphere, in contrast with type 2 alloys. These properties become enhanced as the nickel content increases, and in carburizing and nitriding atmospheres the life of the casting increases with the nickel content. Austenitic ferrous alloys are used extensively under conditions of severe temperature fluctuations like those encountered by fixtures used in quenching and by parts that are not heated uniformly or that are heated and cooled intermittently. In addition, these alloys

have characteristics of electrical resistance that make them suitable for electrical resistor heating elements.

4. Nickel-base alloys that contain about 50% Ni and appreciable amounts of molybdenum. The nickel-base alloys that contain high percentages of molybdenum as a primary alloying element were developed for their resistance to chemical corrosion. However, these alloys also have good mechanical properties at high temperature and are used in the form of castings for certain high-temperature applications that warrant the cost of the alloys — higher than that of iron-base heat-resistant alloys.

5. Cobalt-base alloys that contain about 50% or more cobalt. The cobalt-base high-temperature alloys were first used

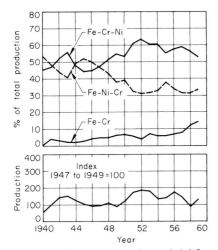

Fig. 1. (Upper) *Percentage of total heat-resistant alloy production contributed by the three principal groups.* (Lower) *Trend of total production of heat-resistant alloy castings.*

Table 1. Designations and Compositions of Heat-Resistant Casting Alloys

ACI	AISI(b)	SAE	ASTM(c)	C	Cr	Ni	Si (max)
HA	...	.....	A217, A199, A200	0.20 max	8 to 10	....	1.00
HB	...	.....	....	0.30 max	18 to 22	2 max	2.00
HC	446	70446	A297	0.50 max	26 to 30	4 max	2.00
HD	327	70327	....	0.50 max	26 to 30	4 to 7	2.00
HE	312	70312	A297	0.20 to 0.50	26 to 30	8 to 11	2.00
HF	302B	70308	A297	0.20 to 0.40	19 to 23	9 to 12	2.00
HH	309	70309	A297, B190	0.20 to 0.50	24 to 28	11 to 14	2.00
HI	...	.....	A297	0.20 to 0.50	26 to 30	14 to 18	2.00
HK(d)	310	70310	A297	0.20 to 0.60	24 to 28	18 to 22	2.00
HL	...	70310A	....	0.20 to 0.60	28 to 32	18 to 22	2.00
HN	...	.....	A297	0.20 to 0.50	19 to 23	23 to 27	2.00
HT	330	70330	A297, B207	0.35 to 0.75	13 to 17	33 to 37	2.50
HU	...	70331	A297	0.35 to 0.75	17 to 21	37 to 41	2.50
HW	...	70334	A297	0.35 to 0.75	10 to 14	58 to 62	2.50
HX	...	70335	A297	0.35 to 0.75	15 to 19	64 to 68	2.50

(a) Manganese is 0.35 to 0.65% for HA, 1% for HC, 1.5% for HD and 2% for the other alloys. Phosphorus and sulfur are 0.04% max. Molybdenum is not intentionally added except in type HA, which has 0.90 to 1.20% Mo; maximum for other alloys is set at 0.5% Mo. Type HH also contains 0.2% max N. (b) Wrought alloy designation. All others are cast alloy designations. These are listed for identification only. Castings should be ordered under cast alloy designation. (c) ASTM specification numbers; alloy designations are the same as ACI. (d) AMS 5365.

Table 2. Compositions of Some Nickel-Base and Cobalt-Base Heat-Resistant Casting Alloys

Alloy designation	C	Mn	Si	Cr	Ni	Co	Mo	Al	Fe	Others
Nickel-Base Alloys										
Hastelloy B	0.12 max	0.8	0.7	1.0	rem	....	28	....	5	....
Hastelloy C	0.15 max	0.8	0.7	16.5	rem	2.5 max	17	....	5	4 W
Hastelloy X	0.20 max	1.0 max	1.0 max	22	rem	1.5	9	....	18.5	1 W
Inconel 713C	0.12	0.50	0.50 max	13	rem	1.0 max	4.5	6.0	2.50 max	2.00 Cb + Ta, 0.6 Ti
Udimet 500	0.12	1.0 max	0.25 max	20	rem	10	4	2.75	2.0 max	3.0 Ti
Thetaloy	....	2.5	....	25	rem	12.5	3.0	....	....	7.0 W
GMR-235	0.15	0.25 max	0.60 max	15.5	rem	....	5.25	3.0	10	2.0 Ti, 0.05 B
Cobalt-Base Alloys										
HS 21	0.25	0.60	0.60	27	3	62	5.5	....	1	....
HS 31	0.50	0.60	0.60	25	10	55	....	....	1	8 W
HA 36	0.40	1.2	0.50	18.5	10	54	....	2.0 max	14.5 W	

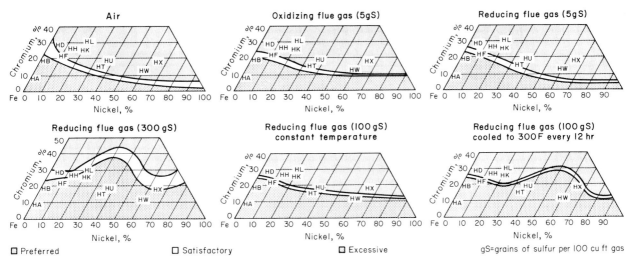

Fig. 2. *Relative performance of cast alloys in air and sulfur-bearing flue gas, based on 100-hr tests at 1800 F* (E. A. Schoefer, Machine Design, April 2, 1959)

Table 3. Approximate Rates of Corrosion of Commercial Heat-Resistant Castings in Air and in Flue Gas(a)

(Data, based on 100-hour tests, expressed in inches penetration per year)

Alloy	Ni	Cr	Oxidation in air 1600 F ipy	1800 F ipy	2000 F ipy	Sulfur-containing gas at 1800 F — Oxidizing(b) 5 gS ipy	100 gS ipy	Reducing(b) 5 gS ipy	100 gS ipy
HB	2 max	20	0.025—	0.250—	0.500—	0.100+	0.250—	0.500—	0.500
HC	4 max	28	0.010	0.050	0.050	0.025—	0.025	0.025+	0.025—
HD	5	28	0.010—	0.050—	0.050—	0.025—	0.025	0.025—	0.025—
HE	10	28	0.005—	0.025—	0.035—	0.025—	0.025	0.025—	0.025—
HF	10	20	0.005—	0.050+	0.100	0.050+	0.050+	0.100+	0.250—
HH	12	26	0.005—	0.025—	0.050	0.025—	0.025	0.025	0.025—
HI	15	28	0.005—	0.010+	0.035—	0.025—	0.025	0.025—	0.025—
HK	20	26	0.010—	0.010—	0.035—	0.025—	0.025	0.025—	0.025—
HL	20	30	0.010+	0.025—	0.035	0.025—	0.025	0.025—	0.025—
HN	25	20	0.005	0.010+	0.050—	0.025—	0.025	0.025	0.025
HP	30	30	0.025—	0.025+	0.035—	0.025—	0.025	0.025—	0.025—
HS	30	10	0.100	0.250	1.000+	0.250	0.500—	0.160	0.500+
HT	35	15	0.005—	0.010+	0.050	0.025	0.025	0.025—	0.100
HU	39	19	0.005—	0.010—	0.035—	0.025—	0.025	0.025—	0.025
HW	60	12	0.005—	0.010—	0.035	0.025	0.050—	0.025—	0.250
HX	67	17	0.005—	0.010—	0.035—	0.025—	0.025	0.025—	0.025—

(a) From Alloy Casting Institute, and A. Brasunas, J. T. Gow and O. E. Harder, Resistance of Fe-Ni-Cr Alloys to Corrosion in Air at 1600 to 2200 F, ASTM Symposium on Materials for Gas Turbines, June 1946, p. 129-52. (b) gS, grains of sulfur per 100 cu ft of flue gas.

for gas-turbine blading in 1941. Initial application was at temperatures no higher than those at which the heat-resistant alloys of the Fe-Cr-Ni and Fe-Ni-Cr groups were used. However, the alloys served under conditions of high stress and far surpassed the older alloys in rupture strength and creep properties.

The first three types of alloys are more widely used than the nickel-molybdenum and cobalt-base materials. Designations of the Alloy Casting Institute, ASTM, AISI and SAE for these alloys are given in Table 1. Compositions of several nickel and cobalt-base heat-resistant casting alloys are given in Table 2. Tables 3 and 4 show the approximate rates of corrosion and the mean coefficients of thermal expansion, respectively, for the alloys listed in Table 1. In Table 5 are listed the nominal mechanical properties of ACI-type alloys at room temperature.

Figure 2 provides a guide to the selection of castings on the basis of their resistance to air and sulfur-bearing flue gas at 1800 F.

Elevated temperature properties and other data for the cast heat-resistant alloys are given separately in the data compilations directly following this article. Service data for a number of these alloys in high-temperature applications are given in the article on Furnace Parts and Fixtures in this Handbook.

Manufacture. Castings produced in these several compositions may be made from heats melted in an electric-arc furnace that has either an acid or a basic lining, or in a high-frequency induction furnace. However, acid lined furnaces are seldom used because chromium losses are high and silicon content is difficult to control. Heat-resistant alloys containing appreciable amounts of titanium or other reactive metals may be melted under vacuum or a protective atmosphere. Castings may be made by the static method, by the centrifugal method or by the investment process. The centrifugal method is used extensively in the production of tubes for radiant heating furnaces and for other tubular parts.

With the exception of a small tonnage of iron-chromium alloys, the heat-resistant castings are not usually heat treated before shipment. When the 12 and 18% Cr alloys are used, it is sometimes necessary to give the castings a full annealing treatment to remove casting stresses, because these two alloy grades are hardenable.

Metallurgical Structures

Most of the alloys used for industrial heat-resisting service are either Fe-Cr-Ni or Fe-Ni-Cr types. Depending chiefly on the chromium and nickel contents, the basic structures may be ferritic, martensitic, or austenitic. Ferrite is soft, relatively weak and ductile at high temperatures. Martensite is unstable. Austenite is strong and relatively tough. Thus, the stronger industrial high-temperature alloys depend on an austenitic matrix. Use of the various chromium-nickel combinations is motivated by such factors as strength at elevated temperature, resistance to carburization and hot gas corrosion, and cost of the part.

Table 4. Thermal Expansion of Commercial Heat-Resistant Castings

ACI type	Ni	Cr	Mean coefficient of linear thermal expansion(a) Temp, F	Per °F
HC	4 max	28	32 to 1000	6.3
			32 to 1850	7.6
HE	10	28	120 to 1650	10.5
HF	10	20	80 to 1600	10.4
HH	12	26	70 to 1300	9.9
			70 to 1650	10.1
			70 to 1850	11.0
HK	20	26	32 to 950	9.2
			32 to 1850	10.6
HT	35	15	70 to 1850	9.7
HU	39	19	32 to 1850	9.5
HW	60	12	32 to 1850	9.0
HX	67	17	32 to 1850	8.5

(a) Micro-in. per in. per °F

Table 5. Mechanical Properties at Room Temperature

Property	As cast	Aged
HA Alloy		
(See first alloy in data compilations immediately following this article)		
HC Alloy(a)		
(2.0% min Ni, 0.15% min N)		
Tensile strength, 1000 psi...	110	115
0.2% yield strength, 1000 psi	75	80
Elongation in 2 in., %......	19	18
Hardness, Bhn	223	...
HD Alloy		
Tensile strength, 1000 psi...	85	...
0.2% yield strength, 1000 psi	48	...
Elongation in 2 in., %......	16	...
Hardness, Bhn	190	...
HE Alloy(a)		
Tensile strength, 1000 psi...	95	90
0.2% yield strength, 1000 psi	45	55
Elongation in 2 in., %......	20	10
Hardness, Bhn	200	270
HF Alloy(a)		
Tensile strength, 1000 psi...	85	100
0.2% yield strength, 1000 psi	45	50
Elongation in 2 in., %......	35	25
Hardness, Bhn	165	190
HH Alloy(a)		
Type I. Partially Ferritic		
Tensile strength, 1000 psi...	80	86
0.2% yield strength, 1000 psi	50	55
Elongation in 2 in., %......	25	11
Hardness, Bhn	185	200
Type II. Wholly Austenitic		
Tensile strength, 1000 psi...	85	92
0.2% yield strength, 1000 psi	40	45
Elongation in 2 in., %......	15	8
Hardness, Bhn	180	200
HI Alloy(a)		
Tensile strength, 1000 psi...	80	90
0.2% yield strength, 1000 psi	45	65
Elongation in 2 in., %......	12	6
Hardness, Bhn	180	200
HK Alloy(b)		
Tensile strength, 1000 psi...	75	85
0.2% yield strength, 1000 psi	50	50
Elongation in 2 in., %......	17	10
Hardness, Bhn	170	190
HL Alloy		
Tensile strength, 1000 psi...	82	...
0.2% yield strength, 1000 psi	52	...
Elongation in 2 in., %......	19	...
Hardness, Bhn	192	...
HN Alloy		
Tensile strength, 1000 psi...	68	...
0.2% yield strength, 1000 psi	38	...
Elongation in 2 in., %......	17	...
Hardness, Bhn	160	...
HT Alloy(b)		
Tensile strength, 1000 psi...	70	75
0.2% yield strength, 1000 psi	40	45
Elongation in 2 in., %......	10	5
Hardness, Bhn	180	200
HU Alloy(c)		
Tensile strength, 1000 psi...	70	73
0.2% yield strength, 1000 psi	40	43
Elongation in 2 in., %......	9	5
Hardness, Bhn	170	190
HW Alloy(d)		
Tensile strength, 1000 psi...	68	84
0.2% yield strength, 1000 psi	36	52
Elongation in 2 in., %......	4	4
Hardness, Bhn	185	205
HX Alloy(c)		
Tensile strength, 1000 psi...	65	73
0.2% yield strength, 1000 psi	36	44
Elongation in 2 in., %......	9	9
Hardness, Bhn	176	185

(a) The aging treatment for this alloy consists of holding at 1400 F for 24 hr and furnace cooling. (b) Aging treatment, 24 hr at 1400 F, air cooled. (c) Aging treatment, 48 hr at 1800 F, air cooled. (d) Aging treatment, 48 hr at 1800 F, furnace cooled.

Fine dispersions of carbides or intermetallic compounds in the austenitic matrix will increase high-temperature strength considerably. The fact that the carbon content of the heat-resistant castings is significantly higher than that of the corrosion-resistant castings is predicated primarily on the improvement in high-temperature strength provided by the additional carbon.

Castings will develop considerable segregation as they freeze. As cast, or after rapid cooling from a temperature near the melting point, much of the carbon in the standard grades will be in supersaturated solid solution. Subsequent reheating will precipitate excess carbides. The lower the reheating temperature, the slower the reaction and the finer the precipitated carbides are likely to be. The fine carbides increase creep strength and decrease ductility. Intermetallic compounds such as Ni₃Al, if present, can be made to play the same role.

Reheating the precipitated carbides in the range from 1800 to 2200 F will agglomerate and spheroidize them, reducing creep strength and increasing ductility. Above 2000 F so many of the fine carbides are dissolved or spheroidized that this strengthening mechanism loses its importance. By holding at temperatures where carbon diffusion is rapid, as above 2200 F, and then rapidly cooling, a high and uniform carbon content is established and retained in the austenite.

Aging at a lower temperature, such as 1400 F, where a fine, uniformly dispersed carbide precipitate will form, will confer a high level of strength up to temperatures where carbide agglomeration changes its character (overaging). This pattern of solution heat treatment or quench annealing followed by aging is a precipitation-hardening mechanism that may be invoked to attain maximum creep strength.

Ductility is usually reduced when strengthening occurs, but in some alloys, when the treatment corrects an unfavorable grain boundary network of brittle carbides, both properties may benefit.

However, such treatment is costly and may warp the castings excessively. Hence, this treatment is applied to only a small fraction of the total production of heat-resistant castings where premium performance justifies the high cost.

Carbides and solid solutions of carbide and nitride sometimes appear as platelets or as lamellar aggregates with austenite. This form has a lesser strengthening effect than the finely dispersed carbides.

Carbide networks at grain boundaries are generally undesirable. They usually occur in very high carbon alloys or in those that have cooled slowly through the high-temperature ranges where excess carbon in the austenite is rejected at grain boundaries rather than as dispersed particles. These networks confer brittleness in proportion to their continuity.

Carbide networks also provide paths for selective attack in some atmospheres and in certain molten salts. While chromium usually confers its protective qualities on the surface oxide whether it is present in carbides or in the matrix, it is advisable in some salt bath applica-

tions to sacrifice high-temperature strength and gain resistance to intergranular corrosion by specifying carbon of 0.08% max.

Iron-Chromium Alloys

"Straight chromium" alloys are made with 9, 18 and 28% Cr. Alloy types HC and HD are included among the straight chromium grades, although they contain 4% max Ni and 4 to 7% Ni, respectively.

Type HA (9% Cr, 1% Mo), a heat treatable ferritic material, contains enough chromium to provide good resistance to oxidation at temperatures up to about 1200 F. A molybdenum content of 1% contributes desirable strength. Castings of HA alloy are widely used in oil-refinery service. A modification of this alloy containing more chromium (12 to 14%) is widely used in the glass industry.

The alloy has a structure that is essentially ferritic with carbides in pearlitic areas or agglomerated particles, depending on prior heat treatment. Hardening of the alloy occurs on cooling in air from temperatures above 1500 F. In the normalized and tempered condition, the alloy

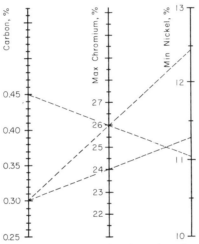

Fig. 3. Nomograph of carbon, chromium and nickel contents for wholly austenitic type HH (26% Cr, 12% Ni) heat-resistant alloy (J. T. Gow and O. E. Harder, ASM Trans, 30, 855, 1942)

exhibits satisfactory toughness throughout its useful temperature range.

Type HB (18% Cr). This alloy is only slightly hardenable and is ferritic. It contains enough chromium to make it suitable for castings that are to resist oxidation up to 1500 F. Castings are usually air cooled from 1350 to 1450 F, after which they are machinable, and are usually furnished to the consumer in this condition.

Type HC (28% Cr) provides excellent resistance to oxidation and high-sulfur flue gases at temperatures to 2000 F. Its use is limited to applications where strength is not a consideration, or for moderate-load service around 1200 F. It is also used where appreciable nickel cannot be tolerated, as in very high sulfur atmospheres, or where nickel may crack hydrocarbons through catalytic action.

The alloy is ferritic at all temperatures and for this reason is not hardened by heat treatment. Its ductility and impact strength are very low at room temperature. Its creep strength is very low at elevated temperature unless some nickel is present. In the HC-type alloy containing more than 2% Ni, substantial improvement in these properties is obtained by increasing the nitrogen content to 0.15% or more.

The HC alloy is embrittled when heated for prolonged periods at temperatures between 750 and 1025 F, and it shows low resistance to impact. The alloy is magnetic and has a low coefficient of thermal expansion, comparable to carbon steel. Electrical resistivity is about eight times that of carbon steel and thermal conductivity half. Heat conduction, however, is roughly double the value for the iron-chromium-nickel alloys.

Type HD (28% Cr, 5% Ni) is very similar in general properties to the HC grade, except that the nickel content imparts somewhat greater strength at high temperatures. Its high chromium content makes this grade suitable for use in high-sulfur atmospheres.

The alloy has a two-phase, ferrite-plus-austenite structure that is not hardenable by conventional heat treating procedure. Long exposure in the range from 1300 to 1650 F, however, may result in considerable hardening and severe loss of room temperature ductility through formation of sigma phase. Ductility may be restored by heating the alloy uniformly to a temperature of 1800 F or higher, and then cooling it rapidly to below 1200 F.

Iron-Chromium-Nickel Alloys

Ferrous alloys in which the chromium content exceeds the nickel content are made in compositions ranging from 28% Cr, 10% Ni to 30% Cr, 20% Ni.

Type HE (28% Cr, 10% Ni) has excellent resistance to corrosion at elevated temperatures. Because of the higher chromium content, it can be used at higher temperatures than HF alloy and is suitable for applications up to 2000 F. The material is stronger and more ductile at room temperature than the straight-chromium types. In recent years, a great deal of work on the HH-type alloy has provided a basis for evaluating the effect of variations in composition within the group of Fe-Cr-Ni alloys similar to type HH. Type HE belongs to this group.

In the as-cast condition the alloy has a two-phase, austenite-plus-ferrite structure containing carbides. Type HE castings cannot be hardened by heat treatment but, as with the HD grade, long exposure to temperatures around 1500 F will promote formation of the sigma phase with consequent embrittlement of the alloy at room temperature. The ductility of this grade can be improved somewhat by quenching the alloy from about 2000 F.

Castings of type HE have good machining and welding properties. Thermal expansion is about 50% greater than for carbon steel or the iron-chromium HC alloy. Thermal conductivity is much lower than for types HD or HC, but electrical resistivity is about the same. The alloy is weakly magnetic.

Type HF (20% Cr, 10% Ni) is essentially the same as the well-known 18% Cr, 8% Ni composition widely selected for its resistance to corrosion. The HF alloy is also suitable for use at temperatures up to 1600 F. When used for resistance to oxidation at elevated temperatures, it is not necessary to keep the carbon content at the low level specified for corrosion-resistant castings. Molybdenum, tungsten, columbium and titanium are sometimes added to the basic type HF composition to impart greater strength at elevated temperature.

As cast, the alloy has an austenitic matrix containing interdendritic eutectic carbides and, occasionally, an unidentified lamellar constituent. Aging at service temperatures usually promotes the precipitation of finely dispersed carbides. This results in higher room temperature strength and some loss of ductility. Improperly balanced, the alloy may be partially ferritic as cast. Such materials are susceptible to embrittlement from sigma phase formation after long exposure at 1400 to 1500 F.

Type HH (26% Cr, 12% Ni). Alloys of this nominal composition comprise about one third of the total production of heat-resistant castings. The alloy is used extensively in high-temperature applications, because of its strength and resistance to oxidation at temperatures up to 2000 F.

The composition balance of this type of alloy is critical. The range of nickel and chromium contents lies at the junction of the alpha-plus-gamma, the gamma, and the gamma-plus-sigma regions of the Fe-Cr-Ni system at 1650 F. Consequently, relatively large amounts of ferrite-promoting elements, in comparison with the amounts of austenite-promoting elements, may result in the presence of substantial amounts of ferrite in the metal. The ferrite decreases the strength at high temperatures, and if the ferrite transforms to sigma at intermediate temperatures, castings may suffer from brittleness or cracking when held within certain temperature ranges or when heated and cooled repeatedly through these ranges.

To achieve maximum strength at high temperature and to retain good ductility, this alloy must be wholly austenitic. If a balance is maintained between ferrite-promoting elements, such as chromium and silicon, and austenite-promoting elements, such as nickel, carbon and nitrogen, the desirable austenitic structure can be obtained. In commercial alloys of the HH type, with the usual carbon, nitrogen, manganese and silicon contents, Gow and Harder found that the ratio of chromium to nickel necessary for a stable austenitic alloy is expressed by the formula:

$$\frac{Cr\ (\%) - 16\ C\ (\%)}{Ni\ \%} = \text{less than } 1.7$$

The work of Gow and Harder provides strong evidence of the effect of silicon and molybdenum on the formation of sigma. Silicon in excess of 1% has the effect on structure of three times its weight in chromium, and molybdenum has the effect of four times its weight in chromium.

After the metal has been heated between 1200 and 1600 F, loss in ductility results from two changes within the alloy: *(a)* precipitation of carbides and *(b)* transformation of ferrite to sigma. When the composition of the alloy is balanced so that the structure is wholly austenitic, only the carbide precipitation normally occurs. In partially ferritic alloys, both carbides and sigma phase may form.

Figure 3, showing the relationship of carbon, chromium and nickel in the wholly austenitic HH alloy, will be found useful in balancing the composition to obtain an austenitic structure.

Before selecting the HH alloy as material for heat-resistant castings, it is advisable to consider the relation between the chemical composition and the temperature range in which the castings must operate. If castings are to operate continuously appreciably above 1600 F, there is little danger of serious embrittlement from either carbide precipitation or sigma phase. In this instance, the composition of the alloy may be 0.50% C max (0.35 to 0.40% preferred), 10 to 12% Ni and 24 to 27% Cr. On the other hand, if the operating temperature is to be from 1200 to 1600 F, the composition should be 0.40% C max, 11 to 14% Ni and 23 to 27% Cr. For such service the composition should be balanced to provide an austenitic alloy; for instance, for the 1200 to 1600 F application, the use of 11% Ni with 27% Cr is likely to produce sigma phase and its associated embrittlement, which occurs most rapidly around 1600 F. It is better, therefore, not to use the maximum chromium content with the minimum nickel content. The mechanical properties of the HH alloy at room temperature vary considerably with the composition, as shown in the data compilations following this article.

Short-time tests of properties of fully austenitic HH alloys at elevated temperatures show that the tensile strength and elongation depend on the carbon and nitrogen contents. The nomographic charts in Fig. 4 may be used to determine the properties of these alloys at 1600 and 1800 F. Nominal mechanical properties for this alloy, as determined from short-time tests at elevated temperatures, are given in the data compilations following this article.

In the application of heat-resistant alloys, the considerations include *(a)* resistance of the alloy to corrosion at elevated temperature; *(b)* the stability of the alloy — its resistance to warping, cracking or thermal fatigue; *(c)* the resistance of the alloy to plastic flow — its creep strength. For maximum creep strength the HH alloy should be fully austenitic in structure. Creep data are given in the data compilations.

In the design of load-carrying castings, data concerning creep stresses should be used with an understanding of the limitations of these data. An extrapolated limiting creep stress for 1% elongation in 10,000 hr cannot necessarily be sustained for that length of time without structural damage. Stress-rupture tests are a valuable adjunct to creep tests and are useful in the selection of design stresses. A comparison of creep rates and rupture times at various stresses and temperatures is shown for a wholly austenitic alloy in the data compilations for HH alloy following this article.

As pointed out before, HH alloys of wholly austenitic structure have greater strength at high temperatures than partially ferritic alloys of similar composition. Although a ratio factor of less than 1.7, calculated from the nickel, chromium and carbon contents, is an indication of a wholly austenitic material, ratio factors greater than 1.7 do not give a quantitative indication of the amount of ferrite present. It is possible, however, to measure the amount of ferrite in the material by magnetic analysis, after quenching the alloy from about 2000 F.

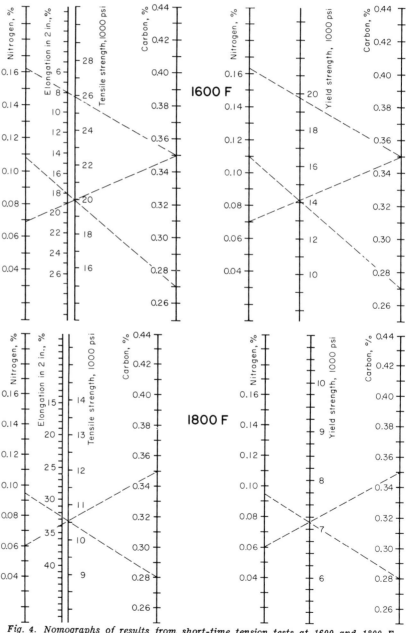

Fig. 4. Nomographs of results from short-time tension tests at 1600 and 1800 F showing elongation, tensile strength and yield strength in relation to carbon and nitrogen contents of type HH (26% Cr, 12% Ni) heat-resistant alloy (J. T. Gow and O. E. Harder, Trans ASM, 30, 855, 1942)

The magnetic permeability of HH alloys increases with the amount of ferrite. Thus, preferably after holding the alloy 24 hr at 2000 F and then quenching it in water, measurement of magnetic permeability can be related to the creep strength, since this property depends also on the structure of the material.

Alloys of the HH type are often evaluated by comparing the percentage elongation of sample bars in a tension test at room temperature after the bars have been held 24 hr at 1400 F. Such a test may be misleading, since there is a tendency to favor compositions that produce the greatest elongation after this particular heat treatment. High ductility values are likely to be related to low creep resistance.

Type HI (28% Cr, 15% Ni) is similar to type HH but contains more nickel and chromium. The increased chromium content makes this grade more resistant to oxidation than HH, and the additional nickel serves to maintain good strength at high temperature. Exhibiting adequate strength, ductility and corrosion resistance, this alloy has been used extensively for retorts operating with an internal vacuum at a continuous temperature of 2150 F. The alloy has an essentially austenitic structure containing carbides and, depending on the exact composition balance, may or may not contain small amounts of ferrite. Aging at 1400 to 1600 F is accompanied by precipitation of finely dispersed carbides, which may increase the strength and decrease the ductility at room temperature. At service temperatures above 2000 F, however, such carbides remain in solution and room temperature ductility is not impaired.

Type HK (26% Cr, 20% Ni) is somewhat similar to a wholly austenitic HH alloy in general characteristics and mechanical properties. Although less resistant to oxidizing gases than HC, HE or HI, the HK alloy has chromium content high enough to insure good resistance to corrosion by hot gases, including sulfur-bearing gases, in both oxidizing and reducing conditions. The high nickel content helps to make the HK grade one of the strongest heat-resistant casting alloys at temperatures above 1900 F. Accordingly, HK-type castings are widely used for stressed parts in structural applications to 2100 F. As normally produced, the HK alloy is stably austenitic over its entire temperaure range of application. The as-cast microstructure consists of an austenitic matrix containing relatively large carbides as scattered islands or networks. After aging at service temperature, the alloy exhibits precipitation of fine, granular carbides within the grains of austenite, with subsequent agglomeration if the temperature is high enough. These fine, dispersed carbides contribute to the creep strength of the alloy. A lamellar constituent resembling pearlite also is frequently observed in HK alloys, but its exact nature is not known.

Unbalanced compositions are possible within the stated ranges of this grade; hence, some ferrite may be present in the austenitic matrix. Such ferrite will transform to the brittle sigma phase if the alloy is held for more than a short time at about 1500 F, with consequent weakening at this temperature and embrittlement at room temperature. Formation of sigma phase in HK alloy can occur directly from austenite in the range from 1400 to 1600 F, particularly at the lower carbon level (0.20 to 0.30%), and for this reason a considerable scatter in properties at intermediate temperatures is observed.

Figure 5 indicates the spread in room-temperature mechanical properties obtained statistically for an HK alloy. These data were obtained by a single foundry and are based on 183 heats of the same alloy.

Type HL (30% Cr, 20% Ni) is similar to HK; its higher chromium content gives this grade greater resistance to corrosion by hot gases, particularly those containing appreciable amounts of sulfur.

Because essentially equivalent high-temperature strength can be obtained with either HK or HL, the improved corrosion resistance of HL makes it especially useful for service where excessive scaling must be avoided. The as-cast and aged microstructures of the HL alloy, as well as its physical properties and fabricating characteristics, are similar to those of HK.

Iron-Nickel-Chromium Alloys

In general these alloys have a more stable structure than the compositions in which chromium is the predominant alloying element. There is no evidence of a phase change that would cause brittleness within certain ranges of temperature. Experimental data indicate that the composition limits are not critical, and therefore the production of castings from these alloys does not require such close composition control as when alloys in the preceding Fe-Cr-Ni group are used.

In many respects there is no sharp line of demarcation among the Fe-Ni-Cr alloys as far as service applications are concerned. The follow-

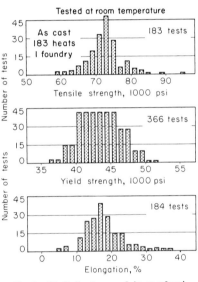

Fig. 5. Statistical spread in mechanical properties of HK alloy

ing general observations, however, should be considered when the most suitable alloy is being selected for any given service: (*a*) as the nickel content is increased, the tendency for the alloys to carburize is decreased; (*b*) as the nickel content is increased, the tensile strength of the alloys at elevated temperatures is decreased somewhat but the alloys are more resistant to thermal shock and thermal fatigue; (*c*) as the chromium content is increased, the alloys become more resistant to corrosion and oxidation; (*d*) as the carbon content is increased, the tensile strength of the alloys at elevated temperatures is increased; and (*e*) as the silicon content is increased, the tensile strength of the alloys at elevated temperatures is decreased, but the resistance to carburization is increased somewhat.

Type HN (25% Ni, 20% Cr) contains enough chromium for good high-temperature corrosion resistance and nickel content greater than chromium. The alloy has properties somewhat similar to the much more widely used type HT alloy. Although extensive field service data are lacking, the information available indicates that the HN grade may be a potentially useful alternate to alloys of higher nickel content in some highly stressed parts. The alloy is austenitic at all temperatures and lies well within the stable austenite field. In the as-cast condition, carbide areas are present and additional fine carbides precipitate on aging. The alloy is not susceptible to sigma-phase formation, nor is increased carbon content especially detrimental to ductility.

Types HT, HU, HW and HX

Alloys HT, HU, HW and HX as a group normally constitute about one third of the total production of heat-resistant alloys. These four alloys have given excellent service life in fixtures and trays for heat treating furnaces where the castings are subjected to rapid heating or cooling. Since these compositions carburize less readily than the Fe-Cr-Ni types, they are used extensively for parts in carburizing furnaces. Because they form an adherent scale that does not flake off, castings of high-nickel alloys are also made for the enameling industry, for applications where loose scale would be detrimental to the process.

In many respects there is no sharp line of demarcation between the HT, HU, HW and HX alloys as far as service applications are concerned. The following general observations, however, should be considered when the most suitable alloy is being selected for any given service.

Type HT (35% Ni, 15% Cr) contains about equal amounts of iron and alloying elements. The high nickel content makes this grade useful in resisting the thermal shock of rapid heating and cooling. In addition, the alloy is resistant at high temperature to oxidation and carburization and has good strength at heat treating temperatures. Except in high-sulfur gases, it performs satisfactorily to 2100 F in oxidizing atmospheres and up to 2000 F in reducing atmospheres, provided limiting creep stress values are not exceeded.

However, in considering maximum operating temperatures for a heat-resistant alloy the allowable ranges of nickel and chromium must be considered. For example, an HT alloy having both nickel and chromium on the high side of the allowable range is essentially the same as HU having these elements at the low side. These variations will influence the maximum operating temperatures they can withstand.

The HT grade is widely used for general heat-resistant applications in highly stressed parts. The alloy has an austenitic structure containing varying amounts of carbides, depending on its carbon content and thermal history. In the as-cast condition, large carbide areas are present at the interdendritic boundaries, but fine carbides precipitate within the grains after exposure at service temperatures, with subsequent decrease in room-temperature ductility. Increased carbon content may affect the high-temperature ductility of the alloy. Additional protection against carburization is obtained with silicon contents above about 1.6%, but at some sacrifice of strength at elevated temperatures. The alloy can be made still more resistant to thermal shock by the addition of up to 2% Cb.

Type HU (39% Ni, 19% Cr) is similar to type HT but its higher chromium and nickel contents give it greater resistance to corrosion by either oxidizing or reducing hot gases including those containing up to 100 grains of sulfur per 100 cu ft of gas (Fig. 2). High-temperature strength, resistance to thermal fatigue and resistance to carburization of the alloy are essentially the same as for HT; hence its improved corrosion resistance makes HU especially suited for severe service conditions involving high stress and rapid thermal cycling.

Type HW (60% Ni, 12% Cr), in which the high nickel content contributes toward excellent resistance to carburization, is especially useful in applications where wide, rapid and severe temperature fluctuations are encountered. In addition, the alloy is resistant to oxidation at high temperature and, although not as strong as HT, has good strength at heat treating temperatures. It performs satisfactorily up to about 2050 F in strongly oxidizing atmospheres and to 1900 F in oxidizing or reducing products of combustion, provided sulfur is not present in the gas. The generally adherent nature of its oxide scale makes the HW alloy suitable for enameling furnace service where even small flakes of dislodged scale could ruin the work in process.

This grade is widely used for intricate heat treating fixtures that are quenched with the load and for many other applications involving thermal shock and steep temperature gradients under conditions of high stress, such as furnace retorts and muffles.

The alloy has an austenitic structure containing varying amounts of carbides, depending on carbon content and thermal history. In the as-cast alloy, the microstructure consists of a continuous interdendritic network of elongated eutectic carbides. After aging at service temperatures, the austenitic matrix becomes uniformly peppered with small carbide particles except in the immediate vicinity of the eutectic carbides. This change in structure is accompanied by an increase in room-temperature strength without change in ductility.

Type HX (66% Ni, 17% Cr) is similar to HW but contains more nickel and chromium. The increased chromium content confers substantially improved resistance to corrosion by hot gas, even sulfur-bearing, which permits this grade to be employed for severe service applications at temperatures up to 2100 F. However, it has been reported that this alloy decarburizes rapidly in the temperature range from 2000 to 2100 F. High-temperature strength, resistance to thermal fatigue and resistance to carburization of the alloy are essentially the same as for HW; hence it is suitable for the same general applications where corrosion must be minimized. The as-cast and aged microstructures of HX alloy, as well as its mechanical properties and fabricating characteristics, are similar to HW.

Nickel-Base Alloys

Nickel-base high-temperature alloys containing 1% Al or more are strengthened primarily by the precipitation of Ni$_3$Al in a solid solution hardened by molybdenum and chromium. Virtually all commercial alloys of this category also contain titanium, which partially replaces aluminum up to an atomic ratio of 3 Ti to 1 Al in the intermetallic phase without altering its crystallographic structure. A very effective precipitation hardening is achieved in this system because both the matrix and precipitate are face-centered cubic and their lattice parameters are quite close, differing, generally, by less than 0.5%. Kinetic studies have shown the rejection of Ni$_3$Al from solid solution to be extremely rapid. In addition to this precipitation, the alloys derive supplementary benefit from boron or zirconium additions in small amounts.

The nickel-base alloys that contain a high percentage of molybdenum as a primary alloying element were developed for their resistance to chemical corrosion. However, these alloys also have good mechanical properties at high temperature and are used in the form of castings where the cost of the alloys can be justified.

Hastelloys. The toughness, high-temperature strength, and machinability of Hastelloy B (61% Ni, 28% Mo), combined with its corrosion resistance, make it particularly useful in the design of chemical plant equipment. Solution treatment serves to develop its maximum ductility and corrosion resistance.

Hastelloy C (51% Ni, 17% Mo, 16 Cr) castings have greater strength than wrought low-carbon and medium-carbon steels. The alloy has considerable toughness, although it is less ductile than steel. It can be readily ground to a good finish and can be highly polished. This is important for applications such as valve seating surfaces.

Hastelloy B contains no more than 1% Cr, which limits its application under oxidizing conditions to a maximum temperature of about 1400 F. The alloy is particularly satisfactory with reducing atmospheres. Hastelloy C, having a considerable chromium content, is useful for its resistance to oxidation at temperatures to 2100 F and is also satisfactory in atmospheres that are reducing or otherwise corrosive. This alloy is used for tracks and trays in heat treating furnaces operating to 2000 F, for chain conveyors in furnaces to 1750 F, and for carburizing equipment.

Hastelloy X (45% Ni, 22% Cr, 9% Mo) has high strength and oxidation resistance to 2200 F. Since it has a relatively low strategic alloy content, it is suitable for use in furnace applications involving oxidizing, reducing, and neutral atmospheres. It has been used in a variety of jet-engine parts including tailpipes, afterburner components, turbine heads and nozzle vanes.

Inconel 713 C (Ni-base, 13% Cr, 4% Mo, 7% Al, Ti) is an investment-casting alloy used in the jet-engine field. It is currently the strongest cast turbine blade material commercially available. Designed for 1700 F service, it has a tensile strength of slightly under 80,000 psi at this temperature with an elongation of about 10%.

Udimet 500, Thetaloy and R-235 are other nickel-base casting alloys currently used in jet-engine service.

Cobalt-Base Alloys

The cobalt-base heat-resistant alloys were first used for highly stressed gas-turbine blading, in 1941. The first alloy used was a modification of Vitallium, a dental alloy, employed for a number of years before the second World War. The application and modification of the alloy and the production of gas-turbine blading by precision casting on an industrial scale were important wartime developments in the field of metallurgy. The wartime application of this alloy was at temperatures no higher than those at which the heat-resistant alloys of the Fe-Cr-Ni and Fe-Ni-Cr types were used, but the alloy served under conditions of high stress and far surpassed the older alloys in creep and rupture properties.

Most of the high-temperature cobalt-base alloys contain an appreciable amount of carbon and derive their strength not only from solid-solution hardening by such elements as tungsten and chromium, but also from carbide precipitation. Nickel is commonly employed to stabilize the high-temperature form of cobalt (face-centered cubic).

Chromium carbide precipitation is predominantly Cr$_{23}$C$_6$ with some substitution of chromium by other elements such as iron and tungsten. In contrast to the precipitation of Ni$_3$Al in nickel-base alloys, precipitation of Cr$_{23}$C$_6$ is relatively sluggish. Aging in this type of alloy can come from two sources, one being the precipitation of Cr$_{23}$C$_6$ from supersaturated solid solution, the other involving the decomposition of the metastable Cr$_7$C$_3$ (present as massive primary carbides in alloys with sufficient carbon) to Cr$_{23}$C$_6$ plus carbon. The liberated carbon atoms diffuse into the carbon-lean matrix and combine with chromium atoms to form Cr$_{23}$C$_6$, which precipitates as discrete particles.

HS 21 (Co-base, 27% Cr, 5.5% Mo) is resistant to oxidizing and reducing atmospheres at temperatures to 2100 F. The alloy has excellent strength at elevated temperatures, and this strength is maintained through aging that occurs in the temperature range from 1300 to 2100 F. At 1500 F, for example, it has an average as-cast tensile strength of 62,500 psi with an elongation of 16%. The alloy is used in applications requiring resistance to thermal shock, such as in the turbine nozzle diaphragms of a jet engine. These vanes, which guide the hot gases from the combustion chambers to the turbine buckets, are subject to severe thermal shock. The alloy has also been used successfully in buckets for turbosuperchargers. Normally used in the as-cast condition, it ages at operating temperatures. This increases its strength and slightly lowers ductility.

HS 31 (X-40) (Co-base, 25.5% Cr, 10.5% Ni, 7.5% W). At 1500 F, the alloy has an average as-cast tensile strength of 63,000 psi with an elongation of 15%. The average 100-hr rupture stress at this temperature is 27,000 psi. At 1500 F, the average stress to produce 1% elongation in 100 hr is 18,000 psi.

The alloy has been investment-cast into turbine blading used in a number of operational jet engines. It has also been used for turbine wheels in other types of high-temperature gas turbine engines and turbosupercharger units. In common with HS 21, this alloy is generally used in the as-cast condition and ages at service temperatures.

HS 36 (Co-base, 18% Cr, 14.5% W, 10% Ni) is a cobalt-base cast alloy comparable to wrought HS 25. Suitable for investment casting, it has excellent creep and rupture properties along with good hot ductility at rupture. Its elevated temperature properties are improved by aging for 16 hr at 1350 F, although ductility is decreased.

Applications

The greatest tonnage of heat-resistant castings is used in metallurgical and other industrial furnaces. A detailed discussion of furnace applications is presented in the article on furnace parts and fixtures, in this Handbook. Other widespread applications for heat-resistant castings include turbosuperchargers, gas turbines, power plant equipment, and equipment involved in the manufacture of glass, cement, synthetic rubber, chemicals and petroleum products.

Service Example 1. Data pertaining to the use of cast alloys in the manufacture of glass fiber are shown in Fig. 6. In this application, cast alloy bushings are mounted in the forehearth of a glass-melting tank. Molten glass is fed by gravity to each bushing and flows through the forming tips at the bottom of the bushing as it is mechanically drawn. The diameter of each glass fiber is determined by the size of the hole in the bushing tip, the speed of pull, and the temperature and type of glass used. Since the bushing must be kept hot, it is heated by resistance through water-cooled clamps attached to a terminal at each end.

The cast alloy bushings are subjected to corrosion and erosion resulting from the passage of molten glass at high temperature and the oxidizing effects of the surrounding air. At 1830 F, HF with lower chromium and nickel content had an average life of only 46 days. It was replaced by HK alloy, thereby increasing the average bushing life to 77 days. This improvement was effected with only a moderate increase in alloy content. When the equipment was used at 1850 F, however, it was necessary to go to an alloy rich in nickel; average life was extended to 229 days, as shown in Fig. 6. (HT and HW alloys were tried experimentally but were not used in production, so that no performance data are available.)

Service Example 2. The selection of cast heat-resistant alloys for a burning layout in a cement mill is shown in Fig. 7. Significantly, more than one alloy can be used successfully in most of the twenty components listed. Environmental conditions that provide the basis for selection are listed for each component.

Chemical processing and petroleum refinery applications are discussed in the articles on these subjects in this book.

Cost

Cost per hour of service life is the ultimate criterion in selection of a heat-resistant alloy. On this basis an alloy that is more expensive in initial cost frequently provides lower cost per hour of service than a less expensive type. Several examples of this relationship, as well as studies of the distribution in service life results among a large number of identical cast parts, are included in the article on furnace parts and fixtures, in this Handbook.

Machinability

Machinability data for ACI alloys and for nickel-base and cobalt-base aircraft alloys are summarized in

terms of speed, feed and depth of cut in Tables 6 and 7, respectively. These are recommended starting points for machining, rather than actual data for a specific tool life.

Casting Design

For the majority of applications, the ACI types of heat-resistant alloys are sand cast. Section thicknesses of $\frac{3}{16}$ in. and more can be cast satisfactorily and somewhat lighter sections may also be possible, depending on casting design and pattern equipment. Dimensional tolerances for rough castings are influenced by the quality of the pattern equipment. In general, over-all dimensions and locations of cored holes can be held to $\frac{1}{16}$ in. per foot.

More specific information regarding casting design details is to be found in Fig. 8. In Fig. 9 a comparison is made of dimensional varia-

Table 6. Machining Data for ACI Heat-Resistant Alloys

ACI designation	Max hardness, Bhn	Rough turning(a) Speed, sfm	Rough turning(a) Feed, ipr	Finishing Speed, sfm	Finishing Feed, ipr	Drilling speed, sfm(b)
HA	220	40 to 50	0.010 to 0.030	80 to 100	0.005 to 0.010	35 to 70
HC	220	40 to 50	0.025 to 0.035	80 to 100	0.010 to 0.015	40 to 60
HD	190	40 to 50	0.025 to 0.035	80 to 100	0.010 to 0.015	40 to 60
HE	270	30 to 40	0.020 to 0.025	60 to 80	0.005 to 0.010	30 to 60
HF	190	25 to 35	0.020 to 0.025	50 to 70	0.005 to 0.010	20 to 40
HH	200	25 to 35	0.015 to 0.020	50 to 70	0.005 to 0.010	20 to 40
HI	200	25 to 35	0.015 to 0.020	50 to 70	0.005 to 0.010	20 to 40
HK	190	25 to 35	0.020 to 0.025	50 to 70	0.005 to 0.010	20 to 40
HL	190	30 to 40	0.020 to 0.025	60 to 80	0.005 to 0.010	30 to 60
HN	160	35 to 45	0.020 to 0.025	70 to 90	0.005 to 0.010	40 to 60
HT	200	40 to 45	0.025 to 0.035	80 to 90	0.005 to 0.010	40 to 60
HU	190	40 to 45	0.025 to 0.035	80 to 90	0.010 to 0.015	40 to 60
HW	200	40 to 45	0.025 to 0.035	80 to 90	0.010 to 0.015	40 to 60
HX	185	40 to 45	0.025 to 0.035	80 to 90	0.010 to 0.015	40 to 60

(a) Single-point high speed steel tools are usually ground to 4 to 10° side and back rake, 4 to 7° side relief, 7 to 10° end relief, 8 to 15° end cutting-edge angle, 10 to 15° side cutting-edge angle and $\frac{3}{32}$ to $\frac{1}{8}$-in. nose radius. (b) Recommended drilling feeds are as follows: For a drill diameter up to $\frac{1}{8}$ in., 0.001 to 0.002 ipr; $\frac{1}{8}$ to $\frac{1}{4}$-in. diam, 0.002 to 0.004 ipr; $\frac{1}{4}$ to $\frac{1}{2}$-in. diam, 0.004 to 0.007 ipr; $\frac{1}{2}$ to 1-in. diam, 0.007 to 0.015 ipr; over 1-in. diam, 0.015 to 0.025 ipr. *Tapping* speeds recommended for alloys HA, HC, HD, HE and HL are 10 to 25 sfm; HF, HH, HI and HK, 10 to 20 sfm; HN, HT, HU, HW and HX, 5 to 15 sfm.

Table 7. Machining Data for Nickel-Base and Cobalt-Base Heat-Resistant Alloys

Alloy (investment cast)	Maximum Rockwell C hardness	Rough turning (carbide)	Finish turning (carbide)	Precision finishing and boring (carbide)	All operations (high speed steel)
HS 21 (AMS 5385)	35	40	60	100	10 to 15
HS 31 (AMS 5382)	35	40	60	100	10 to 15
Hastelloy C (AMS 5388)	21	40	60	100	10 to 15
Hastelloy C (AMS 5389)	15	40	60	100	10 to 15
Hastelloy X (AMS 5390), solution treated	23	50	75	100	10 to 15

(a) Ratings are based on the use of positive rake angles. For maximum economy, negative rake should be used wherever possible. Speeds given above are for depths and feeds as follows:

	Depth of cut	Feed
Roughing	$\frac{1}{4}$ in. or more	0.015 to 0.025 in.
Finishing	1/32 to $\frac{1}{4}$ in.	0.010 to 0.015 in.
Precision finishing and boring	Less than 1/32 in.	Less than 0.010 in.

(b) Approximate machinability rating for the Stellites and Hastelloy C was 15% (20% for Hastelloy X), based on 100% for free-machining steel bar stock under each of the following conditions:

	Carbide	High speed steel
Rough turning	300 sfm	100 sfm
Finish turning	400 sfm	125 sfm
Precision finishing and boring	500 sfm	225 sfm
Milling	350 sfm	125 sfm
Drilling	125 sfm	100 sfm

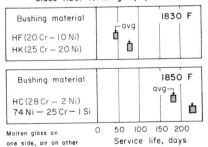

Glass fiber forming equipment

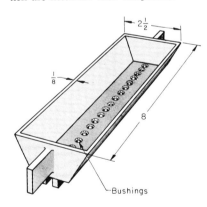

Fig. 6. Effect of composition on service life of cast bushings used in equipment for forming glass fibers. Corrosion and erosion of the orifices were the main criteria of failure. (See Service Example 1 in text.)

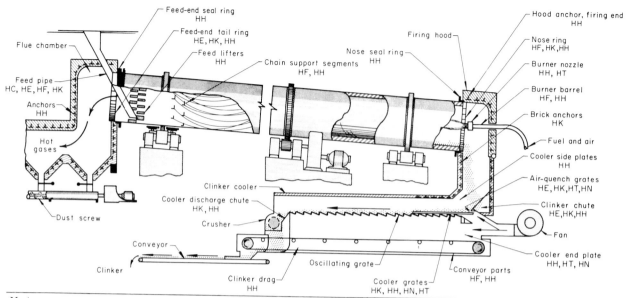

Maximum operating temp, F	Part name	Alloys used(a)	Environmental conditions	Service life, yr
1200	Conveyor parts	HF, HH₂	Severe abrasion and oxidation	Indefinite
1200	Cooler discharge chute	HH₂, HK	Severe abrasion and oxidation	3 to 5
1200	Clinker drag	HH₂	Severe abrasion and oxidation	5 to 10
1400	Feed-end seal ring	HH₂	Some abrasion and oxidation	Indefinite
1500	Brick anchors	HK	Even temperature	Indefinite
1500	Burner barrel	HF, HH₂	Slight abrasion and oxidation	5 to 10
1500	Hood, anchor firing end	HH₂	Even temperature, oxidation	Indefinite
1500	Clinker chute	HE, HH₂, HK	Severe abrasion, impact, oxidation	Indefinite
1500	Air-quench grates	HE, HK, HN, HT	Severe abrasion and oxidation	3 to 7
1700	Anchors	HH₂	Even temperature	Indefinite
1800	Feed pipe	HC, HE, HF, HK	Moderate abrasion inside feed and dust particles outside, thermal shock, oxidation and sulfur gases	2 to 7
1800	Feed-end tail ring	HE, HH₂, HK	Abrasive dust particles, thermal shock and oxidation	10 to 15
1800	Feed lifters	HH₂	Some abrasion, thermal shock, oxidation and sulfur gases	5 to 10
1800	Chain support segments	HF, HH₂	Intermittent temperature surges, light abrasion, sulfur gases	Indefinite
1800	Cooler end plates	HH₂, HN, HT	Severe abrasion and oxidation	1 to 5
1800	Cooler grates	HH₂, HK, HN, HT	Severe abrasion and oxidation	1 to 5
1800	Cooler side plate	HH₂	Severe abrasion and oxidation	1 to 5
2000	Nose seal ring	HH₂	Some abrasion, oxidation and sulfur gases	3 to 10
2000	Burner nozzle	HH₂, HT	Some abrasion, oxidation and sulfur gases	1 to 3
2200	Nose ring	HF, HH₂, HK	Extreme abrasion, oxidation and sulfur gases	3 to 5

(a) HH₂ is the type 2, wholly austenitic (26% Cr, 12% Ni) alloy. See data compilation following this article for discussion.

Fig. 7. A typical burning layout for manufacturing cement, showing location of specific components and cast heat-resistant alloys that have been used. Operating conditions and service life for 20 components are tabulated above.

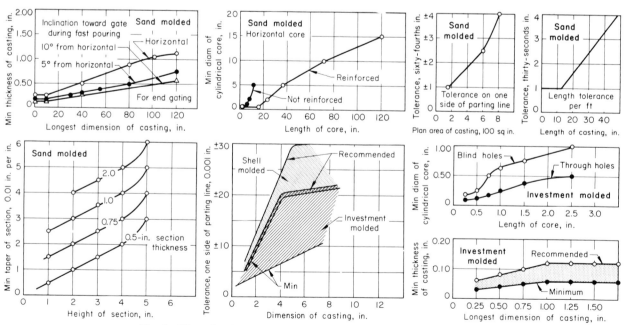

Fig. 8. Dimensional relations for sand, shell and investment molding

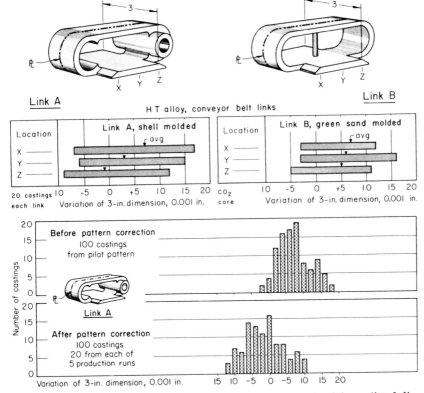

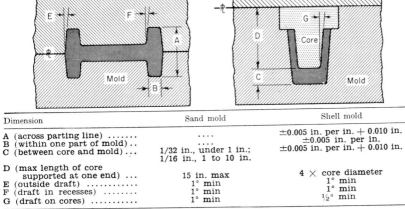

Fig. 9. Charts at top give the dimensional variations encountered in casting belt links in shell and green sand molds. Below are shown the dimensions before and after pattern correction of the shell molded link.

Dimension	Sand mold	Shell mold
A (across parting line)	....	±0.005 in. per in. + 0.010 in.
B (within one part of mold)..	....	±0.005 in. per in.
C (between core and mold)...	1/32 in., under 1 in.; 1/16 in., 1 to 10 in.	±0.005 in. per in. + 0.010 in.
D (max length of core supported at one end) ...	15 in. max	4 × core diameter
E (outside draft)	1° min	1° min
F (draft in recesses)	1° min	1° min
G (draft on cores)	1° min	½° min

Fig. 10. Recommended minimum tolerances for casting the above shapes

17% Cr steel, shell molded

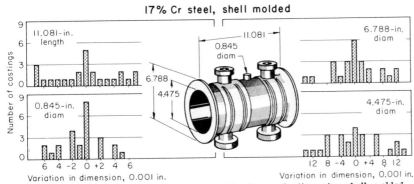

Fig. 11. Dimensional variations encountered in the production of a shell molded valve body. These data illustrate the magnitude and frequency of dimensional variations involved in casting a relatively complex form. The gating and risering needed to solidify a casting of this over-all shape are shown in Fig. 12. Certain section changes were made to avoid hot tears.

tions encountered in casting belt links in green sand and shell molds. Dimensional variation is of about the same magnitude for both types of molds. The greater number of measurements on the plus side of the tolerance range is largely a matter of pattern correction. Figure 9 (bottom) shows two stages in the development of Link A, with pattern correction to bring the variation in casting dimension from the high side to the middle of the ±0.015-in. tolerance range. In this example, removing approximately 0.010 in. from the pilot pattern cost $25. If a production pattern had been made to the drawing dimension, it would have been necessary to build a complete new pattern at a cost of approximately $2000.

Figure 10 provides another comparison of minimum tolerances for sand and shell molds used in making two typical shapes. The composition of the alloy does not significantly affect these tolerances.

The relations between process variables and casting shape are of fundamental importance for all castings. High-alloy castings are similar to steel castings in many of these design-process relations. For a detailed discussion of such matters, the reader may refer to the design section of the article on steel castings in this Handbook. One example relating to design changes in a complex shape is given below:

Design-Process Example. A shell molded valve body is shown in Fig. 11, along with a record of dimensional variations encountered in its production. These data illustrate the magnitude and frequency of dimensional variations likely to be met in casting a relatively complex form. This valve body is an example of a casting in which design revisions were necessary to suit processing requirements. Two of the revisions involved blending at section changes, one dealt with ribs, and the fourth required the reduction of mass in a chilled boss. Each change was related to the mode of solidification applicable to the casting shape. The gating and risering needed to solidify a casting of this over-all shape are shown in the top part of Fig. 12.

The section changes shown at View A in Fig. 12 were not expected to give trouble but nevertheless resulted in hot tears. Why hot tears would develop here can be understood by studying the direction of freezing, which could be predicted by the location of the six risers. Since the freezing progresses generally from points E1 and E2 (Fig. 12) diagonally toward the four side risers and the two end risers, concentrated shrinkage stresses are across these section increases, and a localized hot zone exists at the 1/16-in. fillet. Had these section changes been oriented parallel to the freezing direction, stress would not have developed and hot tears would not have occurred. The blend was redesigned with a 15° taper as shown, and no further defects were observed during subsequent production of 245 castings.

The four flanged outlets shown as View B obviously required risering with directional freezing toward each riser. The original design provided for the sharply changing section detailed, which resulted in hot tears in 20% of the castings. It was recognized that the hot tears were caused by the sharpness of the section change closest to the riser and that the blend should be more gradual. The temperature gradient between the two diameters of the outlet was too great; a reduction in the temperature gradient was needed to improve producibility.

The importance of the temperature gradient can be appreciated when it is realized that the cylindrical part of the outlet was freezing considerably in advance of the flanged portion. This, combined with the predictable existence of retarded cooling in the corner at the junction of the cylinder and flange, produced an insufficient thickness of solidified skin at the corner to resist rupture (hot tears) from the stress by differential shrinkage of these adjacent portions.

The flange was redesigned with a blend as shown and no more castings were rejected for hot tears in this area.

Inspection of the eight ribs (View C) revealed about 5% rejects due to cold shuts. One of the common foundry solutions to the problem of cold shuts is to pour the metal at a higher temperature in order to keep it fluid long enough to run into the narrow openings in the mold. Although the pouring temperature was already high, it was decided that pouring at a still higher temperature was worth the calculated risk. But this resulted in excessive gas evolution from the mold and from gas dissolved in the metal, which produced new defects at random throughout the casting. Thus, the alternative was to enlarge the rib in the areas where cold shuts were observed.

It is evident in comparing these two designs that the increase in width of the rib was very near the minimum required to avoid the defect. For this reason, the 0.187-in. rib width probably reflects a minimum for ribs of this height in a 17% Cr shell molded steel casting of comparable design and rigging.

The boss shown at section X-X in Fig. 12 is an example of the conflict that can exist between part function and required foundry procedures. The chilled face of this boss was to receive a hole and a butterfly valve. This required that the face be acceptable as cast.

Thus, a machining operation would have been required to remove either a riser or a rib added to feed from the nearby riser. The boss was at first believed small enough to respond to chilling. The total mass of metal involved nevertheless cooled more slowly than anticipated and therefore tended to freeze after surrounding areas had solidified, thus isolating the boss from feed metal. This resulted in shrinkage cracks at the base of the boss, as shown. The solution was to reduce the mass of the boss so that chilling would be effective and freezing would begin in this boss and continue into adjacent areas.

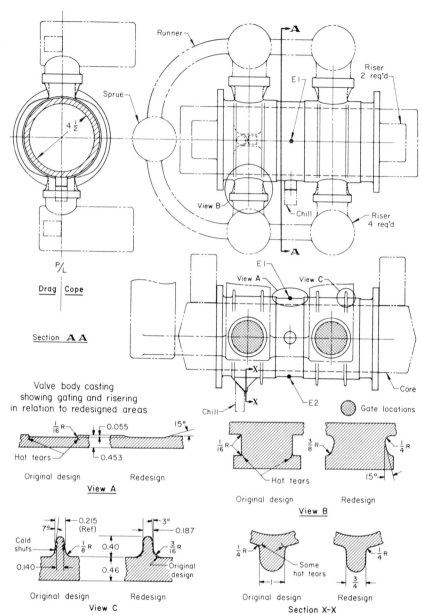

Fig. 12. Details of a shell molded valve body. Dimensional variations involved in producing this casting are given in Fig. 11.

PROPERTIES OF HEAT-RESISTANT ALLOY CASTINGS

*Compiled by the ASM Committee on Heat-Resistant Castings**

HA alloy

(9% Cr, 1% Mo)

A 3 *ASTM number.* A217, grade C12
C 1 *Density* at 70 F. 7.72 g per cu cm (0.279 lb per cu in.)
D 6 *Thermal expansion,* micro-in./in./°F

68 to 212 F (20 to 100 C) 6.1
68 to 1000 F (20 to 538 C) 7.1
68 to 1200 F (20 to 650 C) 7.5

11 *Specific heat* at 70 F. 0.11 Btu/lb/°F

*Most of the data on the H-alloys (HA through HX) are from Alloy Casting Institute data, compiled by E. A. Schoefer.

J 6 *Microstructure.* See the article on heat-resistant castings.
K 1 *Mechanical properties.* Modulus of elasticity in tension: 29,000,000 psi
M 1 *Chemical composition limits.* 0.20 max C, 0.35 to 0.65 Mn, 1.00 max Si, 0.04 max P, 0.04 max S, 0.90 to 1.20 Mo, 8 to 10 Cr, no Ni, rem Fe
N 1 *Melting temperature.* 2750 F (1510 C)
   9 *Annealing temperature.* 1625 F (885 C)
  15 *Heat treatment.* For maximum softness, castings should be annealed by heating to 1625 F, or slightly higher, and slowly cooled in the furnace at about 50 °F per hr to below 1300 F. For improved strength, castings are normalized by heating to 1825 F and air cooling to below 1300 F, followed by tempering at about 1250 F.

HA alloy	Annealed	Normalized, tempered(a)
Tested at Room Temperature		
Tensile strength, . . . 1000 psi	95	107
0.2% yield strength, . 1000 psi	65	81
Elongation in . . . . . . 2 in., %	23	21
Reduction of . . . . . . area, %	. . .	56
Hardness, Bhn	180	220
Impact strength, ... Charpy keyhole, ft-lb	. . .	32

(a) Normalized 1825 F, tempered 1250 F

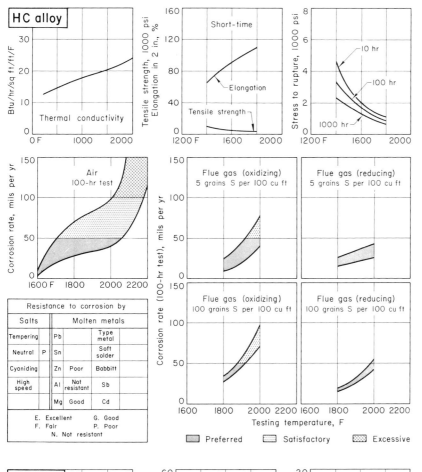

HC alloy

(28% Cr)

A 3 *ASTM number.* A297, grade HC
   5 *Other designations.* SAE 70446
B 1 *Typical uses.* Boiler baffles, electrodes, furnace grate bars, gas outlet dampers, kiln parts, lute rings, rabble blades and holders, recuperators, salt pots, soot blower tubes, support skids, tuyeres
C 1 *Density* at 70 F. 7.53 g per cu cm (0.272 lb per cu in.)
   5 *Patternmaker's shrinkage.* 7/32 in./ft
D 6 *Thermal expansion,* micro-in./in./°F

   68 to 1000 F (20 to 538 C) 6.3
   68 to 1200 F (20 to 650 C) 6.4
   68 to 1400 F (20 to 760 C) 6.6
   68 to 1600 F (20 to 871 C) 7.0
   68 to 1800 F (20 to 982 C) 7.4
   68 to 2000 F (20 to 1093 C) 7.7
   1200 to 1600 F (650 to 871 C) 8.7
   1200 to 1800 F (650 to 982 C) 9.3

   11 *Specific heat* at 70 F. 0.12 Btu/lb/°F
E 2 *Electrical resistivity* at 70 F. 77.0 microhm-cm
G 3 *Magnetic permeability.* Ferromagnetic
J 6 *Microstructure.* Ferritic at all temperatures and for this reason is not hardenable by heat treatment. Ductility and impact strength are very low at room temperature, and the creep strength very low at elevated temperature, unless some nickel is present.
K 1 *Mechanical properties.* Modulus of elasticity, in tension: 29,000,000 psi
M 1 *Chemical composition limits.* 0.50 max C, 1.00 max Mn, 2.00 max Si, 0.04 max P, 0.04 max S, 0.5 max Mo, 26 to 30 Cr, 4 max Ni, rem Fe. Molybdenum is not intentionally added.
N 1 *Melting temperature.* 2725 F (1495 C)
   15 *Heat treatment.* Castings are normally used in the as-cast condition.

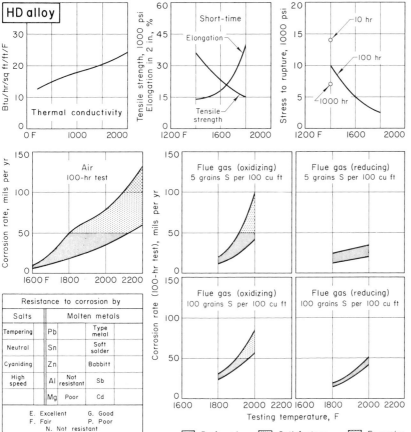

HD alloy

(28% Cr, 5% Ni)

A 5 *Other designations.* SAE 70327
B 1 *Typical uses.* Brazing furnace parts, cracking equipment, furnace blowers, gas burner parts, holding pots, kiln parts, pouring spouts, rabble shoes and arms, and recuperator sections.
C 1 *Density* at 70 F. 7.58 g per cu cm (0.274 lb per cu in.)
   5 *Patternmaker's shrinkage.* 7/32 in./ft
D 6 *Thermal expansion,* micro-in./in./°F

   68 to 1000 F (20 to 538 C) 7.7
   68 to 1200 F (20 to 650 C) 8.0
   68 to 1400 F (20 to 760 C) 8.3
   68 to 1600 F (20 to 871 C) 8.6
   68 to 1800 F (20 to 982 C) 8.9
   68 to 2000 F (20 to 1093 C) 9.2
   1200 to 1600 F (650 to 871 C) 10.3
   1200 to 1800 F (650 to 982 C) 10.6

   11 *Specific heat* at 70 F. 0.12 Btu/lb/°F
E 2 *Electrical resistivity* at 70 F. 81.0 microhm-cm
G 3 *Magnetic permeability.* Ferromagnetic
J 6 *Microstructure.* Two-phase: ferrite plus austenite structure nonhardenable by customary heat treating procedure. Long exposure to temperatures at 1300 to 1500 F, however, may result in hardening of the alloy accompanied by severe loss of room temperature ductility through the formation of sigma phase. Ductility may be restored by heating the alloy to a uniform temperature of 1800 F, or higher, and then cooling rapidly to below 1200 F.
K 1 *Mechanical properties.* Modulus of elasticity in tension: 27,000,000 psi
M 1 *Chemical composition limits.* 0.50 max C, 1.50 max Mn, 2.00 max Si, 0.04 max P, 0.04 max S, 0.5 max Mo, 26 to 30 Cr, 4 to 7 Ni, rem Fe. Molybdenum is not intentionally added to this alloy.
N 1 *Melting temperature.* 2700 F (1482 C)
   15 *Heat treatment.* Castings are normally used in the as-cast condition.

HE alloy

(28% Cr, 10% Ni)

A 3 *ASTM number.* A297, grade HE
  5 *Other designations.* SAE 70312
C 1 *Density at* 70 F. 7.67 g per cu cm
     (0.277 lb per cu in.)
  5 *Patternmaker's shrinkage.* 9/32 in./ft
D 6 *Thermal expansion,* micro-in./in./°F

    68 to 1000 F (20 to 538 C) 9.6
    68 to 1400 F (20 to 760 C) 10.2
    68 to 1800 F (20 to 982 C) 10.8
    68 to 2000 F (20 to 1093 C) 11.1
    1200 to 1600 F (650 to 871 C) 12.2
    1200 to 1800 F (650 to 982 C) 12.5

 11 *Specific heat at* 70 F. 0.14 Btu/lb/°F
 17 *Thermal conductivity.* Btu/hr/sq ft/
    ft/°F
    1500 F (816 C)
    ....10.0

E 2 *Electrical resisitvity at* 70 F. 85.0
    microhm-cm
G 3 *Magnetic permeability.* 1.3 to 2.5
J 6 *Microstructure.* As cast, austenite plus
    ferrite and carbides.

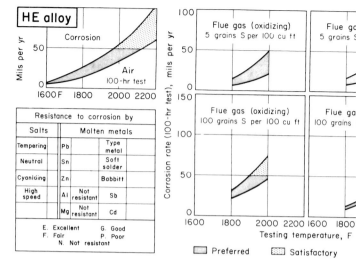

HF alloy

(20% Cr, 10% Ni)

A 3 *ASTM number.* A297, grade HF
  4 *Government numbers.* MIL-S-17509
    (ships)
  5 *Other designations.* SAE 70308
B 1 *Typical uses.* Arc furnace electrode
    arms, annealing boxes and trays,
    baskets, brazing channels, burner
    tips, burnishing rolls, conveyor belts.
C 1 *Density at* 70 F. 7.75 g per cu cm
    (0.280 lb per cu in.)
  5 *Patternmaker's shrinkage.* 9/32 in./ft
D 6 *Thermal expansion,* micro-in./in./°F

    68 to 212 F (20 to 100 C) 7.2
    68 to 1000 F (20 to 538 C) 9.9
    68 to 1200 F (20 to 650 C) 10.1
    68 to 1400 F (20 to 760 C) 10.3
    68 to 1600 F (20 to 871 C) 10.4
    68 to 1800 F (20 to 982 C) 10.5
    68 to 2000 F (20 to 1093 C) 10.9
    1200 to 1600 F (650 to 871 C) 11.1

 11 *Specific heat at* 70 F. 0.12 Btu/lb/°F
E 2 *Electrical resistivity.* 80.0 microhm-
    cm
G 3 *Magnetic permeability.* 1.00
J 6 *Microstructure.* As cast, austenitic
    matrix containing interdendritic
    eutectic carbides and occasionally an
    unidentified lamellar constituent.
K 1 *Mechanical properties.* Modulus of
    elasticity in tension: 28,000,000 psi
M 1 *Chemical composition limits.* 0.20 to
    0.40 C, 2.00 max Mn, 2.00 max Si, 0.04
    max P, 0.04 max S, 0.5 max Mo, 19 to
    23 Cr, 9 to 12 Ni, rem Fe. Molybdenum
    is not intentionally added.
N 1 *Melting temperature.* 2550 F (1400 C)
 15 *Heat treatment.* Castings are nor-
    mally used in the as-cast condition.
    The alloy cannot be hardened by heat
    treatment but if service conditions
    involve repeated heating and cooling,
    improved performance may be ob-
    tained by heating castings at 1900 F
    for 6 hr followed by furnace cooling
    prior to placing in service.

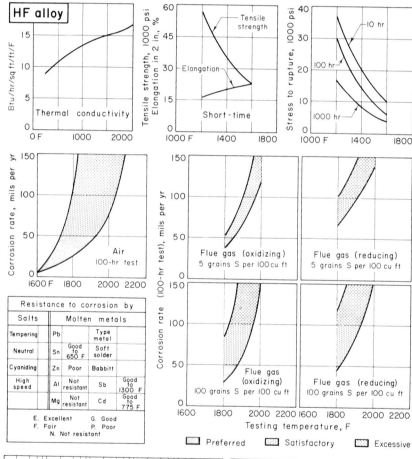

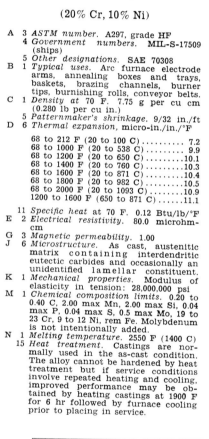

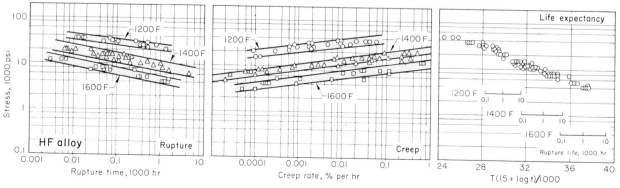

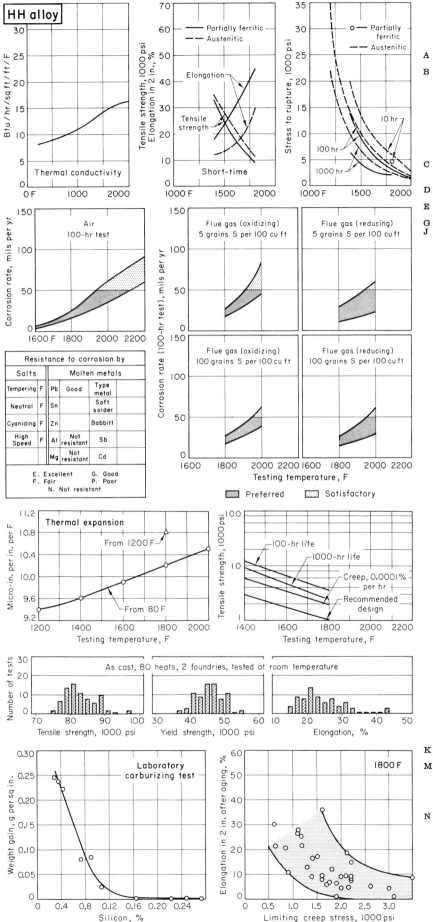

HH alloy

(26% Cr, 12% Ni)

A 3 *ASTM number.* A297 and B190, gr HH
 5 *Other designations.* SAE 70309
B 1 *Typical uses.* Annealing trays, billet skids, burner nozzles, carburizing boxes, convection tube supports, dampers, exhaust manifolds, flue gas stacks, grate supports, hardening trays, kiln nose ring segments, muffles, normalizing disks, pier caps, quenching trays, rabble arms and blades, radiant tubes and supports, refractory supports, retorts, roller hearths and rails, and stoker parts.
C 1 *Density* at 70 F. 7.72 g per cu cm (0.279 lb per cu in.)
 5 *Patternmaker's shrinkage.* 5/16 in./ft
D 6 *Thermal expansion.* See graphs.
 11 *Specific heat* at 70 F. 0.12 Btu/lb/°F
E 2 *Electrical resistivity* at 70 F. 75 to 85 microhm-cm
G 3 *Magnetic permeability.* 1.0 to 1.9
J 5 *Microstructure.* Sufficient nickel is present in this alloy to maintain austenite as the major phase but the alloy is borderline in character and its microstructure is sensitive to composition balance. Two distinct grades of material can be obtained within the stated chemical composition range of the type HH alloy. These grades are defined as type I and type II in ASTM B190.

HH alloy is basically austenitic and holds considerable carbon in solid solution, but carbides, ferrite (soft, ductile and magnetic) and sigma (hard, brittle and nonmagnetic) may also be present in the microstructure. The amounts of the various structural constituents present depend on composition and the thermal history of the sample under consideration. Near 1600 F the partially ferritic alloys tend to embrittle from the development of sigma phase, while around 1400 F carbide precipitation may cause a comparable loss of ductility. Such possible embrittlement suggests that 1700 to 2000 F is the best service temperature range, but this is not critical for steady temperature conditions in the absence of unusual thermal or mechanical stresses.

A serious cause of embrittlement is absorption of carbon from the service environment. Hence, type HH alloy is seldom used for carburizing applications. High silicon content (over 1.5%) will fortify the alloy against carburization under mild conditions but will promote ferrite formation and possible sigma embrittlement.

The partially ferritic (type I) HH alloy is adapted to operating conditions which are subject to changes in temperature level and applied stress. A plastic extension in the weaker, ductile ferrite under changing load tends to occur more readily than in the stronger austenitic phase, thereby reducing unit stresses and stress concentrations and permitting rapid adjustment to suddenly applied overloads without cracking. Where load and temperature conditions are comparatively constant, the wholly austenitic (type II) HH alloy provides the highest creep strength and permits use of maximum design stress. The stably austenitic alloy is also favored for cyclic temperature service that might induce sigma phase formation in the partially ferritic type.
K 1 *Mechanical properties.* Modulus of elasticity in tension: 27,000,000 psi
M 1 *Chemical composition limits.* 0.20 to 0.50 C, 2.00 max Mn, 2.00 max Si, 0.04 max P, 0.04 max S, 0.5 max Mo, 0.2 max N, 24 to 28 Cr, 11 to 14 Ni, rem Fe. Molybdenum is not intentionally added.
N 1 *Melting temperature.* 2500 F (1370 C)
 15 *Heat treatment.* Castings are used in the as-cast condition. The alloy cannot be hardened by heat treatment. For alloys of medium carbon content (about 0.30%), in applications involving thermal fatigue from rapid heating and cooling, improved performance sometimes may be obtained by heating castings at 1900 F for 12 hr followed by furnace cooling prior to placing in service.

HI alloy

(28% Cr, 15% Ni)

A 3 *ASTM number.* A297, grade HI
C 1 *Density* at 70 F. 7.72 g per cu cm (0.279 lb per cu in.)
  5 *Patternmaker's shrinkage.* 5/16 in./ft.
D 6 *Thermal expansion,* micro-in./in./°F

68 to 1000 F (20 to 538 C) 9.9
68 to 1200 F (20 to 650 C) 10.0
68 to 1400 F (20 to 760 C) 10.1
68 to 1600 F (20 to 871 C) 10.3
68 to 1800 F (20 to 982 C) 10.5
68 to 2000 F (20 to 1093 C) 10.8
1200 to 1600 F (650 to 871 C) 11.0
1200 to 1800 F (650 to 982 C) 12.0

  11 *Specific heat* at 70 F. 0.12 Btu/lb/°F
G 3 *Magnetic permeability.* 1.0 to 1.7
J 6 *Microstructure.* Austenitic containing carbides and, depending on the composition balance, small amounts of ferrite. Aging at 1400 to 1600 F is accompanied by precipitation of finely dispersed carbides which tend at room temperature to increase the mechanical strength and to decrease the ductility. At service temperature above 2000 F, however, such carbides remain in solution.
K 1 *Mechanical properties.* Modulus of elasticity, in tension: 27,000,000 psi
M 1 *Chemical composition limits.* 0.20 to 0.50 C, 2.00 max Mn, 2.00 max Si, 0.04 max P, 0.04 max S, 0.5 max Mo, 26 to 30 Cr, 14 to 18 Ni, rem Fe. Molybdenum not intentionally added.
N 1 *Melting temperature.* 2550 F (1400 C)

Resistance to corrosion by

Salts		Molten metals	
Tempering	F	Pb	Good	Type metal
Neutral	P	Sn		Soft solder
Cyaniding	G	Zn	Poor	Babbitt
High speed	P	Al	Not resistant	Sb
		Mg	Not resistant	Cd

G. Good F. Fair P. Poor

HK alloy

(26% Cr, 20% Ni)

A 3 *ASTM number.* A297, grade HK
  5 *Other designations.* SAE 70310, AMS 5365
B 1 *Typical uses.* Billet skids, brazing fixtures, calcining tubes, cement kiln nose segments, conveyor rolls, furnace door arches and lintels, heat treating trays and fixtures, pier caps, rabble arms and blades, radiant tubes, retorts, rotating shafts, skid rails.
C 1 *Density* at 70 F. 7.75 g per cu cm (0.280 lb per cu in.)
  5 *Patternmaker's shrinkage.* 5/16 in./ft.
D 6 *Thermal expansion,* micro-in./in./°F

68 to 1000 F (20 to 538 C) 9.2
68 to 1200 F (20 to 650 C) 9.4
68 to 1400 F (20 to 760 C) 9.6
68 to 1600 F (20 to 871 C) 9.7
68 to 1800 F (20 to 982 C) 10.0
68 to 2000 F (20 to 1093 C) 10.1
1200 to 1600 F (650 to 871 C) 10.5
1200 to 1800 F (650 to 982 C) 11.1

  11 *Specific heat* at 70 F. 0.12 Btu/lb/°F
E 2 *Electrical resistivity* at 70 F. 90.0 microhm-cm
G 3 *Magnetic permeability.* 1.02
J 6 *Microstructure.* As cast: austenite containing massive carbides as scattered islands or networks. After aging at service temperature, the alloy exhibits precipitation of fine, granular carbides within the austenite grains, with subsequent agglomeration if the temperature is high enough. These fine, dispersed carbides contribute to the creep strength of the alloy. A lamellar constituent tentatively identified as an austenite-carbonitride eutectoid resembling pearlite also is frequently observed in HK alloys but its exact nature is in doubt. Except when present in excessive amount, however, it is not associated with loss of hot strength.
K 1 *Mechanical properties.* Modulus of elasticity, in tension: 29,000,000 psi
M 1 *Chemical composition limits.* 0.20 to 0.60 C, 2.00 max Mn, 2.00 max Si, 0.04 max P, 0.04 max S, 0.5 max Mo, 24 to 28 Cr, 18 to 22 Ni, rem Fe. Molybdenum is not intentionally added.
N 1 *Melting temperature.* 2550 F (1400 C)
  15 *Heat treatment.* Castings are normally used in the as-cast condition. HK alloy cannot be hardened by heat treatment.

Resistance to corrosion by

Salts		Molten metals	
Tempering	G	Pb		Type metal
Neutral	G	Sn		Soft solder
Cyaniding	G	Zn		Babbitt
High speed	G	Al	Not resistant	Sb
		Mg	Not resistant	Cd

E. Excellent G. Good
F. Fair P. Poor
N. Not resistant

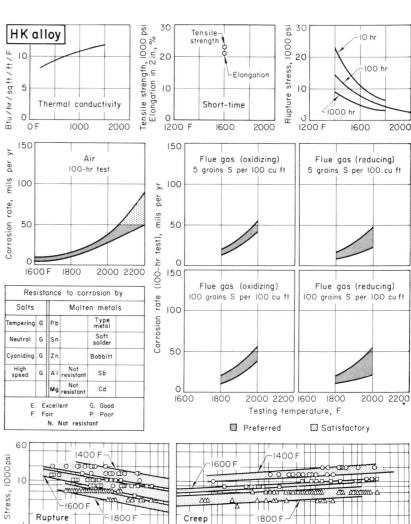

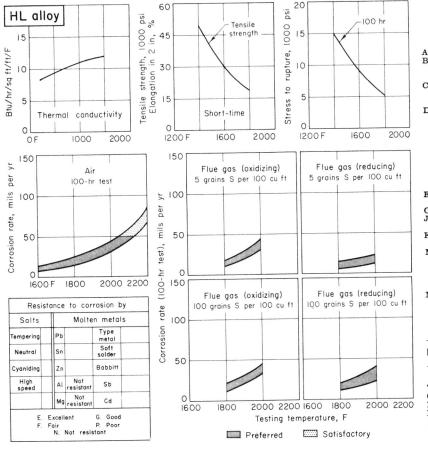

HL alloy

HL alloy

(30% Cr, 20% Ni)

A 5 *Other designations.* SAE 70310A
B 1 *Typical uses.* Carrier fingers, enameling furnace fixtures, furnace skids for slabs and bars, and radiant tubes.
C 1 *Density* at 70 F. 7.72 g per cu cm (0.279 lb per cu in.)
　 5 *Patternmaker's shrinkage.* 5/16 in./ft
D 6 *Thermal expansion,* micro-in./in./°F

　　68 to 1000 F (20 to 538 C) 9.2
　　68 to 1400 F (20 to 760 C) 9.6
　　68 to 1600 F (20 to 871 C) 9.7
　　68 to 1800 F (20 to 982 C) 9.9
　　68 to 2000 F (20 to 1093 C) 10.1
　　1200 to 1600 F (650 to 871 C) 10.5
　　1200 to 1800 F (650 to 982 C) 10.7

　 11 *Specific heat* at 70 F. 0.12 Btu/lb/°F
E 2 *Electrical resistivity* at 70 F. 94.0 microhm-cm
G 3 *Magnetic permeability.* 1.01
J 6 *Microstructure.* The as-cast and aged microstructure is similar to type HK.
K 1 *Mechanical properties.* Modulus of elasticity, in tension: 29,000,000 psi
M 1 *Chemical composition limits.* 0.20 to 0.60 C, 2.00 max Mn, 2.00 max Si, 0.04 max P, 0.04 max S, 0.5 max Mo, 28 to 32 Cr, 18 to 22 Ni, rem Fe. Molybdenum is not intentionally added.
N 1 *Melting temperature.* 2600 F (1425 C)
　 15 *Heat treatment.* Castings are normally used in the as-cast condition. The alloy cannot be hardened by heat treatment.

HL alloy	As cast

Tested at Room Temperature

Tensile strength, 1000 psi............	82
0.2% yield strength, 1000 psi........	52
Elongation in 2 in., %................	19
Hardness, Bhn	192

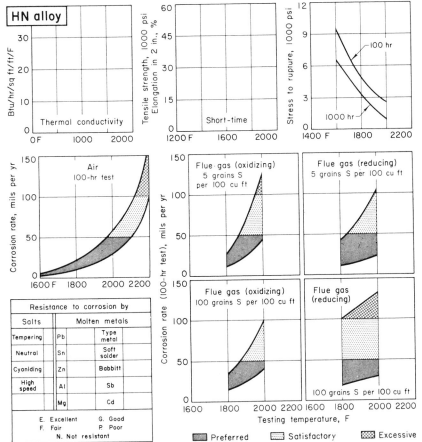

HN alloy

HN alloy

(25% Ni, 20% Cr)

A 2 *Trade name.* Type HN alloy
B 1 *Typical uses.* Brazing fixtures, chain, furnace beams and parts, pier caps, radiant tubes and tube supports.
C 1 *Density* at 70 F. 7.83 g per cu cm (0.283 lb per cu in.)
　 5 *Patternmaker's shrinkage.* 5/16 in./ft
D 6 *Thermal expansion.* No data.
　 11 *Specific heat* at 70 F. 0.11 Btu/lb/°F
G 3 *Magnetic permeability.* 1.10
J 6 *Microstructure.* Austenitic at all temperatures, well within the stable austenite field. In the as-cast condition carbide areas are present and additional fine carbides precipitate on aging. The alloy is not susceptible to sigma phase formation.
K 1 *Mechanical properties.* Modulus of elasticity, in tension: 27,000,000 psi
M 1 *Chemical composition limits.* 0.20 to 0.50 C, 2.00 max Mn, 2.00 max Si, 0.04 max P, 0.04 max S, 0.5 max Mo, 19 to 23 Cr, 23 to 27 Ni, rem Fe. Molybdenum is not intentionally added.
N 1 *Melting temperature.* 2500 F (1370 C)
　 15 *Heat treatment.* Castings are normally used in the as-cast condition.
　 17 *Suitable joining methods.* Castings can be welded by metal-arc, inert-gas, and oxyacetylene gas methods. Oxyacetylene welding is more satisfactory for high-temperature applications of this alloy. Type 330 bare filler rods should be used for gas welding, and the flame should be rich in acetylene.

HN alloy	As cast

Tested at Room Temperature

Tensile strength, 1000 psi...........	68
0.2% yield strength, 1000 psi........	38
Elongation in 2 in., %	17
Hardness, Bhn	160

HT alloy

(35% Ni, 15% Cr)

A 3 *ASTM number.* A297 and B207, gr HT
5 *Other designations.* SAE 70330

B 1 *Typical uses.* Air ducts, brazing trays, carburizing containers, chain, cyanide pots, dampers, dippers, door frames, enameling bars and supports, fan blades, feed screws, gear spacers, glass molds, glass rolls, hearth plates, heat treating fixtures and trays, idler drums, kiln nose rings, lead pots, malleablizing baskets, muffles, oil burner nozzles, point bars.

C 1 *Density* at 70 F. 7.92 g per cu cm (0.286 lb per cu in.)
5 *Patternmarker's shrinkage.* 5/16 in./ft

D 6 *Thermal expansion,* micro-in./in./°F

68 to 1000 F (20 to 538 C)	8.5
68 to 1200 F (20 to 650 C)	8.9
68 to 1400 F (20 to 760 C)	9.2
68 to 1600 F (20 to 871 C)	9.3
68 to 1800 F (20 to 982 C)	9.8
68 to 2000 F (20 to 1093 C)	9.8
1200 to 1600 F (650 to 871 C)	10.4
1200 to 1800 F (650 to 982 C)	11.5

11 *Specific heat* at 70 F. 0.11 Btu/lb/°F

E 2 *Electrical resistivity* at 70 F. 100.0 microhm-cm

J 6 *Microstructure.* Austenite containing various amounts of carbides depending on the carbon content and thermal history. In the as-cast condition large carbide areas are present at the grain boundaries, but fine carbides precipitate within the grains after exposure at service temperatures with subsequent decrease in room temperature ductility. Increased carbon content does not significantly affect the high temperature ductility of the alloy; this characteristic makes it especially useful for carburizing fixtures or containers. Additional protection against carburization is obtained with silicon contents above about 1.6%, but at some sacrifice of hot strength.

K 1 *Mechanical properties.* Modulus of elasticity in tension: 27,000,000 psi

M 1 *Chemical composition limits.* 0.35 to 0.75 C, 2.00 max Mn, 2.50 max Si, 0.04 max P, 0.04 max S, 0.5 max Mo, 13 to 17 Cr, 33 to 37 Ni, rem Fe. Molybdenum is not intentionally added

N 1 *Melting temperature.* 2450 F (1345 C)
15 *Heat treatment.* Castings are normally used in the as-cast condition. This alloy cannot be hardened by heat treatment but for applications involving thermal fatigue from repeated rapid heating and cooling, improved performance may be obtained by heating castings at 1900 F for 12 hr followed by furnace cooling prior to placing them in service.
17 *Suitable joining methods.* Castings of this type have good welding properties if proper techniques are employed. Thermal expansion is about one third greater than for carbon steel or Fe-Cr alloy types HC or HD. Electrical resistance is over six times that of carbon steel and is characterized by a low temperature coefficient of resistivity (0.00017 per °F in the range from 68 to 930 F). Welding can be accomplished by metal-arc, inert-gas arc, and oxyacetylene gas methods. It is generally considered that oxyacetylene welding is more satisfactory than arc welding for high-temperature applications of this alloy. Type 330 bare filler rods should be used for gas welding, and the flame should be adjusted to be very rich in acetylene.

HT alloy	Condition	
	As cast	Aged(a)
Tested at Room Temperature		
Tensile strength, 1000 psi	70	75
0.2% yield strength, 1000 psi	40	45
Elongation in 2 in., %	10	5
Hardness, Bhn	180	200
Impact strength, Charpy keyhole, ft-lb	4	...

(a) Aged 24 hr at 1400 F, air cooled

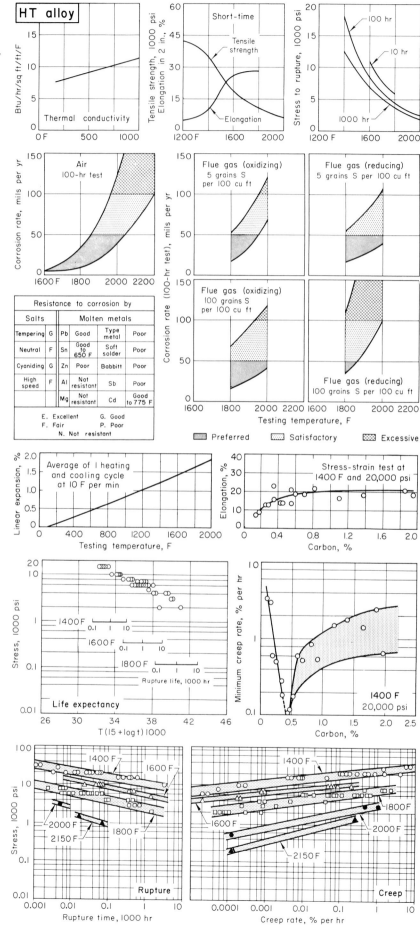

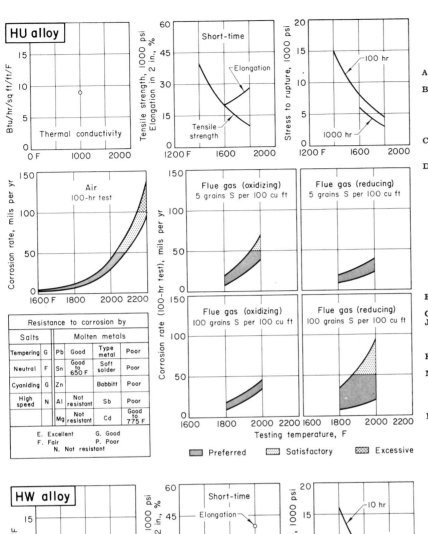

HU alloy

(39% Ni, 19% Cr)

A 3 *ASTM number.* A297, grade HU
  5 *Other designations.* SAE 70331
B 1 *Typical uses.* Trays, burner tubes, carburizing retorts, conveyor screws and chains, cyanide pots, dipping baskets, furnace rolls, lead pots, muffles, pouring spouts, radiant tubes, resistor guides
C 1 *Density* at 70 F. 8.04 g per cu cm (0.290 lb per cu in.)
  5 *Patternmaker's shrinkage.* 5/16 in./ft
D 6 *Thermal expansion,* micro-in./in./°F

68 to 1000 F (20 to 538 C)	8.8
68 to 1200 F (20 to 650 C)	9.0
68 to 1400 F (20 to 760 C)	9.2
68 to 1600 F (20 to 871 C)	9.4
68 to 1800 F (20 to 982 C)	9.6
68 to 2000 F (20 to 1093 C)	9.7
1200 to 1600 F (650 to 871 C)	10.5
1200 to 1800 F (650 to 982 C)	10.6

  11 *Specific heat* at 70 F. 0.11 Btu/lb/°F
  17 *Thermal conductivity,* Btu/hr/sq ft/ft/°F

1000 F (538 C)	8.9

E 2 *Electrical resistivity* at 70 F. 105.0 microhm-cm
G 3 *Magnetic permeability.* 1.10 to 2.00
J 6 *Microstructure.* Austenite with varying amounts of carbides depending on carbon content and thermal history.
K 1 *Mechanical properties.* Modulus of elasticity in tension: 27,000,000 psi
M 1 *Chemical composition limits.* 0.35 to 0.75 C, 2.00 max Mn, 2.50 max Si, 0.04 max P, 0.04 max S, 0.5 max Mo, 17 to 21 Cr, 37 to 41 Ni, rem Fe. Molybdenum is not intentionally added.
N 1 *Melting temperature.* 2450 F (1345 C)
  15 *Heat treatment.* Castings are used in the as-cast condition.

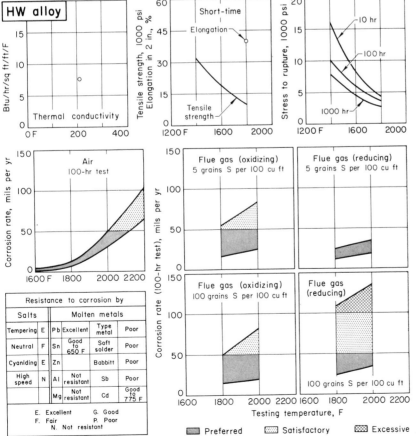

HW alloy

(60% Ni, 12% Cr)

A 2 *Trade name.* Type HW alloy
  3 *ASTM number.* A297, grade HW
  5 *Other designations.* SAE 70334
B 1 *Typical uses.* Cyanide pots, electric heating elements, enameling tools, gas retorts, hardening fixtures, hearth plates, lead pots, muffles
C 1 *Density* at 70 F. 8.14 g per cu cm (0.294 lb per cu in.)
  5 *Patternmaker's shrinkage.* 9/32 in./ft
D 6 *Thermal expansion,* micro-in./in./°F

68 to 1000 F (20 to 538 C)	7.8
68 to 1200 F (20 to 650 C)	8.1
68 to 1400 F (20 to 760 C)	8.4
68 to 1600 F (20 to 871 C)	8.6
68 to 1800 F (20 to 982 C)	8.8
68 to 2000 F (20 to 1093 C)	9.2
1200 to 1600 F (650 to 871 C)	10.0
1200 to 1800 F (650 to 982 C)	10.1

  11 *Specific heat* at 70 F. 0.11 Btu/lb/°F
  17 *Thermal conductivity,* Btu/hr/sq ft/ft/°F

212 F (100 C)	7.7

E 2 *Electrical resistivity* at 70 F. 112.0 microhm-cm
G 3 *Magnetic permeability.* 16.0
J 6 *Microstructure.* Austenite with varying amounts of carbides depending on the carbon content and thermal history. In the as-cast alloy, the microstructure consists of a continuous interdendritic network of massive and elongated eutectic carbides.
K 1 *Mechanical properties.* Modulus of elasticity in tension: 25,000,000 psi
M 1 *Chemical composition limits.* 0.35 to 0.75 C, 2.00 max Mn, 2.50 max Si, 0.04 max P, 0.04 max S, 0.5 max Mo, 10 to 14 Cr, 58 to 62 Ni, rem Fe. Molybdenum is not intentionally added.
N 1 *Melting temperature.* 2350 F (1290 C)
  15 *Heat treatment.* Castings are normally used in the as-cast condition.

HX alloy

(67% Ni, 17% Cr)

C 1 *Density* at 70 F. 8.14 g per cu cm (0.294 lb per cu in.)
 5 *Patternmaker's shrinkage.* 9/32 in./ft
D 6 *Thermal expansion,* micro-in./in./°F

68 to 1000 F (20 to 538 C)	7.8
68 to 1200 F (20 to 650 C)	8.1
68 to 1400 F (20 to 760 C)	8.5
68 to 1600 F (20 to 871 C)	8.8
68 to 1800 F (20 to 982 C)	9.2
68 to 2000 F (20 to 1093 C)	9.5
1200 to 1600 F (650 to 871 C)	10.7
1200 to 1800 F (650 to 982 C)	11.3

 11 *Specific heat* at 70 F. 0.11 Btu/lb/°F
K 1 *Mechanical properties.* Modulus of elasticity in tension: 25,000,000 psi
M 1 *Chemical composition limits.* 0.35 to 0.75 C, 2.00 max Mn, 2.50 max Si, 0.04 max P, 0.04 max S, 0.5 max Mo, 15 to 19 Cr, 64 to 68 Ni, rem Fe. Molybdenum is not intentionally added.
N 1 *Melting temperature.* 2350 F (1290 C)
 15 *Heat treatment.* Castings are normally used in the as-cast condition.

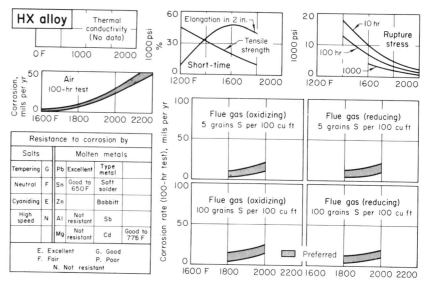

Resistance to corrosion by				
Salts			Molten metals	
Tempering	G	Pb	Excellent	Type metal
Neutral	F	Sn	Good to 650 F	Soft solder
Cyaniding	E	Zn		Babbitt
High speed	N	Al	Not resistant	Sb
		Mg	Not resistant	Cd Good to 775 F

E. Excellent   G. Good
F. Fair   P. Poor
N. Not resistant

Nickel-Base Super-Strength Alloys

Thetaloy (50 Ni — 25 Cr — 12.5 Co — 7 W — 3 Mo — 2.5 Mn), as-cast test bars

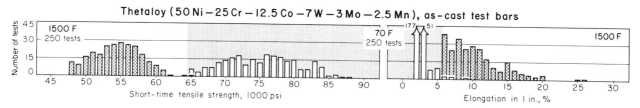

Hastelloy B (65 Ni — 28 Mo — 5 Fe), investment cast test specimens

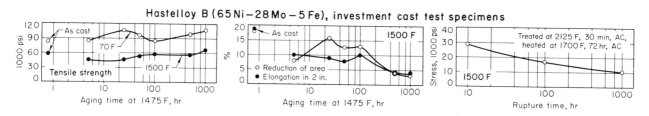

Hastelloy C (55 Ni — 17 Mo — 16 Cr — 5 Fe — 4 W), as investment cast

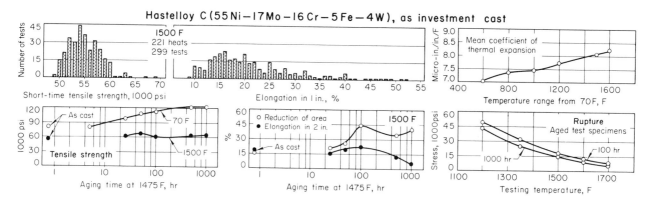

Hastelloy X (45 Ni — 22 Cr — 18 Fe — 9 Mo — 2 Co — 1 W), as investment cast

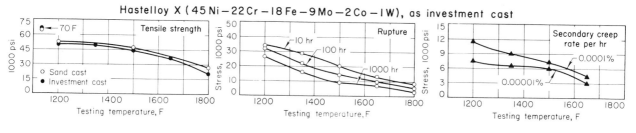

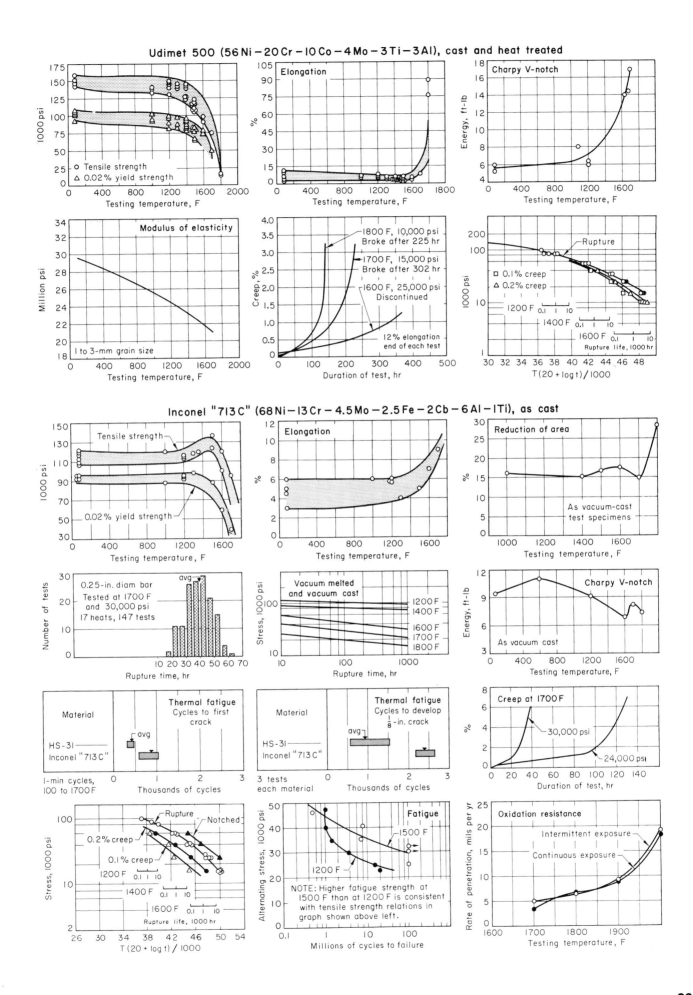

Udimet 500 (56 Ni − 20 Cr − 10 Co − 4 Mo − 3 Ti − 3 Al), cast and heat treated

Inconel "713 C" (68 Ni − 13 Cr − 4.5 Mo − 2.5 Fe − 2 Cb − 6 Al − 1 Ti), as cast

GMR-235 (65Ni – 15Cr – 10Fe – 5Mo – 3Al – 2Ti)

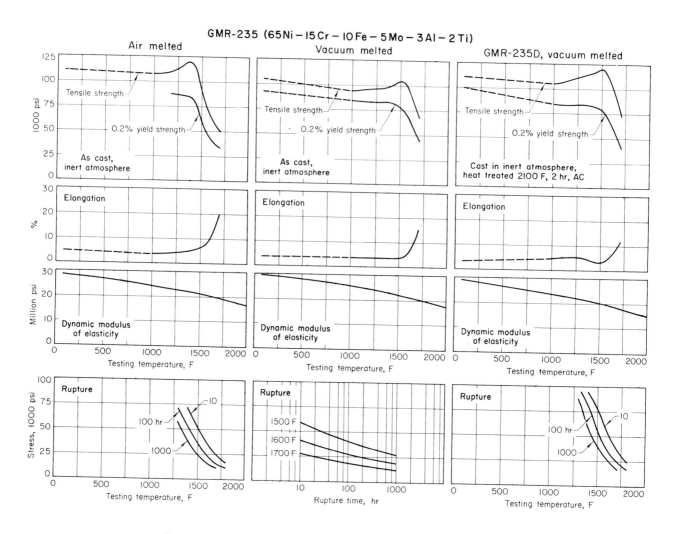

Cobalt-Base Super-Strength Alloys

HS-36 (L-251) (54Co – 19Cr – 15W – 10Ni), aged 16 hr at 1350 F

HS-31 (X-40) (55Co – 25Cr – 10Ni – 8W)

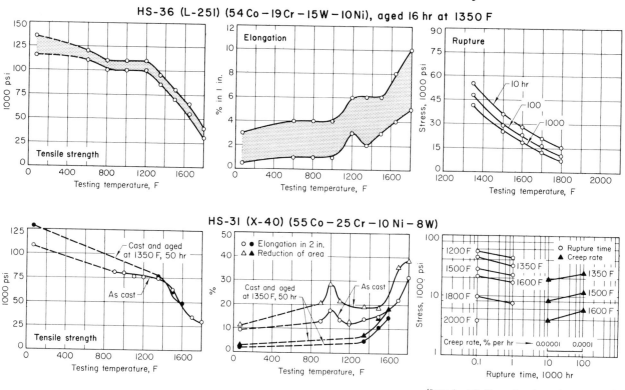

(Data for HS-31 continued on the next page)

HS-31 (continued), as cast

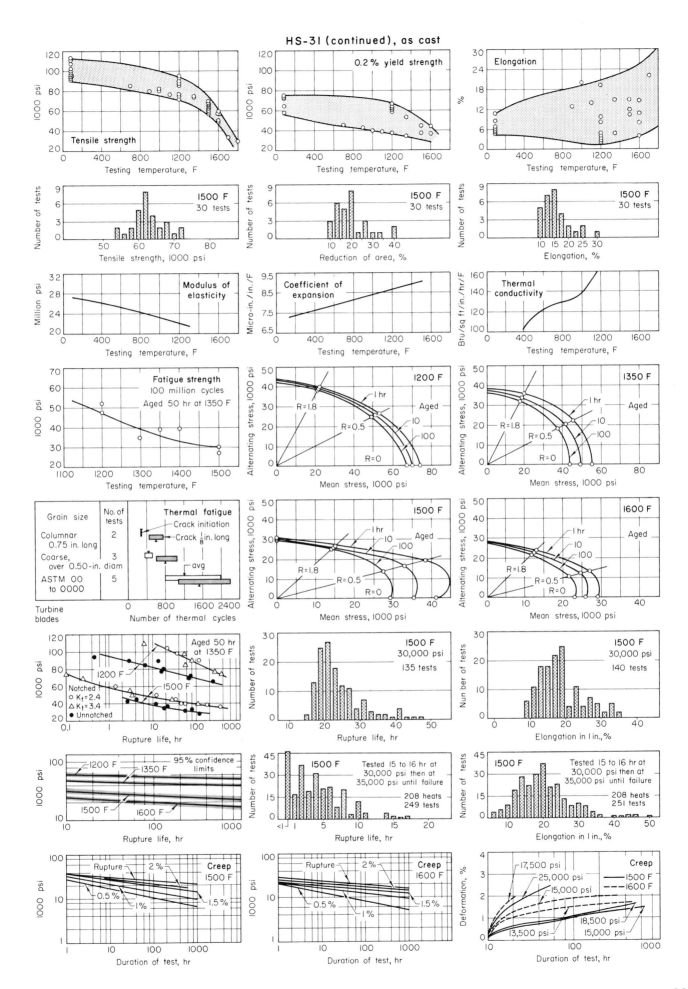

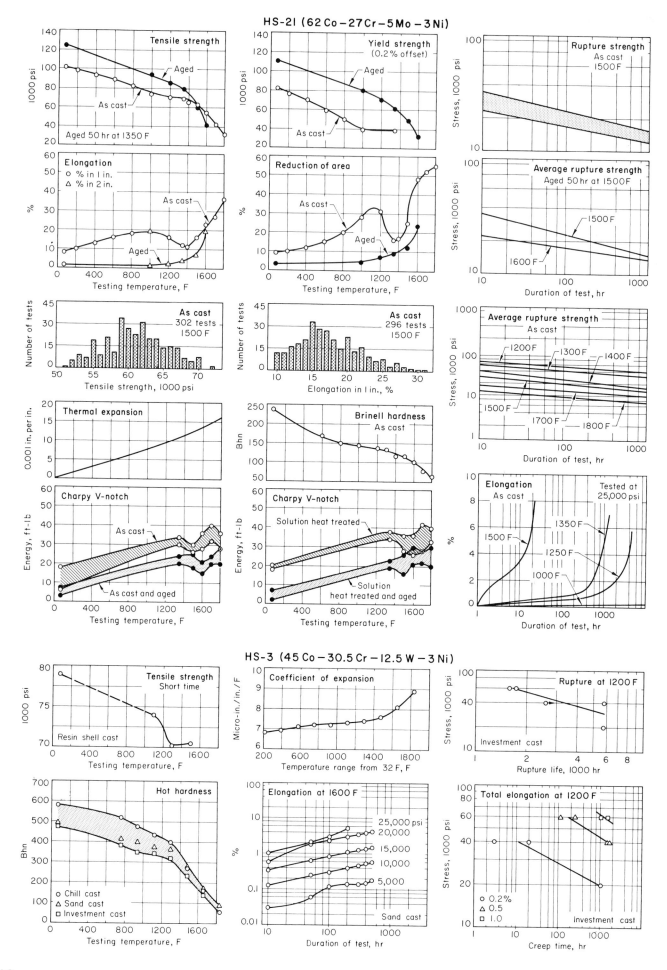

HS-21 (62 Co − 27 Cr − 5 Mo − 3 Ni)

HS-3 (45 Co − 30.5 Cr − 12.5 W − 3 Ni)

SECTION VII:
Iron-Base and Iron-Containing Superalloys

Microstructures and Mechanical Properties of Iron-Base (-Containing) Superalloys

C. P. Sullivan and M. J. Donachie Jr.

This *review* is primarily concerned with microstructural characteristics, and the bearing they have on mechanical property response. Two sections provide a background, with some general aspects of physical metallurgy and the various classifications of alloys. Precipitation hardening by the formation of the intermetallic compounds γ' [Ni$_3$(Al, Ti)] and γ'' (Ni$_3$Cb) are given a relatively detailed treatment; and also included is information on the types of phases associated with overaging. Subsequent sections deal with topologically close packed phases; carbides (both as a general precipitate for matrix strengthening and at grain boundaries); boron and trace elements; environment; and processing operations. The review closes with some remarks concerning areas for further exploration, together with an assessment of the future of iron-base (-containing) superalloys.

Introduction

Iron-base superalloys are nominally defined as those alloys with iron as the major constituent, and which are hardened by a carbide or intermetallic precipitate. Because many of the iron-base superalloy principles can be applied to other iron-containing systems, information will be included on those superalloys containing iron in relatively large amounts (defined for the present purposes as >10 wt %). Furthermore, the increased emphasis on iron as a lower-cost substitute for nickel- and cobalt-base superalloys has resulted in development of moderate strength, solid-solution-hardened formulations; these are also considered.

Physical Metallurgy

Three papers (1-3) have dealt with the relationship between the type of alloy base (iron, nickel or cobalt) and the phases that are precipitated within it. Results show that (1) iron-base alloys with a face-centered cubic (FCC) matrix (often referred to as austenite) have a strong tendency to form so-called topologically close packed (TCP) phases such as sigma, mu, Laves and chi; (2) FCC nickel-base alloys have a tendency towards precipitation of ordered geometrically-close-packed (GCP) phases such as γ' [Ni$_3$(Al,Ti)] and η (Ni$_3$Ti); and (3) FCC cobalt-base alloys have a strong potential for solid solutioning. Iron

Dr. Sullivan and Dr. Donachie are associated with the Materials Engineering and Research Laboratory, Pratt & Whitney Aircraft, Div. of United Aircraft Corp., East Hartford, Conn.

can tolerate, more readily than nickel, the formation of phases with abnormally short interatomic distances such as those which occur in Laves and σ. This characteristic is of extreme practical importance.

A variety of elements are added to perform one or more desirable functions. The most potent strengthening for alloys with a FCC matrix is provided by elements such as nickel, aluminum, titanium, and columbium. These precipitate intermetallic phases such as ordered FCC γ' [Ni$_3$(Al,Ti),L1$_2$] and ordered body-centered tetragonal γ'' (Ni$_3$Cb, DO$_{22}$) from the matrix by a suitable heat treating schedule. Elements such as iron and chromium may also be found in γ' and γ''. FCC alloys can be hardened by the addition of carbon in relatively large amounts ($\sim$0.5 wt %) to form a general carbide precipitate; nitrogen and phosphorous are sometimes added to enhance this effect. Carbon may promote the formation of grain boundary carbide phases such as M$_{23}$C$_6$ and M$_6$C to provide strength in these regions.

Elements in solid solution add some limited degree of strengthening; the most potent are interstitial in nature. These elements also alter the lattice parameter of the FCC matrix (4). As a result, the degree of hardening produced by a precipitate can be affected through the magnitude of coherency strains developed. Oxidation resistance is provided by judicious use of elements such as chromium and manganese, while nickel additions retard formation of the generally undesirable TCP phases. Use of boron or lanthanum in relatively limited amounts can greatly improve elevated temperature properties.

Iron-base alloys of most importance for application temperatures >1,000 F (540 C) are those with a FCC matrix, since a close-packed lattice is more resistant to time-dependent deformational processes. Under many circumstances, the microstructural/property interactions in FCC iron-base superalloys can be considered analogous to nickel-superalloys.

Table I lists several elements used in iron-base alloys together with their nominal functions. These effects are not necessarily the same for each alloy, nor are they necessarily produced simultaneously in a specific alloy. Also, under some circumstances deleterious effects may result from a given element.

Alloy Types

The most important class of iron-base (-containing) alloys are those which are strengthened by intermetallic compound precipitation in a FCC matrix. The most common precipitate is γ', as typified by A-286 (5), V-57 (6), and Pyromet 860 (7). On the other hand, Inconel 718 is hardened by γ'' (8, 9). HNM (10) and the CRMD series (11) represent the alloy class hardened by carbides, nitrides and carbonitrides; elements such as tungsten and molybdenum also may be added to give solid solution strengthening. Hastelloy X is an example of a third alloy class which is essentially solid solution hardened, but which may derive some strengthening from carbide precipitation through working plus aging.

Iron-base superalloys hardened by intermetallic compound precipitation have found primary usage in gas turbine engines as blades, disks, casings, and fasteners (12). For example, some Pratt & Whitney gas turbine engines use A-286 forgings for turbine disks and Incoloy 901 for turbine hubs. A-286 also finds application as a turbine case material. Alloys relying on the precipitation of carbides have not been widely employed in the United States as yet, but the CRM alloys such as 18D may eventually be broadly applied in automotive gas turbines. Solid solution hardened Hastelloy X is used in sheet form for applications such as burner cans in gas turbine engines (13).

Representative alloy compositions are given in Table II. The first three groups are γ' hardened listed roughly in the order of increasing nickel content, next are γ'' hardened alloys, then carbide and carbonitride compositions, and finally solid solution formulations. Table III presents rupture strengths, together with those of representative nickel- and cobalt-base superalloys for comparison.

Strengthening by Intermetallic Compounds

Gamma prime precipitation in FCC matrix

The magnitude of hardening in this type of system can be related to several factors in a fashion similar to that for nickel-base superalloys (14): anti-phase boundary (APB) and fault energy of γ', γ strength, γ' strength, coherency strains, volume fraction γ', γ' particle size, diffusivity in γ and γ', and possibly $\gamma - \gamma'$ modulus mismatch. These effects are not necessarily additive.

Strengthening is a function of γ' particle size. Before the age hardening peak is reached, strength increases with size at constant volume fraction of precipitate (15, 16). In this condition, dislocations cut through the γ' precipitates, and are paired owing to the high anti-phase-domain boundary energy (15, 17). (Since γ' is ordered, a superlattice dislocation consisting of two regular lattice dislocations is required to penetrate the precipitate if order is to be maintained.) The unusual temperature dependence of γ' strength, wherein there is an increase with temperature up to approximately 1,300 F (700 C), is of importance in this regime (18). After the age hardening peak is reached, strength decreases with continuing particle growth because dislocations largely bypass γ' particles by looping around them (15, 17). As the volume fraction of precipitate is generally less than 0.2 (15), optimum strengthening occurs for γ' sizes of 100 to 500 Å. The γ' particles are usually spherical, although other morphologies have been reported (18, 19).

The degree of strengthening produced by $\gamma - \gamma'$ coherency strains is open to some question. Separate investigations on Pyromet 860 (20) and an alloy similar to Incoloy 901 (21) indicated a mixture of random and preferred orientations of the spherical γ' It appears that co-

Table I Effects of Elements in Iron-Base (-Containing) Superalloys

Element	Effects
Aluminum	Form γ' [Ni₃(Al,Ti)]; retard formation of hexagonal η(Ni₃Ti)
Titanium	Form γ' [Ni₃(Al,Ti)], carbide (MC)
Columbium Tantalum	Form body-centered tetragonal γ'', carbide (MC)
Carbon	Form carbides (MC, M₇C₃, M₂₃C₆, M₆C); stabilize FCC matrix
Phosphorous	Promote general precipitation of carbides
Nitrogen	Form carbonitrides [M(C,N)]
Chromium	Oxidation resistance; solid solution strengthening
Molybdenum	Solid solution strengthening; carbide former (M₆C)
Tungsten Nickel	Stabilize FCC matrix; form γ', inhibit information of deleterious phases
Boron Zirconium	Improve creep properties; retard formation of grain boundary η(Ni₃Ti)
Lanthanum	Oxidation resistance

herency was not a primary source of strength in these two cases. Similarly, studies on A-286 and PE 16 plus four other alloys of varying hardener (aluminum + titanium) content have shown that there is little effect on strength of misfits up to at least 0.3% (15, 16).

Proper heat treatment is critical to satisfactory alloy performance. In general, if low temperature creep strength or high short term properties are required, a low solution temperature is used prior to aging. For high creep and rupture strengths (accompanied by reduced ductility), a higher solution temperature is used prior to aging. With A-286 for example, the former requirement is met by 1,650 F (900 C) solution, and the latter, by 1,800 F (980 C) solution (22). If A-286 is solutioned at 2,050 F (1,120 C) even higher creep properties result, but ductility is reduced to the point where notch sensitivity becomes a problem.

In several alloys, a two step aging sequence is employed to achieve a balance of properties. The aging temperature(s) must be kept below the level where significant amounts of γ' have transformed to hexagonal Ni₃Ti.

Alloy strength also depends upon volume fraction of precipitate. This can be increased by adding more hardener elements (aluminum and titanium). If optimum properties are to be achieved, there are limits to the amounts of the individual elements, their ratio and their sum. The desired end is to form γ' and prevent other types of phase formation; addition of titanium is effective for this purpose (23). However, aluminum additions produce a synergistic effect, and inhibit the formation of the undesirable η (Ni₃Ti) phase (4, 24). When aluminum and titanium contents are comparable, Ni₂AlTi tends to form; even greater aluminum levels promote β phase [Ni(Al,Ti) or NiAl] (4). In both instances, there is a decrease in properties. With increasing volume fraction of precipitate (degree of supersaturation), both the rate and degree of over-aging increase (24). It has been suggested that the optimum composition contains approximately 2.5% Ti and 1.0% Al (4). Strength was correlated with titanium to aluminum (Ti/Al) ratios ranging from about one to eight. Since APB energies for γ' increase with Ti/Al ratio, it follows that the optimum amounts of titanium and aluminum will be principally determined by achieving the maximum APB energy and volume fraction of γ' consistent with maintaining alloy stability together with adequate tensile and creep ductility.

Table II Nominal Compositions of Some Iron-Base (-Containing) Superalloys

Designation	C	Mn	Si	Cr	Ni	Fe	Ti	Al	B	Mo	W	Cb + Ta	Other
						Composition, Wt %							
Tinidur	0.10	0.8	0.8	14.0	30.0	Bal	1.8	—	—	—	—	—	—
A-286 (G-68)	0.05	1.3	0.6	15.0	25.0	Bal	2.15	0.15	—	1.0	—	—	0.20 V
Discaloy	0.03	0.8	0.8	13.5	26.0	Bal	1.6	0.10	—	3.0	—	—	—
W-545	0.05	1.3	0.3	13.0	26.0	Bal	2.85	0.20	0.08	1.5	—	—	—
V-57	0.08	0.25	0.55	14.8	26.0	Bal	3.0	0.25	0.008	1.25	—	—	0.30 V
Unitemp 212	0.08	0.05	0.15	16.0	25.0	Bal	4.0	0.15	0.06	—	—	0.5	0.05 Zr
M-308	0.08	0.03	0.70	14.0	33.0	Bal	2.0	0.25	0.005	4.0	6.5	—	0.25 Zr
CG-27	0.05	—	—	13.0	38.0	Bal	2.5	1.6	0.01	5.75	—	0.6	0.7 Co
Incoloy 901	0.05	0.45	0.40	13.5	Bal	34.0	2.5	0.25	0.01	6.1	—	—	—
PE 16	0.05	—	—	16.4	Bal	35.0	1.24	1.1	0.003	3.4	—	—	—
D-979	0.05	0.50	0.50	14.0	Bal	28.0	3.0	1.0	0.01	4.0	4.0	—	—
Pyromet 860	0.05	1.0	1.0	14.0	Bal	28.0	3.0	1.1	0.01	6.0	—	—	4.0 Co
Inconel 706	0.02	0.05	0.05	16.0	Bal	39.0	1.7	0.3	0.004	—	—	2.75	—
Inconel 718	0.04	0.20	0.20	19.0	Bal	18.5	0.9	0.50	0.005	3.05	—	5.30	—
René 62	0.05	0.25	0.25	15.0	Bal	22.0	2.5	1.25	0.01	9.0	—	2.25	—
HNM	0.30	3.50	0.50	18.5	9.5	Bal	—	—	—	—	—	—	0.23 P
G-18B	0.40	0.80	1.0	13.0	13.0	Bal	—	—	—	2.0	2.5	3.0	10.0 Co
AF 71	0.30	18.0	0.30	12.5	—	Bal	—	—	0.20	3.0	—	—	0.2 N, 0.9 V
HTX	0.45	8.50	0.45	21.0	8.0	Bal	—	—	—	1.5	—	—	0.2 N, 0.23 P
CRM-18D	0.75	5.0	0.50	23.0	5.0	Bal	—	—	0.003	—	1.0	2.0	0.25 N
Hastelloy X	0.10	0.5	0.5	22.0	Bal	18.5	—	—	—	9.0	0.6	—	1.5 Co
René Y	0.15	1.0	1.0	22.0	Bal	18.5	—	—	—	9.0	2.0	—	0.2 La, 2.5 Co

Precipitation of γ'' in FCC Matrix

The type of alloy which relies on a columbium-rich precipitate for strengthening is typified by Inconel 718. In addition to the high strength levels obtainable at a moderate temperature [1,200 F (650 C)], the alloy is reported to be less susceptible to strain age cracking in the heat-affected zone (HAZ) of welds (25) due to its slower aging response. Another potentially important alloy of this type is Inconel 706. Although similar to Inconel 718 in aging response, it is reportedly more machinable (26, 27). The effective hardening in Inconel 718 is associated with coherency strains produced by precipitation of a body-centered tetragonal phase, γ'' (28, 29). The γ'' possesses a disk shaped morphology bearing the relationship with the FCC matrix: $[001]_{\gamma''} \parallel <001>_{\gamma}$ and $\{100\}_{\gamma''} \parallel \{100\}_{\gamma}$ (9, 30). The alloy also contains lesser amounts of spherical γ'. Above 1,300 F (700 C), strength decreases markedly due to: rapid coarsening of γ''; some solutioning of γ'' and γ'; and formation of orthorhombic δ Ni_3Cb phase (31). In the fully heat treated condition slip is planar at room temperature (31) which suggests that dislocations are cutting through the γ'' precipitate in a manner similar to γ' hardened alloys when aged for optimum strength (17). The degree of slip planarity has significant implications regarding the creep and fatigue behavior of superalloys (32).

A lack of notch ductility in Inconel 718 has been associated with grain boundary zones denuded in γ''; these can be effectively eliminated and alloy ductility restored by appropriate heat treatment (33). The presence of PFZ (precipitate free zones) along grain boundaries is not always harmful to properties. In fact, they may be quite beneficial, depending upon the size of the zones and their strength relative to the grain boundary and matrix proper (34).

Inconel 718 reportedly can be subjected to stressed exposure at 1,200 F (650 C) for at least 10,000 hr without

Table III Typical 1,000 Hr Rupture Strengths of Some Superalloys

Designation	1,200F* (650C)	1,400F* (760C)	1,500F* (815C)
	Iron Base		
A-286	46 (32)	15 (11)	7.7 (5.4)
W-545	65 (46)		
CG-27	77 (54)		22 (15)
Pyromet 860	81 (57)		17 (12)
inconel 718	86 (60)	25 (18)	
CRM-18D†	54 (38)		19 (13)
Hastelloy X	31 (22)	15 (11)	10 (7)
	Nickel Base		
Udimet 700	102 (72)	62 (44)	43 (30)
MAR-M246†		86 (60)	62 (44)
	Cobalt Base		
X-40†	51 (36)	33 (23)	22 (15)
S-816	46 (32)		18 (13)

* Values are 1,000 psi (Kgf per sq mm)
† Cast alloys

formation of significant amounts of δ (35). However, the same work suggested that creep stress accelerated aging reactions. It has been shown that prior creep exposure of Inconel 718 for 500 hr at 50,000 psi (35.2 kgf per sq mm) at either 1,250 or 1,300 F (675 or 700 C) produces γ'' coarsening and formation of δ plates, with a consequent increase in rupture ductility and decrease in life upon sub-

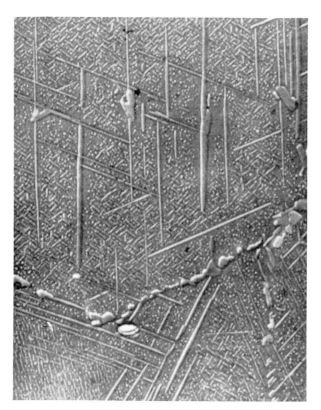

Fig. 1. This electron replica micrograph shows Inconel 718 which has been heated at 1,300F (700C) for 6,048 hr under a stress of 37,000 psi (26 kgf per sq mm). Precipitates are disks of γ'', spheres of γ', and plates of δ. Note precipitation depletion around δ plates. 5,000X. (Source: Ref. 35)

sequent creep testing at 1,200 F (650 C) and 110,000 psi (77.3 kgf per sq mm) (36). Figure 1 shows γ', γ'', and δ phases.

Overaging

In iron-base (-containing) superalloys, the strengthening precipitate usually gives way in a moderate temperature range (1,200 to 1,400 F, 650 to 760 C), forming another structure and precipitate morphology which is much less effective in strengthening the alloy. This behavior can be related to the greater tolerance of iron for other structural forms, and sets an upper temperature limit on the utility of this superalloy class. In alloys hardened by titanium and aluminum, with the former being preponderant, hexagonal η (Ni$_3$Ti, DO$_{24}$) can eventually form and replace γ' (1, 3). Unlike γ', this phase has no significant solubility for other elements.

The precipitation of η may occur in two forms: at the grain boundaries as a pearlitic (cellular) product, or intragranularly as plates (Widmanstätten form) (37). Cells appear at a lower temperature than plates. In both situations, there is a matching of close packed planes and directions of the η with the γ matrix with which it is in contact: $(0001)_\eta \parallel \{111\}_\gamma$; $<1210>_\eta \parallel <110>_\gamma$ (37). The cellular precipitation (also referred to as discontinuous precipitation and the recrystallization reaction) consists of alternating lamellae of γ and η having a random orientation relationship to the grain in which they are growing. Its presence is often associated with a loss of mechanical properties

due to the coarse interlamellar spacing of η. For example, notched stress/rupture strength and creep ductility may decrease (4, 23). The formation mechanism of Widmanstätten η appears to be temperature dependent (23, 39). Its appearance is accompanied by a loss of strength, but not ductility. Denudation of γ' around growing Widmanstätten plates has been observed (37). It is suspected that this would create planes of weakness along the $\{111\}$ slip planes since plates have a $\{111\}$ habit. Both cold work and increasing titanium enhance the driving force for η formation (37-39). Additions of aluminum, on the other hand, inhibit η precipitation for two possible reasons. First, η has no solubility for this element while γ' does; thus it must be removed from η nucleation sites before η can nucleate and grow (39). Second, aluminum may decrease γ'/γ mismatch, thereby reducing the driving force for η precipitation (24).

For alloys in which Al/Ti $\geq$ 1, elevated temperature exposure can lead to replacement of γ' by Ni$_2$AlTi or β [NiAl or Ni(Al,Ti)] (40). Rapid overaging occurs at a moderate temperature because of the large lattice disregistry of these body-centered cubic type phases with the γ matrix (23). No cellular form of precipitation is observed in this Al/Ti range, only the formation of massive plates due to rapid overaging (4).

In an alloy like Inconel 718, plates of orthorhombic δ phase will form at sufficiently long times and elevated temperatures (9, 14). Its appearance suggests that there is a matching of close packed planes and directions as in η precipitation. The general features of δ formation are closely analogous to those of η formation. There is a deterioration of properties when excessive amounts of δ form because of its relatively coarse morphology. Recent data (41) indicate that the precipitation of δ occurs much more rapidly than previously reported (42). This is important because δ may form during a "solutioning" treatment of relatively short time and cause undesirable changes. In contrast to Inconel 718, the kinetics of δ formation in Inconel 706 are significantly slower (26). The transformation to δ phase is speeded by high silicon and columbium and low aluminum (14), while the substitution of tantalum for columbium appreciably retards δ formation (43).

Minor Intermetallic Phases

As noted earlier, iron-base (-containing) superalloys are prone to precipitation of a number of minor phases typified by sigma and Laves. The presence of these phases can alter mechanical properties by virtue of their own mechanical characteristics together with the concomitant changes in matrix and grain boundary chemistry. The degree of change is also related to the amount, morphology, and distribution of the phases in question, and it is not always clear that their presence will be deleterious.

Sigma phase

This tetragonal phase is found over a wide range of compositions and an atom size ratio spread of 0.93 to 1.15. A continuous sigma layer at grain boundaries can severely reduce ductility; in contrast, a blocky intragranular form can actually increase creep ductility (23). Small amounts of this phase in Inconel 718 have not proven harmful to stress/rupture properties (35, 36). When sigma precipitates in platelets, it is expected that the interface with the matrix will often provide a path of relatively easy fracture.

The relationship of various composition modifications to the resultant propensity for sigma phase precipitation provides some guidelines to the alloy designer. Analysis of the phase in A-286 shows that nickel tries to avoid participation in its formation (44). Thus, the formation of sigma (and Laves also) will be promoted when the nickel in the alloy is tied up in other phases, or when the total nickel content is reduced. Increasing aluminum favors presence of the sigma phase, and in very high concentrations can make an alloy quite brittle and difficult to forge (45, 46). Part of the aluminum effect is associated with a removal of nickel from the matrix to form γ' or β; however, not all of the effect is necessarily due to this (46). Increasing titanium assists sigma phase precipitation because both the displacement of $M_{23}C_6$ by TiC and precipitation of more Ni_3Ti increase matrix chromium concentration (47).

Work on alloys which contain neither titanium or aluminum has provided additional insight. Several studies (48, 49) indicate that $M_{23}C_6$ does not nucleate sigma. In fact, separate competitive precipitation sequences are involved. Two favored sites for sigma phase are grain and annealing twin boundaries (48-50), while cold deformation accelerates the precipitation kinetics (48, 49). Precipitation of carbides may either retard the precipitation of sigma or accelerate it, depending upon the elements involved in the formation (50, 51). Increasing silicon in an iron-nickel-chromium system promotes sigma, and for amounts of sigma $\gtrsim 5\%$ room temperature impact is severely reduced (52, 53).

Alpha prime

This phase, a chromium-rich ferrite, may be of importance in alloys of high nickel plus chromium content by acting as a transition phase for sigma formation. (54). Alpha prime has been reported to be harmless when present in minor amounts at grain boundaries in Inconel 718 (35). Initially, this phase precipitates intergranularly, and then intragranularly; prior precipitation of $M_{23}C_6$ in these locations inhibits α' formation because iron and chromium are involved in both reactions (54).

Laves phase

In these iron-base alloys, Laves phase is usually found as the $MgZn_2$ hexagonal type with a general formula: (Fe, Mn, Cr, Si)$_2$ (Mo, Ti, Cb). Increases in certain elements such as silicon and columbium directly promote Laves phase formation (55). As with sigma phase, elemental additions can affect phase stability in an indirect, but highly significant manner. For example, increasing aluminum in Inconel 718 promotes instability (55), and the level of boron in Incoloy 901 determines whether Laves or another phase appears (1). Attempts have been made to develop an alloy hardened by Laves phase precipitation (56, 57). Although strengthening was achieved, growth of the phase was rapid. Tensile tests at 1,290 F (700 C) showed that premature cracking was associated with Laves phase at grain boundaries. Though Inconel 706 is reported to form Laves phase (26), the annealing schedule, aging sequence, and use temperature are such as to minimize its formation. Therefore, deleterious mechanical property effects do not seem likely.

The presence of 2 to 3% Laves in CG-27 reduces stress/ rupture life (45), and in Inconel 718, room temperature yield strength and ductility are reduced (42). In contrast, Laves phase at grain boundaries in A-286 does not affect 1,100 F (595 C) properties (23). An interesting attempt was made to employ a heavy, blocky precipitation of Laves phase at grain boundaries to promote ductility by excluding carbide film formation at this location (58). The alloy, intended for welded sheet applications, was successful in its initial objective, but weld repair capability was compromised due to acicular Laves formation during subsequent exposure (59).

G phase

Several compositions have been reported for this complex FCC phase, which is usually located at grain boundaries in globular form. A common elemental make-up (based on work on A-286) involves nickel, titanium and silicon (60). Cobalt and iron can substitute for nickel in this phase (61), and silicon plays a key role in its development. One important consequence of G phase formation is the removal of nickel from the matrix, and a resultant increase in the tendency to form sigma and Laves phases. Although G phase has little effect on room temperature impact energy and short time 1,200 F (650 C) tensile properties of A-286, 1,200 F (650 C) stress/ rupture properties show deterioration associated with cracking at G phase/grain boundary interfaces (23). Pitler (62) has pointed out that this stress/rupture deterioration is accompanied by substantial increases in ductility which might prove useful if an application requires a trade-off.

Mu phase

This trigonal phase takes the form of platelets parallel to $\{111\}_\gamma$, and has been reported to form in Incoloy 901 containing 0.1% B (1) and Pyromet 860 (63). The reduction in room temperature strength observed in the latter alloy when it contains mu phase is not attributable to its presence, also neither room temperature nor elevated temperature tensile fractures occur along mu platelets.

Carbides

The function of carbides in the iron-base superalloys can be considered as three-fold. First, if sufficient levels of carbon and other elements are present, a general carbide precipitation occurs in the matrix, providing a primary source of strength (precipitate size is on the order of 300 Å for effective hardening). Second, carbides at grain boundaries promote strength in this region, and are often essential for optimum overall alloy performance. Third, carbon, in combining with certain elements, prevents them from contributing to phase instability. Types of carbides which have been reported are MC, M_6C, M_6C', $M_{23}C_6$ and to a lesser extent M_7C_3 and M_4C_3 (M represents one or more types of metal atom).

Carbides (carbonitrides)—Matrix Hardening

In considering hardening of austenitic matrices by precipitation, alloys may employ carbon alone, carbon plus nitrogen (carbonitride), carbon plus phosphorous, and carbon plus nitrogen plus phosphorous. The addition of phosphorous promotes a precipitate distribution which is more general, more finely dispersed, and smaller than is produced by carbon alone (64, 65). It appears, from lattice parameter measurements, that phosphorous is a substitutional replacement in $M_{23}C_6$, producing $(M,P)_{23}C_6$. This precipitate has an improved matching with the matrix lattice, thereby assisting in forming more general precipitation. The element also retards the overaging tendency of the general form of precipitate. Phosphorous does not produce any additional phases in performing its alloying function. Precipitates involved in this class of alloys may also form as an intragranular $M_{23}C_6$ Widmanstätten type, or in a cellular reaction nucleated at grain boundaries (66). The cellular reaction decreases room temperature yield strength and ductility, and is also deleterious with respect to elevated temperature strength. As carbon plus nitrogen, or carbon plus nitrogen plus phosphorous increase, elevated temperature properties improve, but the tendency for cellular reaction to occur becomes stronger (4, 65, 66). Thus, it is necessary with element levels which produce reasonable strengths $(C + N + P \approx 0.5\%)$ to find an optimum aging temperature; this is generally around 1,300 F (700 C), and thus limits the application range.

HNM and HTX are examples of alloys which use phosphorous to promote carbide precipitation, while G-18B relies on carbon alone. G-18B contains the strong primary carbide former, columbium. This element produces, in addition to M_6C and $M_{23}C_6$, an MC phase which has greater high-temperature stability. Precipitation of primary CbC carbides occurs on matrix dislocations in G-18B (67), and precipitation on dislocations in other iron-containing alloys has been shown as well. After chromium-nickel steels with 1% Cb or Ti are aged in the 1,022 to 1,562 F (550-850 C) range, precipitation takes place predominantly on dislocations (68). Some of the dislocations dissociate during precipitation to form large bands of stacking faults which are the sites of very fine CbC or TiC precipitation (9). The onset of stacking fault precipitation is associated with a marked increase in strength and moderate decrease in ductility, but this effect is much reduced at lower solution temperatures (69). Thus, the degree of ductility obtained in a carbide hardened alloy will depend on the heat treatment used.

CRM-6D, -15D and -18D alloys are relatively recent alloy developments which are intended for use in automotive gas turbines. As columbium is used, primary carbides are present in addition to the chromium-rich M_7C_3 and $M_{23}C_6$ carbides produced by aging heat-treatments.

In opposition to matrix hardening, it should be noted that, under certain conditions, carbides in superalloys are prone to cracking, and thus cause stress concentration effects. Carbide-associated cracking has been observed in Inconel 718 due to severe machining (70). A critical amount of surface deformation is necessary for carbide cracking, and it can be prevented by a suitable machining sequence. Prevention of cracking is important because iron (iron-nickel) containing superalloys are frequently used in fatigue-limited applications.

Carbides—Grain Boundary Effects

The microstructural evidence for the role played by carbides in the overall performance of iron-base superalloys is not as well documented as it is for nickel-base superalloys. Yet, with $< 0.01\%$ C, stress/rupture life and ductility can be drastically reduced, and an increased tendency towards the formation of undesirable phases can occur (71). Although $M_{23}C_6$ has been reported in some iron-base alloys, there are several instances where little or none of this carbide type occurs. In Inconel 718, for example, only M_6C precipitates have been found, and these are located on the boundaries of coarse grains (72). In an alloy closely approximating the A-286 composition, the only grain boundary phase reported was TiC (73). The effect on mechanical properties of carbide precipitation at grain boundaries may be associated with an accompanying PFZ formation rather than the carbide itself (74, 75). The formation of TiC grain boundary films in V-57 and A-286 with exposure at 2,000 F (1,090 C) and higher has been reported to be detrimental to forging and welding characteristics (76). It appears that, in alloys of this type, MC is not so stable as to preclude some breakdown followed by a reprecipitation of TiC at grain boundaries (77). In a recent study of V-57 (78) it was found possible to improve its cyclic rupture capability by eliminating a nearly continuous TiC grain boundary film through a heat treatment which permitted cellular precipitation of Ni_3Ti to occur first. Although it appears that carbide films in general reduce properties in some manner or other, it has been stated that they are not detrimental to the creep/rupture and tensile properties of Inconel 718 (36, 42).

In the past, notch sensitivity of iron-base alloys was mistakenly attributed to a well-developed cellular precipitate, but it now is known that alloys like A-286 and V-57 in this condition do not necessarily exhibit this behavior (79). [An excessive amount of transformation to η by discontinuous precipitation will reduce stress-rupture properties, however (80).] The notch sensitive condition in A-286 has been identified with a poorly-defined cellular precipitate of $M_{23}C_6$ (81).

Characteristics of grain boundary $M_{23}C_6$ precipitation have been carefully examined in a noncommercial alloy (82). There is a sensitivity of precipitate distribution to boundary misorientation. Triple points are favorable carbide precipitation sites, and the grain boundaries immediately adjacent to these locations are denuded of carbides. This behavior probably results from a denudation of solute atoms in these areas due to the rapid growth of carbides at the triple points.

Fig. 2. Blocky Laves phase and acicular δ phase are associated with casting segregation in Inconel 718. The as-cast alloy was heated for 1 hr at 1,800F (980C) and then for 16 hr at 1,325F (720C). (Source: Ref. 104)

As mentioned earlier, the formation of PFZ adjacent to grain boundaries can have an important effect on mechanical properties. It has been observed, in an austenitic steel containing 4% Ti, that the tendency of grain boundary $M_{23}C_6$ precipitation to be accompanied by γ'-precipitate-free zones depends upon aging temperature (83). A complication can arise regarding PFZ formation in those circumstances where the $M_{23}C_6$ precipitate is replaced by TiC, which in turn begins to dissolve upon continued aging (47). In this instance, PFZ are initially present, but the dissolution of TiC releases titanium with the consequence that γ' forms in the formerly denuded zones.

Boron and Trace Elements

The beneficial effect of boron (and to a lesser extent, zirconium) on creep properties is a well-documented phenomenon in nickel-base superalloys (84). Although it is generally agreed that boron exerts its influence at grain boundaries, the details of the mechanism(s) for property improvement remain obscure (14).

Boron and Zirconium

Boron additions can markedly improve stress/rupture properties. One way in which this improvement can occur is through inhibition of the η cellular precipitation reaction, which, in excessive amounts, reduces creep strength (39, 85). The fact that boron retards precipitation of η, but not the Widmanstätten form, which is intragranular, is evidence for its effectiveness being grain-boundary associated (39). Zirconium has been reported to act similarly to boron with regard to η precipitation (86). The practical consequence of this behavior is that more titanium can be added to produce a stronger alloy since the boron will counteract the tendency of titanium to promote the property-degrading cellular reaction.

In Discaloy, boron additions lead to a breaking-up of the grain boundary precipitate with a subsequent increase in ductility (87). This, in turn, allows the titanium content to be increased (within limits), providing that boron is suitably increased. Exploitation of an effect of this sort led to the development of W-545.

Boron also reduces the solubility of carbon in austenite. As a result, it can reduce the size of TiC and $M_{23}C_6$ precipitates while increasing their number (67). Grain size refinement has also been reported (38), and this may be of value in applications with a high ductility requirement, such as weldability. If sufficient quantities of boron are added, borides will form. These may cause deleterious effects such as nucleation of cellular η (80), formation of a low melting point grain boundary eutectic (88), or initiation of fatigue (84).

Trace Elements

The unintentional incorporation of trace elements such as lead, antimony, arsenic, tin, bismuth, gallium, tellurium, and selenium is a topic of importance in the superalloy field (89). Concentrations as low as 1 to 25 ppm may affect the creep behavior of nickel- and cobalt-base superalloys, and it seems logical to presume that these elements will also embrittle iron-base (-containing) superalloys. In nickel superalloys, for example, bismuth is embrittling at >1 ppm. Many of these elements are already known to participate in temper embrittlement reactions when they are present in concentrations from 10 to 700 ppm (90). Substantive data are generally lacking for trace element embrittlement in iron-base (-containing) superalloys. Russian work (91) on stainless steels indicates that lead and antimony in amounts from 0.02 to 0.12% decrease the rupture life at 1,110 and 1,290 F (600 and 700 C).

Environment

The importance of environment on the performance of nickel- and cobalt-base superalloys is becoming well-documented because surface reactions occur relatively rapidly at the temperatures to which many of them are exposed. With iron-base superalloys intended for use in the 1,000 to 1,400 F (540 to 760 C) range this aspect of environment is less critical. Also, chromium content is generally about 15% or more, ensuring reasonable protection from oxidation. However, if an alloy such as Hastelloy X is exposed to 1,800 F (980 C) and higher, oxidation reactions assume greater importance, and protection must be provided.

Uncoated Alloys

Both molybdenum and tungsten adversely affect oxidation resistance, the former producing a blistering effect presumably associated with formation of the volatile molybdenum oxide (92). Under certain conditions of engine operation involving a salt- and sulfur-containing environment, the alloy N-155 can undergo a severe attack similar to the hot corrosion (sulfidation) produced in nickel- and cobalt-base superalloys. Another aspect of environmental interaction is the oxidation enhancement in certain localized regions produced in A-286, V-57, and M-308 by low-cycle fatigue strain (93). This effect can be significant for component application. During cycling, regions of intense local oxidation develop. The appearance of these regions as surface markings is due to the repeated rupture of oxide film by the highly localized deformation associated with fatigue bands. Localized oxidation, in turn, leads to notching at the surface, further localization of the strain, and hence further concentration of oxidation. This cycle of events repeats itself until local fracture occurs.

Hastelloy X is used in sheet components of gas turbine engines which may be exposed to oxidative environments up to 1,800 F (980 C) and even higher. At such temperatures, the affected substrate of this alloy consists of two layers. The inner one involves alloy depletion, and the outer one, an internal oxidation product forming in a previously depleted area (94). After elevated temperature exposure, mechanical properties will be adversely affected, particularly in thin sections, thus limiting its use temperature. The oxidation behavior of Hastelloy X and other alloys whose primary oxide scale is Cr_2O_3 is complicated by the significant amount of Cr_2O_3 volatilization which occurs at 1,800 F (980 C) and above by reaction with oxygen to form gaseous CrO_3 (95). Under dynamic exposure conditions, the rate of oxidation becomes catastrophic if the temperature is high enough (96) because CrO_3 is continually swept from the alloy surface by the high-velocity gas stream. The addition of small amounts of rare earth elements, such as lanthanum and cerium, together with about 1% Mn to Hastelloy X, results in an adherent nonvolatile oxide being formed upon exposure (97). Although these alloying additions improve serviceability, a coating of the aluminide type will provide even greater protection if this is required.

A surface effect, which may be of some importance, can occur as a result of oxidation during solution heat treatment. It may involve a decrease in chromium and iron (86) or deboronization (80), but in either case the effect is to hasten the discontinuous precipitation of η either during aging or subsequent service. It is conceivable that surface dependent properties, notably fatigue, would show some degradation if the affected layer is not removed prior to use.

Coated Alloys

As indicated above, Hastelloy X must be used with a protective coating for long term service above a certain temperature range. An aluminide type coating is used because Al_2O_3 is the most protective oxide which can be formed. Coating formation and degradation mechanisms on nickel-base superalloys have been described in some detail (98). In large measure, the same principles are applicable to Hastelloy X, although the absence of aluminum in the alloy substrate must be taken into consideration.

Application of a coating can affect mechanical properties. For thin sections, creep behavior will be modified because the coating, which consumes some of the substrate during its formation, possesses less load carrying capability than the alloy itself. Fatigue behavior will be modified by the coating regardless of section thickness as cyclic behavior is generally sensitive to surface condition. Since ductility decreases and protectivity increases with increasing aluminum content, it is clear that serious consideration must be given to the details of designing a coated system if maximum systems effectiveness is to be realized. Some of the ramifications of coatings as regards mechanical properties have been recently discussed (99).

Processing

Processing techniques will probably represent the most significant factor in materials systems performance advances which are made in the near future.

Casting

Although most of the alloys that have been discussed are used in the wrought form, casting is of critical importance in determining final wrought properties. Also, some alloys, such as those in the CRM-D family, are expected to find use in the cast condition. Chemical segregation effects produced in ingot material may lead to unwanted phase formation and other undesirable effects in wrought products. For example, areas of "freckle" segregation in A-286, associated with Laves phase and incipient melting as low as 2,100 F (1150 C), can produce adverse results in subsequent forging (100). Detailed studies of the nature and cause of freckling in superalloys have been reported (101, 102). Segregation in Incoloy 901 castings may lead to the formation of patchy areas containing η; the alloy then has lower yield strength and ductility. The use of homogenization cycles has been suggested as a procedure to reduce Laves formation in A-286 (103) and δ formation in Inconel 718 (104). If segregation is severe enough, homogenization treatments of themselves may not be sufficient to remove all the associated adverse effects. Figure 2 shows unwanted phases associated with casting segregation.

The use of consumable-electrode vacuum-arc melting procedures produces an improved product over air induction melted ingot because: (1) vacuum refining can occur; (2) formation of oxides, nitrides and carbonitrides is minimized; (3) no crucible contamination occurs; and (4) rapid solidification rate decreases elemental segregation. These factors have led to such improvements as: enhanced forgeability; production of larger ingots; superior wrought tensile, fatigue, ductility, and stress/rupture properties; and more uniform properties in the final wrought product (105-107). Recent work on Inconel 706 has suggested that even better properties may accrue from the use of electroflux remelting of iron-base (-containing) superalloys (26). Microcleanliness, macrostructure and mechanical properties were equally good for electroflux and consumable melted Inconel 706 except that the former material had superior stress/rupture ductility at 1,200 F (649C).

Forging Practice

As iron-base (-containing) superalloys are generally used in the wrought condition, a forging operation is normally involved. Billets for forging are generally furnished as press-forged squares or as hot-rolled rounds, depending on size requirements. Some forging companies use as-cast ingots for die forging (108). Alloy cleanliness has probably the greatest effect on hot forgeability. Nitride and carbonitride segregation found in castings ultimately shows up as stringers in wrought products. Increased deformation rate tends to narrow the hot working range for A-286 and similar iron-base (-containing) alloys (108). Factors which restrict the range of forging temperatures may significantly alter the properties attainable in these alloys.

One of the most significant microstructural variables in applications for iron-base (-containing) superalloys is grain size. Production of a given grain size in a wrought product is strongly connected with forging practice. Grain size variations are generally accompanied by other changes which result in significant property effects.

As a rule of thumb, a larger grain size produces better creep/rupture properties with less ductility, while a fine grain size yields superior time-independent mechanical properties with greater ductility. A uniform grain size is a better guarantee of uniform properties; if a structure is duplex, deficiencies of both grain sizes may be inherited. Therefore, it is desirable to tailor the forging practice to achieve as uniform a grain size as possible which will respond well to subsequent heat treatment (72, 109, 110). It has been suggested that a grain size of ASTM 6 to 7 in Inconel 718 provides optimum stress/rupture properties; larger sizes lead to notch brittleness, and finer to a loss in strength (36).

It has been shown that orthorhombic δ phase in Inconel 718 (36, 72, 111, 112) and hexagonal η phase in Pyromet 860 (113) and Incoloy 901 (112, 114) can be used to refine grain size during hot working to enhance the final properties. Quantity, size and dispersion of the δ and η phases must be controlled to produce optimum properties. The principle behind such processing is that the equilibrium precipitating phase (δ or η) is the only one present in the forging range, and that subsequent aging is then applied to achieve matrix hardening through the precipitation of γ'' in Inconel 718, and γ' in Pyromet 860 and Incoloy 901. [It has also been suggested that a beneficial dislocation substructure may be involved, particularly if some γ' or γ''

is present during working (115).] In Incoloy 901, grain size refinement can be achieved through carefully tailored forging/heat treating cycles, improving fatigue strength. Laves phase has also been used to produce fine grain sizes in a similar manner (116).

The effects of going from the Inconel 718 to Inconel 706 composition illustrate the relationship between finish properties and composition. Lowered columbium in Inconel 706 reduces its tendency to freckling, and so a more homogeneous distribution of phases and alloy hardeners is expected (26). Because of the compositional changes, Inconel 706 has a lower flow stress and works better than Inconel 718, which should lead to better structure control. Inconel 706 has a lower recrystallization temperature (higher iron, lower columbium), and the forging process must take this factor into account to achieve microstructure control. Grain sizes of ASTM 9 to 10 can be produced by standard forging practice (27).

Heat Treatment

Creep strength of W-545 can be significantly increased by a two-step aging treatment in which the first step is at the lower temperature (87). This creep strength can be translated into useful life if boron is added to the alloy to simultaneously promote ductility. The enhanced strength is due to the inverted aging sequence, which promotes more precipitate nuclei by employing the lower aging temperature first. Another interesting case of heat treating for a specific application is the use of Pyromet 860 as steam turbine cylinder bolts (117). In the standard heat treated condition, the alloy contracts during exposure due to further γ' precipitation, raising stresses to an undesirable level. A heat treatment has been devised which promotes additional γ' precipitation by a high-temperature intermediate age prior to final age, thereby decreasing the degree of contraction and the amount of stress build-up during subsequent exposure. The two aging steps are also tailored so that creep and tensile strengths remain at an acceptable level.

A recent study on D-979 (118) has shown that the coupling of a new heat treating schedule with an improved forging sequence leads to a significant improvement of the alloy as a disk material. The reason for this exceptional improvement in properties is associated with three factors: fine grain size; a finer γ' distribution (yield stress improvement); and more optimum grain boundary strengthening by a closely spaced network of μ and χ phases (stress rupture improvement).

Closing Remarks

In many of the applications for iron-base superalloys, notch sensitivity is of critical importance. Resistance to the deleterious effect of notches should be achieved without large compromises in other mechanical properties. Therefore, great attention has to be given to suitable alloy selection and processing, particularly with regard to fracture toughness. The task is difficult, because generally as strength increases, toughness decreases, while operation at a high stress to take advantage of the strength increase, decreases critical crack length.

The extent and consequences of property degradation produced by the formation of TCP phases requires further definition. If large amounts form, particularly at grain boundaries, deleterious effects are to be expected. Because parts are being required to remain in service for longer and longer periods of time, the chances for the occurrence of TCP phase formation may become greater.

The effect of grain boundary carbide precipitation has not received detailed attention. Precipitation of $M_{23}C_6$ carbides occupies a less significant position than in the nickel-base systems, and it appears that MC, M_6C, and other phases are contributors to grain boundary mechanical characteristics as well. A better understanding of the manner in which grain boundary phase formation affects deformation and fracture modes is required.

Processing details will assume increasing prominence in system performance. Improvement of existing alloys or new alloys will be forthcoming by exploitation of the processing route. Less thought has been given to the application of new processing techniques to iron-base superalloys because of their limited temperature range of application, but the possibility of marked intermediate temperature improvement exists.

What is the future for iron-base (-containing) superalloys? The reason for using iron centers primarily on economy and conservation of strategic materials commensurate with obtaining reasonable properties. It appears that most of the iron-base (-containing) superalloys have an upper use temperature limit of 1,300 to 1,500 F (700 to 815 C). As gas turbine performance increases and temperatures go up, alloys of this type are being replaced by nickel-base superalloys. Because of weight (or strength/density) considerations, a push by titanium-base alloys is underway at the other end of the temperature spectrum. On the other hand, hundreds of small and large parts for gas turbines still use A-286. The CRM-D series of carbide/nitride strengthened alloys, specifically designed for use in automotive gas turbines, may find extensive use based on both relatively good performance to 1,500 F (815 C) and economic considerations.

REFERENCES

1. H. J. Beattie, Jr. and W. C. Hagel: Intragranular Precipitation of Intermetallic Compounds in Complex Austenitic Alloys, *Trans. TMS-AIME*, Vol. 221, 1961, p. 28.
2. H. J. Beattie, Jr. and W. C. Hagel: Compositional Control of Phases Precipitating in Complex Austenitic Alloys, *Trans. TMS-AIME*, Vol. 233, 1965, p. 277.
3. H. J. Beattie, Jr. and W. C. Hagel: Intermetallic Compounds on Titanium-Hardened Alloys, *Trans. TMS-AIME*, Vol. 211, 1957, p. 911.
4. K. J. Irvine, D. T. Llewellyn, and F. B. Pickering: High-Strength Austenitic Stainless Steels, *JISI*, Vol. 199, 1961, p. 153.
5. G. Mohling and W. W. Dyrkacz: Ferrous Base Chromium-Nickel-Titanium Alloy, U. S. Patent #2,641,540, June 9, 1953.
6. G. A. Aggen: Austenitic Alloy, U. S. Patent #3,065,067, November 20, 1962.
7. G. B. Heydt and C. R. Whitney: High Temperature Austenitic Alloy, U. S. Patent #3,183,084, May 11, 1965.
8. H. L. Eiselstein: Age Hardenable Nickel Alloy, U. S. Patent #3,046,108, July 24, 1962.
9. I. Kirman and D. H. Warrington: The Precipitation of Ni_3Nb Phases in a Ni-Fe-Cr-Nb Alloy, *Met. Trans.*, 1, 1970, p. 2667.
10. R. Schempp, P. Payson, and J. G. Chow: Age Hardening Austenitic Steel, U. S. Patent #2,686,116, August 10, 1954.
11. A. Roy and W. E. Jominy: Castable Heat Resisting Iron Alloy, U. S. Patent #3,165,400, January 12, 1965.
12. W. F. Simmons and H. J. Wagner: Where You Can Use Today's Superalloys, *Metal Progress*, Vol. 91, June 1967, p. 87.
13. E. F. Bradley, D. G. Phinney, and M. J. Donachie Jr.: The Pratt & Whitney Gas Turbine Story, *Metal Progress*, Vol. 97, March 1970, p. 68.

14. R. F. Decker: Strengthening Mechanisms in Nickel-Base Superalloys, "Symposium: Steel Strengthening Mechanisms," Zurich, 1969, Climax Molybdenum Co., Greenwich, Conn., 1970, p. 147.
15. D. Raynor and J. M. Silcock: Strengthening Mechanisms in γ'-Precipitating Alloys, *Metal Science Jul.*, Vol. 4, 1970, p. 121.
16. R. K. Ham: Discussion to paper by R. F. Decker, "Symposium: Steel Strengthening Mechanisms," Zurich, 1969, Climax Molybdenum Co., Greenwich, Conn., 1970, p. 179; R. F. Decker: author's reply, ibid., p. 180.
17. F. G. Wilson and F. B. Pickering: Some Aspects of the Deformation of an Age-Hardened Austenitic Steel, *JISI*, Vol. 207, 1969, p. 490.
18. P. H. Thornton, R. G. Davies, and T. L. Johnston: The Temperature Dependence of the Flow Stress of the γ' Phase Based Upon Ni_3Al, *Met. Trans.*, Vol. 1, 1970, p. 207.
19. G. N. Maniar, J. E. Bridge, Jr., H. M. James, and G. B. Heydt: Correlation of Gamma-Gamma Prime Mismatch and Strengthening in Ni Fe-Ni Base Alloys Containing Aluminum and Titanium as Hardeners, *Met. Trans.*, Vol. 1, 1970, p. 31.
20. G. N. Maniar and H. M. James: Electron Microstructure Study of an Iron-Nickel Base Heat-Resistant Alloy Containing Cobalt, "Fifty Years of Progress in Metallographic Techniques," ASTM STP 430, Am. Soc. Testing Mats., 1968, p. 262.
21. C. M. Hammond and G. S. Ansell: Gamma-Prime Precipitation in an Fe-Ni-Base Alloy, *ASM Trans Quart*, Vol. 57, 1964, p. 727.
22. A. J. Lena: Precipitation Reactions in Iron-Base Alloys, "Precipitation From Solid Solution," ASM, Cleveland, Ohio, 1959, p. 244.
23. R. F. Decker and S. Floreen: Precipitation from Substitutional Iron-Base Austenitic and Martensitic Solid Solutions, "Precipitation From Iron-Base Alloys," Eds. Gilbert R. Speich and John B. Clark, Gordon and Breach, New York, 1965, p. 69.
24. F. G. Wilson and F. B. Pickering: Effects of Composition and Constitution on the Ageing of Austenitic Steels Containing Aluminum and Titanium, *JISI*, Vol. 204, 1966, p. 628.
25. R. M. Evans: The Welding and Brazing of Alloy 718, DMIC Report 204, June 1, 1964.
26. H. L. Eiselstein: Properties of Inconel Alloy 706, presented at Materials Engineering Congress, Cleveland, Oct. 1970.
27. J. H. Moll, G. N. Maniar, and D. R. Muzyka: Heat Treatment of 706 Alloy for Optimum Stress Rupture Ductility, presented at TMS-AIME Spring Meeting, Las Vegas, May 1970.
28. I. Kirman and D. H. Warrington: Identification of the Strengthening Phase in Fe-Ni-Cr-Nb Alloys, *JISI*, Vol. 205, 1967, p. 1264.
29. P. S. Kotval: Identification of the Strengthening Phase in "Inconel" Alloy 718. *Trans. TMS-AIME*, Vol. 242, 1968, p. 1764.
30. H. J. Wagner and A. M. Hall: Physical Metallurgy of Alloy 718, DMIC Report 217, June 1, 1965.
31. D. F. Paulonis, J. M. Oblak, and D. S. Duvall: Precipitation in Nickel-Base Alloy 718, *Trans. ASM*, Vol. 62, 1969, p. 611.
32. M. Gell, G. R. Leverant, and C. H. Wells: The Fatigue Strength of Nickel-Base Superalloys, "Achievement of High Fatigue Resistance in Metals and Alloys," ASTM STP 467, Philadelphia, Penn., 1970, p. 113.
33. E. L. Raymond: Effect of Grain Boundary Denudation of Gamma Prime on Notch-Rupture Ductility of Inconel Nickel-Chromium Alloys X-750 and 718, *Trans. TMS-AIME*, Vol. 239, 1967, p. 1415.
34. C. P. Sullivan: A Review of Some Microstructural Aspects of Fracture in Crystalline Materials, Welding Research Council Bulletin 122, New York, May 1967.
35. J. F. Barker, E. W. Ross, and J. F. Radavich: Long Time Stability of Inconel 718, *Jnl. of Metals*, Vol. 22, 1970, p. 31.
36. J. P. Stroup and R. A. Heacox: Effect of Grain Size Variations on the Long-Time Stability of Alloy 718, *Jnl. of Metals*, Vol. 21, 1969, p. 46.
37. B. R. Clark and F. B. Pickering: Precipitation Effects in Austenitic Stainless Steels Containing Titanium and Aluminum Additions, *JISI*, Vol. 205, 1967, p. 70.
38. D. Dulieu and B. Aronsson: Structure and Mechanical Properties of Pure 15% Cr-20% Ni Austenites Hardened with Titanium, and some Observations on the Influence of Boron Additions, *Jernkontorets Annaler*, Vol. 150, 1966, p. 787.
39. J. R. Mihalisin and R. F. Decker: Phase Transformations in Nickel-Rich Nickel-Titanium-Aluminum Alloys, *Trans. TMS-AIME*, Vol. 218, 1960, p. 507.
40. H. Hughes: Precipitation in Alloy Steels Containing Chromium, Nickel, Aluminum, and Titanium, *JISI*, Vol. 203, 1965, p. 1019.
41. W. J. Boesch and H. B. Canada: Precipitation Reactions and Stability of Ni_3Cb in Inconel Alloy 718, *Jnl. of Metals*, Vol. 22, 1969, p. 34.
42. H. L. Eiselstein: Metallurgy of a Columbium-Hardened Nickel-Chromium-Iron Alloy, "Advances in the Technology of Stainless Steels," ASTM STP 369, Philadelphia, Penn., 1965, p. 62.
43. P. S. Kotval: Superalloy by Powder Metallurgy for Use at 1000-1400° F, Final Report, NASA CR-72644, Stellite 70-7691, December 31, 1969.
44. H. J. Beattie, Jr., and F. L. VerSnyder: A-286 Alloy Phases, G. E. Technical Information Report Series, No. DF 55 SE 26, July 5, 1955.
45. P. Lillys, M. Kaufmann, A. M. Aksoy, and R. C. Gibson: Hot Strength Iron-Base Alloys, U. S. Patent #3,243,287, March 29, 1966.
46. J. D. Jones and W. Hume-Rothery: Constitution of Certain Austenitic Stainless Steels, With Particular Reference to the Effect of Aluminum, *JISI*, Vol. 204, 1966, p. 1.
47. W. C. Hagel and H. J. Beattie Jr.: Cellular and General Precipitation During High-Temperature Ageing, "Precipitation Processes in Steels," Iron and Steel Institute, London, 1959, p. 98.
48. P. Duhaj, J. Ivan, and E. Makovický: Sigma-Phase Precipitation in Austenitic Steels, *JISI*, Vol. 206, 1968, p. 1245.
49. P. A. Blenkinsop and J. Nutting: Precipitation of the Sigma Phase in an Austenitic Steel, *JISI*, Vol. 205, 1967, p. 953.
50. J. J. Irani and R. T. Weiner: Ageing Behavior of an Austenitic Steel Containing Vanadium, *JISI*, Vol. 203, 1965, p. 913.
51. F. E. Asbury and D. G. Harris: Precipitation in a 19% Cr, 13% Ni, 10% Co, Nb, W, Mo Steel, *JISI*, Vol. 204, 1966, p. 497.
52. F. B. Foley and V. N. Krivobok: Sigma Formation in Commercial Ni-Cr-Fe Alloys, *Metal Progress*, Vol. 71, May 1957, p. 81.
53. A. M. Talbot and D. E. Furman: Sigma Formation and its Effect on the Impact Properties of Iron-Nickel-Chromium Alloys, *Trans. ASM*, Vol. 45, 1953, p. 429.
54. L. K. Singhal and J. W. Martin: The Formation of Ferrite and Sigma Phase in Some Austenitic Stainless Steels, *Acta Met.*, Vol. 16, 1968, p. 1441.
55. F. J. Rizzo and J. D. Buzzanell: Effect of Chemistry Variations on the Structural Stability of Alloy 718, *Jnl. of Metals*, Vol. 22, 1969, p. 24.
56. A. W. Denham and J. M. Silcock: Precipitation of Fe_2Nb in a 16 wt-% Ni 16 wt-% Cr Steel, and the Effect of Mn and Si Additions, *JISI*, Vol. 207, 1969, p. 585.
57. M. C. Chaturvedi and R. W. K. Honeycombe: Precipitation of Laves Phases in Fe-Mn-Nb Austenites, *JISI*, Vol. 207, 1969, p. 593.
58. H. T. McHenry, J. F. Barker, and R. J. Stuligross: René 62 —A Strong Superalloy for Welded Structures, *Metal Progress*, Vol. 84, December 1963, p. 86.
59. S. F. Sternasty: Reported in A Summary of the Symposium on High-Temperature Alloys, *Jnl. of Metals*, Vol. 18, 1966, p. 606.
60. H. J. Beattie Jr. and F. L. VerSnyder: A New Complex Phase in a High-Temperature Alloy, *Nature*, Vol. 178, 1956, p. 208.
61. R. Leveque and A. Mercier: Study of a Creep Resistant Austenitic Steel with Cobalt-Tungsten-Titanium, *Mem. Sc. Rev. Metal.*, Vol. 65, 1968, p. 691.
62. R. K. Pitler: Discussion in reply to paper by R. F. Decker and S. Floreen, "Precipitation from Iron-Base Alloys," Eds. Gilbert R. Speich and John B. Clark, Gordon and Breach, New York, 1965, p. 129.
63. G. N. Maniar, D. R. Muzyka, and C. R. Whitney: Microstructural Stability of Pyromet 860 Iron-Nickel-Base Heat Resistant Alloy, *Trans. TMS-AIME*, Vol. 245, 1969, p. 701.
64. B. R. Banerjee, E. J. Dulis, and J. J. Hauser: Role of Phosphorous in $M_{23}C_6$ Precipitation in an Austenitic Stainless Steel, *Trans. ASM*, Vol. 61, 1968, p. 103.
65. A. G. Allten, J. G. Y. Chow, and A. Simon: Precipitation Hardening in Austenitic Chromium-Nickel Steels Containing High Carbon and Phosphorous, *Trans. ASM*, Vol. 46, 1954, p. 948.
66. E. J. Dulis: Age-Hardening Austenitic Stainless Steels, "Metallurgical Developments in High-Alloy Steels," ISI Special Report No. 86, 1964, p. 162.
67. M. G. Gemmill: The Technology and Properties of Ferrous

Alloys for High-Temperature Use, CRC Press, Cleveland, Ohio, 1966, p. 133.

68. J. S. T. Van Aswegen and R. W. K. Honeycombe: Segregation and Precipitation in Stacking Faults, *Acta Met.*, Vol. 10, 1962, p. 262.

69. R. W. K. Honeycombe: The Microstructure of Steels, "Metallography 1963," ISI Report No. 80, London, 1964, p. 245.

70. E. E. Brown: private communication, Materials Development Laboratory, Pratt & Whitney Aircraft, East Hartford, Connecticut, 1968.

71. J. P. Stroup and L. A. Pugliese: How Low-Carbon Contents Affect Superalloys, *Metal Progress*, Vol. 93, February 1968, p. 96.

72. D. R. Muzyka and G. N. Maniar: Effects of Solution Treating Temperature and Microstructure on the Properties of Hot Rolled 718 Alloy, *Met. Eng. Quart.*, Vol. 9, 1969, p. 23.

73. J. M. Silcock and N. T. Williams: Precipitation During Aging at 700° C of a Commercial Austenitic Steel Containing Titanium and Aluminum, *JISI*, Vol. 204, 1966, p. 1100.

74. T. Oda, S. Ueda, and M. Nakamura: Effect of Carbide Precipitation on the Creep-Rupture Characteristics of Austenitic Heat-Resisting Steels, Mitsubishi Heavy Industries Ltd., Tokyo, Japan, Rpt. No. MTB 41, Dec. 1966; U. S. Govt. Document PB 186264.

75. T. Oda, S. Ueda, and M. Nakamura: Effect of Carbide Precipitation on the Creep-Rupture Elongation of Austenitic Heat Resisting Steels, *Trans. Japan Inst. Metals*, Vol. 5, 1964, p. 53.

76. J. F. Radavich: A High-Temperature Study of Phase Precipitation in Superalloys, "Advances in X-ray Analysis," Vol. 3, ed. by W. M. Mueller, Plenum Press, New York, 1960, p. 365.

77. M. J. Fleetwood and C. A. P. Horton: The Use of the Microprobe for the Identification of Precipitate Particles in a Nickel-Chromium-Iron Alloy, *Jnl. Royal Mic. Soc.*, Vol. 83, 1964, p. 245.

78. A. Vinter and L. G. Wilbers: Advancing the Engineering Properties of a Ni-Fe Base Superalloy Through Conventional Control Procedures, *Jnl. of Metals*, Vol. 22, 1970, p. 46.

79. G. B. Heydt: Investigation of Notch Sensitivity in A-286 Alloy, *Trans. ASM*, Vol. 54, 1961, p. 220.

80. J. E. Coyne Jr.: Some Factors Influencing Precipitation of Lamellar Constituent in A-286 Stainless Steel, Master of Science Thesis, Hartford Graduate Center, Rensselaer Polytechnic Institute, East Windsor Hill, Conn., May 15, 1958.

81. G. N. Maniar and H. M. James: Notch Sensitivity in A-286 Alloy, *Trans. ASM*, Vol. 57, 1964, p. 368.

82. L. K. Singhal and J. W. Martin: The Growth of $M_{23}C_6$ Carbide in Grain Boundaries in an Austenitic Stainless Steel, *Trans. TMS-AIME*, Vol. 242, 1968, p. 814.

83. F. G. Wilson and F. B. Pickering: A Study of Zone Formation in an Austenitic Steel Containing 4% Titanium, *Acta Met.*, Vol. 16, 1968, p. 115.

84. C. P. Sullivan and M. J. Donachie Jr.: Some Effects of Microstructure on the Mechanical Properties of Nickel-Base Superalloys, *Met. Eng. Quart.*, Vol. 7, 1967, p. 36.

85. K. Metcalfe: Solve Lamellar Phase Problem in A-286, *Iron Age*, Vol. 182, 1958, p. 72.

86. F. M. Richmond: Advanced Concepts of Phase and Structure Control in Fe-Cr-Ni-Ti Alloys, Doctor of Philosophy Thesis, University of Pittsburgh, 1966.

87. J. T. Brown and J. Bulina: W545—A New Higher-Temperature Turbine Disk Alloy, "High Temperature Materials," Eds. R. F. Hehemann and G. Mervin Ault, John Wiley and Sons, Inc., New York, 1959, p. 38.

88. H. Brown: Metallurgical Characteristics of A-286 Alloy, DMIC Memorandum 59, July 26, 1960.

89. J. L. McCabe: General Electric Co., Evendale, Ohio, as referenced by Defense Metals Information Center, Battelle Memorial Institute, Columbus, Ohio, Nov. 1969.

90. C. J. McMahon: private communication, University of Pennsylvania, Philadelphia, Pa., 1970.

91. M. V. Pridantsev: Influence of Impurities and Rare-Earth Elements on the Properties of Alloys, Feb. 1966, available as U. S. Govt. Document AD 634697.

92. J. F. Radavich: High-Temperature Oxidation of Gas-Turbine Alloys, "High Temperature Materials" Eds. R. F. Hehemann and G. Mervin Ault, John Wiley and Sons, Inc., New York, 1959, p. 520.

93. L. F. Coffin Jr.: Cyclic-Strain-Induced Oxidation of High-Temperature Alloys, *Trans. ASM*, Vol. 56, 1963, p. 339.

94. S. T. Wlodek: The Oxidation of Hastelloy Alloy X, *Trans. TMS-AIME*, Vol. 230, 1964, p. 177.

95. C. S. Tedmon Jr.: The Effect of Oxide Volatilization on the Oxidation Kinetics of Cr and Fe-Cr Alloys, *Jnl. Electrochemical Soc.*, Vol. 113, 1966, p. 766.

96. P. Lane Jr., and N. M. Geyer: A Critical Look at Superalloy Coatings for Gas Turbine Components, *Jnl. of Metals*, Vol. 18, 1966, p. 186.

97. S. T. Wlodek: Nickel-Base Alloy and Article, U. S. Patent #3,383,206, May 14, 1968.

98. G. W. Goward, D. H. Boone, and C. S. Giggins: Formation and Degradation Mechanisms of Aluminide Coatings on Nickel-Base Superalloys, *Trans. ASM*, Vol. 60, 1967, p. 228.

99. D. H. Boone and G. W. Goward: The Use of Nickel-Aluminum Intermetallic Systems as Coatings for High Temperature Nickel Base Alloys, "Ordered Alloys—Structural Applications and Physical Metallurgy, Eds. Bernard H. Kear, Chester T. Sims, Norman S. Stoloff, and Jack H. Westbrook, Claitor's Publishing Division, Baton Rouge, La., 1970, p. 545.

100. G. C. Gould: Freckle Segregation in Vacuum Consumable-Electrode Ingots, *Trans. TMS-AIME*, Vol. 233, 1965, p. 1345.

101. A. F. Giamei and B. H. Kear: On the Nature of Freckles in Nickel Base Superalloys, *Met. Trans.* Vol. 1, 1970, p. 2185.

102. S. M. Copley, A. F. Giamei, S. M. Johnson, and M. F. Hornbecker: The Origin of Freckles in Unidirectionally Solidified Castings, *Met. Trans.* Vol. 1, 1970, p. 2193.

103. F. L. VerSnyder and H. J. Beattie Jr.: The Laves Phase in A-286 Alloy, G. E. Technical Information Report Series No. R-55TL028, March 1, 1955.

104. J. F. Barker: A Superalloy for Medium Temperatures, *Metal Progress*, Vol. 81, May 1962, p. 72.

105. R. K. Pitler, E. E. Reynolds, and W. W. Dyrkacz: Consumable-Electrode Vacuum Remelting of High-Temperature Alloys, "High-Temperature Materials," Eds. R. F. Hehemann and G. Mervin Ault, John Wiley and Sons, Inc., New York, 1959, p. 378.

106. J. Bulina and J. T. Brown: Evaluation of Properties Obtained from an Air-Induction and Vacuum-Arc Melted High-Temperature Alloy, *Trans. TMS-AIME*, Vol. 215, 1959, p. 571.

107. W. W. Dyrkacz: Combined Arc and Vacuum-Melting Process Offers Higher Quality Superalloys, *Iron Age*, Vol. 176, 1955, p. 75.

108. Heat Resistant Superalloys, Chapter 11, "Forging Materials and Practices," Eds. A. M. Sabroff, F. W. Boulger, and H. J. Henning, Reinhold, New York, 1968, p. 253.

109. D. M. Gadsby: Forging and Solution Treating Alloy 718, *Metal Progress*, Vol. 90, November 1966, p. 85.

110. L. J. Hull: How to Fabricate A-286, *Metal Progress*, Vol. 76, December 1959, p. 76.

111. R. S. Cremisio, H. M. Butler, and J. F. Radavich: The Effect of Thermomechanical History on the Stability of Alloy 718, *Jnl. of Metals*, Vol. 21, 1969, p. 55.

112. E. E. Brown, R. C. Boettner, and R. A. Sprague: The Effect of Grain Size in the Mechanical Properties of Nickel-Base Alloys, presented at Metallurgical Society of AIME, Fall Meeting, Cleveland, October 1970.

113. C. R. Whitney: The Effects of Ti and Al Variations on Pyromet 860 Alloy Mechanical Properties and on Eta and Gamma Prime Solvus Relationships, Master of Science Thesis, Lehigh University, Bethlehem, Pa., 1970.

114. H. A. Hauser, R. A. Sprague, and E. F. Bradley: Thermomechanical Working of Superalloys for Fatigue Life and Tensile Strength, presented at NATO Materials Conference, London, England, April 29-30, 1970.

115. J. F. Slepitis: Method of Fabricating and Heat Treating Precipitation-Hardenable Nickel-Base Alloys, U. S. Patent #3,420,716, January 7, 1969.

116. Preliminary Data Sheets—Special Metals, Inc., New Hartford, New York, 1964.

117. D. R. Muzyka, C. R. Whitney, and M. L. Wolpert: Heat Treating Pyromet 860, an Iron-Nickel Base Superalloy, for Use As Steam Turbine Cylinder Bolts, TMS paper selection, paper No. F69-5, October, 1969.

118. B. A. Ewing, R. P. Dalal, and P. B. Mee: D979—An Upgraded Gas Turbine Material, *Jnl. Matls.*, Vol. 5, 1970, p. 651.

SECTION VIII:
Nickel-Base Superalloys

Some Effects of Microstructure on the Mechanical Properties of Nickel-Base Superalloys

C. P. Sullivan and M. J. Donachie, Jr.

The relation between microstructure and resulting mechanical properties is developed by a consideration of specific wrought and cast nickel-base superalloys. Particular attention is given to two or three wrought alloys as they illustrate many of the basic principles common to both wrought and cast products. The effects of microstructural changes produced prior to, and during testing are discussed. Included are gamma-prime and carbide hardening, sigma-phase formation and the coarsening of various phases. The effects of these on creep-rupture strength, tensile strength and ductility, together with several other properties are given.

THE PRESENT review is not intended to be an exhaustive summary of the effects microstructure can have on mechanical properties; rather, it is planned to introduce several examples of the changes brought about by variations in microstructure in order to show general trends. However, there can be exceptions to these trends, which is understandable in view of the microstructural and compositional complexity of nickel-base superalloys. A general section, to provide background, is followed by a discussion of the effects which microstructural changes produced prior to testing have on resultant properties. The paper continues with a section on the effects which microstructural changes occurring during testing can produce and ends with some critical comments.

General Background

Nickel-base superalloys contain a variety of elements in a large number of combinations to produce several desired effects. Some elements go into solid solution either to provide strength (molybdenum and tungsten), oxidation resistance (chromium), or decrease the solubility of elements which form a precipitate (cobalt) (1, 2). Other elements, such as aluminum and titanium, are added to form the ordered intermetallic face-centered-cubic γ' phase $Ni_3(Al,Ti)$. It is this γ' phase which is responsible for the useful high-temperature properties of most nickel-base superalloys. Other elements, such as chromium and cobalt, can also be present in the γ'. The mismatch between the lattices of γ' and the solid solution

The authors are associated with the Advanced Materials Research and Development Laboratory and the Materials Development Laboratory, respectively Pratt and Whitney Aircraft, North Haven, Conn. This paper was presented November 3, 1966 at the National Metal Congress, Chicago.

matrix, γ is often quite small accounting in large degree for the relative elevated-temperature stability of this precipitate (3). The γ' precipitate is also unique in that its strength increases with temperature up to about 1400 F (4). Other types of intermetallic precipitate, such as orthorhombic Ni_3Cb, are used in some alloys for specialized purposes (5). Regardless of the nature of the precipitate, strengthening by precipitation requires some form of heat treatment in order to optimize the particular properties required.

Carbides are also an important constituent in nickel-base superalloys because they must be present at grain boundaries to produce desired strength and ductility characteristics. Carbides are also essential for grain size control in wrought alloys. Some carbides are almost unaffected by heat treatment while others require such a step in order to be present. Various types of carbides are possible, depending on the alloy composition and heat treatment; some of the important ones are: MC, M_6C, $M_{23}C_6$ and M_7C_3, where M stands for one or more types of metal

atom. Those carbides exist jointly to some extent; however, they are usually formed by sequential reactions. The normal carbide reaction for titanium-aluminum hardened nickel-base alloys is MC to $M_{23}C_6$ to M_6C (6).

In addition to the carbides, boron and zirconium have been found to increase the rupture life of nickel-base alloys when added in small amounts. Boron has the greater effect on properties but maximum strengthening is achieved by a proper combination of both elements. Table 1 summarizes some effects of various elements.

The evolution of nickel-base superalloys has closely paralleled that of the aircraft gas-turbine engine because of its requirements for materials with high-temperature strength (7). The first major advance was the development of Nimonic 80 by the British which achieved its strength by the precipitation of a titanium-containing intermetallic phase in what was essentially a Nichrome matrix (80 wt % Ni, 20 wt % Cr). A subsequent advancement came about from the discovery that aluminum additions increased the strength even further (Nimonic 80A). Cobalt additions raised the solvus temperature of γ' (Nimonic 90); increased amounts of titanium and aluminum were added (Nimonic 95); molybdenum increased solid-solution strength while aluminum became a more prominent element than titanium (Nimonic 100). Development of alloys in the United States has followed a similar pattern. The Nimonics are wrought alloys, but as the level of high-temperature performance increases forging operations become more difficult. This has led to the use of the most advanced type of alloys available today in cast form because such alloys are essentially unworkable. In these alloys, the most recent trend in composition has been to lower the chromium content to approximately 10% in order to be able to add more aluminum for hardening by γ' precipitation. The lowered chromium content has been balanced by the increased aluminum content which

Table 1. Effects of Several Elements in Nickel-Base Superalloys

Element	Effect
Chromium	Oxidation resistance, carbide former $M_{23}C_6$ and M_7C_3
Molybdenum Tungsten	Solid-solution strengtheners, carbide former M_6C, MC
Aluminum Titanium	Formation of γ' [$Ni_3(Al,Ti)$] hardening precipitate, titanium forms MC type carbide as well.
Cobalt	Raises solvus temperature of γ'
Boron Zirconium	Increase rupture strength, boron also forms borides when present in large enough amounts
Carbon	Formation of various carbides such as MC, M_6C, $M_{23}C_6$, M_7C_3
Columbium	Formation of Ni_3Cb hardening precipitate, carbide former MC
Tantalum	Solid-solution strengthener, carbide former MC

acts to retain oxidation resistance roughly equivalent to the 80-20 nickel-chromium composition. However, the highest normal operating temperatures of today's engines are such that even 20% Cr does not offer sufficient oxidation resistance. In retrospect, it is also apparent that sulfidation (hot corrosion) resistance of these alloys is low due to the lower chromium content. The result of high operating temperatures and impaired sulfidation resistance is a dependence upon coatings, and their technology is now an important part of alloy development. Table 2 lists the nominal compositions of some typical nickel-base superalloys.

Effect of Prior Microstructure on Mechanical Properties

Gamma-Prime Precipitation

The elevated-temperature properties of nickel-base superalloys are dependent on the nature and distribution of the γ' phase and hence on alloy composition (Al + Ti content; Al/Ti ratio) and on the heat treatment employed to precipitate the γ' phase. It is not always possible to separate out the specific effects which changes in γ' precipitation exert because of concurrent changes which take place in other microstructural constituents; however, certain trends are evident. In general, alloys with higher Al/Ti ratio have found use in applications at the higher temperatures (8). To achieve the greatest hardening, however, heat treatment is required and it is usually necessary to heat treat at a temperature somewhat above the solubility limit of γ'; (solution heat treatment) this ensures that upon subsequent aging heat treatments at lower temperatures will produce the maximum amount of precipitate. In some alloys more than one aging treatment is employed to form a duplex precipitate size; one reason for this procedure is to provide a reasonable degree of strength over a wide range of operating temperature. An essential feature of age hardening in nickel-base superalloys is that a temporary over-

heat does not necessarily produce permanent property damage because subsequent cooling to normal operating conditions reprecipitates γ'.

To achieve effective strengthening by the precipitation of an intermetallic phase the following conditions have been listed as being essential (9):
1. Volume fraction of at least 30%
2. Interparticle spacing on the order of 500A
3. A strength greater than that of the matrix to minimize slip cutting through
4. Good lattice match (within one percent) with matrix to promote stability
5. Sufficient ductility to prevent formation of an easy fracture path

These requirements are certainly desirable, but in any specific situation some may not be as critical as others. The spacing requirement may not be stringent in cases where the volume fraction of precipitate is 40% or greater. For example, it has been found that the creep resistance at 1600 F is a sensitive function of particle size and spacing in M-252 (small volume fraction of γ') but not in Inconel 700 (large volume fraction of γ') (10). It was concluded that this resulted from the relative insensitivity of certain geometric parameters to particle spacing when volume fraction was on the order of 40% or greater. These geometric parameters were considered to be critical in determining the alloy strength produced by a γ' dispersion.

René 41 provides an example of the use of varied heat treatments to achieve optimum properties under different test conditions (11). Solutionizing is carried out at 1975 F and results in the dissolution of γ' and $M_{23}C_6$ carbides but retention of M_6C carbides. In order to maximize short-time strength subsequent aging is carried out at 1400 F to produce a fine precipitate; the same treatment is also suitable for creep applications below 1400 F. For long-time service at more elevated temperatures an aging temperature of 1650 F is employed to provide a greater degree of stability.

An extensive study has been carried out on an alloy similar to Udimet 700 to determine the effects of size and distribution of the γ' on creep properties (12). Separate treatments which resulted in a finely dispersed general precipitation led to essentially equal responses at 1600 F under 25,000 psi. Creep life under the same test conditions was reduced however, when material was furnace cooled from the solutionizing temperature because the precipitate was in the overaged condition. Detrimental effects also occurred in material ice-brine quenched and aged prior to testing because the γ' precipitate was partially in the form of cells at grain boundaries.

It is not certain in all cases whether the strength produced by the γ' precipitate is associated just with its size and spacing. Coherency strains resulting from a certain degree of mismatch between precipitate and matrix may be important in determining the strength of some alloys such as Waspaloy (13). This suggestion is based on the observation that room-temperature hardness variations cannot be always ascribed to the size and dispersion of the γ' precipitate. In the case of alloys where the mismatch is very small coherency strains are insufficient to explain strengthening. Recent work on MAR-M200 suggests that the tensile yield behavior is controlled by the fact that two dislocations in the matrix must constrict into a pair before they enter the ordered precipitate (14). In such a case the degree of order is an important factor in strength considerations.

In the final analysis it is not possible to judge alloy performance by just considering the strength characteristics of the γ' hardened γ matrix, because it is the relationship between this strength and that of the grain boundaries which really determines the performance of the alloy being tested. For example, if the matrix becomes much stronger relative to the grain boundaries, premature failure occurs either because relaxation of stresses becomes difficult or

Table 2. Nominal Compositions of Some Nickel-Base Superalloys

Alloy	C	Cr	Al	Ti	Mo	W	Co	Cb	B	Zr	Other	Ni
Nimonic 75	0.12	20	—	0.5	—	—	—	—	—	—	—	Bal.
Nimonic 80A	0.08	20	1.5	2.4	—	—	—	—	—	—	—	Bal.
Nimonic 90	0.10	20	1.6	2.4	—	—	17.5	—	—	—	—	Bal.
Nimonic 95	0.12	20	2.0	3.0	—	—	17.5	—	—	—	—	Bal.
Nimonic 100	0.20	11	5.0	1.3	5.0	—	20.0	—	—	—	—	Bal.
Waspaloy	0.08	19	1.3	3.0	4.4	—	13.5	—	0.008	0.08	—	Bal.
Udimet 700	0.10	15	4.3	3.5	5.2	—	18.5	—	0.03	—	—	Bal.
René 41	0.09	19	1.5	3.1	10.0	—	11.0	—	0.005	—	—	Bal.
IN-100 (cast)	0.18	10	5.5	5.0	3.0	—	15.0	—	0.015	0.05	—	Bal.
MAR-M200 (cast)	0.15	9.0	5.0	2.0	—	12.5	—	1.0	0.015	0.05	—	Bal.
B-1900 (cast)	0.11	8.0	6.0	1.0	6.0	—	10.0	—	0.015	0.07	4.3 Ta	Bal.
INCO-713 (cast)	0.14	13.0	6.0	0.75	4.5	—	—	2.3	0.01	0.1	—	Bal.
								Cb + Ta				
M-252	0.15	19.0	1.0	2.5	9.8	—	10.0	—	0.005	—	5.0 Fe (max.)	Bal.

because of an actual weakness in the grain-boundary zones. In a study of the creep-rupture behavior of Nimonic 80A, the aging temperature for γ' precipitation was found to be of secondary importance in determining creep properties at 1380 F under 38,000 psi compared to the condition of the grain boundaries (15). The effect of precipitation at grain boundaries on alloy performance is discussed below.

Carbide Precipitation

Carbides exert a profound influence on properties by their precipitation on grain boundaries because this affects the strength of such regions. In most alloys $M_{23}C_6$ is precipitated at the grain boundaries at some point in the heat treatment cycle. If it precipitates as a continuous film properties can be severely degraded. Specifically, this was found to be the case for room-temperature impact resistance in M-252 (16). The presence of a continuous film provides an easy fracture path. Also grain-boundary sliding would be restricted leading to excessive stress build-up. Lowered rupture lives and ductility in forged Waspaloy have also been attributed to overheating during forging and resultant $M_{23}C_6$ continuous carbide film formation upon aging (17). Aging the M-252 alloy at a higher temperature produces a globular form of carbide which alleviates the problem, probably because such a configuration tends to restrict grain-boundary cracks to the intercarbide spaces (18). On the other hand, the absence of carbides at the grain boundaries can lead to premature failure. The low-cycle fatigue resistance at 1400 F of Udimet 700 heat treated so as to produce γ' precipitation both within grains and at grain boundaries but no $M_{23}C_6$ precipitation was drastically reduced compared to normally heat treated material containing grain-boundary carbides (19). Therefore, for a given matrix strength, optimum alloy properties are achieved by a certain distribution of carbides at the boundary. If no carbides are present, excessive grain-boundary sliding leads to premature failure; on the other hand, if a continuous film exists, stress relaxation becomes a problem and failure again occurs prematurely. The carbide distribution required is one such that stresses can be relieved by a restricted degree of sliding, but where a continuous fracture path is avoided. The condition where carbides are relatively uniform and small seems to lead to the best combination of properties (1).

Another effect produced by grain-boundary $M_{23}C_6$ carbide precipitation is the occasional formation of a zone on either side of the boundary depleted in γ' precipitate. Electron-probe microanalyses of grain boundaries in Nimonic 80A has shown that the depletion is apparently associated with a decrease in chromium content (20).

The decrease is supposedly sufficient to allow complete solubility of the titanium and aluminum, thus accounting for the absence of γ', Ni_3 (Al,Ti). In addition, the precipitation of certain grain-boundary carbides may deplete the zones of molybdenum, tungsten or columbium. The net result is the presence of narrow regions which are somewhat weaker but more ductile than the matrix. These are thought to be able to relieve stress concentrations arising during creep, thus leading to improved lives even when the matrix is highly strengthened by a large volume fraction of γ' (15, 21). If such zones should become wide or much weaker than the matrix, deformation would concentrate there resulting in early failure. This behavior has been noted in the sense that failure has been observed in γ' depleted zones adjacent to creep-agglomerated $M_{23}C_6$ carbides (22).

In related effects, the morphology of carbides has been connected to the weldability of René 41 (11). If the alloy in the mill annealed condition (solution heat treat at 2150 F) is welded and subsequently heat treated, serious cracking problems are encountered. If, however, an anneal at 1975 F preceeds welding little difficulty is encountered. Such behavior has been postulated to be associated with carbides. At a temperature of 2150 F the M_6C carbide normally present in this alloy goes into solution and subsequent heat treatment between 1300 and 1800 F brings out $M_{23}C_6$ as a continuous film along grain boundaries. As noted above, this has a deleterious effect on some properties. However, by annealing at 1975 F the M_6C does not go into solution (although $M_{23}C_6$ does go into solution) and no $M_{23}C_6$ films form with further heat treatment.

The $M_{23}C_6$ carbide can also precipitate in a cellular form under certain conditions with a resultant embrittling effect (1, 16). The cells are composed of rods or plates of the carbide phase in a matrix which has an orientation different from that of the grain into which it is growing. It appears that some degree of matrix supersaturation is required for cellular growth hence a high-temperature or intermediate treatment to precipitate carbides reduces this supersaturation and prevents subsequent $M_{23}C_6$ carbide cell formation at lower temperatures.

The ductility of a nickel-base superalloy has been observed to be impaired by a Widmanstätten precipitation of M_6C carbide at grain and twin boundaries (23). This form of precipitate can be suppressed by a heat treatment at a temperature between the normal temperatures for solution and aging treatments. However, formation of Widmanstätten M_6C molybdenum-rich carbides after creep exposure of a complex cast alloy, B-1900,

has not appeared to degrade ductility. Thus, whether such a carbide morphology is harmful may depend upon several factors such as alloy composition and the conditions of test. The relatively massive MC type carbides which are present in almost all nickel-base superalloys can be potential sources of trouble. A recent low-cycle fatigue study on Udimet 700 over a range of temperatures showed that cracking can be initiated at such particles (24). Usually the cracks are not very large, but it is conceivable that such cracking could be important under some circumstances. Also, if the material is worked in a specific direction, stringers containing the same type of carbides are produced which provide a path for fracture.

Boron and Zirconium Effects

The main effect of boron and zirconium is thought to reside in grain boundaries. One suggestion is that the boron and zirconium atoms migrate to grain boundaries and fill open spaces present there. This then strengthens the region by slowing down grain-boundary diffusional processes which contribute to creep deformation (2). Another hypothesis, which is supported by experimental data, is that the presence of the two elements in the grain boundary suppresses early agglomeration of $M_{23}C_6$ carbides and the subsequent formation of creep cracks in these regions (25). The retardation is believed to be due to the reduction of cabon, titanium and aluminum segregation, to grain boundaries, by boron and zirconium. An additional effect of boron in this study (25) was the increase of intragranular carbide precipitation which might also have increased alloy creep resistance. The enhanced matrix precipitation is consistent with a reduction in the grain-boundary segregation tendency of carbon.

The presence of boron and zirconium at grain boundaries may change interfacial energy relationships sufficiently to cause a second phase to assume a different morphology. Recent Russian work indicates that these two elements favor coalescence and spheroidization of secondary phase precipitates along grain boundaries (26). It is felt that coalesced spheroids enhance grain-boundary strength.

If boron is added in excessive amounts complex borides, which are hard and brittle, form at grain boundaries. These can also be a source of difficulty in forging due to their relatively low melting points. Another possible case of harmful effects arising from excess boron was that involving two heats of Udimet 700 which had comparable properties at room temperature, but not at 1400 F (19). The material with the poorer characteristics contained an excess of M_3B_2 type borides. It is possible that the inferior 1400 F properties were associated with the presence of a brittle boride phase at the grain boundary. Conversely, the complete absence of borides in thin sections of Udimet 700, heat treated in atmospheres which promoted deboronization, can result in low thermal (low-cycle) fatigue strength and possible premature rupture.

Segregation and Defects

As the compositions of nickel-base superalloys have become more complex the presence of segregation in wrought material has assumed greater importance. The source of segregation is the original casting, and the manifestation of a carry-over of this segregation in the wrought product is termed banding. Banding consists of alternate layers of lighter and darker material which reflects a change in alloying element composition. This segregation leads to forging difficulties and mechanical property variability (27). It is possible to eliminate banding to some extent by thermal treatments and controlled forging deformation; however, the degree of success is dependent upon the amount of segregation present in the starting casting. When banding is severe, the alloy-rich regions contain a massive form of γ' which is a remnant of a eutectic structure formed in pouring the ingot. Figure 1 gives an example of the massive γ' structure. If maximum rupture properties are to be achieved this form of γ' must be dissolved otherwise the full precipitation potential of the system is not achieved.

Defects can affect properties to a marked extent. The presence of pores in cast material can be troublesome, particularly if they are large or located at a strategic location in a part. One reason for a lowering of properties is that load-bearing area is reduced, leading to an acceleration of deformation. A second reason for a decrease in properties is associated with the fact that an initiation site for crack propagation is provided, thus shortening the time to fracture. A striking example of this behavior has been found in a nickel-base alloy subjected to cyclic creep (alternating stress superimposed on a mean stress with their ratio being less than one) at 1700 F. Internal cavities produced by the casting process serve as fatigue-crack initiators with the result that failures originate in the interior. Figure 2 shows an example. In this particular case the fatigue-crack changed to a tensile failure upon reaching the specimen surface. Forging defects will also affect resultant properties, their elimination requires careful control of the working process which becomes more difficult as alloy complexity increases.

Although not considered segregation in the accepted sense, the presence of external coatings on parts to prevent their rapid deterioration in aircraft gas turbine environments represents an intentional segregation of a most pronounced sort. With today's ever-increasing engine operating temperatures, coatings are being used more frequently on parts. The subject is large in scope, and it is merely intended here to point out that the presence of a coating introduces a new group of problems to mechanical behavior. There are problems involving the mechanical characteristics of the coating itself and those dealing with the compatibility of substrate and coating. In addition, composition gradients (and therefore structural gradients) of considerable magnitude are present in a small zone near the interface between coating and substrate. When heat treating coated parts the most important consideration may well be the effects produced in the coating or at coating-base metal interfaces rather than those produced in the underlying material.

Grain Structure

Grain size is one important aspect of grain structure. A fine-grained material has superior hardness, yield strength, tensile strength, fatigue strength and impact resistance at room temperature compared to a coarse-grained material. At elevated temperatures creep properties are generally better for coarse-grained materials, but this advantage can be offset by other inferior properties under other conditions of loading. One problem frequently encountered in wrought materials, particularly in large disk forgings which are not uniformly worked, is abnormal grain growth wherein some grains grow to an extreme size. The resultant duplex structure is not desirable because the material possesses the weaknesses associated with both coarse and fine-grained conditions. It has been shown that if nickel-base superalloys receive more than 5 to 10% reduction, abnormal grain growth will be prevented. Nonuniform metal flow during forging is a common source of this trouble. Creep testing is now being undertaken, with aim periods of up to 30,000 hr, on Waspaloy from disk forgings to ascertain the degree of mechanical and metallurgical instability to be expected from such duplex structures (29).

The presence of a fine grain size does not always lead to superior fatigue properties. The initiation of low-cycle fatigue cracking in Udimet 700 is transcrystalline at room temperature (30), but intercrystalline at 1400 F (31). The form of cracking at 1400 F does not appear to be related to a creep phenomenon. A finer grain size in the same material gave superior results at room temperature in low-cycle fatigue, but, if anything, was inferior at 1400 F (19). This result is probably due to the change in cracking path with temperature. At 1400 F the ease of crack initiation, which is intergranular, is not affected much by the change

Fig. 1. Massive γ' constituent in TRW-1800 alloy after 4-hr heat treatment at 2200 F. Arrows point to some examples. Etched in 62% H_2O: 15% HF, 15% H_2SO_4, and 8% HNO_3. X100. (Ref. 35).

in grain size; if anything, the finer grained structure might present a somewhat more favorable cracking path. When cracking is transgranular however, the effectiveness of microcracks produced is a direct function of grain size, being less harmful when grains are smaller.

An indirect effect of grain size results from the extent of precipitation possible at grain boundaries. The larger the grain size the smaller is the grain boundary area available for precipitate formation. A given volume of precipitate, such as $M_{23}C_6$, will therefore form a thicker and more continuous film as grain size increases. In Udimet 500 a grain size greater than ASTM 3 coupled with a 1550 to 1600 F heat treatment results in a continuous $M_{23}C_6$ carbide film at grain boundaries (32). This film can be deleterious to fatigue, ductility and impact properties. For a grain size of ASTM 4 or finer there is not sufficient $M_{23}C_6$ to cause brittleness from this source. On the other hand, if the grain size is too fine, the density of $M_{23}C_6$ becomes so small that a loss in rupture life occurs.

Fig. 2. Example of fatigue crack initiated at internal void in cast nickel-base superalloy subjected to cyclic creep at 1700 F. Arrows outline the fatigue zone while X marks the origin of failure. X10.

Another aspect of grain structure is the presence of annealing twins in wrought material. In the low-cycle fatigue of Udimet 700 at room temperature and 1400 F surface cracks initiate along the coherent boundaries of annealing twins as well as in slip bands or grain boundaries (30, 31). These cracks are relatively long because the twins generally run across the grains rather than ending within them; Fig. 3 illustrates this behavior in a specimen tested at room temperature. It is not known whether the absence of annealing twins would result in an increased life, but it is clear that they do provide a path for initial cracks to propagate relatively long distances. It has also been observed that interior cracks form at coherent annealing twin boundaries in low-cycle fatigue at temperatures somewhat below 1400 F (24). The same type of boundary is a site of enhanced shear in creep specimens of Udimet 700 (33). Figure 4A illustrates the offset in a grain boundary which has resulted from this type of shear. If severe enough, the shear leads to grain boundary cracking as shown in Fig. 4B. The contribution of twin boundary shear to the over-all process of deformation and fracture is not known at present.

In the broad sense casting processes can be considered as producing one form of grain structure while the working of such a structure produces another. Not only are there grain size differences, but compositional gradients (with resultant microstructural gradients) are more prevalent in castings. The dendritic structure of castings can also be an important consideration in their performance. It is often stated that castings provide superior creep-rupture properties above the equicohesive temperature, but in many cases castings and forgings are not compared on a similar

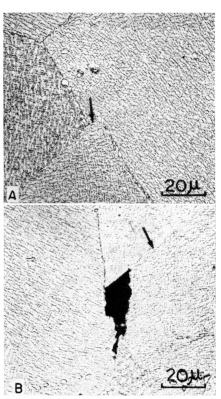

Fig. 4. Micrographs showing shear along twin boundaries in creep specimens of Udimet 700. (A) After cycling under 36,600 psi with ¼ hr at each temperature (1640 and 1690 F) every cycle for 35.2 hr. Arrow indicates shear displacement. (B) After isothermal creep under 36,600 psi at 1690 F for 14.5 hr. Crack produced at grain-boundary shear. Arrow points to twin boundary. (G. A. Webster and C. P. Sullivan, to be published in J. Inst. Met.). Etched in 50% lactic acid 33% HCl, 17% HNO_3; X600.

basis. There is no doubt that the newer cast alloys such as MAR-M200 and B-1900 have superior creep properties compared with the older wrought alloys such as Udimet 700, but this is to be expected since the

former two are alloyed to give a higher hardening potential. Some research has been carried out on cast and wrought conditions of the same alloy. Udimet 700 has been evaluated in both conditions and it was found that castings had superior rupture times at 1800 F while the wrought product had superior ductility at the same temperature (34). B-1900, a more complex casting alloy, exhibits similar behavior when forged. Figure 5 shows stress-rupture properties for wrought and cast Udimet 700 at 1400, 1600 and 1800 F. It will be noted that the cast material has a coarser grain size which probably enhances its creep strength at 1800 F, hence the comparison between cast and wrought conditions is not exactly on the same terms. Inconel 713 also has shown superior creep strength in the cast form compared to the wrought form when tested at 1800 F under 18,000 psi (34). The difference in grain sizes is not known, but the important point is that working of a type which would be used to produce hardware of Inconel 713 results in lowered creep properties. The same statement holds for Udimet 700.

A recent study has been undertaken wherein cast and double-extruded compositions of the most advanced nickel-base superalloys were evaluated in creep tests at 1800 F and tensile tests at 1400 F (35). In all cases the double-extruded material was superior in creep strength to Udimet 700, which is about the best commercial wrought alloy available today; but the cast strength of each material, although approached, was never exceeded in the wrought form. However, at 1400 F the wrought materials displayed superior or comparable tensile strengths and ductilities to their cast counterparts, an effect which may be due to the more regular structure in the former.

In summary, it can be said that a cast alloy is superior in creep strength to its wrought counterpart above a

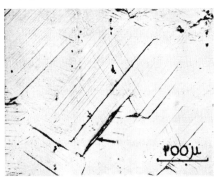

Fig. 3. Surface cracking along coherent annealing twin boundaries (arrows) in Udimet 700 room-temperature low-cycle fatigue sample, stress axis horizontal. Electropolished surface, metallographic replica (extrusion-like marking due to replicating material entering crack and then being pulled free). X160. (C. H. Wells and C. P. Sullivan, Pratt and Whitney Aircraft, Advanced Materials Research and Development Laboratory, Report 66-002, April 1964.)

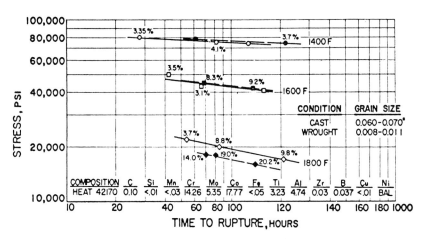

Fig. 5. Stress-rupture properties of wrought and cast Udimet 700. (W. H. Sharp, High Temperature Materials II, 18 (1963) 189.)

certain temperature. However, other properties such as tensile strength and fatigue life may be inferior. The reason for creep-strength superiority of castings is associated both with grain size and the microstructure peculiar to the alloying element heterogeneities present. From a practical viewpoint it would seem that the extra expense involved in working of the most advanced alloy compositions would only be justified if such a process produced an improvement in some specific property which was considered absolutely essential for proper part performance.

Several other aspects of grain structure have recently been developed by more stringent control of grain size, porosity, and eutectic γ' in a conventional polycrystalline casting (36) and by a controlled directional casting process (directional solidification). The latter produces grain boundaries running parallel to each other over the entire length of a part (37) or produces single crystals (38). Figure 6 illustrates, in cast MAR-M200, the aligned grain-boundary structure and contrasts it to the conventional structure. By

Fig. 6. Macroetched turbine blades of directionally solidified MAR-M200 on the left and conventionally cast MAR-M200 on the right. Note parallel grain boundaries in the former material running along the blade length. (B. J. Piearcey and F. L. Ver-Snyder, Pratt & Whitney Aircraft, Advanced Materials Research and Development Laboratory Report 65-007, April 1965).

eliminating grain boundaries that lie perpendicular to the principal stress, the ductility and thermal fatigue resistance are improved considerably because grain boundaries are paths of easy crack propagation. Thermal fatigue resistance of B-1900 processed to controlled fine polycrystalline grain sizes also is improved over conventional castings but not to the degree to

which MAR-M200 is improved by the directional solidification process. An important consideration in the controlled directional casting process is that the growth direction of the columnar crystals is <100>. This is the direction of maximum creep strength properties (38) and accounts in part for the superior creep strength compared to conventionally cast MAR-M200 where the grains are not oriented in the maximum strength direction. Another advantage of directional solidification is that it produces parts that are freer of shrinkage porosity; Fig. 2 is just one example of the effect which porosity can produce on mechanical behavior. The properties of the directionally solidified product improve, up to a point, with increasing solidification rate due to changes in grain size; amount of porosity; and carbide content, size and shape (40).

The directional solidification process resulting in aligned grain boundaries has been applied to three cast alloys in addition to MAR-M200 (40). It was concluded that the process is most beneficial in cases where a severe grain-boundary cracking problem exists as in conventionally cast MAR-M200; but if a cast alloy has grain boundaries with strength comparable to the matrix, directional solidification no longer leads to large increases in creep life. More important, however, is that the use of a strongly hardened matrix is not curtailed by premature grain-boundary failure when the boundaries are aligned.

The production of single crystals is a natural outgrowth of the concept of grain-boundary alignment. If grain boundaries are a source of failure, then maximum improvement results when none are present. For example, single crystals of MAR-M200 have outstanding thermal fatigue resistance (38). The creep strength and rupture lives are considerably improved over those of the aligned grain-boundary structure. The differences are attributable to effects of grain boundaries, orientation and thermal history, but the specific contribution of each has not been determined in detail as yet.

Dislocation Arrangements

One possible way in which dislocation structure can be affected is by the presence of solute atoms that lower the stacking fault energy, thereby increasing the distance between partial dislocations which, in turn, makes dislocation climb and cross slipping more difficult (41). The net result is an increase in creep resistance. Recent work on pure metals has shown that the steady-state creep rate increases with stacking fault energy in a marked fashion (42). Solute atoms may also segregate to dislocations, effectively pinning them and thus increasing creep resistance in this manner as well.

Dislocation structure can also be affected by a controlled elevated-temperature deformation process applied prior to testing. In addition to changing dislocation structure such a high-temperature thermomechanical treatment (HTMT) can also change the course of subsequent precipitation and therefore the resultant properties. The object of HTMT is to produce a prior dislocation substructure that will be relatively stable at the contemplated operating temperature range, and which will also be more deformation resistant than the structure in normally heat treated material. Recent Russian work on an alloy of the Nimonic type shows that the initial deformation, primary creep rate, and primary creep strain are all reduced by HTMT (43). This is attributed to a reduction in the number of mobile dislocations and to the formation of a relatively stable dislocation substructure. Improvements produced in one property may not be accompanied by similar improvements in others, hence the type of treatment would have to be varied depending upon which property is being tailored. The potential of HTMT in nickel-base superalloys is an exciting one (44) and it promises to be a fruitful avenue of research in the years to come.

Effect of Exposure on Mechanical Properties

Gamma Prime

Changes in the γ' precipitate during testing will produce an effect on subsequent properties, but this may be overshadowed by other microstructural changes. Common changes in γ' morphology are its growth and agglomeration with time at temperature in a creep test. The latter process is illustrated in Fig. 7. A structure such as shown in Fig. 7B can result in a large reduction in strength and increase in ductility upon subsequent tensile testing at room temperature (45). In Nimonic 80A, coarsening of the precipitate also leads to an increase in ductility in subsequent tensile testing at elevated temperatures (46). It is also likely that agglomeration is a significant contributor to the accelerated drop in strength of Udimet 700 observed in long-time rupture testing, although data on engine-operated turbine blades of this same alloy usually do not show any strength or creep ductility degradation after short-time engine exposures of up to 100 hr at turbine inlet temperatures of 1900 F. In some nickel-base superalloys with a relatively low volume fraction of γ', it is possible to reduce the creep rate by periodic overheating (without stress) (10). The coarsened and agglomerated γ' is taken back into solution, and upon subsequent cooling a finer dispersion of γ' is produced which is more creep resistant. Tests on Nimonic 80A annealed at the

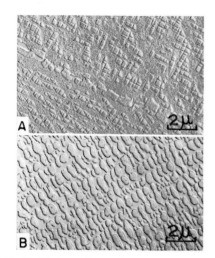

Fig. 7. Electron micrographs illustrating change in γ' morphology in creep testing of Udimet 700. Etched in 50% lactic acid, 33% HCl, 17% HNO$_3$; X400. (A) As heat-treated. (B) 91.2 hr under 36,600 psi at 1640 F. (G. A. Webster and C. P. Sullivan, Pratt & Whitney Aircraft, Advanced Materials Research and Development Laboratory Report 65-017, September 1965).

creep temperature of 1380 F after various amounts of creep show that complete recovery from the effects of creep elongation occurs after comparatively short annealing times (47). Prolonged lives can apparently be obtained from single specimens by repetitive creep/anneal cycles.

In René 41 creep testing at 1500 and 1600 F of material in the condition yielding optimum tensile properties (1975 F/30 min plus 1400 F/16 hr) produced an increase in particle size and a heavy concentration of precipitate along slip planes (11). This condition resulted in lowered room-temperature tensile strength and a brittle behavior. However, exposure to creep conditions does not always lead to a decrease in strength. Sometimes an increase occurs due to further precipitation of the hardening phase (46). Such an occurrence depends upon the initial condition of the material as well as the subsequent test conditions since both affect the resultant precipitate condition.

An interesting example of the effect which prior heat treatment has on subsequent precipitation during creep tests occurs in the superalloy Inconel 713C (48). This alloy was solution heat treated at 2225 F to dissolve the coarse γ' and then air cooled to produce a fine γ' dispersion. The treatment dissolve a carbide of the MC type and during subsequent creep tests at 1700 F a heavy agglomerate embrittled the grain boundaries. The agglomerate appeared to be carbon-containing γ' which owed its existence to the prior solution of the MC carbide at 2225 F. What appears to be a rather subtle effect of change in γ' morphology has been observed in Udimet 700 cycled between 1640 and 1690 F under

constant stress (33). If equal periods of time were spent at each temperature a strengthening effect was observed. This was attributed to the fact that, for equal strains, the dislocation substructure produced by interaction with the precipitate present at the lower temperature was more deformation resistant to creep at 1690 F than the dislocation substructure produced by straight creep at 1690 F.

Carbides

As exposure time increases there is a change in morphology of carbides and there can also be a change in the types or relative amounts of the carbides present. An example of such changes occurs in René 41. Initially the primary carbide phase present is M_6C with lesser amounts of $M_{23}C_6$ and Ti(C,N) titanium carbonitrides. As the time and temperature increase in the intermediate temperature range (1300 - 1600 F) the relative amount of $M_{23}C_6$ increases while the other two decrease in amount (45). This increase in $M_{23}C_6$ corresponds to a thickening of grain boundaries and leads to a reduction in tensile ductility below about 1400 F, but above this temperature the boundary thickening is reportedly responsible for an increase in tensile ductility.

A recent study of MAR-M200 in the directionally solidified aligned grain boundary form has made an attempt to isolate the microstructural change(s) responsible for the weakening of the alloy in creep testing (49). A number of specimens were deliberately overaged under conditions which were optimum for separately promoting precipitation of $M_{23}C_6$, M_6C and γ' agglomeration. Creep tests at 1800 F under 25,000 psi showed that considerable weakening was only associated with M_6C precipitation, suggesting, in turn, that the removal of the hardening element tungsten from solid solution was the critical factor.

An overtemperature during exposure could prove to be quite serious if $M_{23}C_6$ were the primary grain-

boundary carbide. This carbide goes back into solution around 1900 F in many of the nickel-base superalloys (50). Without the initial presence of $M_{23}C_6$ it is quite likely that subsequent properties would be drastically altered if cracking is grain-boundary initiated. As noted in the previous section on effects of prior microstructure, $M_{23}C_6$ in cellular form is sometimes deleterious to properties, so that if a cellular precipitate formed under operational conditions part performance could be adversely affected.

Formation of Transition Element Phases

Because nickel-base superalloys contain a variety of transition elements the formation of transition element phases such as sigma, mu and related intermetallic compounds during service can occur with sometimes deleterious effects (51). Sigma and mu phases generally precipitate in the form of long thin plates with a definite orientation relationship to the matrix; these plates presumably provide crack-propagation paths through the grains. Also these phases are usually brittle. Figure 8 shows that fracture follows the pattern of sigma phase precipitation in the weld metal of a Udimet 700 sample. The presence of sigma phase generally is accompanied by a loss in low-temperature ductility. In addition, rupture strengths can be reduced as shown in the cases of the cast alloy IN-100 (51) and the wrought alloy Udimet 700 (52).

On the other hand there are circumstances in which sigma phase formation does not appear to be significant. Results of tests on Inconel-713 may be analyzed for the combined effects of mechanical creep damage (void formation, etc.) and sigma phase formation (53). Data (Table 3) was generated by post-exposure creep tests at 1400 F, 85,000 psi after significant amounts of sigma phase had formed during 1500, 1600, and 1700 F prior creep exposure. Although the amount of sigma phase was about the same in all cases, the rupture lives decreased in order of in-

Fig. 8. Crack propagation path in room-temperature bend specimen of Udimet 700 weld metal heat treated at 1550 F for 16 hr. Stress axis vertical. Fracture path follows sigma phase precipitation pattern. ×350. (W. A. Owczarski and C. P. Sullivan, Welding J. Res. Supp., 44 (1965) 241-S).

Table 3. Effects of Prior Creep Exposure On Inconel-713 (after Ref. 52)

Temp, F	Prior creep exposure			Sigma phase	Post-exposure rupture at 1400 F, 85,000 psi	
	Stress, psi	Approx. creep %	Life, hr/hr ratio		Life hr	R.A. %
1500	30,000	0.5	0.04	Yes	1.6	4.0
1600	20,000	1.0	0.14	Yes	0.1	1.7
1700	13,000	1.0	0.22	Yes	<0.01	1.6
Base	Line	Data		No	5.5	2.0

creasing prior creep life use ratio (0.04 to 0.22). Thus, sigma is not always the important factor in determining resultant properties. The major factor in determining strength and ductility may be creep-life ratio in cases where the base-line life is short. Results on Udimet 700 (54), subsequently tested in stress-rupture and 70 F tensile tests, after long-time annealing at elevated temperatures produced sigma phase, were claimed to indicate that sigma phase caused minor effects. IN-100 does not always deteriorate in elevated-temperature creep properties when sigma phase is produced by long-time (1000-hr) creep exposures although deleterious effects on room-temperature properties may exist (55).

The degree of damage attributable to sigma phase formation is variable. Part of this behavior can be ascribed to the fact that other microstructural changes are occurring simultaneously. A proper evaluation, therefore, requires the use of statistical analysis. It has been possible to avoid the formation of sigma phase during long-time creep exposure by a judicious balancing of alloying elements using the principles of electron vacancy theory (51, 52). The same technique with suitable modifications should be able to avoid premature formation of the related intermetallic compounds.

Environment

The environment to which a material is exposed can be the source of severe property changes. Oxidation, for example, can remove elements from the surface and also change the rate of crack initiation and propagation. Atmospheres containing sulfur-bearing compounds can lead to a form of rapid attack whose precise mechanism is not fully understood as yet, but which may be catalyzed by the presence of chlorides. Exposure to air results in depletion of aluminum and boron from the surface layers of nickel-base superalloys and a resultant degradation in strength properties. The loss of aluminum is revealed either by the absence of the γ' precipitate or a decrease in its amount. The fatigue characteristics in the presence of such a layer may be markedly different in the initial cracking stages. In a recent study on the effect of air on creep-rupture life, it was shown that nickel, un-

der certain conditions of stress and temperature, had a longer life in air than in vacuum, but that the lives of superalloys were always less in air (56). Strengthening in air can occur because oxide formation acts to partially heal cracks by taking up some of the load thus offsetting the accelerated development of cracks. In the case of superalloys however, the stresses at which they are tested are high enough to overshadow any possible strengthening due to oxide formation so that the accelerated crack development predominates. Less complex superalloys (solid-solution strengthened) do not necessarily conform to this pattern (57).

Coatings

The composition of a coating continually changes during high-temperature exposure due to loss of elements. The element is aluminum in the case of the commonly employed aluminide coatings. The loss of aluminum will result in a gradual shift from β (NiAl) to mixtures of β and γ' to γ', to mixtures of γ' and γ (aluminum in solid solution). The mechanical characteristics which depend on the amount, distribution and nature of the phases will reflect these changes. Little information is now available on such changes, but it is clear that investigations into this area are essential. Another aspect of the exposure of coated material is the formation of unwanted phases in the substrate material. While aluminum is diffusing out through the surface it also diffuses inwards, altering the composition in the matrix. The resultant composition, in turn, may be prone to the formation of sigma phase or a phase related to it. The precipitation of a needle-like phase in the substrate adjacent to an aluminide coating has been observed after long-time exposure (45, 58). The presence of such a phase is undesirable as it may directly degrade the matrix strength either by its own characteristics or indirectly by removal of certain elements from the surroundings.

Defects

Defects, generally in the form of microcracks, are produced during the course of a test. These in turn can have a pronounced effect if stresses of a different nature are applied to a

part. One such case is the application of an impact at a relatively low temperature in the presence of creep cracks. If the material is notch-sensitive, catastrophic failure may ensue, depending upon the geometry of the cracks and the magnitude of the impact stresses. The importance of adequate fracture toughness in a material is an important consideration in its selection. Another example of interaction is the propagation of creep cracks by an alternating stress. Recent work on Udimet 700 has shown that in cyclic creep at 1400 and 1700 F, a surface creep crack of grain size dimensions can propagate rapidly to failure in fatigue under a sufficiently high alternating stress, leading to a drastic reduction in creep life (59). Figure 9 shows a typical case of a fracture surface produced in such a way. The intergranular origin shown by the arrow gives way to a fatigue zone consisting of a smooth portion 'A' and a rougher portion 'B'. Final separation occurs due to overloading and constitutes the remainder of the fracture surface. The reason why surface creep cracks figure so prominently in the testing of Udimet 700 is that creep cracks form on the surface before they do so in the interior and hence are available there first for propagation in fatigue.

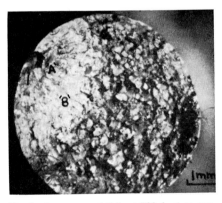

Fig. 9. Appearance of Udimet 700 fracture surface produced in specimen cycled under 80,000 ± 10,000 psi at 1400 F and 10 cycles/sec. Fracture initiated at surface creep crack (arrow), spread in fatigue (areas 'A' and 'B') and terminated by overload. (E. G. Ellison and C. P. Sullivan, Pratt & Whitney Aircraft, Advanced Materials Research and Development Laboratory Report 66-024, July 1966).

Conclusion

It can be appreciated that the effects of microstructure on properties in nickel-base superalloys are extremely complex due both to the compositional make up of the alloys and to the time dependent nature of the distribution and types of phases present. The treatment in this review can only be considered as one which serves to introduce the newcomer to the mechanical metallurgy of high-temperature alloys by presenting a few examples of mechanical property changes induced by variations in microstructural parameters. Even the specific examples cannot always be considered as being totally accurate since results in nickel-base superalloys usually require sophisticated interpretation. It is very difficult, for example, to change just one microstructural variable at a time. Another problem in interpretation involves taking into account microstructural changes which occur during testing when attempting to evaluate the specfic effect of changes in the starting microstructure.

Often, a serious limitation is the use of one, or at the most, two types of testing procedures to determine the results of microstructural variations. Many times creep tests are the sole source of information. Furthermore, the degree of creep testing may be limited in itself and this can prejudice alloy evaluation procedures. Generally speaking, the use of this test is not always relevant because the pertinent conditions are often not static in nature. The use of various types of fatigue tests (low-cycle, high-cycle, high frequency) both alone, or in combination, and in combination with creep, must assume more importance in mechanical property evaluation. Fracture toughness should be determined because of its relevance to all types of crack propagation problems. More testing should be carried out on structures that have been artifically aged to a condition representing the state of the material after some extended period of service exposure. Mechanical problems may not be evident until these later stages, hence testing of as-heat-treated parts cannot reveal the real deficiencies of an alloy.

It appears that the study of mechanical property variation, by microstructural means, in superalloys has barely been touched. The opportunity for fruitful research is therefore almost unlimited.

REFERENCES

1. J. Chander, Hardening Mechanism and Corrosion Resistance of Nickel-Base Alloys, A Review, Canadian Met. Quart., 3 (1964) 57.
2. C. H. Lund, Physical Metallurgy of Nickel-Base Superalloys, DMIC Report 153 (1961).
3. D. McLean, Mechanical Properties of Metals, John Wiley and Sons, Inc., New York (1962) 323.
4. R. G. Davies and N. S. Stoloff, On the Yield Stress of Aged Ni-A1 Alloys, Trans AIME, 233, (1965) 714.
5. H. J. Wagner and A. M. Hall, Physical Metallurgy of Alloy 718, DMIC Report 217 (1965).
6. H. Beattie and F. VerSnyder, The Influence of Molybdenum on the Phase Relationships of a High Temperature Alloy, Trans. ASM, 49 (1956) 883.
7. W. Betteridge, The Nimonic Alloys, Edward Arnold Ltd., London (1959).
8. J. Barker, E. Dunn and L. Woodyatt, 1400° F Ultra High Strength Alloy Development Program, Tech. Report AFML-TR-65-278, I (Aug. 1965).
9. R. F. Decker and R. R. DeWitt, Trends in High-Temperature Alloys, J. Metals, 17 (1965) 139.
10. John P. Rowe and James W. Freeman, Relation Between Microstructure and Creep Resistance in Nickel-Base Alloys, Joint Int. Conf. on Creep 1963, Inst. Mech. Eng., London (1963) 1-65.
11. Donald P. Moon, James F. Barker and Ward F. Simmons, High-Temperature Properties of René 41 and Astroloy, Metals Eng. Quart., 1, (1961) 3.
12. R. F. Decker, John P. Rowe, W. C. Bigelow and J. W. Freeman, Influence of Heat Treatment on Microstructure and High-Temperature Properties of a Nickel-Base Precipitation-Hardening Alloy, NACA Tech. Note 4329, (July 1958).
13. W. C. Bigelow, J. A. Amy, C. L. Corey and J. W. Freeman, An Electron Metallographic Study of the Precipitation Hardening Process in Commercial Nickel-Base Alloys, Symposium on Advances in Electron Metallography, ASTM STP No. 245, Philadelphia, (1958) 73.
14. S. M. Copley, B. H. Kear and B. J. Piearcey, Coherent Precipitation Hardening in Ni-Base Superalloys, Pratt & Whitney Aircraft, Advanced Materials Research and Development Laboratory, Report 66-016, (May 1966).
15. W. Betteridge and A. W. Franklin, The Effect of Heat-Treatment and Structure on the Creep and Stress-Rupture Properties of Nimonic 80A, J. Inst. Met. 85 (1956-57) 473.
16. W. C. Hagel and H. J. Beattie, Jr., Aging Reactions in Udimet 500 and M-252, General Electric Co., Gas Turbine Department and Metallurgical Products Dept. Report DF-57-SL-349, (November 1958).
17. W. Danesi, J. Radavich and M. Donachie, The Effects of Forging Practice on Rupture Properties of Waspaloy, unpublished research Pratt and Whitney Aircraft, to be submitted to Trans. Quart., ASM.
18. C. W. Weaver, Intergranular Cavitation, Structure, and Creep of a Nimonic 80A-Type Alloy, J. Inst. Met. 88 (1959-60) 296.
19. C. H. Wells, Advanced Materials Research and Development Laboratory, Pratt and Whitney Aircraft, unpublished research.
20. M. J. Fleetwood, The Distribution of Chromium Around Grain-Boundary Carbides in Nimonic 80A, J. Inst. of Met., 90 (1961-62) 429.
21. J. Heslop, Wrought Nickel-Chromium Heat-Resisting Alloys Containing Cobalt, Cobalt, 24, (1964) 128.
22. R. F. Decker, J. P. Rowe and J. W. Freeman, Boron and Zirconium from Crucible Refractories in a Complex Heat-Resistant Alloy, NACA Report 1392, (1958).
23. J. R. Mihalisin and J. S. Iwanski, Microstructural Study of the Response of a Complex Superalloy to Heat Treatment, Trans. AIME, 215 (1959) 912.
24. C. H. Wells and C. P. Sullivan, Cyclic Strain Ageing of Udimet 700, Pratt and Whitney Aircraft, Advanced Materials Research and Development Laboratory, Report 66-019, (June 1966).
25. R. F. Decker and J. W. Freeman, The Mechanism of Beneficial Effects of Boron and Zirconium on Creep Properties of a Complex Heat Resistant Alloy, Trans. AIME, 218 (1960) 277.
26. B. S. Natapov, V. E. Ol'shanetskii and E. P. Ponomarenko, Effect of Alloyed Elements on the Form of Secondary Precipitates in Refractory Nickel Alloys, Metal Science and Heat Treatment, (January 1966) 11.
27. A. J. DeRidder and R. W. Koch, Controlling Variations in Mechanical Properties of Heat Resistant Alloys During Forging, ASM Report System, Tech. Report No. W 12-5-65 (February 1965).
28. R. F. Decker, A. I. Rush, A. G. Dano, and J. W. Freeman, Abnormal Grain Growth in Nickel-Base Heat-Resistant Alloys, NACA Tech. Note 4082 (December 1957).
29. Private communication, G. Tlewellyn and K. Chamberlin, Bristol-Siddley Aircraft Co., England to M. Donachie, Pratt and Whitney Aircraft, July 5 1966.
30. C. H. Wells and C. P. Sullivan, The Low-Cycle Fatigue Characteristics of a Nickel-Base Superalloy at Room Temperature, ASM Trans. Quart., 57 (1964) 841.
31. C. H. Wells and C. P. Sullivan, Low-Cycle Fatigue Damage of Udimet 700 at 1400 F, ASM Trans. Quart., 58 (1965) 391.
32. M. Kaufman and A. E. Palty, The Relationship of Structure to Mechanical Properties in Udimet 500, Trans. AIME, 218 (1960) 107.
33. G. A. Webster and C. P. Sullivan, Effects of Temperature Cycling on the Creep Behavior of a Nickel-Base Alloy, to be published in J. Inst. Met.
34. W. H. Sharp, Status and Future of Cobalt- and Nickel-Base Alloys in High Temperature Materials II, Ed. by G. M. Ault, W. F. Barclay and H. P. Munger, Interscience Publishers, New York, (1963) 189.
35. S. T. Scheirer and R. J. Quigg, Extrusion to Cast Nickel-Base Superalloys, ASM Trans. Quart., 59 (1966) 49.
36. A Technical Report on a New Cast Nickel-Base Superalloy PWA 663 (B-1900), Pratt and Whitney Aircraft, Report No. PWA-2617 (June 18 1965).
37. B. J. Piearcey and F. L. VerSnyder, A New Development in Gas Turbine Materials—The Properties and Characteristics of PWA 664, Pratt and Whitney Aircraft, Advanced Materials Research and Development Laboratory Report 65-007 (April 1965).
38. B. J. Piearcey and F. L. VerSnyder, Monocrystaloys—A New Concept in Gas Turbine Materials—The Properties and Characteristics of PWA 1409, Pratt and Whitney Aircraft, Advanced Materials Research and Development Laboratory Report 66-007 (February 1966).
39. B. J. Piearcey and R. W. Smashey, The Structure of a Unidirectionalloy Solidified Nickel-Base Superalloy, Pratt and Whitney Aircraft, Advanced Materials Research and Development Laboratory Report 65-018 (October 1965).
40. B. E. Terkelsen and B. J. Piearcey, The Effect of Unidirectional Solidification on the Properties of MAR-M200, B-1900, IN-100 and TRW 1900, Pratt and Whitney Aircraft, Advanced Materials Research and Development Laboratory Report 66-008 (April 1966).
41. J. Nutting and J. M. Arrowsmith, The Metallography of Creep-Resistant Alloys in Structural Processes in Creep, Iron Steel Instit., London (1961) 136.
42. Craig R. Barrett and Oleg D. Sherby, Influence of Stacking-Fault Energy on High-Temperature of Pure Metals, Trans. AIME, 233 (1965) 1116.
43. Y. E. N. Sokolkov, M. G. Gaydukov and S. N. Petrova, Features of the Initial Stage of Creep in an Alloy of the Nimonic Type After High-Temperature Thermomechanical Treatment, Phys. Metals Metallography, 19 (1965) 91.
44. C. Slunder and A. Hall, Thermal and Mechanical Treatments for Nickel and Selected Nickel-Base Alloys and Their Effect on Mechanical Properties, NASA Report TM X-53443 (April 20 1966); AD 633040.
45. George J. Wile, Materials Considerations for Long Life Jet Engines, see SAE Journal, 74 (1966) 81.

46. W. Betteridge and J. A. Towers, Precipitation During Creep of Titanium-Hardened Nickel-Chromium Alloys in Precipitation Processes in Steels, Iron and Steel Institute, London (1959) 235.

47. P. Davies, J. Dennison and H. Evans, Recovery of Pоperties of a Nickel-Base High-Temperature Alloy After Creep at 750 C, J. Inst. Met., 94 (1966) 270.

48. R. F. Decker and C. G. Bieber, Microstructure of a Cast Age-Hardenable Nickel-Chromium Alloy in Symposium on Electron Metallography, ASTM STP No. 262, Philadelphia, (1959) 120.

49. G. A. Webster and B. J. Piearcey, The Effects of Load and Temperature Cycling on the Creep Behavior of PWA 664, Pratt and Whitney Aircraft, Advanced Materials Research and Development Laboratory Report 66-020 (June 1966).

50. C. H. Lund and H. J. Wagner, Identification of Microconstituents in Superalloys, DMIC Memorandum 160, 1962.

51. L. R. Woodyatt, C. T. Sims, and H. J. Beattie, Jr., Prediction of Sigma-Type Phase Occurrence from Compositions in Austenitic Superalloys, Trans. AIME, 236 (1966) 519.

52. William J. Boesch and John S. Slaney, Preventing Sigma Phase Embrittlement in Nickel-Base Superalloys, Metals Prog., 86 (1964) 109.

53. D. Jensen, A. Pinkowish and M. Donachie, Effects of Prior Creep Exposure on 1400° F Strength of INCO 713 Cast Nickel-Base Alloy, Trans. ASME, 88D (1966) 109.

54. W. Smith, M. Donachie and J. Johnson, Relationship of Prior Creep Exposure to Strength of Wrought Udimet 700 Nickel-Base Alloy, Trans. ASME, 88D (1966) 4.

55. D. Jensen, A. Pinkowish and M. Donachie, Effects of Prior Creep Exposure on Strength of IN-100, unpublished research, Pratt and Whitney Aircraft, to be submitted to Trans. ASME.

56. P. Shahinian, Creep-Rupture Behavior of Unnotched and Notched Nickel-Base Alloys in Air and in Vacuum, Trans. ASME, J Basic Eng., 87 (1965) 344.

57. R. Shepheard and M. Donachie, Creep-Rupture Behavior of Hastelloy-X in Vacuum, Matl. Res. and Std., 4 (1964) 495.

58. W. Boesch and J. Slaney, Phase Stability in Wrought Precipitation Hardened Nickel-Base Alloys, presented at AIME Annual Meeting, New York, February, 1966; to be published.

59. E. G. Ellison and C. P. Sullivan, The Effect of Superimposed Fatigue Upon the Creep Behavior of the Nickel-Base Alloy Udimet 700, Pratt and Whitney Aircraft, Advanced Materials Research and Development Laboratory Report 66-024 (July 1966).

Effects of Solution Treating Temperature and Microstructure on the Properties of Hot Rolled 718 Alloy

Donald R. Muzyka and G. N. Maniar

THERE IS relatively limited information in the literature relating the properties and microstructure of hot rolled bars of 718 alloy. Data of this type are invaluable in understanding and/or predicting the behavior of 718 alloy in this product form in service.

It has been observed qualitatively at the Carpenter Research Laboratory and by others (1) that certain amounts of grain boundary precipitates are necessary for stress-rupture notch ductility, while lesser or larger amounts can lead to notch brittleness. It is also well known that a uniform fine grained microstructure will yield best all-around properties for 718 alloy.

More recently, qualitative results have been published (2) indicating

The authors are associated with the Research and Development Center, The Steel Division of The Carpenter Technology Corp., Reading, Pa. 19603. This paper was presented 11 March at the 1969 Westec Conference, Los Angeles.

that increasing Ni_3Cb plate size, associated with increasing grain size, decreases the stress-rupture ductility of this alloy. These investigations (2) have also indicated that a decreasing volume of Ni_3Cb platelets appears to decrease stress-rupture ductility. The investigators (2) feel that the platelet size effect overshadows the platelet volume effect. There also appears to be a consensus (2, 3) that "carbide films" are not detrimental to the creep-rupture and tensile properties of 718 alloy.

The majority of recent structural work on this alloy has been performed on disc-type forgings. This is in line with the fact that much of the current usage of this alloy is in this type forging. Also, recent literature confines itself to "specification" heat treatments for the alloy.

The present work was designed to develop data on 718 alloy that would be useful in promoting further under-standing of the relationships between properties and microstructure of hot rolled bars of this alloy. It is felt that this product form will be more and more used as 718 alloy wins its way into fastener applications now enjoyed by other, lower strength, high-temperature alloys. Some of the solution heat treatments employed were out of the ordinary range of those recommended for the alloy. This was done to develop a range of structures and properties for study.

Test Program

A 12-ft length of 2¾-in. round 718 alloy bar was obtained from inventory. This bar had been forged and hot rolled from a 14-in. diam vacuum induction plus vacuum consumable remelted ingot. The chemistry of this bar is shown in Table 1. Macroscopic and microscopic evaluation showed the bar to be essentially uniform with a grain size of about ASTM 4.5.

Table 1. Heat Analysis of 718 Alloy Bar

| | | | | | | Composition, % | | | | | | | | |
C	Mn	Si	P	S	Cr	Ni	Mo	Cu	Co	Ti	Al	B	Cb + Ta	Fe
0.07	0.09	0.09	0.006	0.003	18.68	52.43	3.14	0.01	0.04	0.96	0.56	0.0051	5.42	bal

| | | | | | | Typical Specifications | | | | | | | | |
| 0.08 max | 0.035 max | 0.035 max | 0.015 max | 0.015 max | 17.00/ 21.00 | 50.00/ 54.00 | 2.80/ 3.30 | 0.30 max | 1.00 max | 0.65/ 1.15 | 0.20/ 0.80 | 0.006 max | 4.75/ 5.50 | bal |

Hot Rolling

In order to obtain a range of hot worked structures for study, sections of the 2¾-in. round bar were rolled on a hot mill to 1³⁄₁₆-in. rcs bar using the sequences described in Table 2. Furnace temperatures were varied from 1900 to 2000 F. Various optical pyrometer temperature readings, taken during the processing, are given in Table 2 as are as-rolled grain size data.

Heat Treatment

Sections of the as-rolled bars were solution heat treated 1 hr/A.C. at 1725, 1750, 1775, 1800, 1850, or 1900 F and given a common age of 1325 F/8 hr/cool 100 F/hr to 1150 F/8 hr/A.C. Previous work (4) had shown that for usual 718 alloy, only the 1850 and 1900 F treatments would produce any appreciable grain growth. This was desirable in the present work so that a broad range of structures and properties could be compared.

Mechanical Tests

Heat treated sections of all bars were then tested as follows:
1. 70 F tensiles (0.252-in. gage diam × 1-in. gage length).
2. 1200 F/100 ksi stress ruptures (0.178-in. gage diameter × K_t 3.8).
3. Hardness, R_c.

Limited testing of a few of the hot rolled bars was also done, for two heat treatments, 1750 F plus age and 1850 F plus age, as follows:
4. 1200 F tensiles (0.357-in. gage diam × 1.2-in. gage length).
5. 1300 F/75 ksi stress ruptures (0.178-in. gage diam × K_t 3.8).
6. 1100 F/135 ksi stress ruptures (0.178-in. gage diam × K_t 3.8).

All test specimens conformed to ASTM requirements.

Metallography

As-rolled and heat treated sections of all bars were examined via optical metallography, electron metallography, and X-ray diffraction.

For optical microscopy, the samples were etched in glyceregia (15 ml HCl, 5 ml HNO_3, 10 ml glycerol). For fine-structure study by electron microscopy, chromium shadowed parlodion replicas were prepared from the micros etched in glyceregia. Extraction replicas were made for selected area electron diffraction analysis of grain boundary and matrix precipitates. For this purpose, a mixture of 50 ml HCl, 48 ml H_2O and 2 ml H_2O_2 was found to be satisfactory in that it selectively attacks the fine matrix gamma-prime precipitate particles but not the grain boundary carbides or the Ni_3Cb needles.

Table 2. Sequences Used to Hot Roll Sections of 2¾-In. Round 718 Alloy Bar to 1³⁄₁₆-In. Rcs.

I. Initial Processing (Furnace temp 2000 F)

Bar No.	Reduction, %	Finish temp, F	Reheat	Reduction, %	Finish temp, F	Finish size, in.	Grain size (ASTM)
1	50	1840	"	67.5	1810	1 rcs.	9
2	50	1860	"	58.5	1850	1⅛ rcs.	8
3	50	1865	"	49.0	1880	1¼ rcs.	6.5/7
4	50	1880	"	38.5	1900	1⅜ rcs.	6.5/7
5	50	1860	"	67.5	1710	1 rcs.	9
6	50	1870	"	38.5	1870	1⅜ rcs.	7.5

II. Final Processing* (Bars 1, 2, 3 and 5 cut in half prior to this step due to excessive length)

Bar No.	Furnace Temp, F	Reduction, %	Finish temp, F	Estimated hot rolled grain size†, ASTM
1A	1900	34	1710	4/5
1B	1900	34	1740	4/5
2A	1900	48	1580	5/6 (95%) + 10⁺ (5%)
2B	1900	48	1710	5/6 (90%) + 10⁺ (10%)
5A	2000	34	1800	5/6 (90%) + 10⁺ (10%)
5B	2000	34	1810	5/6 (80%) + 10⁺ (20%)
6‡	2000	47	1580	5 (60%) + 10⁺ (40%)
3A	1900	58	1730	5/6 (30%) + 10⁺ (70%)
3B	1900	58	1760	5/6 (20%) + 10⁺ (80%)
4	1900	65	1770	5/6 (20%) + 10⁺ (80%)

* Bars listed in order of decreasing average hot rolled grain size.
† 10⁺ = Grains finer than ASTM 10. Numbers in parenthesis are the percentages of the grain sizes present in the structure.
‡ This bar only finished at 1 in. rcs due to trouble entering passes.

Source: *Metals Engineering Quarterly*, Nov 1969

The precipitates from several samples were electrolytically extracted in 1:10 HCl:methanol solution. To determine the effect of heat treatment on dissolution of the various phases, a quantitative estimate of the extracted residue was obtained in each case. The phases were identified by X-ray diffraction analysis. In some cases, the residues were analyzed by X-ray fluorescence methods using a mini-probe attachment.

Mechanical Properties

Tensile Tests

The 70 F tensile test results are listed in Table 3. The data show very little, if any, consistent effect of the hot rolling variables on properties. Differences between tests on identically processed bars were often greater than differences between bars with widely separated processing. This effect is believed to be due to the realtively fine, and/or mixed grain sizes of all of the bars tested. The main differences in the tensile properties developed were due to variations in solution treating temperature. Figures 1 and 2 show these data in graphical form. Figure 1 shows that, on the average, all tensile strength properties appear to be reduced for increasing solution treatment temperatures from 1725 to 1850 F. The limited data for 1900 F solution treating show that this trend continues to 1900 F. At the same time, Fig. 2 shows that 70 F tensile ductility is apparently increased as solution temperature is increased. This is especially true of 70 F tensile reduction in area. These results agree with those of previous investigators (1,3,4).

Results of the 1200 F tensile tests are given in Table 4. These limited data indicate that the trend of generally decreasing tensile strength properties with increasing solution treating temperature established for 70 F tests, also holds for 1200 F tensile tests. However, no significant changes in 1200 F tensile ductility with increasing solution treating temperature are obvious. These observations agree with those of Eiselstein (3).

Table 3. 70 F Tensile Properties of Hot Rolled 718 Alloy Bars

Bar No.	Solution temp, F*	0.2% YS, ksi	UTS, ksi	Elong, %	RA, %	Hardness, (R_C)
1A	1725	177.9	210.9	18.2	34.6	44
1B		183.8	213.1	16.8	32.7	44
2A		171.3	209.4	20.0	35.8	44
2B		180.4	215.4	18.5	33.3	45.5
3A		184.4	220.4	18.1	33.3	44.5
3B		180.9	217.4	17.6	34.6	44.5
4		177.9	216.4	22.9	35.8	44
4		179.4	218.4	19.1	31.9	45
5A		176.4	212.4	16.9	31.9	44
5B		179.9	217.4	17.6	33.3	43.5
6		181.4	220.4	16.5	31.9	45
6		175.4	217.4	15.6	30.6	44.5
1A	1750	170.3	206.4	22.4	40.8	44
1B		171.5	206.8	18.2	37.6	44
2A		172.8	209.4	21.1	38.3	44.5
2B		176.4	213.9	20.0	35.8	45
3A		181.4	217.4	17.7	34.6	45
3B		179.4	215.9	20.0	35.8	45
4		176.4	214.9	18.3	35.8	44
4		176.4	216.4	19.0	35.8	44
5A		174.3	208.9	18.7	35.8	44
5B		177.9	213.4	18.3	37.1	44
6		171.8	213.4	15.4	26.8	44.5
6		172.8	212.9	16.4	31.9	44
1A	1775	167.7	204.1	22.5	42.0	44
1B		165.8	204.4	22.5	39.6	44
2A		170.3	207.9	23.0	40.8	44
2B		169.3	209.9	21.8	38.3	45
3A		177.4	215.9	20.2	35.8	44
3B		176.4	214.9	21.3	38.3	45
4		177.4	214.4	20.8	37.1	44
4		177.4	215.4	20.3	33.3	44
5A		174.3	210.4	19.9	39.6	44
5B		177.8	212.6	18.8	37.8	44
6		171.3	210.9	20.4	38.3	44.5
6		172.8	210.4	19.1	35.8	43
1A	1800	171.3	204.4	23.6	43.2	44
1B		169.7	203.6	24.4	42.2	44
2A		177.8	212.1	21.0	42.8	44
2B		177.3	214.1	18.9	39.1	44.5
3A		175.4	213.9	21.6	37.1	44
3B		174.8	212.4	19.1	39.6	44
4		177.9	213.9	23.5	38.3	44
4		177.9	214.4	22.1	38.3	44
5A		172.8	208.9	19.6	40.8	45
5B		175.4	210.4	20.2	39.6	44.5
6		171.0	210.3	20.1	38.1	44.5
6		172.3	209.9	20.2	37,1	44
1A	1850	172.1	201.4	22.9	46.8	43.5
1B		172.8	202.4	21.4	45.8	43.5
2A		174.3	206.9	22.2	45.6	44
2B		172.7	206.6	21.7	44.6	44
3A		174.8	208.4	22.7	44.4	44
3B		174.3	205.9	20.0	44.4	44
4		172.3	206.4	21.0	44.4	44
4		173.3	206.4	20.0	43.2	43
5A		173.3	204.9	20.6	45.6	44
5B		170.3	203.4	21.8	45.6	43
6		174.3	206.9	22.5	45.6	44
6		173.0	205.3	22.0	44.8	43.5

(Table 3. continued next page)

Table 3. (continued)

Bar No.	Solution temp, F*	0.2% YS, ksi	UTS, ksi	Elong, %	RA, %	Hardness, (Rc)
1A	1900	167.3	198.4	25.9	48.0	42.5
1B		170.7	197.6	24.0	48.1	41.5
2A		169.8	200.4	21.4	48.0	43
2B		171.3	200.4	23.1	48.0	43
3A		167.7	200.4	26.5	48.7	41.5
3B		166.7	200.4	27.8	48.7	43
5A		169.8	198.0	24.2	49.4	42.5
5B		164.3	199.3	22.5	49.6	42
Typical Specifications						
	1750/ 1800	150 min	185 min	12 min	15 min	36† min

* 1 hr./A.C., all bars aged 1325 F/8 hr/Cool 100 F/hr. to 1150 F/8 hr/A.C.
† Rc 36 ≃ Bhn 331

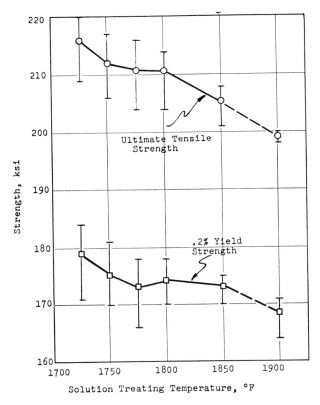

Fig. 1. Effect of solution treating temperature on the average and ranges of 70 F tensile strength properties of fully heat treated, hot rolled bars of 718 alloy.

Stress-Rupture Tests

Stress-rupture data for tests at 1200 F/100 ksi are given in Table 5. Again, there appear to be no significant, systematic differences in properties due to the hot rolling variables. When these data are observed as plotted in Fig. 3, however, it becomes evident that in general, stress-rupture life increases and ductility decreases with increasing solution treating temperature. These results also agree with trends previously shown (3-5). In fact, use of the 1900 F solution treatment leads to stress-rupture notch sensitivity with attendant considerable scatter in rupture lives. Retesting of smooth bar sections of several of the notch-sensitive bars shows that ductility values are generally less than 5% for elongation and 12% or lower for reduction in area.

A limited number of stress-rupture tests at 1100 F/135 ksi and 1300 F/75 ksi, two other conditions of concern in the aircraft industry, Tables 6 and 7, show that the trends established at 1200 F/100 ksi hold for these conditions also. By comparing data for 1750 and 1850 F solution treating in Tables 5, 6 and 7, a not unexpected result is that for decreasing stress-rupture test temperature, stress-rupture ductility also decreases. This corresponds to an increasing "severity" of test. The data from these tables for 1750 F solution treating are plotted in Fig. 4 to further demonstrate this effect.

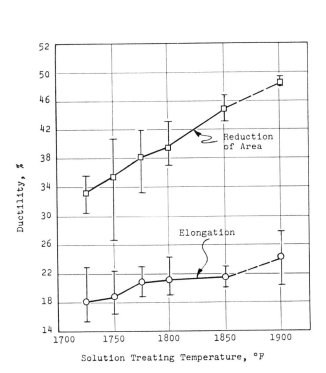

Fig. 2. Effect of solution treating temperature on the average and ranges of 70 F tensile ductility properties of fully heat treated, hot rolled bars of 718 alloy.

Table 4. 1200 F Tensile Properties of Hot Rolled 718 Alloy Bars

Bar No.	Solution temp, F*	0.2% YS, ksi	UTS, ksi	Elong, %	RA, %
1A	1750	141.2	165.0	18.0	41.6
1B		142.5	166.4	20.0	43.6
2A		143.1	167.8	18.0	43.4
2B		144.1	169.0	17.4	41.2
3A		152.9	175.9	15.3	38.8
3B		149.9	176.4	16.8	38.8
5A		143.7	166.9	17.8	43.4
5B		141.4	164.2	19.1	42.8
1A	1850	138.9	156.3	17.8	38.9
1B		138.4	154.9	18.8	42.8
2A		139.4	158.9	17.8	39.6
2B		138.1	158.6	18.5	39.1
3A		136.6	156.3	18.5	41.2
3B		135.8	156.8	17.6	40.0
5A		138.2	156.8	17.6	42.4
5B		139.2	155.1	18.7	40.0
Typical Specifications					
	1750/	125.0	145.0	12.0	15.0
	1800	min	min	min	min

*1 hr/A.C., all bars aged 1325 F/8 hr/cool 100 F/hr to 1150 F/9 hr/A.C.

Hardness

Hardness data, obtained from the thread ends of all 70 F tensile test specimens, are given in Table 3. The results are predominantly R_c 44-45. However, a slight decrease in hardness occurs for 1850 and 1900 F solution treating. This is attributed to increased grain size and the accompanying dissolution of the various phases for these treatments, as will be discussed below.

Optical Microstructure

Hot Rolled

As-rolled grain size data for all the bars are given in Table 2. The structures, for the most part, were moderate to fine, and of mixed grain sizes. The working sequences in Table 2 were designed to develop a wide

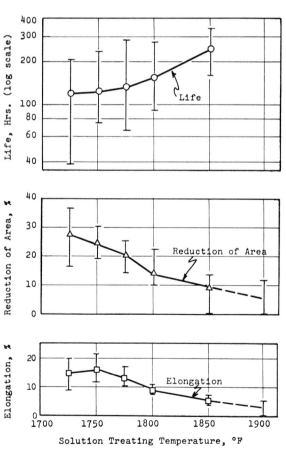

Fig. 3. Effect of solution treating temperature on the average and ranges of 1200 F/100 ksi stress-rupture properties of fully heat treated, hot rolled bars of 718 alloy.

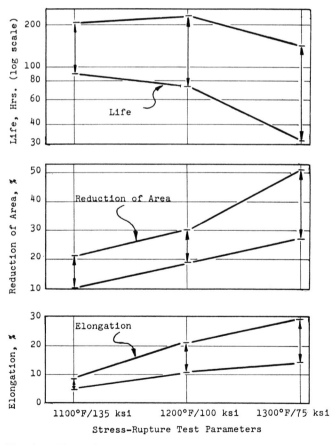

Fig. 4. Effect of test parameters on the stress-rupture properties of hot rolled bars of 718 alloy, heat treated: 1750 F/1 hr/A.C. + 1325 F/8 hr/cool 100 F/hr to 1150 F/8 hr/A.C.

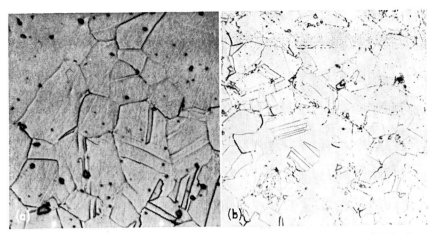

Fig. 5. Effect of heat treatment on microstructure of hot rolled bar of 718 alloy. (a) Bar 1A, as-rolled; (b) bar 1A, heat treated, 1800 F/1 hr/A.C. + aged. Etchant, glyceregia; 250×.

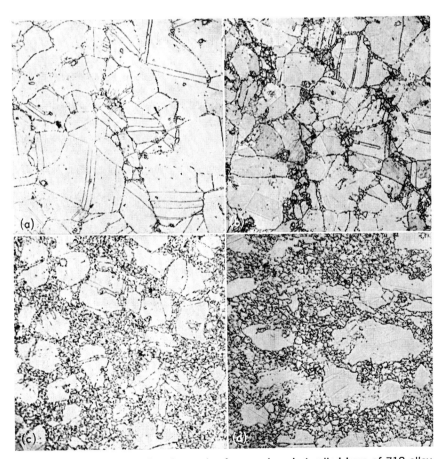

Fig. 6. Optical micrographs of samples from various hot rolled bars of 718 alloy showing fully aged microstructures. All samples solution treated at 1725 F/1 hr/A.C. (below grain growth temperature). (a) Bar 1A; (b) bar 2A; (c) bar 3A; (d) bar 4. Etchant, glyceregia; 250×.

Table 5. 1200 F/100 ksi Stress Rupture Properties of Hot Rolled 718 Alloy Bars

Bar No.	Solution temp, F*	Life, hr	Elong, %	RA, %
1A	1725	180.4	12.9	16.4
1B		207.4	8.4	16.4
2A		170.6	12.5	26.5
2B		112.3	14.6	30.9
3A		88.8	18.3	32.9
3B		39.9	15.4	30.9
4		125.8	15.4	35.5
4		89.2	19.5	30.9
5A		161.6	13.2	23.3
5B		153.2	12.0	26.1
6		112.3	15.2	28.9
6		117.0	13.6	28.9
1A	1750	150.8	15.5	19.1
1B		193.5	13.3	19.2
2A		134.1	14.8	22.4
2B		100.8	20.9	27.3
3A		76.0	18.5	25.8
3B		106.7	20.0	29.3
4		138.7	11.6	26.5
4		91.8	21.2	29.8
5A		136.0	14.0	24.1
5B		230.2	11.8	24.3
6		98.8	13.7	30.1
6		130.0	15.3	21.3
1A	1775	147.3	12.7	15.4
1B		286.6	12.4	16.2
2A		143.9	16.8	14.2
2B		107.4	16.2	20.3
3A		91.1	11.2	19.5
3B		81.2	11.9	22.3
4		120.3	13.3	21.5
4		67.8	10.9	22.6
5A		159.4	11.8	21.3
5B		131.7	10.0	21.3
6		192.8	14.7	25.6
6		186.3	12.9	22.1
1A	1800	223.6	7.1	10.7
1B		175.3	4.8	8.8
2A		272.1	10.8	12.0
2B		154.6	8.0	13.2
3A		138.0	9.9	12.0
3B		137.2	10.3	17.2
4		114.8	7.7	13.2
4		91.9	7.2	10.0
5A		181.5	7.4	15.2
5B		185.5	10.0	19.2
6		150.1	11.2	22.1
6		165.1	10.1	19.2
1A	1850	219.4	5.0	14.0
1B		322.6	6.3	10.1
2A		179.6	3.9	12.0
2B		216.5	6.5	10.9
3A		235.4	5.9	9.0
3B		332.9	3.5	0.1
4		169.3	4.6	12.0
4		192.5	5.0	4.4
5A		238.7	4.2	8.8
5B		349.5	5.5	7.5
6		314.8	6.9	10.0
6		260.5	5.0	12.0

Table 5. (continued next page)

Table 5. (continued)

Bar No.	Solution temp, F*	Life, hr	Elong, %	RA, %
1A	1900	193.9†	Notch break	
1A		224.6†	2.3	7.8
1B		2.2	Notch break	
2A		224.4	4.1	12.0
2B		1.6	Notch break	
3A		15.8†	Notch break	
3A		15.9†	4.5	12.0
3B		1.0	Notch break	
5A		430.5	4.2	12.0
5B		8.6†	Notch break	
5B		8.7†	5.3	10.8
Typical Specification				
1750/ 1800		25 min	5 min	—

* 1 Hr/A.C., all bars aged 1325 F/8 hr/cool 100 F/hr to 1150 F/8 hr/A.C.
† Failed in the notch first; smooth section of bar rehanged to failure.

range of uniform grain structures. Obviously, this was not accomplished.

Bar No. 1, which received a final areal reduction of about 34% from a 1900 F furnace temperature, had a grain size predominantly ASTM ½. Bars 2 and 5 were predominantly ASTM ½ with small amounts (about 20% or less) of fine grain. "Fine grain" refers to a microstructure finer than ASTM 10. Bar No. 6 was inter-

Table 6. 1100 F/135 ksi Stress Rupture Properties of Hot Rolled 718 Alloy Bars

Bar No.	Solution temp, F*	Life, hr	Elong, %	RA, %
1A	1750	170.9	6.7	14.0
1B		186.9	5.5	10.0
2A		182.6	6.9	12.0
2B		205.4	6.9	15.2
3A		163.8	8.4	21.3
3B		111.5	8.3	19.2
5A		91.6	4.8	10.8
5B		145.3	7.4	13.2
1A	1850	19.2	Notch break	
1B		22.8	Notch break	
2A		7.6	Notch break	
2B		8.9	Notch break	
3A		11.2†	Notch break	
3A		82.3†	2.3	10.0
3B		9.2	Notch break	
5A		14.4†	Notch break	
5A		17.3†	3.3	4.4
5B		8.1	Notch break	
(No Applicable Specification.)				

* 1 hr/A.C., all bars aged 1325 F/8 hr/cool 100 F/hr to 1150 F/8 hr/A.C.
† Failed in notch first; smooth section of bar rehanged to failure.

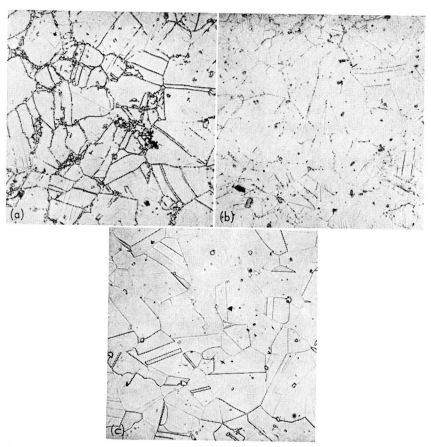

Fig. 7. Optical micrographs of samples from bar 1A, showing solution treated microstructures. All samples treated 1 hr/A.C. at the temperatures indicated. (a) 1725 F; (b) 1800 F; (c) 1900 F. Etchant, glyceregia; 250×.

mediate with about 40% fine grain. Conversely, bars 3 and 4 were predominantly (70% or more) fine grained.

Mixed structures are relatively common in hot worked 718 alloy. It is well known that in the forging of 718 alloy, large amounts of deformation (75-90%) from a relatively high hot working temperature (>2000 F) will produce an essentially uniform structure. In the present work, the data from the Initial Processing section of Table 2 confirm, for hot rolled bars, that large amounts of deformation from relatively high hot working temperatures produce relatively fine, uniform microstructures. Other experience at Carpenter (4) indicates that for such a process, finish temperatures (optical) in the range of 1800-1850 F are ideal to produce uniform structures in the range ASTM 7/7.5. For such a process and finish forging

Table 7. 1300 F/75 ksi Stress Rupture Properties of Hot Rolled 718 Alloy Bars

Bar No.	Solution temp, F*	Life, hr	Elong, %	RA, %
1A	1750	90.4	15.6	28.9
1B		146.9	14.6	27.3
2A		99.7	16.6	37.3
2B		86.8	16.9	37.3
3A		33.2	27.9	46.7
3B		31.8	29.7	51.4
5A		77.1	16.1	30.1
5B		104.5	14.3	30.1
1A	1850	255.7	8.3	18.0
1B		215.5	8.0	18.0
2A		197.8	9.1	17.2
2B		213.5	9.4	18.0
3A		236.3	6.2	15.2
3B		191.3	8.0	15.2
5A		206.2	10.3	19.2
5B		247.5	9.8	20.1
Typical Specification				
1750/ 1800		23 min	5 min	—

* 1 hr/A.C., all bars aged 1325 F/8 hr/cool 100 F/hr to 1150 F/8 hr/A.C.

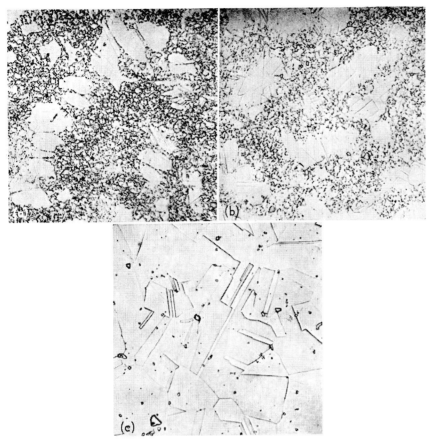

Fig. 8. Optical micrographs of samples from bar 3A, showing solution treated microstructures. All samples treated 1 hr/A.C. at the temperatures indicated. (a) 1725 F; (b) 1800 F; (c) 1900 F.

Table 8. Estimated ASTM Grain Sizes of 718 Alloy Hot Rolled Bars*

Bar No.	As-rolled	Typical for bars heat treated 1725-1800 F	1850 F	1900 F
1A	4/5	5/6	5/6	4/5
1B	4/5	6/7	5/6	
2A	5/6 (95%) + 10⁺ (5%)	6/7 (95%) + 10⁺ (5%)	5/6	4/5
2B	5/6 (90%) + 10⁺ (10%)	6/7 (90%) + 10⁺ (10%)	5/6	
3A	5/6 (30%) + 10⁺ (70%)	6/7 (30%) + 10⁺ (70%)	5/6	4/5
3B	5/6 (20%) + 10⁺ (80%)	6/7 (30%) + 10⁺ (70%)	5/6	
4	5/6 (20%) + 10⁺ (80%)	6/7 (20%) + 10⁺ (80%)	5/6	
5A	5/6 (90%) + 10⁺ (10%)	6/7 (95%) + 10⁺ (5%)	5/6	4/5
5B	5/6 (80%) + 10⁺ (20%)	6/7 (80%) + 10⁺ (20%)	5/6	
6	5 (60%) + 10⁺ (40%)	6/7 (70%) + 10⁺ (30%)	5/6	

* 10⁺ = Grains finer than ASTM 10. Numbers in parentheses are percentages of the grain sizes present in the structures.

temperatures of 1750 F and below, however, a uniform structure is more difficult to obtain. Also, when finish temperatures are below about 1750 F, the average grain size often becomes so fine that the stress-rupture life requirements of the usual aircraft materials specifications are difficult to meet.

The present results indicate that given a relatively uniform starting structure, a relatively smaller amount of reduction (<50%) from a moderate furnace temperature (1900-2000 F), will also produce a uniform or nearly uniform microstructure in rolled bars of 718 alloy. A hot rolling reduction of about 34% from a

1900 F furnace, finishing at an optical temperature below 1750 F, produced the most uniform structures (ASTM 4/5). However, when heated at 1725 F or above, this recrystallized to a fairly uniform ASTM 6/7. These results agree with recent data published in the literature (1,2) for forgings of 718, and some unpublished work at Carpenter (4) which shows that, depending upon prior working sequences, deformations in the range of 8-30% from moderate furnace temperatures (ideally 1800-1900 F) are most desirable as final hot working steps to produce a 718 product with a uniform fine grain size. For processes such as these, involving relatively low furnace temperatures, it is also known that prior hot working steps have a significant effect on final grain sizes.

In general, it is felt (4) that the lower the hot working temperature, the lower the "optimum" amount of deformation that is needed to produce an essentially uniform microstructure. For example, for an 1800 F furnace temperature, 8-10% may be satisfactory, while at 1900 F, 20-25% may be better. Conversely, as hot working temperature increases, larger amounts of deformation are necessary to produce uniform microstructures. For example, 75% reduction may be adequate for 2000 F, but 90% may be necessary for 2050 F. Final grain size will also be related to the finish hot working temperature (2).

Heat Treated

Grain size results on all of the heat treated bars showed no noticeable changes as solution treatment temperature increased from 1725 to 1800 F. Thus, for these temperatures, only typical heat treated grain size results for each bar are given in Table 8. Grain sizes are generally mixed for treatments at 1800 F and below. At 1850 F, all bars had grains in the range ASTM 5/6. At 1900 F, the bars examined had grains in the range ASTM 4/5.

Grain size comparisons of as-hot-rolled vs heat treated bars show that the relatively coarser grained regions in all of the as-rolled bars recrystallize due to 1-hr treatments at 1725 to 1900 F. This effect can be seen by comparing Fig. 5a and 5b. Recrystallization within warm worked grains of 718 can occur as low as about 1700 F, since this is the limit of the gamma-prime solvus band for usual heats of this alloy (4). However, grain growth is inhibited by Ni_3Cb platelets which form in the boundaries of the original as-hot-worked grains. Typical grain size data for each bar, Table 8, show that this leads to a finer reading on the "coarser" grains in each bar for the heat treated vs the as-hot-rolled structure.

Detailed, optical metallographic studies of the present bars were confined to representative examples of the various conditions observed. Bars 1A and 2A (34:1 and 48:1 final rolling reduction) and bars 3A and 4 (58:1 and 65:1 final rolling reduction) were chosen to represent the relatively coarser and finer grained bars, respectively. The micrographs in Fig. 6 show the fully aged microstructures of these bars after 1725 F solution treatment. These micrographs are also typical of the structure of these bars for solution treating to about 1800 F.

Figures 7 and 8 show the optical microstructures of bars 1A and 3A, respectively, as a function of solution treating temperature. No changes in grain size are observed to 1800 F; however, after solution treatment at 1900 F (Fig. 7c and 8c) and 1850 F (not shown), the structures of these bars are uniformly coarser than their as-rolled structures. As will be shown below, the grain boundary and matrix precipitates in 718 alloy dissolve as solution temperature is increased, thereby eventually permitting grain growth.

Properties Versus Optical Microstructure

As indicated above, no systematic correlation between properties and processing variables was established in this work. In general, the properties of all of the bars followed similar trends related primarily to solution treatment temperature. However, a closer look at the rupture data

Table 9. Semi-Quantitative Extractions and XRD Results on Hot Rolled and Heat Treated 718 Alloy Bars

Bar No.	Heat treatment	Wt % residue	X-RD Analysis*
1A	1725 F/1 hr/A.C.	1.34	CbC(S) + Ni₃Cb(S)
1A	1750 F/1 hr/A.C.	1.10	CbC(S) + Ni₃Cb(MS)
1A	1775 F/1 hr/A.C.	1.11	CbC(S) + Ni₃Cb(M)
1A	1800 F/1 hr/A.C.	0.90	CbC(S) + Ni₃Cb(M)
1A	1850 F/1 hr/A.C.	0.59	CbC
3A	1725 F/1 hr/A.C.	2.67	CbC(S) + Ni₃Cb(S)
3A	1750 F/1 hr/A.C.	2.45	CbC(S) + Ni₃Cb(MS)
3A	1775 F/1 hr/A.C.	1.79	CbC(S) + Ni₃Cb(M)
3A	1800 F/1 hr/A.C.	1.11	CbC(S) + Ni₃Cb(M)
3A	1850 F/1 hr/A.C.	0.62	CbC

Legend: S—Strong, M—Medium, MS—Medium-strong.

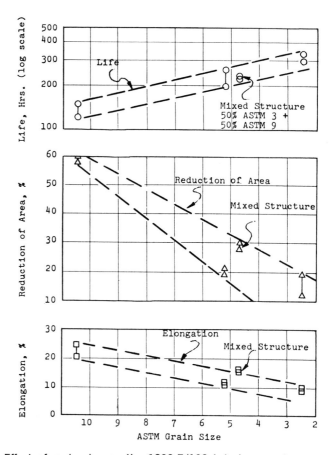

Fig. 9. Effect of grain size on the 1200 F/100 ksi stress-rupture properties of forged sections of a typical 718 alloy heat. Heat treated: 1750 F/1 hr/A.C. plus 1325 F/8 hr/cool 100 F/hr to 1150 F/8 hr/A.C.

in Table 5 will show that bars 3A and 3B are among those with the lowest average rupture lives for solution treatments of 1725, 1750 and 1775 F as compared to the other bars. This trend disappeared for solution treatment at 1800 F and above. Relationships between a few bars and tensile properties were also observed, but these are felt to be relatively unimportant compared to the stress rupture effects.

The stress-rupture data in Table 7, for tests at 1300 F/75 ksi, again seem to indicate that bars 3A and 3B are among those with the lowest average rupture lives and highest average rupture ductilities of all the bars tested, when heat treated at 1750 F.

It is well known that grain size has a pronounced effect on the stress-rupture properties of 718. Data on sections of a typical Carpenter heat, forged to various grain sizes, are shown in Fig. 9 as an example. Reference to Fig. 6 and/or Table 8 will indicate that grain size effects could probably be used as a sufficient reason to explain the behavior of bars 3A and 3B. However, it is also well known that grain size effects are interrelated with fine structure effects.

Based upon these observations, it was decided to confine the more extensive electron metallography in this work to bar 3A as a representative of the finer grained bars and bar 1A as a representative of the coarser grained bars.

Electron Microstructure

Replica Micrographs

For 718 alloy, the phases of interest are the gamma-prime, Ni_3Cb and the various carbides. Figures 10 and 11 show electron micrographs made from chromium shadowed parlodion replicas obtained from bars 1A and 3A, respectively, after a heat treatment of 1725 F/1 hr/A.C. + 1325 F/cool 100 F/hr to 1150 F/8 hr/A.C. The age hardening gamma-prime precipitate is seen in the matrix of both. For the present studies, gamma-prime structure varied relatively little, thus, further comments on fine structure will be confined to Ni_3Cb and carbides. The grain boundaries are heavily precipitated, particularly in the fine grained areas. Grain size data show bar 3A to contain many fine grained areas, while bar 1A contains essentially none (compare Fig. 7 and 8). It follows then that bar 3A has many more heavily precipitated grain boundaries than bar 1A. The major phase precipitated in these grain boundaries, for solution temperatures of 1800 F and below, is Ni_3Cb. As will be shown later, the Ni_3Cb particles in the grain boundaries of bar 3A also appear to be finer than those in bar 1A.

The effect of high solution treating temperatures can be seen in Fig. 12.

Table 10. Comparison of 1200 F/100 ksi Stress-Rupture Elongation Values With Wt % Ni₃Cb for Relatively Coarser Grained (Bars 1A and 1B) and Relatively Finer Grained (Bars 3A and 3B) Hot Rolled 718 Alloy Bars

| Wt % Ni₃Cb* | | 1200 F/100 ksi Elongation | |
Bar 3A (finer grained)	Bar 1A (coarser grained)	Bars 3A & 3B (finer grained)	Bars 1A & 1B (coarser grained)
2.07		15.4-18.3	
1.85		18.5-20.0	
1.19		11.2-11.9	
	0.74		8.4-12.9
0.51		9.9-10.3	
	0.51		13.3-15.5
	0.50		12.4-12.7
	0.30		4.8- 7.1
0		3.5- 5.9	
	0		5.0- 6.3

* Wt % Ni₃Cb = (Wt % Residue − 0.60)

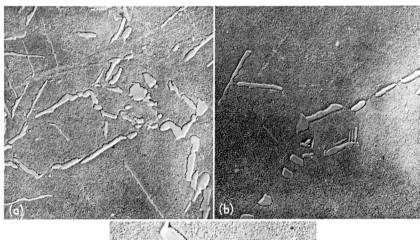

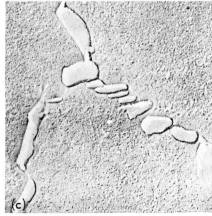

Fig. 10. Electron micrographs from sample of bar 1A, solution treated at 1725 F/1 hr/A.C. plus fully aged. Chromium shadowed parlodion replicas, glyceregia etchant; (a) 7700×; (b) 7700×; (c) 15400×.

For both bars, samples treated at 1800 F followed by aging show many clean grain boundaries due to solutioning of Ni_3Cb as solution temperature is increased. Comparing Fig. 10 and 11 with Fig. 12 shows that Ni_3Cb platelet size also decreases as solution treating temperature is increased from 1725 to 1800 F. For 1900 F solution treating, all grain boundaries become clean except for occasional Cb(C,N) particles.

Extraction Micrographs

The grain boundary and matrix precipitates were extracted on carbon replicas and identified via selected area electron diffraction analysis. Samples from bars 1A and 3A solution treated at 1725 to 1850 F, each followed by the full aging treatment were used.

The following observations are made from the study of numerous extraction replicas and electron diffraction analysis; examples of the structures observed are given in Fig. 13 and 14:

1. The needle-shaped precipitates, both in the matrix (and the grain boundaries at times) of the fine grained areas of both bars, are identified as Ni_3Cb.

2. The replicas from bar 3A showed considerably more of these needles and as a result, more fine grained areas.

3. Increasing solution treating temperature from 1725 to 1750 F for bar 1A seemed qualitatively to have coarsened and partially dissolved the Ni_3Cb needles (Fig. 13a and 13b). Coarsening and dissolution of Ni_3Cb was not evident in bar 3A for 1750 F treating (Fig. 13c and 13d). This effect is attributed to the heavier initial precipitation of Ni_3Cb in bar 3A.

3. For bar 3A, appreciable dissolution of Ni_3Cb occurred only at 1775 F and above. For this bar, no coarsening of Ni_3Cb was noted for any of the treatments studied.

4. The boundaries of the coarser grains contained carbides identified as M_6C type. The morphology of these carbides can be seen in Fig. 14. Since bar 1A contained more coarse grains than 3A, M_6C was more readily seen in bar 1A, Fig. 14. It was also noted that the M_6C in bar 1A coarsened with increasing solution temperature (Fig. 14a and 14b). This was not true of bar 3A (Fig. 14c and 14d).

5. Structural examination also showed that the morphology of the M_6C carbides in bar 1A for treating at 1725 F was very well defined compared to that of the carbides in bar 3A under identical conditions.

6. The structure of bar 2A was very similar to that of bar 1A. Such a similarity was also found between bars 3A and 3B under identical treating conditions.

Properties Versus Electron Microstructure

The above structural observations can be used to partially explain the slightly lower rupture lives for bars 3A and 3B, compared to the other bars, after 1725, 1750, or 1775 F solution treatment. The presence of Ni_3Cb needles is essential in that it controls the grain size. However, it is believed (based on our structural observations) that there is an optimum quantity, platelet size and dispersion of this phase for best properties. This hypothesis is partially supported by the slight improvement in stress-rupture elongation for several bars (e.g. 1A, 1B, 2B, see Table 5) as solution treatment temperature is increased from 1725 to 1750 F. As shown above, in bars 1A and 2A, a dissolution accompanied by coarsening of Ni_3Cb platelets was quite evident after 1750 F solution treatment. A similar effect was not observed in bars 3A or 3B at 1750 F, but only at 1775 F and above.

Another interesting effect can be seen by studying Fig. 15. These plots show the general result that the finer grained bars, such as 3A and 3B, have higher than average 1200 F/100 ksi stress-rupture ductilities for 1725 and 1750 F solution treating; but that for 1775 F solution treating, elongations fall rapidly to slightly below average. Meanwhile, the coarser grained bars, such as 1A and 1B, have

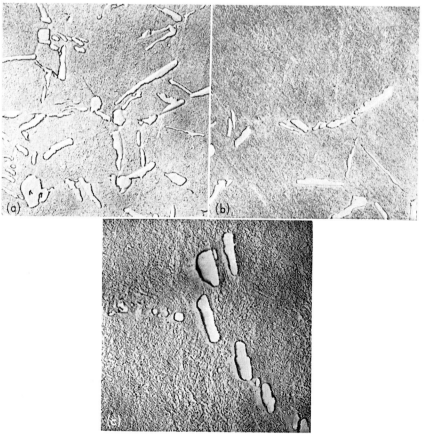

Fig. 11. Electron micrographs from sample of bar 3A, solution treated at 1725 F/1 hr/A.C. plus fully aged. Chromium shadowed parlodion replicas, glyceregia etchant; (a) 7700×; (b) 7700×; (c) 15400×.

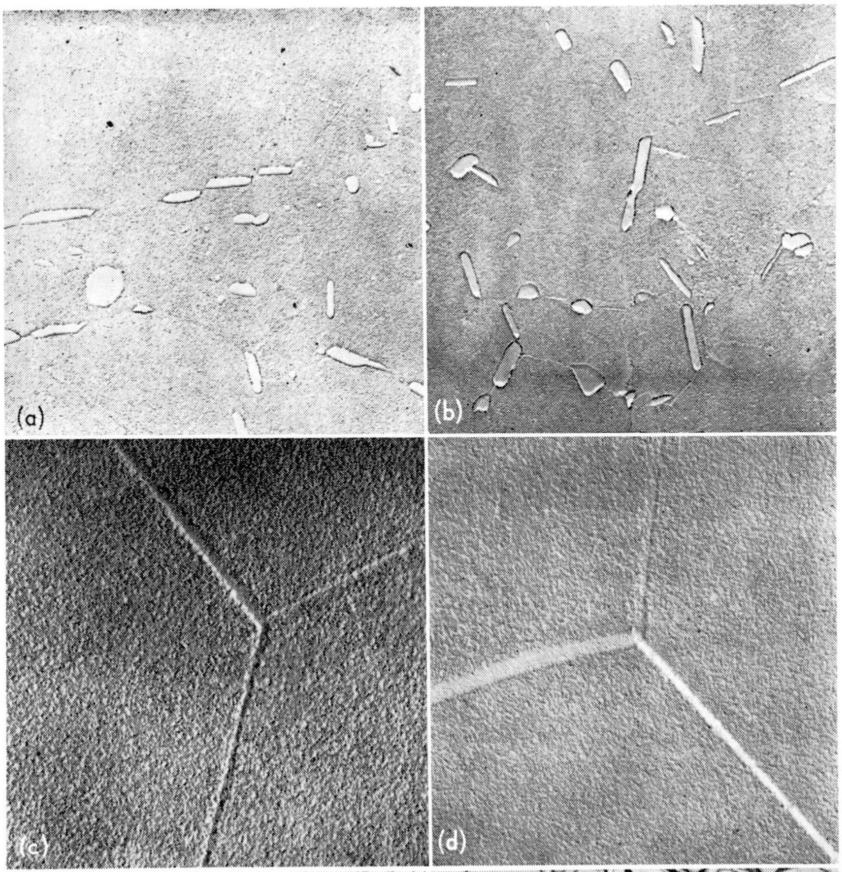

Fig. 12. Electron micrographs from samples of bar 1A, (a and c) and bar 3A, (b and d). Samples heat treated as follows: (a and b) 1800 F/1 hr/A.C. plus fully aged; (c and d) 1900 F/1 hr/A.C. plus fully aged. Chromium shadowed parlodion replicas, glyceregia etchant; (a) 7700×; (b) 7700×; (c) 15400×; (d) 15400×.

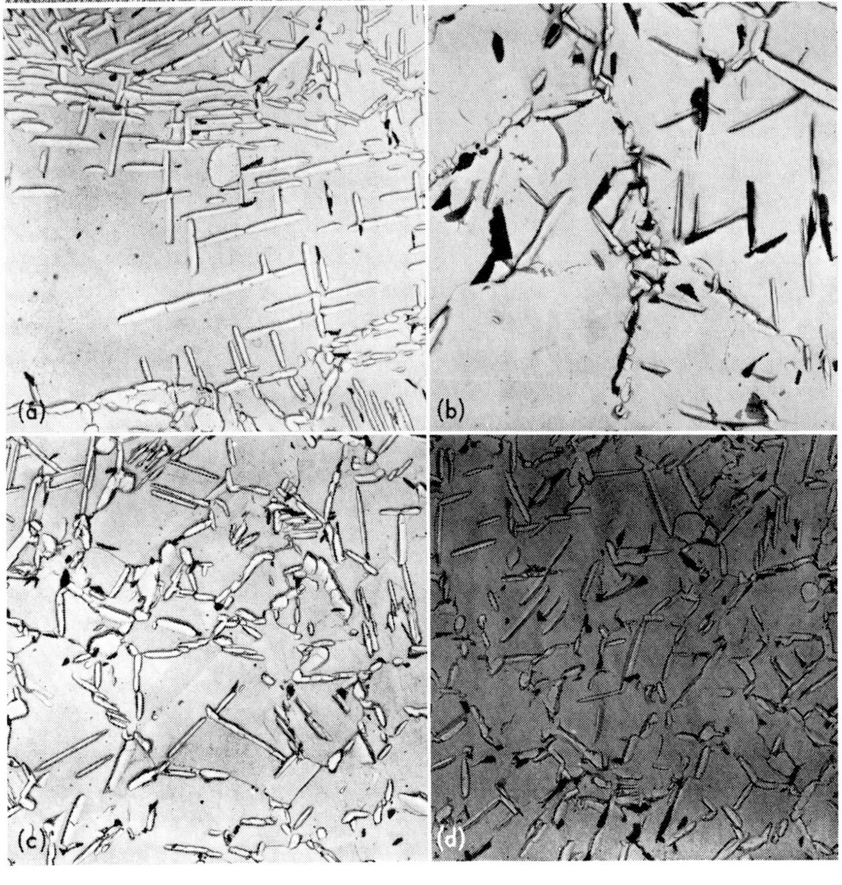

Fig. 13. Extraction replica electron micrographs from samples of bar 1A, (a and b) and bar 3A, (c and d) showing Ni$_3$Cb dispersion. Samples heat treated as follows: (a and c) 1725 F/1 hr/A.C. plus fully aged; (b and d) 1750 F/1 hr/A.C. plus fully aged. Etchant, 50 HCl, 48 H$_2$O, 2 H$_2$O$_2$; 2800×.

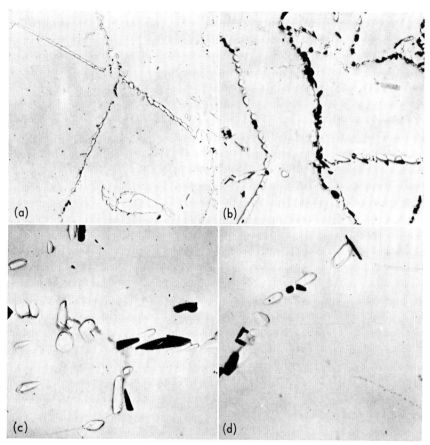

Fig. 14. Extraction replica electron micrographs from samples of bar 1A, (a and b), and bar 3A (c and d) showing grain boundary M₆C carbide distribution. Samples heat treated as follows: (a and c) 1725 F/1 hr/A.C. plus fully aged; (b and d) 1750 F/1 hr/A.C. plus fully aged. Etchant, 50 HCl, 48 H₂O, 2 H₂O₂; (a) 1800×; (b) 1800×; (c) 7700×; (d) 7700×.

the lower 1200 F/100 ksi stress-rupture ductilities for 1725 and 1750 F solution treating, but don't show an appreciable loss in elongation until an 1800 F solution temperature is reached. Figure 15 also shows the life advantage of the coarser grained bars.

Comparing the above ductility effect with the variations in incipient dissolution and coarsening temperatures of Ni₃Cb for bar 1A (1750 F) and bar 3A (1775 F) again leads to the conclusion that the state of the Ni₃Cb in 718 can be optimized for best properties. Obviously, both hot working sequences and heat treatments must be closely controlled.

The above results are further supported by quantitative extractions and XRD results to be discussed below.

Structures of samples solution treated at 1850 or 1900 F, plus fully

aged, show relatively clean grain boundaries with occasional Cb(C,N); a few M₆C type carbides are found in the 1850 F samples, only. These samples, of course, showed the poorest stress-rupture properties; some being quite notch sensitive. It is interesting that no bars showed notched stress-rupture failures at 1200 F/100 ksi for an 1850 F solution treatment but that many did for a 1900 F solution treatment. This difference is tentatively attributed to the presence of M₆C in the grain boundaries for 1850 F solution treating which inhibited grain growth.

Quantitative Extractions and X-RD Results

To verify the above structural observations, semiquantitative extractions were run on samples from bars 1A and 3A. These sections were solu-

tion treated for 1 hr/A.C. at temperatures from 1725 to 1850 F, and were not aged. The results of this work are shown in Table 9. As can be seen, under identical heat treatment conditions, samples from bar 3A show a considerably higher weight percentage of residue than the samples from bar 1A.

The assumption that the amount of Cb(C,N) is constant in these samples, leads to the conclusion that bar 3A does contain more Ni₃Cb, Fig. 16. This assumption is validated by similar quantities of Cb(C,N) seen in both bars after 1850 F treatment (0.59 and 0.62 wt %, respectively). Absence of M₆C in the X-ray diffraction results is attributed to the considerably lower over-all quantity of this carbide in the total residue.

A comparison of 1200 F/100 ksi stress-rupture elongation values with "corrected" residue values, or weight percentage of Ni₃Cb, as in Table 10, suggests correlation between ductility and the weight percentage of Ni₃Cb. These data, plotted in Fig. 17, also suggest that the optimum of weight percentage of Ni₃Cb (for best stress-rupture properties) varies with the grain size. For example, highest ductility was obtained at about 1.85 wt % Ni₃Cb for relatively finer grained bar 3A. Conversely, bar 1A showed highest stress-rupture elongation at about 0.51 wt % Ni₃Cb. Both these values were less than the maximum weight percentages of Ni₃Cb observed among the samples studied. This relationship between optimum weight percentages of Ni₃Cb and grain size is probably a direct result of the greater volume fraction of grain boundaries for a finer grained material as compared to a coarser grained material. Quantitative studies of this effect were precluded by the mixed structures.

Conclusions

The following conclusions are reached as a result of a study of the structure and properties of hot rolled and heat treated 718 alloy bars:

1. Warm worked grains in hot rolled bars of 718 alloy recrystallize due to 1 hr treatments

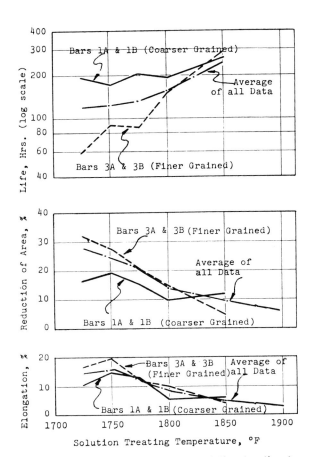

Fig. 15. Relationship of rupture properties to solution treating temperature for relatively coarser grained (bars 1A and 1B) and relatively finer grained (bars 3A and 3B) hot rolled 718 alloy bars. (Fully heat treated.)

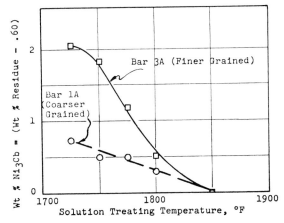

Fig. 16. Effect of solution treating temperature on weight percentage of Ni_3Cb in relatively coarser grained (bar 1A) and relatively finer grained (bar 3A) hot rolled 718 alloy bars.

at 1725 F and above. This effect is believed to be controlled by γ' solvus temperature, about 1700 F for most heats of 718 alloy.

2. Grain growth does not occur in hot rolled 718 alloy bars until a temperature of about 1850 F is reached. For solution temperatures of 1800 F and below, Ni_3Cb appears to control grain size. For temperatures above 1800 F, M_6C or other carbides control grain size.

3. The structures of the rolled bars evaluated, for the most part, were relatively fine and of mixed grain sizes. This may account for the lack of any significant correlation between properties and processing variables.

4. All R.T. tensile strength properties decreased, while ductility values increased, as the solution temperature was increased from 1725 to 1850 F. A similar trend was observed for 1200 F tensile strength data, while 1200 F tensile ductility values did not vary appreciably with solution treating temperature.

5. Stress-rupture life increased and ductility decreased with increasing solution treating temperature.

6. A detailed fine structure study of fully heat treated, representative coarse and fine grained bars, showed the grain boundaries to be heavily precipitated with a needle-shaped phase, particularly in the fine grained areas.

7. This needle-shaped precipitate, identified as Ni_3Cb by electron diffraction, appeared to coarsen as solution treating temperature increased from 1725 to 1750 F in the coarser grained bars, but not in the finer grained bars. Coarsening of Ni_3Cb was not noted in any of the fine grained bars for the heat treatments studied.

8. The 1750 F solution temperature also resulted in a considerable degree of solutioning of Ni_3Cb in the coarser grained bars but not in the finer grained bars.

9. The dissolution of Ni_3Cb occurred only for solution treating temperatures of 1775 F and above for the finer grained bars.

10. The boundaries of the coarser grains showed M_6C type carbides. Thus, M_6C was more readily seen in the coarser grained bar than in the finer grained bar.

11. It was postulated, based on the electron metallographic studies, that there is an optimum quantity, size and dispersion of Ni_3Cb for best properties of hot rolled bars of 718 alloy. These optimums will, of course, vary with the average grain size of the bar.

12. This postulate was partially confirmed by quantitative extractions of phases and XRD analyses. It was found that the optimum quantity of Ni_3Cb increased with decreasing grain size.

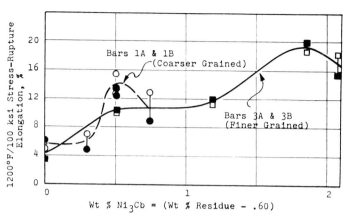

Fig. 17. Effect of weight percentage of Ni_3Cb on 1200 F/100 ksi stress-rupture elongation in relatively coarser grained (bars 1A and 1B) and relatively finer grained (bars 3A and 3B) hot rolled 718 alloy bars.

ACKNOWLEDGMENTS

The authors acknowledge their gratitude to the management of The Carpenter Technology Corporation for permission to publish this work. Thanks are also due to various colleagues at the Research and Development Center who contributed to the experimental work and discussion; particularly to C. R. Whitney for permission to report some of his unpublished work.

REFERENCES

1. D. M. Gadsby, Forging and Solution Treating Alloy 718, Metal Progress (December, 1966) 85-88.
2. J. P. Stroup and R. A. Heacox, Effect of Grain Size Variations on the Long-Time Stability of Alloy 718. Presented at the International Symposium on Structural Stability in Superalloys, Champion, Pa., September 4-6, 1968.
3. H. L. Eiselstein, Metallurgy of a Columbium-Hardened Nickel-Chromium-Iron Alloy, ASTM STP 369, Advances in the Technology of Stainless Steels and Related Alloys, ASTM, (1964) 62-79.
4. Unpublished work, Carpenter Technology Corp.
5. H. L. Eiselstein, U.S. Patent #3,046,-108 (July 24, 1962).

Strengthening Mechanisms in Nickel-Base Superalloys

R. F. DECKER
Paul D. Merica Research Laboratory
The International Nickel Company, Inc.
Sterling Forest, Suffern, New York

Presented at

Steel Strengthening Mechanisms Symposium
Zurich, Switzerland, May 5 and 6, 1969

Sponsored by

Climax Molybdenum Company
A Division of Amax

ABSTRACT

Strengthening mechanisms in nickel-base super-alloys are reviewed and related to composition. The role of the γ matrix and its solid-solution hardeners is weighed. Since γ' plays a predominant role in these alloys, particular attention is given to this phase. The alloy hardening due to γ' is broken down into components due to anti-phase boundary and fault hardening, γ' strength, coherency strains, volume percent γ', particle size of γ', diffusivity in γ' and γ–γ' modulus mismatch. Recent experiments are used to weigh these factors.

Attention is directed toward the recent marked advances in understanding these alloys through phase stability considerations. A modification in electron vacancy calculations is suggested. The current state of knowledge on carbide reactions is summarized. Composition and heat treatment are found to influence the type of carbide formed. Control of grain-boundary stability, thickness/grain size ratio and grain orientation are shown to be important to strength.

Source: *Strengthening Mechanisms in Nickel-Base Superalloys*, International Nickel Co., Inc.

275

INTRODUCTION

Nickel-base superalloys, although known to man but four decades, have assumed an essential role in our advanced transportation and power technology. Applications of these alloys, as components of aircraft, automotive and land-based power systems (Table I) have allowed higher operating temperatures and more efficient operation. It is notable that superalloys make up 50% of the weight of an advanced aircraft gas turbine. The severe aerodynamic heating regimes of space vehicles have been mastered by use of the same alloys in sheet form. Realizing that superalloys are known for their complexity, it comes as no surprise to find that over 40 grades† are in use today. Most are variations of the basic nickel-chromium matrix, hardened by γ′ [Ni₃(Al,Ti)], but with further optional additions of cobalt, iron, tungsten, molybdenum, vanadium, niobium, tantalum, boron, zirconium, carbon and magnesium.

These grades have evolved empirically. However, in the last 15 years, the understanding of mechanisms has reached such a useful level that Professor Bruce Chalmer's term "enlightened empiricism" can now be used to describe the intensive worldwide research and development of superalloys. It is the purpose of this

†See Table II for compositions cited in this paper.

paper to discuss this state of enlightenment on strengthening mechanisms in superalloys. In detail, the analysis will consider the γ matrix, γ′, phase stability, carbides, grain boundaries, grain size and grain orientations.

THE γ MATRIX

Although the nickel base is not endowed with distinctly high modulus of elasticity or low diffusivity, the two basic creep-resisting factors, the matrix is favored by designers for the most severe temperature and time excursions in air atmospheres. Figure 1 illustrates the fact that nickel increases temperature capa-

TABLE I   Uses of Nickel-Base Superalloys

GAS TURBINE	RECIPROCATING ENGINES
discs	superchargers
bolts	exhaust valves
shafts	hot plugs
cases	
vanes	SPACE VEHICLES
blades	aerodynamically
burner cans	heated skins
afterburner	
thrust reversers	STEAM POWER
	bolting

METAL WORKING
hot-work tools and dies

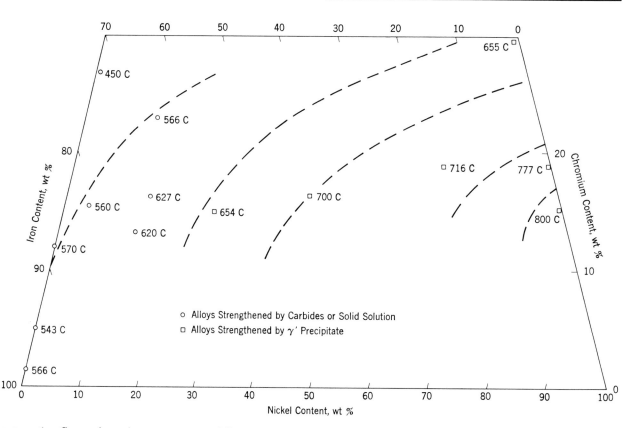

FIGURE 1.   Comparison of temperature capability of some of the strongest available alloys in terms of their basic compositions. Temperature capability is the temperature for rupture in 1000 h under a stress of 31.5 kgf/mm². Richards and Decker[1]*

* See references.

TABLE II Nominal Composition of Superalloys

Alloy Designation	C	Mn	Si	Cr	Ni	Co	Mo	W	Nb	Fe	Ti	Al	B	Zr	Other	C†
Cast Alloys																
MC-102 *	0.04	—	—	20.0	bal	—	6.0	2.5	6.5	—	—	—	—	—	—	850
**	0.20	—	—	23.4	67.4	—	3.8	0.8	4.3	—	—	—	—	—	—	
GMR 235-D *	0.15	—	—	15.5	bal	—	5.0	—	—	4.5	2.5	3.5	0.050	—	—	927
**	0.69	—	—	16.5	65.0	—	2.9	—	—	4.5	2.9	7.2	0.256	—	—	
MAR-M 421 *	0.15	—	—	16.0	bal	9.0	2.0	3.8	2.0	—	1.8	4.3	0.015	0.05	—	975
**	0.70	—	—	17.4	58.5	8.6	0.08	1.2	1.2	—	2.1	9.0	0.078	0.03	—	
Alloy 713LC *	0.05	—	—	12.0	bal	—	4.5	—	2.0	—	0.6	5.9	0.010	0.10	—	980
**	0.23	—	—	12.7	70.4	—	2.6	—	1.2	—	0.7	12.1	0.050	0.06	—	
Alloy 713C *	0.12	—	—	12.5	bal	—	4.2	—	2.0	—	0.8	6.1	0.012	0.10	—	985
**	0.55	—	—	13.2	69.3	—	2.4	—	1.2	—	0.9	12.4	0.060	0.06	—	
IN-738 *	0.17	—	—	16.0	bal	8.5	1.7	2.6	0.9	—	3.4	3.4	0.010	0.10	1.8Ta	901
**	0.80	—	—	17.5	59.4	8.2	1.0	0.8	0.5	—	4.0	7.1	0.052	0.06	0.6Ta	
B-1900 *	0.10	—	—	8.0	bal	10.0	6.0	—	—	—	1.0	6.0	0.015	0.10	4.0Ta	1000
**	0.47	—	—	8.7	62.5	9.6	3.5	—	—	—	1.2	12.6	0.078	0.06	1.3Ta	
IN-100 *	0.18	—	—	10.0	bal	15.0	3.0	—	—	—	4.7	5.5	0.014	0.06	1.0V	1000
**	0.81	—	—	10.4	55.8	13.8	1.7	—	—	—	5.3	11.0	0.070	0.04	1.1V	
IN-731 *	0.15	—	—	10.0	bal	10.0	2.5	—	—	—	4.5	5.6	0.014	0.06	1.0V	1000
**	0.67	—	—	10.4	60.9	9.2	1.4	—	—	—	5.1	11.3	0.069	0.04	1.1V	
MAR-M 200 *	0.15	—	—	9.0	bal	10.0	—	12.5	1.0	—	2.0	5.0	0.015	0.05	—	1020
**	0.74	—	—	10.2	60.8	10.0	—	4.0	0.7	—	2.5	11.0	0.082	0.03	—	
Wrought Alloys																
A-286 *	0.05	1.35	0.50	15.0	26.0	—	1.3	—	—	bal	2.0	0.2	0.015	—	—	780
**	0.23	1.36	0.99	16.0	24.6	—	0.8	—	—	53.3	2.3	0.4	0.076	—	—	
Inconel alloy 718 *	0.04	0.20	0.30	18.6	bal	—	3.1	—	5.0	18.5	0.9	0.4	—	—	—	802
**	0.19	0.21	0.62	20.7	52.2	—	1.9	—	3.1	19.2	1.1	0.9	—	—	—	
IN-120 *	0.04	—	—	21.0	bal	14.0	4.0	—	2.0	—	2.5	0.25	0.005	0.05	—	802
**	0.19	—	—	23.4	55.4	13.8	2.4	—	1.3	—	3.0	0.54	0.026	0.03	—	
Incoloy alloy 901 *	0.05	0.10	0.10	12.5	42.5	—	5.7	—	—	bal	2.8	0.2	0.015	—	—	826
**	0.24	0.11	0.21	14.0	78.0	—	3.5	—	—	—	3.4	0.4	0.080	—	—	
Nimonic 80A *	0.06	0.10	0.70	19.5	bal	1.1	—	—	—	—	2.5	1.3	—	—	—	826
**	0.28	0.10	1.39	20.8	70.8	1.1	—	—	—	—	2.9	2.7	—	—	—	
Inconel alloy X-750 *	0.04	0.70	0.30	15.0	bal	—	—	—	0.9	6.8	2.5	0.8	—	—	—	835
**	0.19	0.26	0.60	16.3	70.7	—	—	—	0.5	6.9	2.9	1.7	—	—	—	
D-979 *	0.05	0.25	0.20	15.0	bal	—	4.0	4.0	—	27.0	3.0	1.0	0.010	—	—	843
**	0.24	0.53	0.41	16.7	44.6	—	2.4	1.3	—	28.0	3.6	2.1	0.053	—	—	
Nimonic 90 *	0.07	0.50	0.70	19.5	bal	18.0	—	—	—	—	2.4	1.4	—	—	—	843
**	0.32	0.50	1.38	20.9	54.3	17.0	—	—	—	—	2.8	2.9	—	—	—	
René 41 *	0.09	—	—	19.0	bal	11.0	10.0	—	—	—	3.1	1.5	0.005	—	—	881
**	0.43	—	—	21.2	54.6	10.1	6.0	—	—	—	3.8	3.0	0.026	—	—	
M-252 *	0.15	0.50	0.50	20.0	bal	10.0	10.0	—	—	—	2.6	1.0	0.005	—	—	884
**	0.72	0.53	1.03	22.2	54.4	9.8	6.0	—	—	—	3.1	2.1	0.026	—	—	
Waspaloy *	0.08	—	—	19.5	bal	13.5	4.3	—	—	—	3.0	1.3	0.006	0.06	—	894
**	0.38	—	—	21.3	56.3	13.1	2.5	—	—	—	3.6	2.8	0.031	0.04	—	
Inconel alloy 700 *	0.12	0.10	0.30	15.0	bal	28.5	3.7	—	—	0.7	2.2	3.0	—	—	—	905
**	0.57	0.10	0.61	16.6	48.1	27.8	2.2	—	—	0.7	2.6	0.6	—	—	—	
Nimonic alloy 105 *	0.20‡	1.0‡	1.0‡	14.6	bal	20.0	5.0	—	—	2.0‡	1.2	4.7	—	—	—	925
**	0.91	0.99	1.9	15.3	46.7	18.5	2.8	—	—	2.0	1.4	9.5	—	—	—	
Udimet 500 *	0.08	—	—	18.0	bal	18.5	4.0	—	—	—	2.9	2.9	0.006	0.05	—	927
**	0.37	—	—	19.4	51.0	17.6	2.3	—	—	—	3.4	6.0	0.030	0.03	—	
Nimonic alloy 115 *	0.15	—	—	15.0	bal	15.0	3.5	—	—	—	4.0	5.0	—	—	—	950
**	0.68	—	—	15.7	53.2	13.8	2.0	—	—	—	4.5	10.1	—	—	—	
Udimet 700 *	0.08	—	—	15.0	bal	18.5	5.2	—	—	—	3.5	4.3	0.030	—	—	960
**	0.37	—	—	16.0	50.3	17.4	3.0	—	—	—	4.0	8.8	0.153	—	—	

* Weight percent.   ** Atomic percent.
† Temperature capability (temperature for 100 h life at 14 kgf/mm²).   ‡ Maximum.

Source: *Strengthening Mechanisms in Nickel-Base Superalloys*, International Nickel Co., Inc.

bility. It is remarkable that these alloys are utilized to 0.8 T_M (melting point) and for times up to 100 000 h at somewhat lower temperatures. This endurance must be attributable to:

1. The high tolerance of nickel for alloying without phase instability due to its nearly filled 3d electron shell.

2. The tendency, with chromium additions, to form Cr_2O_3-rich protective scales having low cation vacancy content, thereby restricting the diffusion rate of metallic elements outward, and oxygen and sulfur inward.

Solid-Solution Hardening of γ

From phase analyses of complex superalloys by Mihalisin and Pasquine,[2] Kriege and Baris,[3] and Loomis,[4] we can list the solid-solution elements in γ as cobalt, iron, chromium, molybdenum, tungsten, vanadium, titanium and aluminum. As seen in Table III, these elements differ from nickel by 1 to 13% in atomic diameter and 1 to 7 in N_v, electron vacancy number.

Using the findings of Pelloux and Grant[5] and Parker and Hazlett,[6] hardening can be related to this atomic diameter oversize as measured by lattice expansion. However, an additional superimposed effect can be attributed to position in the periodic table (or N_v) (see Figure 2).

At least part of this extra effect may result from lowering of stacking fault energy. Such a correlation of N_v with the stacking fault measurements of Beeston, Dillamore and Smallman[7] and Beeston and France[8] is seen in Figure 2. The lowering of stacking fault energy by the alloying elements would make cross-slip more difficult in γ.

The potency of superalloy solid solution elements can be estimated. We shall select a highly alloyed superalloy with γ of composition†:[2,3]

%Co	%Fe	%Cr	%Mo	%W	%V	%Al	%Ti
20	10	20	4	4	1.5	6	1

From Pearson,[9] the change in lattice constants (kX)

† In atomic percent as throughout this paper unless otherwise specifically noted.

TABLE III   Periodic Table of Alloying Elements of Nickel-Base Superalloys

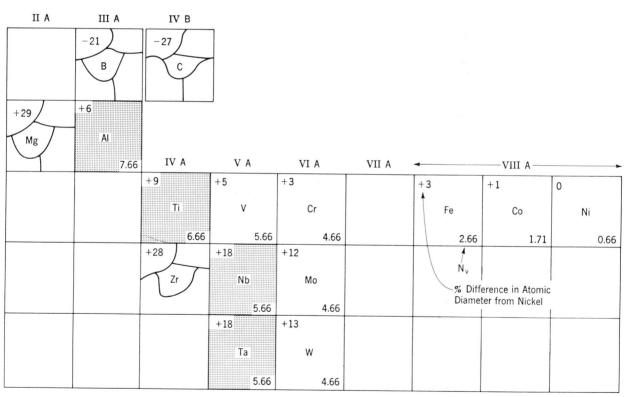

Atomic Diameter of Carbon, Boron, Zirconium, Magnesium – Goldschmidt for CN12
Atomic Diameter of Other Elements from Lattice Parameter Effect in Nickel Binary Alloys

Element Partitions to γ Element Partitions to γ' Element Partitions to Grain Boundary

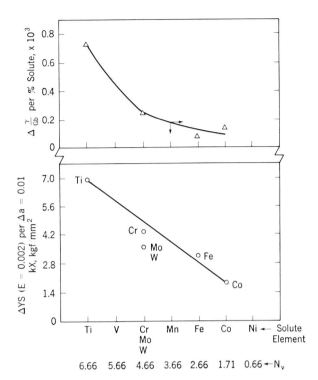

FIGURE 2. Effect of solute position in periodic table and solute N_v on nickel-base binary alloys
  a. change in stacking fault energy of nickel per percent solute[7,8]
  b. change in room-temperature yield strength per unit of lattice constant change[5]

for binary addition to the nickel content of this matrix would be:

Co	Fe	Cr	Mo	W	V	Al	Ti
0.011	0.020	0.033	0.035	0.038	0.006	0.025	0.006

From Figure 2, the change in flow stress at room temperature in kgf/mm², would be:

Co	Fe	Cr	Mo	W	V	Al	Ti
1.8	5.6	16	17	18	3.4	20**	4

Aluminum, usually noted only as a precipitation hardener, rates as a potent solid-solution hardener! Tungsten, molybdenum and chromium also contribute strongly, whereas iron, titanium, cobalt and vanadium serve only as weak solid solution hardeners. That the solid-solution hardening persists to 0.6 T_M (815 C) is supported by the data of Pelloux and Grant[5] in Table IV.

Above 0.6 T_M, the range of high temperature creep, γ strengthening is diffusion dependent. The slow-diffusing elements, molybdenum and tungsten, would be expected to be most potent hardeners. An additional beneficial effect on diffusion was shown by Pridantsev:[10] in a Ni–22%Cr–2.8%Ti–3.1%Al alloy, the presence of molybdenum and tungsten lowered the diffusivity of titanium and chromium at 900 C.

** From extrapolation to N_v of 7.66.

TABLE IV Strengthening Parameter for Alloys with 10 % Solute Element (Pelloux and Grant)[5]

	RT	Rupture Strength			
		650 C		815 C	
	YS	1 h	100 h	1 h	100 h
$\sigma 10\%Cr/\sigma Ni$*	1.5	3.0	2.6	2.0	2.0
$\sigma 10\%Mo/\sigma Ni$	2.4	3.1	3.6	2.4	2.4
$\sigma 10\%W/\sigma Ni$	2.6	3.2	4.6	3.0	2.7

* Ratio of the stress for a 10 at % Cr alloy to the stress for nickel

γ' PHASE, Ni$_3$(Al, Ti)

The precipitation of γ' is a most fortunate phenomenon. By nature its usefulness is limited to high-nickel matrices. First, the lack of compressibility of the nickel atom due to its 3d electron state favors precipitation of γ' rather than more complex phases where atomic size changes are needed. Higher N_v matrices such as iron favor these latter undesirable phases. Then, compatibility of crystal structure (fcc) and lattice constant ($\sim$0–1% mismatch) with γ allows homogeneous nucleation of precipitate with low surface energy and long time stability.

In its own right, γ' is a unique intermetallic phase. γ' has an ordered L1$_2$ structure contributing anti-phase boundary (APB) strengthening to the γ–γ' alloy. Remarkably, the strength of γ' increases as temperature increases. Furthermore, the inherent ductility of γ' prevents severe embrittlement should a massive morphology, such as a grain-boundary film, develop. This is in direct contrast to the severe embrittlement found with comparable grain-boundary morphology of the brittle sigma and Laves phases.

Equilibrium Diagrams

The $\gamma/\gamma + \gamma'$ boundaries of the nickel–chromium–iron–titanium–aluminum system in the aging and service range of 750 C are illustrated in Figure 3a. Replacing about 21% Ni by chromium decreases the solubility for aluminum and titanium. Replacing about 25% Ni in Ni$_3$Cr by iron further reduces solubility. In Figure 3b, we see that substitution of cobalt for nickel also reduces the solubility of the nickel–chromium matrix for aluminum and titanium. Thus, chromium, iron and cobalt can all be used as agents to increase volume percent γ' at a given aluminum plus titanium level.

The use of cobalt in superalloys to ease hot working at high aluminum plus titanium contents may be related to the increased solubility for γ' above 1100 C with cobalt addition. The schematic isothermal section of nickel–aluminum–ternary element alloys of Figure 4 points out how elements substitute and partition in γ'. Cobalt, with its horizontal phase field would substitute for nickel. Titanium, niobium, vanadium, and probably tantalum would substitute for aluminum posi-

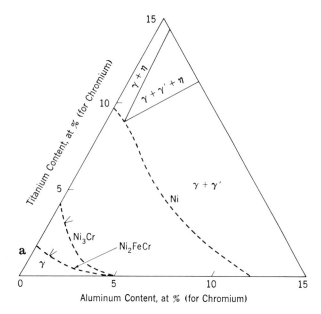

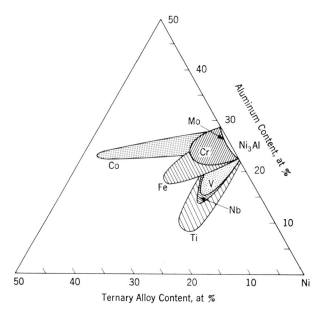

FIGURE 4. Semi-schematic presentation of Ni₃Al solid-solution field at approximately 1150 C for various ternary alloys. Guard and Westbrook;[13] Benjamin, Giessen and Grant[14]

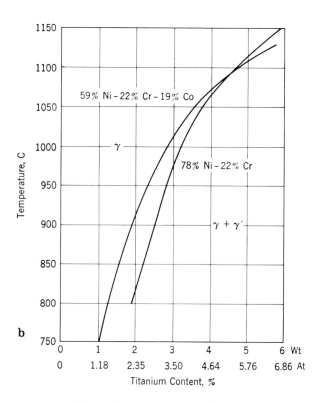

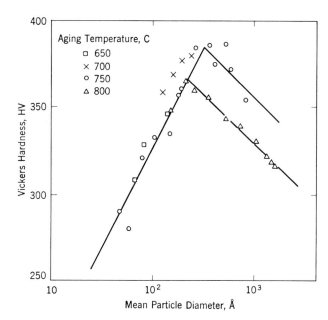

FIGURE 3. Phase diagram for nickel–chromium–cobalt–iron–titanium–aluminum alloys
  a. isothermal section of 750 C. Betteridge[11]
  b. effect of cobalt on solvus (titanium/aluminum ratio 1/1). Heslop[12]

FIGURE 5. Relation between particle diameter and hardness for a nickel-base alloy with 22% Cr, 2.8% Ti, 3.1% Al. Mitchell[15]

tions in the ordered structure because of its phase field running diagonally from Ni₃Al to Ni₃X. Molybdenum, chromium, and iron would substitute for both nickel and aluminum positions due to phase fields intermediate between the two extremes above.

Strengthening of Superalloys by γ′

The strengthening of alloys by precipitates can be expressed in simplest form in terms of alloy hardness vs. particle size. Such a diagram for a Ni–22% Cr–2.8% Ti–3.1% Al alloy is seen in Figure 5 from Mitchell.[15] On the ascending curve, γ′ is cut by dislocations; on the descending curve, dislocations by-pass γ′.

Several basic factors contribute to the magnitude of hardening: anti-phase boundary (APB) and fault energy of γ', γ strength, γ' strength, coherency strains, volume percent γ', particle size of γ', diffusivity in γ and γ' and, possibly, γ–γ' modulus mismatch. They are not additive (Gleiter[16] and Singhal and Martin[17]). These contributing factors will be considered in more detail and will be related to the cutting and by-passing of γ' by dislocations.

APB and Fault Hardening During γ' Cutting

Williams,[18] Merrick,[19] Gleiter and Hornbogen[20] and Copley and Kear[21] developed a theory of APB hardening due to coherent, ordered γ' in disordered γ. Dislocations ($\bar{b} = a/2 <110>$) glide in pairs (see Figure 6) during the shearing stage. The glide motion of the leading dislocation is impeded since it must create APB, whereas the motion of the trailing dislocation is assisted because it annihilates the APB. A net addition to shear strength results. With a nickel–chromium–aluminum alloy as an experimental model, Gleiter and Hornbogen[20] derived a complex mathematical relationship for APB strengthening which would account for the entire aging curve. In the ascending portion of the aging curve, APB hardening increases with:

    a. volume percent γ'
    b. size of γ'
    c. APB energy

In synthesizing the short-time strength of MAR-M 200, Copley and Kear[21] treated the increase in CRSS* due to APB energy, $\Delta\tau_{APB}$, very simply as

$$\Delta\tau_{APB} = \frac{\gamma APB}{2b} \qquad (1)$$

where
$$\gamma APB = \text{APB energy}$$
$$b = \text{Burgers vector}$$

_____
* Critical Resolved Shear Stress.

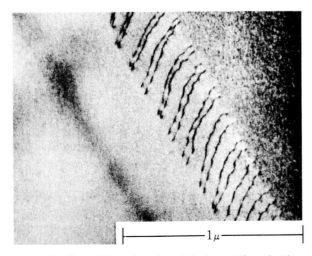

FIGURE 6. Superdislocations found during cutting of γ' in a nickel-base alloy with 20.5% Cr and 8.5% Al strained 0.02 at room temperature. Merrick[19]

The contribution was 33 kgf/mm², over two-thirds of the CRSS of the alloy.

More recently, Kear, Giamei, Silcock and Ham[22] pointed out that γ APB should be replaced by Γ, the fault energy for $\{111\}$-type shear in γ'. Γ would depend upon:

    a. complex fault
    b. APB
    c. superlattice intrinsic fault
    d. superlattice extrinsic fault

Neither the $\frac{\gamma APB}{2b}$ or $\frac{\Gamma}{2b}$ terms provide for effects of variation of volume percent γ', a probable shortcoming. Judging by critical temperatures for ordering of their Ni₃X phases, titanium, niobium and tantalum in γ' would not increase APB energy significantly. However, considering the stacking fault correlation of Figure 2 and the solute contents to be listed under γ' _strength_, titanium and possibly niobium and tantalum could increase the other components of Γ.

γ' Strength During γ' Cutting

During the stage where dislocations cut γ', the drag stress of γ' is important. γ'(Ni₃Al) shows a seven-fold increase in flow stress as temperature is raised from room temperature to 650 C as indicated in Figure 7 (Guard and Westbrook,[13] Flinn,[23] Davies and Stoloff,[24] Copley and Kear[25]). It is believed that this is due to a lattice anisotropy effect, but the exact mechanism is not established as yet.

The γ' of superalloys can be highly alloyed. For example, Mihalisin and Pasquine[2] derived the following compositions for γ' in two nickel-base alloys:

IN-731

$$(Ni_{0.884}Co_{0.070}Cr_{0.032}Mo_{0.008}V_{0.003})_3$$
$$(Al_{0.632}Ti_{0.347}V_{0.013}Cr_{0.006}Mo_{0.002})$$

Alloy 713C

$$(Ni_{0.980}Cr_{0.016}Mo_{0.004})_3$$
$$(Al_{0.714}Nb_{0.099}Ti_{0.048}Mo_{0.038}Cr_{0.103})$$

Taking these analyses and those of Kriege and Baris[3]

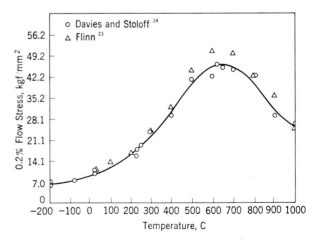

FIGURE 7. Flow stress of Ni₃Al as a function of temperature

Source: _Strengthening Mechanisms in Nickel-Base Superalloys_, International Nickel Co., Inc.

281

on 15 superalloys, γ' can contain up to the following percentages of solute:

Co	Fe	Cr	V	Mo	W	Ti	Nb	Ta
8	3	4	1	2	4	13	3	2

Based on the γ' solid-solution hardening studies of Guard and Westbrook,[13] the following should be the increases in hardness of γ' in kgf/mm² for single additions of each element in the above amounts:

	Co	Fe	Cr	V	Mo	W	Ti	Nb	Ta
RT	30	nil	50	nil	60	no data	130	nil*	nil*
400 C	nil	50	90	nil	100	no data	80	10*	10*
800 C	nil	nil	−20	nil	nil	no data	100	60*	40*

At room temperature and 400 C, chromium, molybdenum and titanium strengthen γ'. Tungsten should be as potent as molybdenum. At 800 C, γ' strengthening from titanium, niobium, tantalum would be significant.

Copley and Kear[21] have related the solid solution component of alloy flow stress, $\Delta\tau_{SSH}$, to γ and γ' drag stress by

$$\Delta\tau_{SSH} = \frac{0.823 \, (\tau_\gamma + \tau_{\gamma'})}{2} \qquad (2)$$

The CRSS appears as a sum independent of volume percent γ' because the slow step in the motion of the dislocation pairs involves one superlattice dislocation moving in γ and the other in γ'. Solution hardening of γ and γ' should be equally potent in hardening the alloy. In Copley and Kear's synthesis of the short-time strength of MAR–M 200, $\Delta\tau_{SSH}$ accounted for 9.8 kgf/mm², a third the flow stress of the alloy.

High Temperature Creep During Cutting Neither the APB, fault nor γ' drag stress treatments above allow for the important diffusion and recovery creep mechanisms. During γ' cutting in high-temperature creep at >0.6 T_M, the creep rate of the alloy may well be determined by diffusion in the γ', through control of the dislocation climb or jog-dragging-screw creep in γ'. The above contents of slow-diffusing molybdenum, tungsten, niobium and tantalum in γ' should then retard creep considerably. More attention needs to be paid to this possibility.

Coherency Strains Coherency strains can play an important strengthening role in superalloys under certain conditions. They can act in both the cutting and the by-pass situations. By critical hypothesis and experiment, Decker and Mihalisin[26] isolated and measured coherency strain effects on nickel–aluminum–ternary element alloys (see Figure 8). Increasing γ–γ' mismatch from <0.2 to 0.8% doubled peak aged hardness. The addition of 200 HV or 70 kgf/mm² by raising mismatch 0.6% was in excellent agreement

* Estimated from 4.5 titanium.

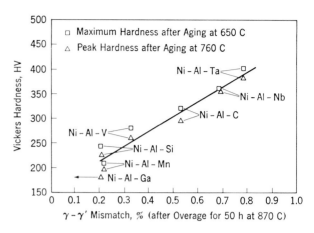

FIGURE 8. Correlation of aged hardness with γ–γ' mismatch of nickel–aluminum–ternary element alloys. Decker and Mihalisin[26]

with the theory of Gerold and Haberkorn[27] on increment of CRSS ($\Delta\tau_{cs}$) arising from coherency strains

$$\Delta\tau_{cs} = 3G\epsilon^{3/2} \left(\frac{hf}{2b}\right)^{1/2} \qquad (3)$$

where   G = γ shear modulus
   ϵ = constrained lattice strain
   h = mean particle diameter
   f = volume fraction of γ'

$\Delta\tau_{cs}$ would increase with coherency strains, volume percent γ' and particle size of γ'.

Practically, due to the following compositional effects, coherency strains would be significant in only certain superalloys (see Figure 9):

1. Titanium and niobium increase coherency strains by partitioning to γ' and by expanding a_o of γ'.
2. Chromium, molybdenum and iron decrease coherency strains by partitioning to the matrix and expanding a_o of γ.
3. Nickel substitution for iron and chromium increases coherency strains by partitioning to the matrix and contracting a_o of γ.
4. Cobalt has little effect.
5. Although data are lacking, tantalum should behave like niobium and tungsten like molybdenum.

The alloys utilizing coherency strains would be molybdenum-free with high titanium/aluminum ratio such as Inconel alloy X-750 or Nimonic 80A. Another type (e.g., Inconel alloy 718) contains high niobium. These alloys are used at <0.6 T_M. Molybdenum is contained in all the more heat-resistant types, perhaps not coincidentally, to lower coherency strains and to reduce γ' growth rate somewhat.

Modulus Mismatch. Extending Fleischer's[28] analysis of solid solution hardening, Phillips[29] has proposed that γ–γ' modulus mismatch should contribute to hardening during dislocation cutting. In nickel–aluminum alloys, Phillips deduced that 40% of the

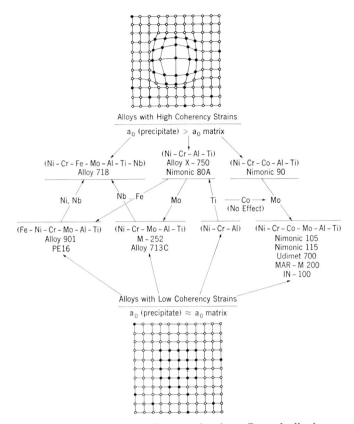

FIGURE 9. Schematic diagram showing effect of alloying additions on coherency strains in superalloys. Bases and commercial alloys having high coherency strains are in the upper half of the diagram; those having low coherency strains are in the lower half. The effect of critical elements in changing coherency strains is given by the direction of the arrows. Decker and Mihalisin[26]

flow stress was derived from modulus mismatch $+ \gamma'$ drag stress. The two effects were not isolated from one another. Because of the complete lack of modulus data on alloyed γ', no estimate can be made of this effect at present.

Dislocation By-Passing of γ'   Beyond a certain γ' particle size, by-passing will occur by either looping or dislocation climb (see Figure 10). The increment of CRSS due to dislocation looping is predicted by the Foreman and Makin[30] modification of the theory of Orowan[31],

$$\Delta\tau_L = 0.2Gb \; \phi \; \frac{2}{\lambda} \ln \frac{h}{2b} \qquad (4)$$

$\phi = $ f (angle between Burgers vector and the tangent to the dislocation)

$\lambda = $ mean planar interparticle separation

Increase of volume percent γ' (through decreasing λ) or increase in γ' particle size, h, would increase hardening. Coherency strain fields would provide additional strengthening by making h appear larger to dislocations.

While the above relationship applies at low temperatures, one must also account for diffusional creep. Lagneborg[32] proposed that γ' strengthened superalloys by decreasing the mobility of climbing dislocations. The retained dislocations contributed to τ, creep shear stress, in proportion to $Gb \sqrt{\rho}$, where $\rho = $ dislocation density. Judicious pre-service introduction of dislocations would strengthen by this relationship.

According to Ansell and Weertman[33] the rate controlling process in high-temperature creep of two-phase alloys would be climb of dislocations over particles. At stresses (σ) above $\frac{Gb}{\lambda}$ dislocations will move past particles by looping. Loops will build up and exert back stresses to prevent more looping until recovery occurs by dislocation climb of the loop nearest γ'. This process is governed by

$$\overset{\circ}{\epsilon} = \frac{2\pi\sigma^4\lambda^2 D}{h \; G \; ^3kT} \qquad (5)$$

where D = self-diffusion coefficient
       k = Boltzmann's constant
       T = temperature

To retard creep, the alloy designer can decrease λ (by increasing volume percent γ'), decrease D by heavy element additions to γ and increase h by aging treatments. At stresses below $\frac{Gb,}{\lambda}$ dislocations climb over γ without pile-up or bowing. Then:

$$\overset{\circ}{\epsilon} = \frac{\pi\sigma b^3 D}{2kTh^2} \qquad (6)$$

Under these conditions of longest service time, the alloy designer can reduce creep by decreasing D of γ and increasing h of γ'.

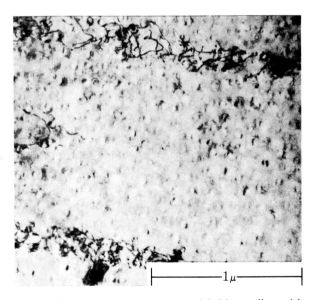

FIGURE 10. Loops around γ' in a nickel-base alloy with 20.5% Cr, 2.52% Ti and 2.83% Al, aged 100 h at 750 C and strained 0.02 at room temperature. Estimated volume percent γ' is about 15. Merrick[19]

Source: *Strengthening Mechanisms in Nickel-Base Superalloys,* International Nickel Co., Inc.

283

The Transition from Cutting to By-Passing As mentioned above, the occurrence of cutting or by-passing depends on particle size. However, the transition size of γ' varies widely in the few documented cases (see Table V). In alloys with about 20% or less γ', dislocation by-passing has been found with particle size greater than 280Å, even with room temperature deformation. Since we shall see that γ' quickly coarsens beyond this size at 750 C or above, the designer must delay by-passing at these service temperatures. As shown by Mitchell[15] in Figure 5, the transition size increases with volume percent γ', suggesting a way to delay by-passing. Evidently this factor, high volume percent γ', accounts for the cutting seen in Table V for the >2500Å particles even in long time creep at 925 C. The highly deformed γ' of Alloy 713C of Figure 11 is testimony to the effect of cutting.

The picture is by no means complete, however. It would be most interesting to observe dislocations after long-time creep above 815 C in alloys with 20 to 60 vol % γ'. At this time, we can only guess that by-pass mechanisms and diffusion in γ would be very important practically, since the major tonnage of superalloys has around 20% γ'. But since the most advanced alloys contain 50% γ', cutting is a very important mechanism in design of new alloys.

Volume percent γ' In the previous sections, several instances have been seen where increasing volume of γ' appeared to be an important design option. Indeed, increase of volume percent γ' increases high temperature strength. In Figure 12, we plot stress-rupture strength vs volume percent γ' in commercial alloys. An increase from 14 to 60% γ' can quadruple the strength of superalloys — an enormously important effect, and the major one used by alloy designers in the last two decades. This, of course, has been accomplished by increasing use of aluminum, titanium, niobium and tantalum, and adjustment of γ' solubility with cobalt, iron and chromium.

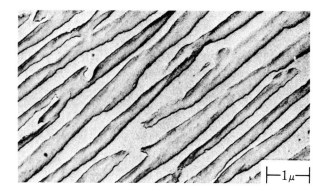

FIGURE 11. Cut γ' in Alloy 713 C after $\epsilon_2 = 0.12$ in 12 182 h at 815 C and 20 kgf/mm². $\mathring{\epsilon}_{ss} = 0.0027\%/h$, $n = \dfrac{\log \epsilon_{ss}}{\log \sigma} = 5.5$

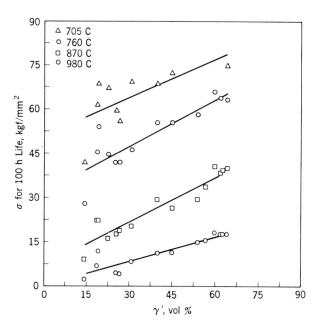

FIGURE 12. Effect of volume percent γ' on strength of superalloys

TABLE V Observations of Dislocation Cutting and By-Passing of γ' in Superalloys

Alloy	Vol % γ'	h, Å	Deformation Temperature, C	Strain	Time, h	Dislocation Motion
Ni–18% Cr–36% Fe–1.8% Mo–1.5% Ti–2.5% Al[15]	~15	2500	815	0.34	5864	by-pass
Fe–23% Ni–18% Cr–2.3% Ti[17]	18	<100	RT	10.0	—	cut
	18	>100	RT	10.0	—	by-pass
Ni–19% Cr–2.15% Ti–1.36% Al[19]	~15	<280	RT	0.02	—	cut
		>280	RT	0.02	—	by-pass
Inconel alloy X-750[34]	14	300	732	0.08	100	by-pass
Udimet 500	33	1500	871	0.012	200	cut
MAR-M 200[21]	56	2500	RT	"Lightly Deformed"	—	cut
Alloy 713LC	48	3000	815	0.037	1074	cut
Alloy 713C	52	3000	815	0.12	12 182	cut
	52	3000	927	0.14	6905	cut

PHASE STABILITY

γ′ Dispersion Stability

Above about 0.6 T_M, γ′ ripens (increases in size) at a significant rate, decreasing $\frac{2}{\lambda} \ln \frac{h}{2b}$ and making dislocation by-passing easier. Measures which minimize ripening will help retain long-time creep resistance. Fleetwood[35] has applied Wagner's theory of Ostwald ripening to γ′:

$$h^3 = \frac{64 \; \gamma_e DC_e V_m^2 t}{9 \; RT} \qquad (7)$$

where   t = time

γ$_e$ = specific γ′/γ interfacial free energy

D = coefficient of diffusion of γ′ solutes in γ

C$_e$ = equilibrium molar concentration of γ′ solute in γ

V$_m$ = molar volume of γ′

R = gas constant

The significant changes with composition come in the terms γ_e, C_e and D. Fleetwood found that the ripening rate of γ′ in nickel–chromium–titanium–aluminum alloys decreased as chromium was increased from 10 to 37%. This arose partly from reduced C_e but also partly from reduction of coherency strains and the consequent reduction of $\gamma_e D$. Increased coherency strains by higher titanium/aluminum ratio (Fell[36]) increased ripening rate.

Mitchell[37] found no effect of cobalt, molybdenum and iron additions and Decker and Freeman[38] no effect of boron and zirconium on growth rate of γ′ in a nickel–chromium–titanium–aluminum alloy. However, increasing niobium from 2 to 5% markedly reduced this rate despite the concurrent increase in coherency strains[37]. Niobium partitions almost completely to γ′ (hence low C_e) and has low D. These reductions in C_e and D from niobium were more influential than the increased coherency strains with increasing titanium/aluminum ratio.

Creep strain has little effect on ripening with 33 or less vol % γ′ (Decker and Freeman,[38] Rowe and Freeman,[39] Mitchell[37]). Perhaps this is characteristic of by-passing. Obviously, the severe cutting of γ′ at 50 vol % γ′ in Figure 11 has accelerated ripening.

Rowe and Freeman,[39] in studies of cyclic overheating of superalloys, found that fine γ′ dispersions were automatically restored at the normal service temperature. However, the loss of creep resistance during γ′ ripening was very dependent on volume percent γ′. M-252 of low volume percent γ′ weakened faster than Inconel alloy 700 of high volume percent γ′. Rowe and Freeman showed that the sensitivity of flow stress in equation (4) to changes in γ′ particle size, h, was much greater in the low volume percent alloy.

Thus, the available design options to retard ripening would be:

(a) increase volume percent γ′

(b) add high partitioning—slow diffusing niobium and tantalum to γ′.

Despite the tolerance of the nickel base for heavy solid-solution and γ′ prime strengthening, a limit still exists beyond which undesirable phases precipitate.

Transformation of γ′ to η and Ni₃Nb

γ′ can transform to other Ni_3X compounds when titanium, niobium or tantalum are sufficiently rich. Pearson and Hume-Rothery[40] related stability of the Ni_3X compounds to size factor. This ranks them in the increasing order of stability: Ni_3Al, Ni_3Ti, Ni_3(Nb or Ta). Although the equilibrium diagrams predict otherwise, almost all the aluminum in Ni_3Al can be replaced by titanium, niobium or tantalum, leaving the possibility of metastable γ′—very important commercially.

The titanium-rich metastable γ′ can transform to Ni_3Ti (η-hcp). The reaction occurs in a Ni–22% Cr alloy when the usual commercial titanium/aluminum ratio of 1/1 is raised to 5/1 (Betteridge[11]). Havalda[41] found that tungsten retarded the transformation in such an alloy. Some intragranular η developed in Incoloy alloy 901 after 2850 h at 730 C, but no undue deterioration in strength or ductility was noted (Clark and Iwanski[42]).

One mode of precipitation, namely cellular precipitation at the grain boundaries, can be detrimental to notched stress–rupture strength. This form is nucleated at grain boundaries and appears over the temperature range of 400 to 980 C. Speich, in a study of an Fe–30% Ni–6% Ti alloy,[43] established the influence of time and temperature on the structure of cellular η and the resulting mechanical properties. The habit of the Ni_3Ti lamellae was such as to leave a semi-coherent η–γ interface with 0.65% mismatch. Surprisingly high hardnesses, over 600 HV, were obtained when the η–γ cells had interlamellar spacings, S, of 10^{-6} cm. Unfortunately, this spacing requires long-time aging at 400 C. Aging at the usual higher temperatures of about 730 to 815 C left coarser spacing and decreased hardness according to the equations

$$S_{cm} = \frac{1.2 \text{ x } 10^{-3}}{982 \; -T} \quad \text{where T} = °C \qquad (8)$$

$$HV \text{ Hardness} = 0.96 S^{-\frac{1}{2}} \qquad (9)$$

S values of η–γ measured in commercial alloys (e.g., A-286 aged at about 815 C) were on the order of 5 x 10^{-5} to 1 x 10^{-4} cm. Equation 9 predicts almost complete loss of hardening when η occurs under these conditions. Thus, the cellular mode leaves a soft zone at grain boundaries backed by γ′ stiffened grains, hence embrittlement.

The second mode, Widmanstätten intragranular, reduces strength, but not ductility. Apparently η is nucleated on stacking faults* in γ′ (see Figure 13a

---

* A stacking fault in γ′ (Ni_3Ti) has the stacking sequence of several layers of hexagonal η (Ni_3Ti).

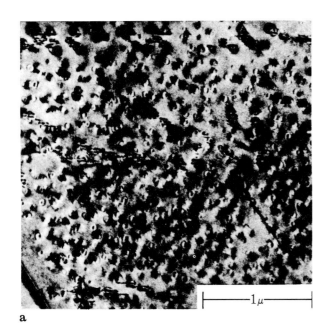

a

from Merrick and Nicholson[44]) on several co-planar γ' particles and on nearly co-planar (111) planes. The η nuclei then join through the γ phase, and eventually devour the mother γ' particles, leaving coarse spacings of 5×10^{-5} to 1×10^{-4} cm.

Several means are available to retard the γ'–η reactions. Trace levels of boron are commonly added to commercial alloys. Equilibrium segregation of boron to grain boundaries retards nucleation of cells (Mihalisin and Decker[45]) with an accompanying increase in notched stress–rupture strength. This effect, plus the retarding effect of aluminum and the accelerating effect of cold work are illustrated in Figure 13b.

Further examples of this use of transition phases are niobium-hardened alloys such as Inconel alloy 718 (Eiselstein[46]), and MC-102 (Haynes[47]). Although there has been considerable disagreement[34, 37, 46, 48–51] it now appears that the precipitating phases take the sequence of

$$\gamma' \to \gamma'_{bct} \to Ni_3Nb \text{ (orthorhombic)} \qquad (10)$$

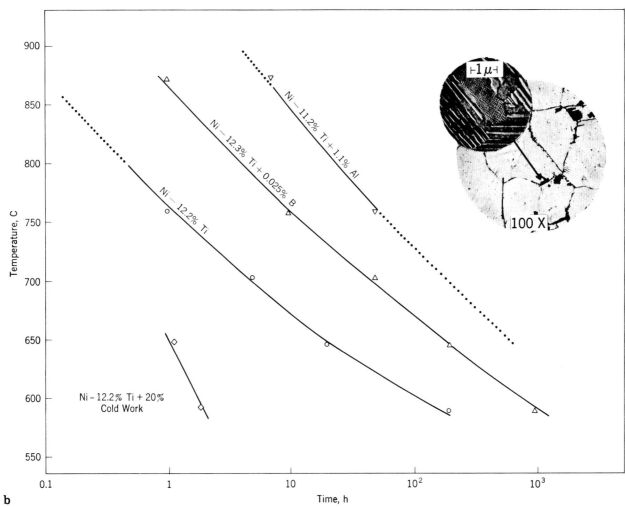

b

FIGURE 13.   The γ' to η transformation

   a. nucleation of η in γ' of nickel–chromium–titanium alloy. Merrick and Nicholson[44]

   b. effect of cold work, boron and aluminum additions on time for the start of cellular precipitation of η. Mihalisin and Decker[45]

with the schematic transformation diagram of Figure 14. The $\gamma' \rightarrow \gamma'_{bct}$ regions are under debate and are not precisely defined. The reaction rate in niobium-hardened alloys is dependent on composition, with increasing titanium and aluminum and decreasing niobium retarding the $\gamma' \rightarrow \gamma'_{bct}$ transformation. The transformation to Ni_3Nb is speeded by high silicon and niobium and low aluminum. During isothermal aging, the early hardening is from γ'. Both γ' and γ'_{bct} can be present at peak hardness. Excessive transformation to Ni_3Nb, which takes a coarse morphology, decreases hardness. As commercially aged for eight hours at 720 C, furnace cool to 18 h at 620 C or for eight hours at 790 C, furnace cool to 620 C, Inconel alloy 718 hardening may result from γ' and/or γ'_{bct}. The structure with γ'_{bct} is seen in Figure 15. During service the $\gamma' \rightarrow \gamma'_{bct}$ transformation proceeds according to Figure 14, retaining the high degree of hardening.

Sigma Phase

Sigma phase can precipitate in superalloys in the temperature region of 650 to 925 C, especially under stress. Loss of stress–rupture strength and room-temperature ductility was the consequence in at least one superalloy, IN-100 (Ross,[52] Mihalisin, Bieber and Grant[53]), due to the plate structure seen in Figure 16.

Based on previous electron theory studies, Boesch and Slaney[54] and Woodyatt, Sims and Beattie[55] derived a quantitative system, PHACOMP, for predicting phase instability from alloy composition. Electron vacancy numbers, N_v, are assigned to the major alloying elements as in Table III. It was demonstrated, for a wide variety of complex alloys, that the tendency for sigma phase formation increases with increase in average electron vacancy number for the matrix, $\overline{N}_v$, given by the following equation:

$$\overline{N}_v = \sum_{i=1}^{n} M_i (N_v)_i \qquad (11)$$

where $\overline{N}_v$ = average electron-vacancy number

M_i = the atom fraction of particular element

N_v = individual electron-vacancy number of particular element

n = number of elements in the matrix

Of course, the effect of γ', carbide and boride precipitation on residual matrix composition must be entered in this quantitative system. The details of the standard system used for calculation of $\overline{N}_v$ at the 1968 International Symposium on Structural Stability are outlined in Table VI.

Woodyatt, Sims and Beattie[55] concluded that sigma could appear in homogeneous wrought alloys at $\overline{N}_v \geq 2.50$. However, some difficulty has been found in application of PHACOMP across wide variation in alloy composition. At least one anomaly has emerged,

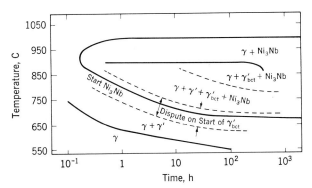

FIGURE 14. Schematic transformation diagram for Inconel alloy 718 drawn from references[34,46,48,49]

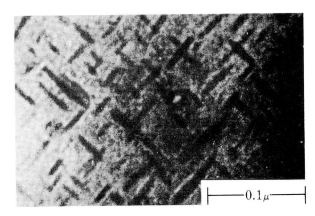

FIGURE 15. γ'_{bct} in Inconel alloy 718. Raymond[34]

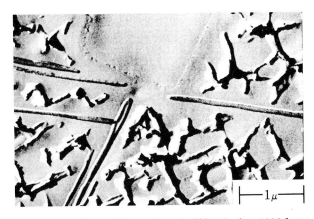

FIGURE 16. Plate of sigma phase in IN-100 after 1006 h at 815 C. Sigma encapsulated in γ'. Cubic particle is MC

sigma phase in Alloy 713C at $\overline{N}_v = 2.07$. Thus, examination and refinement of this useful system now appears appropriate:

1. For sigma precipitation, the critical condition would be long-time exposure at 815 C. To predict carbide effects, the carbide phases actually present at about 5000 h at 815 C should be used to calculate $\overline{N}_v$; not assumption 2 of Table VI. The first regression equation of Table VII can now serve this purpose with high confidence.

2. Mihalisin and Pasquine[2] and Slaney[56] have

Source: *Strengthening Mechanisms in Nickel-Base Superalloys*, International Nickel Co., Inc.

287

pointed out that serious errors existed in assumption 4 of Table VI, especially in aluminum and cobalt partitioning. This assumption can now be replaced by analysis of data from references 2 and 3. This yields more exact γ and γ' compositions by use of a regression equation for volume percent γ' and partitioning ratios for the alloying elements.

3. $\overline{N}_{vz}$ is calculated for both γ and γ' using the expanded equation of Table VIII, accounting for not only the elements in Table VI, but also, niobium, tantalum, titanium and aluminum. As suggested by Mihalisin and Pasquine[2] it is logical to consider $\overline{N}_{vz}$ of γ' because sigma is often found encapsulated in γ' and the electron vacancy number of γ' can be higher than that of γ.

This revised method of computation was reduced to a computer program as outlined in Table VIII. Resulting $\overline{N}_{vz}$ on several commercial alloys are plotted in Figure 17.

In several alloys (e.g., Alloy 713C, IN-100, Udimet

500, Nimonic 115, B-1900, MAR-M 200, René 41), $\overline{N}_{vz}$ of γ' is higher than $\overline{N}_{vz}$ of γ. In fact, sigma phase precipitation correlates better with $\overline{N}_{vz}$ of γ' in these alloys than $\overline{N}_{vz}$ of γ. This is also true of Udimet 700 and IN-738 in which $\overline{N}_{vz}$ is higher than $\overline{N}_{vz}$ of γ'. The sigma boundary is now found from 2.26 to 2.41, a con-

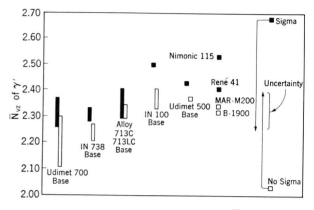

FIGURE 17.   Precipitation of sigma phase vs $\overline{N}_{vz}$ of γ'

TABLE VI   Reference Method for Calculation of Electron Vacancy Number (N_v ref)

1. Convert the composition from weight percent to atomic percent.

2. After long time exposure in the sigma forming temperature range the MC carbides tend to transform to $M_{23}C_6$.
   a. Assume one-half of the carbon forms MC in the following preferential order: TaC, NbC, TiC.
   b. Assume the remaining carbon forms $M_{23}C_6$ of the following composition: $Cr_{21}(Mo, W)_2C_6$ or $Cr_{23}C_6$ in the absence of molybdenum or tungsten.

3. Assume boron forms M_3B_2 of the following composition: $(Mo_{0.5}Ti_{0.15}Cr_{0.25}Ni_{0.10})_3B_2$.

4. Assume gamma prime to be of the following composition: $Ni_3(Al,Ti,Ta,Nb,Zr,0.03Cr^*)$.

5. The residual matrix will consist of the atomic percent minus those atoms tied up in the carbide reaction, boride reaction, and the gamma prime reaction. The total of these remaining atomic percentages gives the atomic concentration in the matrix. Conversion of this on the 100% basis gives the atomic percent of each element remaining in the matrix. It is this percentage that is used in order to calculate the electron vacancy number.

6. The formula for calculation of the electron vacancy number is as follows:
   $(\overline{N}_v \text{ ref}) = 0.66 \text{ Ni} + 1.71 \text{ Co} + 2.66 \text{ Fe} + 3.66 \text{ Mn} + 4.66 (\text{Cr} + \text{Mo} + \text{W}) + 5.66 \text{ V} + 6.66 \text{ Si}$

* (0.03% of the original atomic percent)

TABLE VII   Regression Analyses of Effect of Composition on Carbide Content of Superalloys

f†	Constant	Coefficients									Interaction	$\sigma^{(1)}$	MCC[2]	Fit[3]
		Cr	Ti	Mo	W	Nb	Ta	Al	Co	Fe				
					5000 h at 815 C									
a	0.844	−0.040*	0.0029	0.20*	0.084*	0.21*	0.039*	−0.0073	−0.0027	0.015*	0.0081* CrxMo	0.07	0.996	39.1
					5000 h at 980 C									
b	2.14	−0.067*	0.0022	−0.034	−0.026	0.095	0.18*	−0.10*	−0.011	0.0070		0.21	0.966	4.79
c	−4.12	0.20*	0.14*	−0.26*	−0.20*	0.025	0.17*	0.22*	0.012	−0.0070		0.13	0.989	15.8
d	2.98	−0.13*	−0.14*	0.29*	0.23*	−0.12	−0.35*	−0.12*	0	0		—	—	—

† Fraction of carbide.
(1) Standard error of estimate.
(2) Multiple correlation coefficient.
(3) Goodness of fit.
* T value of coefficient over 0.8.

a–f_{MC} $f_{M_{23}C_6} = 1 - f_{MC}$
b–f_{MC}
c–$f_{M_{23}C_6}$
d–$f_{M_6C} = 1 - f_{MC} - f_{M_{23}C_6}$

TABLE VIII   Method of Calculation for $\overline{N}_{vz}$

1. Convert the composition from weight percent to atomic percent.

2. Assume boron forms M_3B_2 of the following composition $(Mo_{0.5}Ti_{0.15}Cr_{0.25}Ni_{0.10})_3B_2$, subtract M.

3. a. Partition weight percent carbide to MC and $M_{23}C_6$ by first equation of Table VII.

    b. Assume MC is (1) $(Ti_{0.5}Nb_{0.25}Ta_{0.25})C$ with Ti,Nb,Ta in alloy
                           (2) $(Ti_{0.5}Nb_{0.5})C$ with Ti,Nb   in alloy
                           (3) $(Ti_{0.5}Ta_{0.5})C$ with Ti,Ta   in alloy
                           (4) TiC with Ti   in alloy

    c. Assume $M_{23}C_6$ is (1) $Cr_{21}Mo\,W\,C_6$ with Cr,Mo,W in alloy
                              (2) $Cr_{21}Mo_2\,C_6$ with Cr, Mo   in alloy
                              (3) $Cr_{21}W_2\,C_6$ with Cr,W   in alloy
                              (4) $Cr_{23}\,C_6$ with Cr   in alloy

    d. Partition atomic percent carbon to MC and $M_{23}C_6$, subtract M.

4. Calculate volume % γ' from regression equation derived from data on 19 alloys from ref.[2,3]

5. Calculate γ and γ' composition from 4 and partitioning ratios of alloys from ref.[2,3]

6. Calculate $\overline{N}_{vz}$ for γ and γ' from
    $(\overline{N}_{vz}) = 0.66\,Ni + 1.71\,Co + 2.66\,Fe + 3.66\,Mn + 4.66\,(Cr+Mo+W) + 5.66\,(V+Nb+Ta) + 6.66\,(Si+Ti) + 7.66\,Al$

---

siderable improvement over the 2.07 to 2.50 range of the previous method.

Even more striking is the thought that the sequence of calculations and assumptions of Table VI can be completely replaced by *simple chemical analysis of γ' extracts—followed by calculation of $\overline{N}_{vz}$ of γ'*. Of course this could all be done routinely with a modern computer controlled X-ray spectrograph.

An appropriate heat treatment can reduce the rate of sigma formation (Mihalisin, Bieber and Grant[53]). The particular mechanism is not known, but the opportunity to provide further control beyond that offered by composition is enticing. Eventually, such an effect should be handled by electron vacancy calculations.

From the art and science of stainless steels, it is well known that sigma phase need not be detrimental to creep strength or room temperature ductility—if the proper morphology evolves. Apparently, the same situation exists with superalloys. There is a serious lack of data on the effect of volume percent and morphology of sigma phase on properties. Some heats of Alloy 713C and René 41 have contained sigma phase without loss of strength. The no-sigma prescription inherent in PHACOMP may place an unnecessary penalty on the alloy designer, restricting the use of strengthening additions. A critical study of the effect of volume percent and morphology of sigma phase is overdue.

Laves Phase

Laves phase of the general formula $(Fe,Cr,Mn,Si)_2$-(Mo,Ti,Nb) can precipitate in iron-containing superalloys from 650 to 1100 C. Although the morphology can be intragranular, the usual form is coarse intergranular (see Figure 18). Prediction of formation can be treated by PHACOMP, but data are sparse compared to sigma phase predictions. Sims[58] stated that

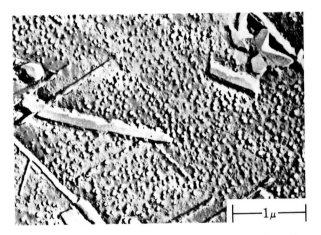

FIGURE 18.   Laves phase as blocky plates in Incoloy alloy 901. Spherical γ' and thin plates of η. Beattie and Hagel[57]

Laves phase was found at $\overline{N}_v \geq 2.30$ if molybdenum and/or tungsten exceed 3% (providing iron is $<20\%$ and cobalt $<30\%$). Silicon is a pronounced stabilizer of Laves phase in alloy 718.[45] The phase has been detected in Incoloy alloy 901.[57] Increasing titanium, niobium and molybdenum and decreasing nickel from that base increase Laves phase.

Laves phase in excess of that in the above alloys can degrade room-temperature ductility[59, 60] with little effect on creep properties.

CARBIDES

The role of carbides in superalloys is curious and the understanding of mechanisms relatively unenlightened. Yet some generalizations can be drawn.

It is the nature of carbides to prefer grain boundaries in nickel, rather than the intragranular sites in high N_v cobalt and iron and the other matrices of higher N_v.

Undoubtedly, many investigators have noted the

Source: *Strengthening Mechanisms in Nickel-Base Superalloys*, International Nickel Co., Inc.

289

detrimental effects of certain grain-boundary carbide morphologies on ductility and taken the next logical step of reducing carbon to very low levels. Surprisingly, studies in this direction found sharply reduced creep life and ductility with <0.03% C in Nimonic 80A[61] and Udimet 500.[62]

Clearly, carbides must be tolerated in superalloys, especially in the grain boundaries. Morphology can influence ductility. The class can influence stability of the matrix through the removal of elements. Therefore, understanding of desirable class and morphology is critical to the alloy designer to save much empiricism in composition and heat treatment variations.

Classes of Carbides and Typical Morphologies

The common classes of carbides are MC, $M_{23}C_6$, Cr_7C_3 and M_6C. MC usually takes a coarse random cubic (see Figure 16) or script morphology. $M_{23}C_6$ shows a marked tendency for grain-boundary forms. At 760 to 870 C, nearly continuous platelet forms predominate (see Figure 19) while at 980 C more blocky and less continuous types are found. On occasion, dislocation decoration by this carbide is seen intragranularly. M_6C can precipitate in blocky form in grain boundaries (see Figure 20), and in Widmanstätten intragranular morphology (see Figure 21). Cr_7C_3 takes a blocky intergranular form. Although data are insufficient for precise correlation, it is apparent that continuous grain boundary $M_{23}C_6$ and Widmanstätten M_6C are to be avoided for best ductility and rupture life.

Carbides after Short-time Heat Treatments

Several of the above types and morphologies can appear in one alloy (e.g., see Figure 22 for data on René 41). Furthermore, MC can transform to M_6C and both MC and Cr_7C_3 can transform to $M_{23}C_6$.

The alloy designer has used short time heat treatments to considerable advantage in creating favorable initial types and morphologies before service. Eleven superalloys (GMR 235-D, M-252, Nimonic 80A, Nimonic 90, Nimonic 115, René 41, Udimet 500, Udimet

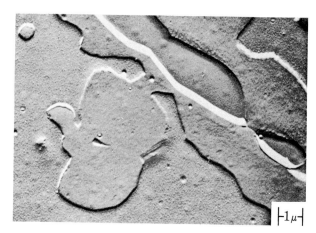

FIGURE 20. Blocky M_6C in grain boundary of Unitemp AF2-1D after 5000 h at 1040 C. Collins[63]

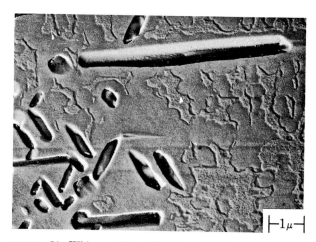

FIGURE 21. Widmanstätten M_6C in nickel-base alloy with 18% Cr, 1% Mo, 0.6% Nb, 0.6% W, 0.6% Ta, 13% Al and 0.1% Ti

700, Waspaloy, IN-738, MAR-M 421) utilize intermediate carbide heat treatments from 1040 to 1100 C before final γ' aging. Six superalloys (alloy X-750, Udimet 500, Udimet 700, Waspaloy, alloy 901, D-979) utilize intermediate treatments at 760 to 850 C.

Suffice it to illustrate the variations in this general treatment with the cases of René 41, Nimonic 80A, Inconel alloy X-750 and Nimonic 115:

René 41 An anneal at 1175 C dissolves M_6C and makes the alloy susceptible to subsequent rapid precipitation of continuous $M_{23}C_6$ film in grain boundaries during γ' aging after forming or welding. Poor ductility and cracking can result. An intermediate treatment at 1040 C leaves the as-worked structure of uniform fine grains with well-dispersed M_6C. Cracking problems and ductility are improved by the delay in formation of $M_{23}C_6$ (Weisenberg and Morris[65]).

Nimonic 80A Betteridge and Franklin[66] found that heat treatment of Nimonic 80A at about 1000 to 1080 C before γ' aging at 700 C produced vastly improved rupture life. At 1080 C, massive Cr_7C_3 formed

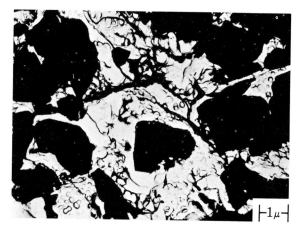

FIGURE 19. Microfractograph of intergranular fracture surface and $M_{23}C_6$ of boron- and zirconium-free Udimet 500 after 0.012 creep in 165 h at 870 C. Decker and Freeman[38]

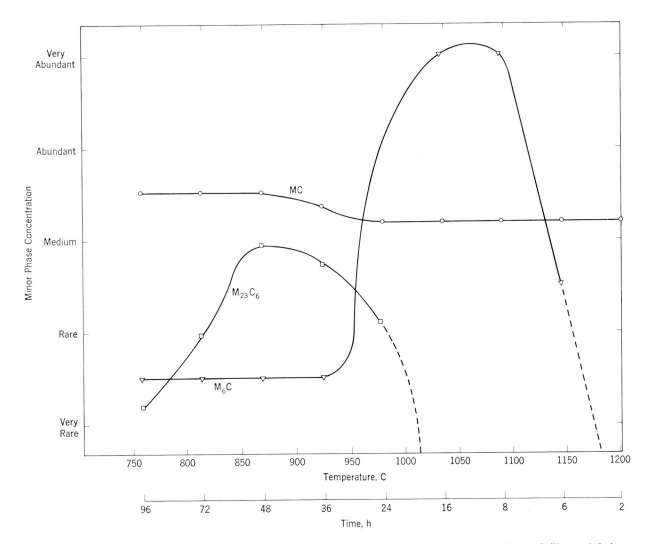

FIGURE 22. Minor phase concentration in René 41 as a function of temperature for short exposure times. Collins and Quigg[64]

at grain boundaries. This pre-existing Cr_7C_3 reduced the initial rate of $Cr_{23}C_6$ precipitation during the standard 16 h at 700 C γ' age.[61] According to Figure 23, about 1% $Cr_{23}C_6$ transformation occurs after intermediate treatment at 1080 C rather than about 10% with a Cr_7C_3-free pretreatment of two hours at 1200 C. Decreasing carbon and titanium plus aluminum speeded the $Cr_7C_3 \rightarrow Cr_{23}C_6$ reaction (Fell[36]). The benefits of the 1080 C treatment were rationalized[66] on the basis of a γ'-denuded grain boundary set up by chromium depletion from Cr_7C_3 precipitation.

Inconel alloy X-750   Raymond[34] came to the opposite mechanistic conclusion in studies of the benefit of double aging used by Bieber[67] for Inconel alloy X-750. Intermediate treatment at 845 C before final 705 C age improved notch ductility (see Figure 24). The intermediate treatment left blocky $M_{23}C_6$ without chromium depletion and allowed fine γ' to precipitate at 705 C up to the grain boundary. In contrast, direct aging at 705 C resulted in cellular $M_{23}C_6$, with regions depleted of chromium and free of γ'. After creep ex-

posure, this denuded zone contained many dislocations, which indicate a weak zone easily crept and ruptured. The notch tough material with intermediate treatment and no denuded zone had intragranular dislocation tangles around γ' and no concentration at grain boundaries.

Nimonic 115   The tensile ductility and impact energy of Nimonic 115 are very sensitive to intermediate heat treatments (Table IX). Furnace cooling or direct transfer treatments to the intermediate-temperature region provide improved ductility and impact energy over material air cooled to room temeprature and reheated. Intergranular-γ' coated blocky $M_{23}C_6$ was associated with the improved properties. A more continuous $M_{23}C_6$ grain-boundary film was found with intermediate air cooling and reheating.

Carbides after Long-Time Service

Some general rules may be derived to relate carbide type to alloy composition after about 5000 h service. From our computer studies of the long-time (700 to

Source: *Strengthening Mechanisms in Nickel-Base Superalloys*, International Nickel Co., Inc.

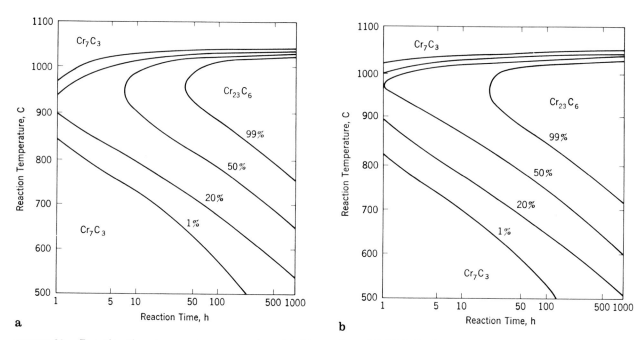

FIGURE 23.   Reaction-time-temperature curves for nickel-base alloy with 22% Cr, 2.8% Ti and 3.1% Al. Fell[36]
a. after 8 h at 1080 C AC   b. after 2 h at 1200 C AC

5000 h) data of Collins,[63] Raymond,[34] Beattie and VerSnyder[69] and Fell[36] on 15 superalloys, we conclude:

1. In the temperature region critical to matrix phase stability (815 C), and where continuous grain-boundary films are likely, only MC and $M_{23}C_6$ are pseudo-equilibrium phases. The fraction of MC of total weight percent MC + $M_{23}C_6$ = 0.844 − 0.0397 Cr + 0.0029 Ti + 0.2021 Mo + 0.0841 W + 0.2130 Nb + 0.0389 Ta − 0.0073 Al − 0.0027 Co + 0.0146 Fe − 0.0081 (Cr x Mo). Low chromium and high niobium, tantalum, molybdenum, tungsten, and iron stabilize MC and should improve ductility. The one exception is M-252 with 10% Mo, 2.6% Ti, and no tungsten, niobium or tantalum, in which M_6C forms. The higher titanium of 3.1% in René 41 removes M_6C and moves the composition into the regime of the above equation.

2. Cr_7C_3 is found only in the basic nickel–chromium–titanium–aluminum superalloy, Nimonic 80A, then at and above 1000 C. Addition of cobalt (Nimonic 90), molybdenum (Waspaloy), tungsten[70] or niobium (Inconel alloy X-750) removes this phase.

3. At the upper design region of 980 C, MC, M_6C and $M_{23}C_6$ are pseudo-equilibrium phases in superalloys other than Nimonic 80A. The equations for partitioning are listed in Table VII. High chromium, titanium, aluminum and tantalum favor $M_{23}C_6$; high molybdenum and tungsten favor M_6C; and high niobium and tantalum favor MC.

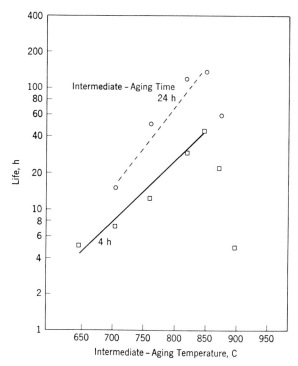

FIGURE 24.   Effect of intermediate-aging treatment on notched-bar rupture life of Inconel alloy X-750 tested at 650 C and 56.25 kgf/mm². Heat treatment: 1150 C (4 h) AC + intermediate aging treatment AC + 705 C (20 h) AC. Raymond[34]

TABLE IX Effect of Heat Treatment on Nimonic 115 (Ward)[68]

Heat Treatment	Stress–Rupture Tests at 11 kgf/mm² and 980 C		Impact Strength at 900 C, kgfm/cm²		Tensile Elongation at 900 C, %	
	Life, h	Elongation, %	As Heat Treated	+ 16 h at 980 C	As Heat Treated	+ 16 h at 980 C
1190 C (1 1/2 h) FC to 1000 C	185	10	2.8	1.4	8	17
1190 C (2 1/2 h) DT to 1125 C (1 h) DT to 1025 C (1/2 h)	158	14	2.2	1.6	14	16
1190 C (1 1/2 h) AC + 1075 C (6 h) AC	203	6	1.0	—	8	6

FC = furnace cooled.
DT = direct transfer.
AC = air cooled.

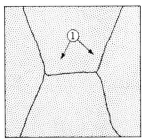

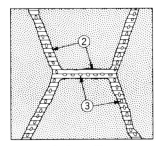

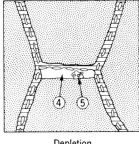

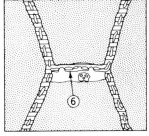

Initial Condition

Agglomeration

Depletion

Microcracking

FIGURE 25. Schematic representation of the steps in the mechanism of stress rupture of a superalloy at 870 C. Axis of tension is vertical. Decker and Freeman[38]
  1. intragranular γ'
  2. agglomerated γ'
  3. agglomerated $M_{23}C_6$
  4. matrix denuded of γ'
  5. γ'-carbide nodule
  6. microcracks at $M_{23}C_6$-denuded zone interfaces

ALLOYING FOR GRAIN BOUNDARIES

Truly the most remarkable alloying effects in superalloys are due to minute additions of boron, zirconium and magnesium.[71–74] The enhancement of creep properties with boron and zirconium additions is shown in Table X. We see that they can increase life 13 times, elongation 7 times, rupture stress 1.9 times and n, stress dependence of creep rate, from 2.4 to 9. Appropriate magnesium addition can supplement boron plus zirconium (see Table XI for boron and zirconium containing IN-120); increasing life three times, elongation two times and reduction of area three times.

In contrast to these unambiguous effects, the mechanisms have resisted clarification. The versatility of these elements probably confounds attempts to assign simple mechanisms. It is widely agreed that boron, zirconium and magnesium serve at grain boundaries, where they segregate because of their odd size (21-29% over- or undersize). However, defining their exact role at grain boundaries is not easy. A severe grain-boundary instability can set into poorly alloyed and/or heat treated superalloys, aided by creep strain (Figure 25). Grain boundaries transverse to stress can undergo a sequence of events leading to bare carbides faced with γ'-depleted matrix (Figure 26a). Massive, widely spaced carbide–γ' nodules in the grain boundaries can serve as sinks for γ' and rob the depleted zone of the finely dispersed γ' and the usual grain boundary carbides of their envelope (Figure 26b). Microcracking sets in at the γ'-carbide interface and rupture life and ductility are cut short (Figure 26b).

Clearly in the case of Udimet 500 (see Table XII), boron and zirconium retarded the evolution of this process and reduced the strain dependency of nodule formation. To illustrate further, in the alloy without boron and zirconium, microcracks developed at the end of first stage (23 h). With boron and zirconium, microcracks did not develop until third stage creep at 214 h.

The sharp effect of boron and zirconium on the exponential dependence, n, of strain rate on stress in second-stage creep (Table X) and the lack of influence on primary creep are notable clues to the mechanism. Apparently n was controlled by denuded zones. When denuded zones were fully developed in second stage creep without boron and zirconium, n was 2.4, characteristic of solid-solution alloys. With few denuded zones, n was 9, typical of γ'-hardened alloys.

The above studies are supported by those of Peter, Müller and Kohlhaas[76] on the role of cobalt and molybdenum in increasing strength and ductility of superalloys. Molybdenum addition to a Ni–22% Cr–2.8%

Source: *Strengthening Mechanisms in Nickel-Base Superalloys*, International Nickel Co., Inc.

293

TABLE X Effect of Boron and Zirconium on Creep of Udimet 500 at 870 C[38]

Alloy	ϵ in Primary Creep	Life, h	Elongation (% in 4D)	σ for $\overset{\circ}{\epsilon}_{ss}$ of 0.004%/h, kgf/mm²	n*
		$\sigma = 17.6$ kgf/mm²			
Base	0.002	50	2	12.0	2.4
+ 0.19 % Zr	0.002	140	6	16.2	4
+ 0.009% B	0.002	400	8	19.7	7
+ 0.009% B + 0.01% Zr	0.002	647	14	22.5	9

$$*n = \frac{\log \overset{\circ}{\epsilon}_{ss}}{\log \sigma}$$

where σ = stress over range of 14.1 to 21.1 kgf/mm²

$\overset{\circ}{\epsilon}_{ss}$ = minimum second stage creep rate

TABLE XI Effect of Magnesium on Creep of IN-120[75]

Alloy	Life, h	Elongation (% in 4D)	Reduction of Area, %
	Stress-Rupture Properties at 70 kgf/mm² and 650 C		
Base	122 to 188	10 to 17	11 to 30
+0.013% Mg	244	24	53
+0.016% Mg	278	31	60
+0.022% Mg	414	30	61
+0.027% Mg	467	29	63
+0.049% Mg	422	28	53

Ti–3.1% Al alloy retarded the appearance of denuded zones as did cobalt addition. Further evidence on the detrimental effect of denuded zones comes from the study of Raymond.[34] Without denuded zones, the replica electron micrographs of Decker and Freeman and transmission electron micrographs of Raymond both showed signs of intragranular γ' deformation. With denuded zones, the intragranular effects were missing and Raymond found high dislocation contents in the denuded zone. Hence highly localized grain boundary sliding, setting up r-type cracking at γ'–carbide interfaces.

Of course, boron and zirconium can benefit γ'-free alloys so that the denuded zone sequence cannot be universal for all nickel alloys. Although there are exceptions, some evidence exists for reduction of carbide precipitation in grain boundaries and shunting of carbon into the grains by boron.[77–80] Magnesium served this purpose in a nickel–chromium–titanium–aluminum alloy by setting up intragranular MC.[81]

To summarize, there is one general mechanism that is consistent with all the above observations. That is, beneficial grain-boundary-segregating elements retard grain-boundary diffusion.

GRAIN CONTROL

The strength of superalloys is very dependent upon grain structure.

Grain Size and Component Thickness

As seen in Figure 27, Richards[82] found that rupture life and creep resistance increased as component thickness to grain-size ratio increased. With this wrought superalloy, provided the ratio was kept constant, life and creep resistance increased with grain size. (The grain size was increased by raising the solution-treating temperature.) A cast superalloy showed the same dependence of life and creep resistance upon thickness to grain size ratio. However, variations in grain size of 2 to 7 mm because of varying casting temperatures failed

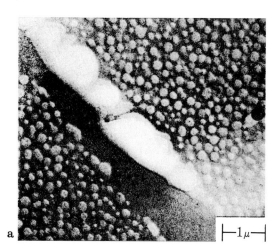

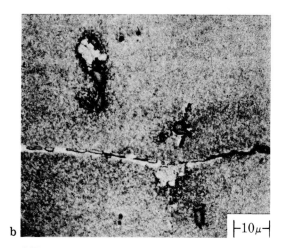

FIGURE 26. Examples of steps in stress rupture of Figure 25. Decker and Freeman[38]

a. bare $M_{23}C_6$ faced with denuded zone

b. bare $M_{23}C_6$ faced with denuded zone, γ'–carbide nodule, microcracks at $M_{23}C_6$–γ' interface

TABLE XII **Effect of Boron and Zirconium on Grain-Boundary Stability of Udimet 500 at 870 C[38]**

| | After $\epsilon = 0.012$ in 200 h | | | After 200 h no σ^* |
Alloy	Denuded Grain Boundaries**	Micro-cracks**	γ'-Carbide Nodules**	γ'-Carbide Nodules**
Base	264	314	418	230
+0.19% Zr	127	78	175	90
+0.009% B	60	30	63	60
+0.009% B +0.01% Zr	23	2	20	20

* No denuded grain boundaries or microcracks in any alloys without stress.

** Number of features detected at 1000 D in 5 mm².

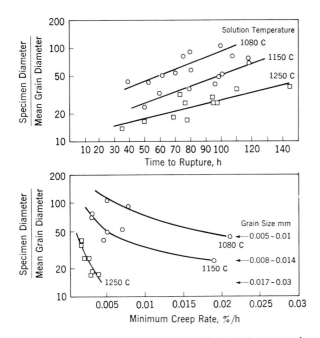

FIGURE 27. Influence of specimen diameter/mean grain diameter ratio and solution-treatment temperature on the creep properties of wrought superalloy tested at 14.2 kgf/mm² at 870 C. Richards[82]

to change properties at a given ratio. These observations deserve serious consideration when dealing with thin sections with large grains. Thin sections of castings can exhibit reduced life and creep resistance as can thin sheet when softened by high temperature annealing for forming.

Directionally Solidified and Single Crystal Superalloys

Reducing the thickness to grain size ratio to the extreme of 1/1 need not be taken in a negative light, however, providing the grains and their boundaries are correctly oriented. VerSnyder and Guard[83] and Piearcey and VerSnyder[84] illustrated this very creatively in directionally solidified alloys. DS 200 (direc-

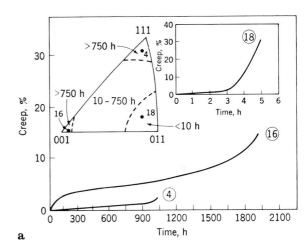

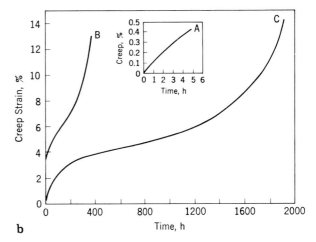

FIGURE 28. Effect of grain orientation on creep properties of MAR-M 200 at 760 C and 70 kgf/mm². Kear and Piearcey[86]
  a. orientation dependence of creep of crystals
  b. creep of
      A. conventional cast random grain specimens
      B. directional grain DS 200 specimens
      C. oriented single grain specimens

tionally solidified MAR-M 200) has superior yield strength to the conventional alloy. Piearcey, Kear and Smashey[85] related this to the preferred cube orientation of the crystals, where advantage is taken of the increase of yield stress for octahedral slip over that for cube slip in γ'. The improved ductility, which also shows up in extended tertiary creep life, is related to the absence of weak, transverse grain boundaries and the greater difficulty of crack initiation and propagation.

Kear and Piearcey[86] extended this research to single-crystal superalloys. At 760 C (Figure 28), there was a marked improvement in creep strength for orientations approaching [001] and [111]. Single crystal MAR-M 200 had life superior to even DS 200. This superiority was related to the more precise <100> orientation and the complete lack of any grain boundaries.

Source: *Strengthening Mechanisms in Nickel-Base Superalloys*, International Nickel Co., Inc.

295

CONCLUSIONS

This review of mechanisms can be summarized in terms of the design steps that can be used to increase the strength of superalloys:

a. solid-solution harden γ

b. increase volume percent γ'

c. increase fault energy of γ'

d. solid-solution harden γ'

e. increase coherency strains for $<0.6\ T_M$

f. decrease ripening rate for $>0.6\ T_M$

g. minimize formation of η, Ni_3Nb, Laves and sigma phases

h. control carbides to prevent denuded zones, $M_{23}C_6$ grain-boundary films and Widmanstätten M_6C

i. stabilize grain boundaries

j. control component thickness/grain size ratio

k. control crystal orientation.

In terms more familiar to the alloy designer, the above factors can be controlled by alloying elements as summarized in Table XIII.

TABLE XIII   Role of Alloying Elements in Superalloys

Element	γ' γ ss	γ' Vol %	γ' Γ	γ' γ'ss	Coherency Strains	Ripening	$\gamma'\to\eta$	$\gamma\to\gamma'_{bct}$	$\gamma'\to Ni_3Nb$	815 C MC	815 C $M_{23}C_6$	980 C MC	980 C $M_{23}C_6$	980 C M_6C	Stable Grain Boundaries
Co	↑W	↑M	—	↑W	nil	nil	—	—	—	—	—	—	—	—	↑
Fe	↑M	↑S	—	↑W	↓	nil	—	—	—	↑W	↓W	—	—	—	—
Cr	↑S	↑S	—	↑S	↓	↓	—	—	—	↓M	↑M	↓M	↑S	↓S	—
Mo	↑S	nil	—	↑S	↓	↓	—	—	—	↑S	↓S	—	↓S	↑S	↑
W	↑S	↑M	—	↑S	↓	↓	↓	—	—	↑M	↓M	—	↓S	↑S	↑
V	↑W	↑M	—	nil	—	—	↓	—	—	↑M	↓M	—	↓S	↑S	—
Nb	—	↑S	↑	↑M	↑	↓	↑	↑	↑	↑S	↓S	—	—	—	—
Ta	—	↑S	↑	↑M	↑	↓	↑	↑	—	↑M	↓M	↑S	↑S	↓S	—
Ti	—	↑S	↑S	↑S	↑	↑	↑	↓	—	—	—	—	↑S	↓S	—
Al	↑S	↑S	↓	↓	↓	↑	↓	↓	↓	—	—	↓S	↑S	↓S	—
B	—	—	—	—	—	nil	↓	—	—	—	↓*	—	—	—	↑
Zr	—	—	—	—	—	nil	—	—	—	—	↓*	—	—	—	↑
C	—	—	—	nil	↑	—	—	—	—	↑	↑	—	—	—	↓
Mg	—	—	—	—	—	—	—	—	—	↑	↓*	—	—	—	↑

*at grain boundaries   ss = solid-solution hardening

W = weak   S = strong   ↓ = decrease

M = medium   ↑ = increase

ACKNOWLEDGMENTS

It is a particular pleasure for the author to acknowledge the helpful discussions of J. S. Benjamin, J. H. Brophy, R. C. Gibson, H. F. Merrick, J. R. Mihalisin, E. L. Raymond and J. W. Schultz.

Particular credit is due to W. G. Wickersty and M. J. Atherton for their valuable assistance in the mathematical and computer treatments of data.

TRADEMARKS

Inconel, Incoloy, and Nimonic are trademarks of International Nickel. MAR-M is a trademark of Martin Marietta Corporation. René 41 is a trademark of Allvac Metals Company (a Teledyne Company). Udimet is a trademark of Special Metals, Incorporated. Waspaloy is a trademark of United Aircraft Corporation.

REFERENCES

1. R. F. Decker and E. G. Richards, International Nickel Limited Technical Publication Paper HT 26, 1967, Inco Power Conference, Switzerland

2. J. R. Mihalisin and D. L. Pasquine, "Phase Transformations in Nickel-Base Superalloys," International Symposium on Structural Stability in Superalloys, Seven Springs, Pa., 1968, I, 134

3. O. H. Kriege and J. M. Baris, Trans. ASM *62*. 195 (1969)

4. W. T. Loomis, The Influence of Molybdenum on the γ' Phase Formed in a Systematic Series of Experimental Nickel-Base Superalloys, PhD Thesis, University of Michigan, 1969

5. R. M. N. Pelloux and N. J. Grant, Trans. AIME *218*. 232 (1960)

6. E. R. Parker and T. H. Hazlett, "Principles of Solution Hardening," Relation of Properties to Microstructure, ASM, 1954, 30

7. B. E. P. Beeston, I. L. Dillamore and R. E. Smallman, Metal Sci. J. *2*. 12 (1968)

8. B. E. P. Beeston and L. K. France, J. Inst. Metals *96*. 105 (1968)

9. W. B. Pearson, Handbook of Lattice Spacings and Structures of Metals and Alloys, Pergamon, New York, 1959, I, II

10. M. V. Pridantsev, Izv. Akad. Nauk SSSR, Metal. *1967*. (5). 115

11. W. Betteridge, The Nimonic Alloys, Edward Arnold Ltd, London, 1959

12. J. Heslop, Cobalt *1964*. (24). 128

13. R. W. Guard and J. H. Westbrook, Trans. AIME *215*. 807 (1959)

14. J. S. Benjamin, B. C. Giessen and N. J. Grant, Trans. AIME *236*. 224 (1966)

15. W. I. Mitchell, Z. Metallk. *57*. 586 (1966)

16. H. Gleiter, Z. Metallk. *58*. 306 (1967)

17. L. K. Singhal and J. W. Martin, Acta Met. *16*. 947 (1968)

18. R. O. Williams, Acta Met. *5*. 241 (1957)

19. H. F. Merrick, "Deformation of Nickel-Chromium Base Alloys," Proc. European Reg. Conf., Electron Microscopy, 3rd, Prague 1964 (A), 171

20. H. Gleiter and E. Hornbogen, Phys. Status Solidi *12*. 235 (1965)

21. S. M. Copley and B. H. Kear, Trans. AIME *239*. 984 (1967)

22. B. H. Kear, A. F. Giamei, J. M. Silcock and R. K. Ham, Scripta Met. *2*. 287 (1968)

23. P. A. Flinn "Solid Solution Strengthening," Strengthening Mechanisms in Solids, ASM, Metals Park, Ohio, 1962, 17

24. R. G. Davies and N. S. Stoloff, Trans. AIME *233*. 714 (1965)

25. S. M. Copley and B. H. Kear, Trans. AIME *239*. 977 (1967)

26. R. F. Decker and J. R. Mihalisin, Trans. ASM *62*. 481 (1969)

27. V. Gerold and H. Haberkorn, Phys. Status Solidi *16*. 675 (1966)

28. R. L. Fleischer, Acta Met. *11*. 203 (1963)

29. V. A. Phillips, Phil. Mag. *16*. 103 (1967)

30. A. J. E. Foreman and M. J. Makin, Phil. Mag. *14*. 911 (1966)

31. E. Orowan, Symposium on Internal Stresses in Metals and Alloys, The Institute of Metals, London, 1948, 451

32. R. Lagneborg, Metal Sci. J. *3*. 18 (1969)

33. G. S. Ansell and J. Weertman, Trans. AIME *215*. 838 (1959)

34. E. L. Raymond, Trans. AIME *239*. 1415 (1967)

35. M. Fleetwood, Private Communication (International Nickel Limited)

36. E. A. Fell, Metallurgia *63*. 157 (1961)

37. W. I. Mitchell, Z. Metallk. *55*. 613 (1964)

38. R. F. Decker and J. W. Freeman, Trans. AIME *218*. 277 (1960)

39. J. P. Rowe and J. W. Freeman, "Relation Between Microstructure and Creep Resistance in Nickel-Base Alloys," Joint International Conference on Creep, The Institution of Mechanical Engineers, London, 1963, Session 1, 65

40. W. B. Pearson and W. Hume-Rothery, J. Inst. Metals *80*. 641 (1951/52)

41. A. Havalda, Trans. ASM *62*. 581 (1969)

42. C. C. Clark and J. S. Iwanski, Trans. AIME *215*. 648 (1959)

43. G. R. Speich, Trans. AIME *227*. 754 (1963)

44. H. F. Merrick and R. B. Nicholson, "Some Features of Precipitation in a Nickel-Chromium-Titanium Alloy," Fifth International Congress for Electron Microscopy, Academic Press, New York, 1962, I, K-8

45. J. R. Mihalisin and R. F. Decker, Trans. AIME *218*. 507 (1960)

46. H. L. Eiselstein, "Metallurgy of a Columbium-Hardened Nickel-Chromium-Iron Alloy," Advances in the Technology of Stainless Steels and Related Alloys, ASTM Special Technical Publication No. 369, Philadelphia, 1965, 62

47. F. G. Haynes, J. Inst. Metals *90*. 311 (1961/62)

48. W. J. Boesch and H. B. Canada, "Precipitation Reactions and Stability of Ni₃Cb in Inconel Alloy 718," International Symposium on Structural Stability in Superalloys, Seven Springs, Pa., 1968, II, 579

49. F. J. Rizzo and J. D. Buzzanell, "Effect of Chemistry Variations on the Structural Stability of Alloy 718," International Symposium on Structural Stability in Superalloys, Seven Springs, Pa., 1968, II, 501

50. I. Kirman and D. H. Warrington, J. Iron and Steel Inst. (London) *205*. 1264 (1967)

51. R. T. Weiner and J. J. Irani, Trans. ASM *59*. 340 (1966)

52. E. W. Ross, Private Communication (General Electric Company, Evandale, Ohio)

53. J. R. Mihalisin, C. G. Bieber and R. T. Grant, Trans. AIME *242*. 2399 (1968)

54. W. J. Boesch and J. S. Slaney, Metal Progress *86*. (1). 109 (1964)

55. L. R. Woodyatt, C. T. Sims and H. J. Beattie, Jr., Trans. AIME *236*. 519 (1966)

56. J. S. Slaney, "A General Method for the Prediction of Precipitate Compositions," International Symposium on Structural Stability in Superalloys, Seven Springs, Pa., 1968, I, 67

Source: *Strengthening Mechanisms in Nickel-Base Superalloys*, International Nickel Co., Inc.

297

57. H. J. Beattie, Jr. and W. C. Hagel, Trans. AIME *221*. 28 (1961)

58. C. T. Sims, J. Metals *18*. 1119 (1966)

59. R. Blower and G. Mayer, J. Iron and Steel Inst. (London) *201*. 933 (1963)

60. K. Bungardt and G. Lennartz, Arch. Eisenhüttenwes. *33*. 251 (1962)

61. E. A. Fell, W. I. Mitchell and D. W. Wakeman, "The Interrelation of Structure and Stress-Rupture Properties of Nickel-Chromium Alloys Strengthened with Titanium and Aluminum," Structural Processes in Creep, Special Report 70, The Iron and Steel Institute (London), 1961, 136

62. R. F. Decker and J. W. Freeman, Private Communication (University of Michigan)

63. H. E. Collins, "Relative Stability of Carbide and Intermetallic Phases in Nickel-Base Superalloys," International Symposium on Structural Stability in Superalloys, Seven Springs, Pa., 1968, I, 171

64. H. E. Collins and R. J. Quigg, Trans. ASM *61*. 139 (1968)

65. L. A. Weisenberg and R. J. Morris, Metal Progress *78*. 70 (Nov. 1960)

66. W. Betteridge and A. W. Franklin, J. Inst. Metals *85*. 473 (1956/57)

67. C. G. Bieber, U. S. Patent No. 2,570,194 (1951)

68. D. M. Ward, Private Communication (International Nickel Limited)

69. H. J. Beattie, Jr. and F. L. VerSnyder, Trans. ASM *49*. 883 (1957)

70. A. Havalda, Trans. ASM *62*. 477 (1969)

71. C. G. Bieber, "The Melting and Hot Rolling of Nickel and Nickel Alloys," Metals Handbook, ASM, Cleveland, 1948, 1028

72. R. W. Koffler, W. J. Pennington and F. M. Richmond, Universal-Cyclops Steel Corporation, Research and Development Department, Report 48, 1956

73. R. F. Decker, J. P. Rowe and J. W. Freeman, "Influence of Crucible Materials on High-Temperature Properties of Vacuum-Melted Nickel-Chromium-Cobalt Alloy," NACA Technical Note 4049, Washington, D. C., June 1957

74. K. E. Volk and A. W. Franklin, Z. Metallk. *51*. 172 (1960)

75. J. Schramm, J. H. Olson and C. G. Bieber, Private Communication (The International Nickel Company, Inc., Paul D. Merica Research Laboratory)

76. W. Peter, H. Müller and E. Kohlhaas, Arch. Eisenhüttenwes. *38*. 329 (1967)

77. B. S. Natapov, V. E. Ol'shanetskii and E. P. Ponomarenko, Metal Science and Heat Treatment *1965*. (1). 11

78. E. G. Richards and P. L. Twigg, "Influence du bore sur les propriétés de tenue au fluage d'un alliage austénitique au nickel-chrome à durcissement structural," Contribution to 11th Creep Colloquium, 1967, Saclay, France

79. F. C. Hull and R. Stickler, "Effects of Nitrogen, Boron, Zirconium and Vanadium on the Microstructure, Tensile and Creep-Rupture Properties of a Chromium-Nickel-Manganese-Molybdenum Stainless Steel," Joint International Conference on Creep, The Institution of Mechanical Engineers, London, 1963, Session 1, 49

80. C. Crussard, J. Plateau and G. Henry, "The Influence of Boron in Austenitic Alloys," Joint International Conference on Creep, The Institution of Mechanical Engineers, London, 1963, Session 1, 91

81. J. W. Freeman and J. W. Schultz, Private Communication (University of Michigan)

82. E. G. Richards, J. Inst. Metals *96*. 365 (1968)

83. F. L. VerSnyder and R. W. Guard, Trans. ASM *52*. 485 (1960)

84. B. J. Piearcey and F. L. VerSnyder, SAE J. *74*. (6). 84 (1966)

85. B. J. Piearcey, B. H. Kear and R. W. Smashey, Trans. ASM *60*. 634 (1967)

86. B. H. Kear and B. J. Piearcey, Trans. AIME *239*. 1209 (1967)

Alloys, Cobalt, No. 39 (June 1968) 63.

7. F. S. Badger, High-Temperature Alloys: 1900-1958, J Metals, 10 (1958) 512.

8. I. Perlmutter, High Performance Jet Engine Design Dependent upon Metallurgical Ingenuity, J Metals, 6 (1954) 113.

9. F. R. Morral, Alloys for the Aircraft Industry – The Role of Cobalt, Cobalt, No. 2 (March 1959) 23.

10. W. F. Simmons, What Alloy Shall I Use for High-Temperature Applications (Above 1200° F.), Metal Progress. 80 (Oct. 1961) 84.

11. W. F. Simmons and H. J. Wagner, Guide to Selection of Superalloys, Metal Progress, 91 (June 1967) 86.

12. J. C. Freche and R. W. Hall, Progress in NASA Programs for Development of High Temperature Alloys for Advanced Engines; paper presented at Sixth International Congress on Aeronautical Sciences AIAA, Munich, Germany, September, 1968.

13. H. J. Wagner and J. Proch, Jr., Where Soviet Super-alloys Stand, Metal Progress, 91 (Mar. 1967) 75.

14. Ward F. Simmons and M. C. Metzger, Compilation of Chemical Compositions and Rupture Strengths of Super-Strength Alloys, ASTM Data Series Publication No. DS 9d, (1967).

15. F. J. Clauss and J. W. Weeton, Relation of Microstructure to High-Temperature Properties of a Wrought Cobalt-Base Alloy Stellite 21 (AMS 5385), NACA TN 3108, (March 1954).

16. E. J. Felten and R. A. Gregg, The Physical Metallurgy and Oxidation Characteristics of a Cobalt-Base Superalloy, SM-302, ASM Trans Quart, 57 (1964) 804.

17. H. J. Beattie, Jr. and W. C. Hagel, Observations on the Formation of Geometrically and Topologically Close-Packed Phases in Commercial Superalloys; in, Precipitation Processes in Steels, ISI Special Rpt. No. 64, London (1959) 108.

18. J. M. Drapier, V. Leroy, C. Dupont, D. Coutsouradis, L. Habraken, Structural Stability of MAR-M509, A Cobalt-Base Superalloy; International Symposium on Structural Stability in Superalloys, held at Seven Springs, Pa., September 4-6, 1968, Vol. 2, 436.

19. G. A. Fritzlen, W. H. Faulkner, B. R. Barrett, Precipitation in Cobalt-Base Alloys; in, Precipitation from Solid Solution, American Society for Metals, Metals Park, Ohio (1959) 449.

20. J. Clauss, written discussion to Effect of Aging Cycle on the Properties of an Iron-Base Alloy Hardened with Titanium, T. W. Eichelberger, Trans, ASM, 50 (1958) 757.

21. N. J. Grant, The Stress-Rupture and Creep Properties of Heat-Resistant Gas Turbine Alloys, Trans ASM, 39 (1947) 281.

22. N. J. Grant, Structural Variations in Gas Turbine Alloys Revealed by the Stress-Rupture Test, Trans ASM, 39 (1947) 335.

23. R. Yoda, T. Watanabe and K. Kawagoe, Study on Cobalt-Base Heat-Resisting Alloys (II), Trans National Res Inst Metals, 3 (1961) 25.

24. M. Gell and G. R. Leverant, The Fatigue of the Nickel-Base Superalloy, MAR-M200, in Single Crystal and Columnar Grained Forms at Room Temperature, Trans AIME, 242, (1968) 1869.

25. D. N. Duhl and M. Gell, private communication, Advanced Materials Research and Development Laboratory, Pratt & Whitney Aircraft, Middletown, Conn., 1968.

26. F. J. Clauss, F. B. Garrett and J. W. Weeton, Effect of Some Selected Heat Treatments on the Operating Life of Cast HS-21 Turbine Blades, NACA TN 3512 (July 1955).

27. G. A. Hoffman and C. F. Robards, Effects of Some Solution Treatments Followed by an Aging Treatment on the Life of Small Cast Gas-Turbine Blades of a Cobalt-Chromium Base Alloy. II-Effect of Selected Combinations of Soaking Time, Temperature, and Cooling Rate, NACA TN, (1951).

28. G. A. Fritzlen, History, Status and Future of Cobalt Alloys; in, High Temperature Materials, John Wiley and Sons, Inc., New York (1959) 56.

29. R. P. De Vries, Jr. and G. Mohling, The Development of Forging and Casting Alloys for Turbine Buckets, AF Tech. Rpt. No. 6615 (August 1951).

30. W. C. Hagel and H. J. Beattie, Jr., Cellular and General Precipitation During High-Temperature Ageing; in, Precipitation Processes in Steels, ISI Spec, Rpt. No. 64, London, (1959) 98.

31. B. L. Lux and W. Bollmann, Precipitation Hardening of Co-Base Alloys by Means of an Intermetallic Phase, Cobalt, No. 11 (June 1961) 4.

32. J. J. Rausch, J. B. McAndrew and C. R. Simcoe, Effect of Alloying on the Properties of Wrought Cobalt; in, High Temperature Materials II, John Wiley and Sons, Inc., New York (1963) 259.

33. R. W. Guard and T. A. Prater, An Austenitic Alloy for High Temperature Use, Trans ASM, 49 (1957) 842.

34. R. W. Guard, Fracture at Elevated Temperatures; in, Fracture of Engineering Materials, American Society for Metals, Metals Park, Ohio, 1964, p. 95.

35. G. S. Lee, Precipitation Hardening Characteristics of a New Cobalt-Base Alloy, Cobalt, No. 37 (1967) 216.

36. R. L. Ashbrook and J. F. Wallace, Modification of Eutectic Alloys for High-Temperature Service, Trans AIME 236, (1966) 670.

37. L. R. Woodyatt, C. T. Sims, and H. J. Beattie, Jr., Prediction of Sigma-Type Phase Occurrence from Compositions in Austenitic Superalloys, Trans AIME, 236 (1966) 519.

38. W. J. Boesch and J. S. Slaney, Preventing Sigma Phase Embrittlement in Nickel-Base Superalloys, Metal Progress, 86 (1964) 109.

39. H. J. Murphy, G. T. Sims and A. M. Beltran, Phacomp Revisited; International Symposium on Structural Stability in Superalloys, held at Seven Springs, Pa., September 4-6, 1968, Vol. 1, 47.

40. C. H. Lund, M. J. Woulds and J. Hockin, Cobalt and Sigma: Participant, Spectator, or Referee?, International Symposium on Structural Stability in Superalloys, held at Seven Springs, Pa., September 4-6, 1968, Vol. 1, p. 25.

41. R. Silverman, W. Arbiter and F. Hodi, Effect of Sigma Phase on Co-Cr-Mo Base Alloys, Trans ASM, (1957) 805.

42. L. Habraken and D. Coutsouradis, Properties of the New Cobalt-Base Alloy UM Co-50, Cobalt Information Center document.

43. M. Urbain, P. Blavier and D. Coutsouradis, Structure, Properties, and Applications of UM Co-50 Alloy, J Metals, 16 (1964) 837.

44. S. T. Wlodek, Embrittlement of a Co-Cr-W (L-605) Alloy, ASM Trans Quart, 56 (1963) 287.

45. J. C. Freche, R. L. Ashbrook and G. D. Sandrock, discussion to Embrittlement of a Co-Cr-W (L-605) Alloy, S. T. Wlodek, ASM Trans Quart, 56 (1963) 971.

46. G. D. Sandrock, R. L. Ashbrook and J. C. Freche, Effect of Silicon and Iron Content on Embrittlement of a Cobalt-Base Alloy (L-605), Cobalt, No. 28 (September 1965) 111.

47. G. D. Sandrock and L. Leonard, Effect of Cold Reduction on Precipitation and Embrittlement of a Cobalt-Base Alloy (L-605), Cobalt, No. 33 (December 1966) 171.

48. G. D. Sandrock and L. Leonard, Cold Reduction as a Means of Reducing Embrittlement of a Cobalt-Base Alloy (L-605), NASA TN D-3528 (August 1966).

49. J. W. Weeton, F. J. Clauss and J. R. Johnston, Performance of As-Forged, Heat-Treated, and Overaged S-816 Blades in a Turbojet Engine, NACA RM E54K17 (March 1955).

50. C. A. Hoffman and C. A. Gyorak, Investigation of Effect of Grain Size upon Engine Life of Cast AMS 5385 Gas-Turbine Blades, NACA RM E53D06 (1953).

51. C. Yaker, F. B. Garrett and P. F. Sikora, Investigation of Engine Performance and High-Temperature Properties of Precision-Cast Turbine Blades of High-Carbon Stellite 21 and Controlled-Grain Size Stellite 21., NACA RM E52D10 (1952).

52. J. R. Johnston, C. A. Gyorak and J. W. Weeton, Engine Performance of Alloy 73J Turbine Blades Cast to Predetermined Grain Sizes, NACA RM E54E05 (July 1954).

53. B. J. Piearcey and B. E. Terkelsen, The Effect of Unidirectional Solidification on the Properties of Cast Nickel-Base Superalloys, Trans AIME, 239 (1967) 1143.

54. R. Yoda, T. Watanabe and K. Kawagoe, Study on Cobalt-Base Heat-Resisting Alloys—III Effects of Ad-

ditional Elements on Their Properties at High-Temperature, Trans National Res Inst Metals, 5 (1963) 31.

55. W. E. Blatz, E. E. Reynolds and W. W. Dyrkacz, Influence of Boron on Cast Cobalt-Base S-816 Alloy; in, Symposium on Metallic Materials for Service at Temperatures Above 1600° F, ASTM STP No. 174 (1956) 16.

56. R. J. Morris, ML-1700 Cast Turbine Bucket Alloy; in, High Temperature Materials, John Wiley and Sons, Inc., New York (1959) 81.

57. G. Llewelyn, Discussion on Engineering Properties of Nickel-Base Alloys, J Inst Metals, 89 (1960-61), 428.

58. J. E. Breen and J. R. Lane, Effect of Rare Earth Additions on the High-Temperature Properties of a Cobalt-Base Alloy; in, Symposium on Metallic Materials for Service at Temperatures Above 1600° F, ASTM STP No. 174 (1956) 57.

59. Haynes Developmental Alloy No. 188, Union Carbide Corp. Materials Systems Division Data Sheet (1967).

60. W. F. Simmons, DMIC Review of Recent Developments-Nickel and Cobalt-Base Alloys (January 1968) 2.

61. R. B. Herchenroeder, Haynes Alloy No. 188 Aging Characteristics; International Symposium on Structural Stability in Superalloys, held at Seven Springs, Pa., September 4-6, 1968, Vol. 2, 460.

62. P. Lane Jr., and N. M. Geyer, A Critical Look at Superalloy Coatings for Gas Turbine Components, J Metals, 18 (1966) 186.

63. P. Galmiche, Chromaluminisation and Tantalisation of Refractory Materials for Gas Turbines, Metals and Matls., 2 (1968) 241.

64. G. W. Goward, D. H. Boone and

G. S. Giggins, Formation and Degradation Mechanisms of Aluminide Coatings on Nickel-Base Superalloys, ASM Trans Quart, 60 (1967) 228.

65. C. H. Wells and C. P. Sullivan, Low-Cycle Fatigue of Udimet 700 at 1700F, ASM Trans Quart, 61 (1968) 149.

66. P. Shahinian, Effect of Environment on Creep-Rupture Properties of Some Commercial Alloys, Trans ASM, 49 (1957) 862.

67. D. C. Ludwigson and F. R. Morral, A Summary of Comparative Properties of Air-Melted and Vacuum-Melted Steels and Superalloys, DMIC Rept. No. 128 (March 1960).

68. M. Woulds, Casting of Cobalt-Base Superalloys, ASM Report System (October 1968).

69. P. A. Clarkin, J. W. Weeton, and P. F. Sikora, Effects of Variations in Carbon Content, Heat Treatment, and Mechanical Working on the Stress-Rupture Properties of a Liquid-Phase Sintered High-Temperature Alloy, Trans AIME, 224 (1962) 116.

70. A. M. Hall and F. R. Morral, From Gas Turbines to Furnace Parts—Cobalt Alloys Beat Heat, Matls. Design Eng, 63 (1966) 76.

71. F. Bollenrath and R. Sonntag, Influence of Cobalt on the Thermal Shock Resistance of Some Heat-Resisting Austenitic Alloys, Cobalt, No. 14 (March 1962) 3.

72. F. L. Muscatell, E. E. Reynolds, W. W. Dyrkacz and J. H. Dolheim, Thermal Shock Resistance of High Temperature Alloys, ASTM, 57 (1957) 947.

73. J. C. Freche and R. O. Hickel, Turbojet Engine Investigation of Thermal Shock Induced by External Water-Spray Cooling on Turbine

Blades of Five High-Temperature Alloys, NACA RM E55J17 (December 1955).

74. D. E. Jensen, private communication, Materials Development Laboratory, Pratt & Whitney Aircraft, East Hartford, Conn., 1968.

75. D. H. Boone, private communication, Advanced Materials Research and Development Laboratory, Pratt & Whitney Aircraft, Middletown, Conn., 1968.

76. F. J. Clauss and J. W. Freeman, Thermal Fatigue of Ductile Materials. I—Effect of Variations in the Temperature Cycle on the Thermal-Fatigue Life of S-816 and Inconel 550, NACA TN 4160 (September 1958).

77. H. L. Wheaton, MAR-M 509, A New Cast Cobalt-Base Alloy for High-Temperature Service, Cobalt, No. 29 (December 1965) 16.

78. D. E. Jensen, private communication, Materials Development Laboratory, Pratt & Whitney Aircraft, East Hartford, Conn., 1967.

79. F. J. Clauss and J. W. Freeman, Thermal Fatigue of Ductile Materials II—Effect of Cyclic Thermal Stressing on the Stress-Rupture Life and Ductility of S-816 and Inconel 550, NACA TN 4165 (September 1958).

80. M. Donachie, R. Brody and E. Bradley, Thermal Fatigue of Turbine Airfoils with Special Reference to B-1900 Nickel-Base Alloy, SAE Paper 660056, (January 1966); Miniature Airfoils Test Thermal Fatigue of Aircraft Gas Turbine Alloys, SAE J, 75, No. 2 (1967) 90-93.

81. F. J. Rizzo and L. W. Lherbier, Research Directed Toward the Development of a Wrought Superalloy, Tech. Rpt. AFML-TR-66-364 (November 1966).

In-Process Metallurgy of Wrought Cobalt-Base Alloys

R. B. Herchenroeder and W. T. Ebihara

IN COMPARISON with the number of wrought nickel-base alloys being used commercially, there are relatively few wrought cobalt-base alloys currently being produced in the United States. Four alloys predominate. The oldest, Haynes Stellite Alloy No. 6B,* was developed in the early 1900's by Elwood Haynes and has been used continuously since that time in applications where wear and abrasion resistance are major considerations. Even though Alloy No. 6B is the oldest of the four dominant wrought cobalt-base alloys, little has been published about its metallurgical characteristics.

Haynes Alloy No. 25* (L-605) is undoubtedly the best known wrought cobalt(base alloy; but even though much has been written about the aging behavior and strength properties of the alloy, only limited information is available regarding its in-process metallurgy.

Two alloys, UMCo-50 and Haynes Alloy No. 188* are new. Both are solid-solution strengthened alloys designed for high-temperature service in oxidizing atmospheres.

This paper deals with Alloy No. 6B, No. 25 and No. 188, with emphasis on cold working characteristics, recrystallization behavior, and precipitation phenomena during recrystallization heat treatments. The objective

The authors are associated with the Union Carbide Corp., Materials System Div., Kokomo, Ind. This paper was presented at the ASM Research Application Program sessions of the 1968 Materials Congress, Detroit.
* Trade names of Union Carbide Corp.

Table 1. Nominal Compositions of Three Haynes Cobalt-Base Alloys

	Alloy No. 188	Alloy No. 25	Alloy No. 6B
Cr	22	20	30
W	14	15	4
Fe	1.5	2	2
C	0.1	0.1	1.0
Si	0.3	0.2	1.0
Co	balance	balance	balance
Ni	22	10	2
Mn	0.75	1.5	1.0
La	0.08	—	—

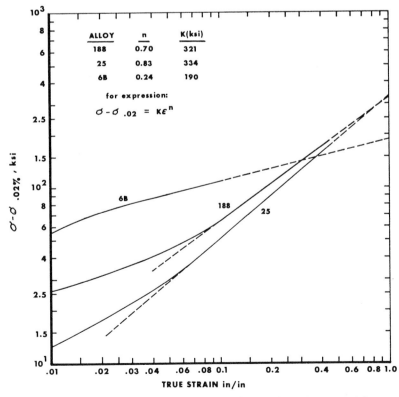

ALLOY	n	K(ksi)
188	0.70	321
25	0.83	334
6B	0.24	190

for expression:

$$\sigma - \sigma_{.02} = K\epsilon^{n}$$

Fig. 1. Strain-hardening characteristics of Alloys No. 188, 25 and 6B.

of this work is to provide fabricators of these alloys with processing data so they can more efficiently handle and form these materials. With such information, better predictions of processing limits are possible, resulting in the reduction of numerous trial-and error processing tests. Knowledge of how variations in composition and processing conditions affect structure and mechanical properties is important to fabricators who furnish material to meet certain design limits.

The nominal compositions of the three alloys to be discussed are reported in Table 1. Alloy No. 188 and No. 25 are somewhat similar in composition, the major difference being that Alloy No. 188 has twice the nickel content of Alloy No. 25 and a controlled lanthanum addition of nominally 0.08 wt %. Alloy No. 6B with its 30% chromium and 1% carbon was originally designed as a cast alloy. However, wide applications have been found for the wrought alloy.

Materials and Treatments

The mill annealed sheet was cut into rectangular blanks and rolled at room temperature on a Stanat four-high mill. Multiple passes were required for reductions of all three of the alloys studied. Total reductions of 10, 20, 30, 40 and 50% could be obtained for Alloy No. 25 and No. 188; however, only 10 and 20% could be achieved for alloy No. 6B. The rolled material was then sectioned into 1-in. by 1-in. coupons for subsequent heat treatment. Only Alloy No. 188 was sectioned for tensile specimens. Rockwell A hardness values were obtained and were correlated with measured DPH (100g load) values.

Metallographic samples of edge sections cut parallel to the rolling direction were prepared by conventional techniques and finished by electropolishing in an H_2SO_4-methanol solution followed by electrolytically etching in either $HCl + H_2O_2$ or $HCl + H Cr_3O_4$ solutions. These samples were used to optically determine recrystallization.

Phase identification was accomplished using extraction and x-ray diffraction techniques. To evaluate the

work hardening characteristics of the three alloys, room-temperature tensile tests were conducted at a strain rate of 0.025 min^{-1}.

Effects of Thermo-Mechanical Treatments

The strain-hardening characteristics observed during tensile loading are helpful in determining the workability of materials. For example, for materials with high stress values and high work hardening exponents would be expected to resist necking, but during rolling, relatively high separation forces would result when large deformations were attempted. This is true of Alloy No. 188 and No. 25, but Alloy No. 6B is the exception.

The strain-hardening characteristics for the three alloys are shown in Fig. 1. Sigma (σ) is the true stress; while $\sigma_{.02}$ represents the 0.02% offset yield stress. K is the extrapolated stress value at unit strain ϵ; n is the work-hardening exponent. Alloy No. 25 and No. 188 show high work hardening exponents of 0.83 and 0.70, respectively. These alloys have a high degree of resistance to necking as would be expected. Alloy No. 25 and No. 188 possess relatively low stacking fault energies (1); therefore, cross-slip would be difficult in these alloys. In this case, the build-up rate of dislocation density is high relative to the strain, resulting in increased strain hardening. Alloy No. 6B ex-

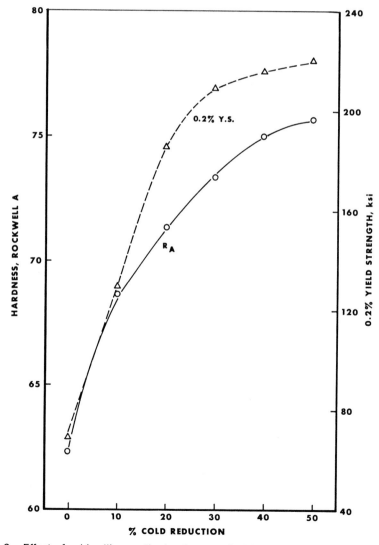

Fig. 2. Effect of cold rolling on the hardness and yield strength of Alloy No. 188.

hibits a value of 0.24 for *n*, but with very little ductility—approximately 10% total tensile elongation. On the other hand, elongations of over 50% can be expected at room temperature for Alloy No. 25 and No. 188. The expression $(\sigma - \sigma_{.02ys} = K\sigma^n)$ does not take into account the ductility of the materials. However, the expression does yield the relative degree of workability of materials. For materials with high *n* and *K* values, one would expect that numerous cold rolling passes with subsequent annealing would be necessary to obtain large reductions. This is true for Alloy No. 188 which was found to require a number of passes to obtain a given amount of reduction on a Sendzimir cluster mill. Alloy No. 6B, on the other hand, with its lower work hardening exponent is difficult to cold work because of the presence of massive Cr_7C_3 carbides, giving rise to a high yield strength.

The yield strength of Alloy No. 188 and No. 25 increases rapidly with increasing cold deformation. An approximation of this increase in strength as a function of cold deformation can be made by checking the hardness of the strained structure, as shown in Fig. 2 for alloy No. 188. This is not an unexpected correlation since the yield stress of a material is approximately proportional to its hardness if the strain introduced by the indentor is taken into account (2). This is an important point since hardness measurements are more easily obtained than yield stress values.

Inasmuch as Alloy No. 25 is often used in the cold worked condition because of the high strengths and hardness obtainable, tests were performed to determine the effects of cold work on the mechanical properties, at room temperature and at selected moderately low temperatures, of Alloy No. 188. The tensile specimens were held approximately 30 min at the testing temperature prior to loading. The room-temperature yield strength is increased two-fold by a cold reduction of 10% and approximately four-fold by a 50% cold reduction. The yield strength of the 10% cold reduced material decreases with tem-

perature as does that of the unworked material. However, no reduction of yield strength relative to that of annealed material is observed. Higher amounts of deformation tend to produce minimums in the yield strength vs temperature curves at 600 to 800° F. Higher temperatures (>800° F) show substantial increases in the yield strength relative to the unworked material. Precipitation of $M_{23}C_6$ carbides from the supersaturated matrix on the slip lines or deformation cell walls is probably responsible for the higher yield strengths. As will be discussed later, hardness peaks are observed at 1000° F for 50% cold reduced material, while the initial hardness peak for the unworked material generally occurs around 1400° F. Apparently, the deformation introduces numerous selective nucleation sites for precipitation which occur at much lower temperatures than in the annealed material.

The effect of cold rolling on the ultimate tensile strength at elevated temperatures was also determined for Alloy No. 188. There was no detectable dip in the ultimate strength curve because of relaxation of the deformed matrix from room temperature up to 1000° F. Ultimate strength was not observed to be as sensitive as yield strength to aging caused by the pre-test soaking period. This is indicated by the fact that only the 40 and 50% cold reduced material showed minimums near 600° F.

The measured elongation as a function of temperature and cold work was found to be consistent with those observed for the yield strength. Materials that suffered deformation greater than 20% never achieved their room-temperature elongation up to 1000° F.

The structure of materials and the morphology of phases can be altered during processing. Hence, to understand what happens within these cobalt-base alloys both during forming and later in actual service conditions, a knowledge of their physical metallurgy is desirable. The phases present in these alloys after varied heat treatments are listed in Table 2.

Table 2. Phases Present in Haynes Cobalt-Base Alloys After Heat Treatment

Alloy No. 188	Alloy No. 25	Alloy No. 6B
2150° F/WQ Fcc matrix M_6C Lanthanum compound	*2200° F/WQ* Fcc matrix M_6C	*2250° F/RAC* Fcc matrix Cr_7C_3 $M_{23}C_6$(?)
Fcc matrix M_6C, M_6C^1 Lanthanum compound	*+2000° F/AC* Fcc matrix M_6C, M_6C^1	Fcc matrix Cr_7C_3 $M_{23}C_6$
Fcc matrix M_6C, M_6C^1 $M_{23}C_6$ A_2B (long term)	*+1800° F/AC* Fcc matrix M_6C, M_6C^1 A_2B	Fcc matrix Cr_7C_3 $M_{23}C_6$
Fcc matrix M_6C, M_6C^1 (disappears) $M_{23}C_6$ A_2B (long term)	*+1600° F/AC* Fcc matrix M_6C, M_6C^1(?) $M_{23}C_6$(?) A_2B	Fcc matrix Hcp(?) Cr_7C_3 $M_{23}C_6$
Fcc matrix M_6C $M_{23}C_6$ A_2B (long term)	*+1400° F/AC* Fcc matrix M_6C $M_{23}C_6$(?) A_2B	Fcc matrix Hcp(?) Cr_7C_3 $M_{23}C_6$

Fig. 3. Microstructures of Alloy No. 188: (A) annealed, (B) 10% cold rolled, (C) 30% cold rolled, (D) 50% cold rolled. 500×.

Several major differences are evident. While both Alloy No. 188 and No. 25 contain fcc M_6C after solution heat treatment at 2150 and 2200° F, respectively, Alloy No. 6B has as its predominant carbide the hexagonal Cr_7C_3. Depending upon the rapidity of the quench, Alloy No. 6B may or may not contain $M_{23}C_6$ in the grain boundaries. Other significant differences are: both Alloy No. 188 and No. 25 precipitate M_6C when exposed at higher temperatures (viz 2000° F); only $M_{23}C_6$ forms in Alloy No. 6B during aging. Part of the Cr_7C_3 in Alloy No. 6B decomposes to $M_{23}C_6$. The degree of this decomposition and precipitation of $M_{23}C_6$ from the carbon enriched matrix depends greatly on the aging treatment.

The hexagonal "Co_2W" (A_2B Laves type), which causes embrittlement, forms readily in Alloy No. 25 during short exposure periods in the temperature range of about 1300 to 1900° F. Precipitation of $M_{23}C_6$ in Alloy No. 25 is dependent upon the silicon content; with low silicon levels $M_{23}C_6$ will form, but with higher levels only the Laves phase forms. In these recrystallization studies, which employed Alloy No. 25 with nominally 0.1 wt % Si, the predominating phase that precipitated at moderate temperatures of 1600-1800° F in the deformed structure was found to be M_6C. There was some evidence of $M_{23}C_6$ being present, but identification was not positive. There was no evidence of Co_2W being present.

In contrast, fcc $M_{23}C_6$ is the predominant precipitate phase when Alloy No. 188 is exposed in the intermediate temperature range of about 1300 to 1800° F. The Laves phase will form in Alloy No. 188, but only after prolonged exposure periods on the order of 100 or more hours.

Pure cobalt has an allotropic transformation from the low-temperature hcp structure to the fcc phase at approximately 785° F. Both alloy No. 188 and No. 25 have stabilized fcc matrices, but Bleil (3) has reported that a partial transformation of the cobalt solid solution matrix of Alloy No. 6B from fcc to hcp can be accomplished by special heat treatments; however, this does not occur with usual processing.

Microstructures are often examined to determine the amount of deformation given a material or to reveal ef-

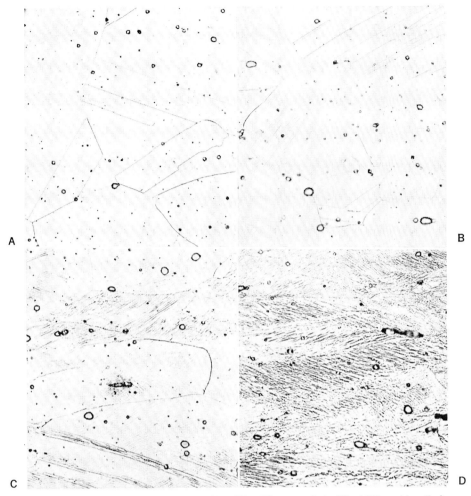

Fig. 4. Microstructures of Alloy No. 25: (A) annealed, (B) 10% cold rolled, (C) 30% cold rolled, (D) 50% cold rolled. 500×.

fects of various heat treatments. The microstructures of Alloy No. 188 with varying amounts of cold reduction are shown in Fig. 3. The solution annealed structure is characterized by primary M_6C carbides in the fcc matrix and by a lanthanum-rich compound, associated with some M_6C carbides. The cold worked structure ranges from one having noticeable slip bands in the equiaxed grains for the 10% reduction to one showing extensive deformation of grains for the 50% cold reduced material.

Typical cold worked microstructures of Alloy No. 25 are shown in Fig. 4. The original material was solution annealed at 2200° F and quenched rapidly. Some primary M_6C carbides are present, as were found for Alloy No. 188. The cold

rolled structures in the 10 to 50% reduced samples are also quite similar to those observed in Alloy No. 188.

The annealed microstructure for Alloy No. 6B, after rapid air cooling from 2250° F, is shown in Fig. 5. The massive primary carbides in the fcc matrix are Cr_7C_3 with some chromium-rich $M_{23}C_6$ at the grain boundaries. After a 20% cold reduction, the microstructure shows some deformed grains.

The hardness recovery characteristics for cold worked Alloy No. 188 are shown in Fig. 6 for 1 hr annealing time at temperature. The initial peak in hardness for the 50% cold reduced material occurs at approximately 1000° F. This is probably due to the precipitation of $M_{23}C_6$ or to some pre-precipitation clustering phenomena.

The peak is deferred until 1200° F for the 30% cold reduced material and until 1400 to 1500° F for the 10% cold reduced material. The secondary peaks for all three curves occur at about 1800° F, which may be attributed to the precipitation of M_6C carbides. Recrystallization has occurred earlier, at approximately 1600° F for the 30 and 50% cold worked material, while the 10% cold worked material did not recrystallize until 1800° F.

The effect of shorter annealing times on hardness recovery characteristics of cold rolled Alloy No. 188 is shown in Fig. 7 for material exposed for 15 min at temperature. Recrystallization begins before 1600° F for the 50% cold reduced material, at approximately 1700° F for the 30% ma-

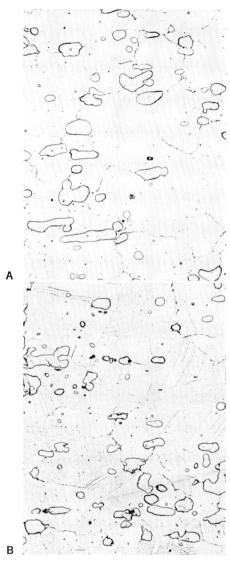

Fig. 5. Microstructures of Alloy No. 6B: (A) annealed, (B) 20% cold rolled. 500×.

1800° F, as shown in Fig. 9C, substantially fewer secondary carbides are formed.

Cold worked Alloy No. 25 exhibits recovery characteristics similar to those of Alloy No. 188 as shown in Fig. 10. However, sharp hardness peaks are not observed for the 10% cold reduced material. Higher recrystallization temperatures are noted for Alloy No. 25 than for Alloy No. 188. Recrystallization is observed after 30 min at 1700° F for the 50% cold reduced material, at 1800° F for 30% and at 2000° F for the 10% deformed samples. The original hardness is not regained until exposure at 2100° F.

Figure 11 shows the microstructures of Alloy No. 25 after 30-min exposures at 1800 and 1900° F. The 30% cold reduced material is observed to be partially recrystallized after the 1800° F exposure. The process is nearly complete after heating to 1900% F. As was the case with unworked Alloy No. 188, the solution annealed Alloy No. 25 shows very little aging after the 30-min dwell at 1800° F.

Cold worked Alloy No. 6B exhibits recovery characteristics quite different from those of alloy No. 188 and No. 25. Substantial hardness reduction is not apparent in Fig. 12. Partial recrystallization is observed after 30 min at 2000° F for the 20% cold worked material and at 2150° F for the 10% material. Increased hardness is evident in the unworked material after exposure at elevated temperatures. Extensive carbide precipitation is manifest in the cold worked material exposed between 1600 and 1800° F.

Figure 13 shows microstructures of Alloy No. 6B after 30-min exposures at 2000 and 2050° F. Partial recrystallization occurs at 2000° F for the 20% cold reduced sample, and the process is nearly complete at 2050° F. Aging is quite evident in the unworked material during the 2000° F heat treatment. Secondary $M_{23}C_6$ carbides are observed, as is the transformation of primary Cr_7C_3 to $M_{23}C_6$.

Annealing treatments utilized merely to achieve recrystallization are

terial and at 1900° F for the 10% cold reduced material. The original hardness is not restored until the 2100° F exposure.

When cold rolled Alloy No. 188 is exposed for 30 min at temperature as shown in Fig. 8, recrystallization begins at about 1500° F for the 50% cold reduced material, at 1600° F for the 30% material, and 1800° F for the 10% cold rolled sample. An intermediate hardness peak, which was not noted for the 15-min exposure, is now observed for the 10% cold worked material. The original hardness is recovered at approximately 2000° F. The alloy exhibited similar

recovery characteristics for 1-hr exposures as for 30-min exposures. Extensive softening occured in the 50% cold reduced material when exposed for 1 hr at 2100° F.

Examples of microstructures for the 30% cold reduced Alloy No. 188 after 30-min exposures at 1700 and 1800° F are shown in Fig. 9. Extensive carbide precipitation is observed at 1700° F where partial recrystallization occurs. At 1800° F, recrystallization is virtually complete but heavy precipitation, which corresponds to the hardness peak shown in Fig. 8, is still evident. When unworked material is exposed for 30 min at

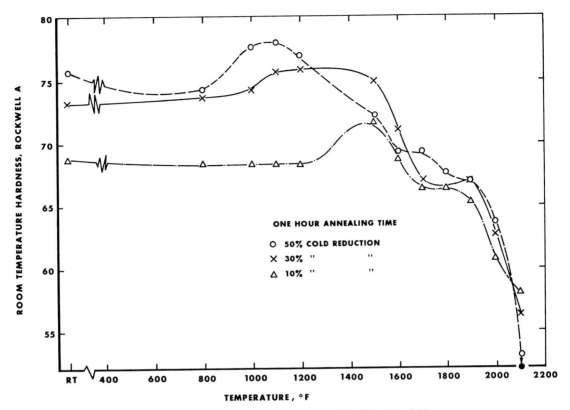

Fig. 6. Isochronal hardness recovery for Alloy No. 188.

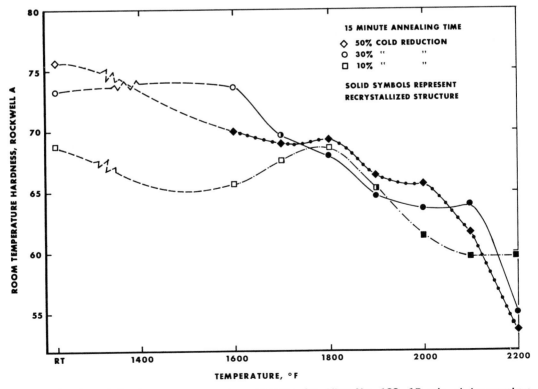

Fig. 7. Isochronal hardness recovery characteristics for Alloy No. 188, 15 min at temperature.

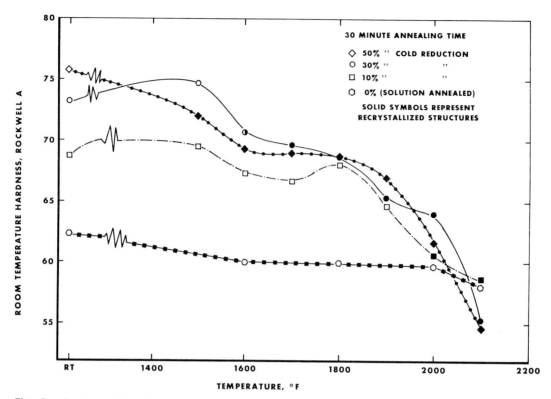

Fig. 8. Isochronal hardness recovery characteristics for Alloy No. 188, 30 min at temperature.

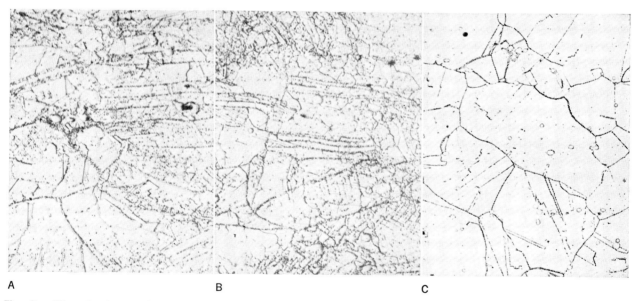

A   B   C

Fig. 9. Microstructures of Alloy No. 188 after 30-min annealing exposures: (A) 30% cold roll + 1700° F, 1000×; (B) 30% cold roll + 1800° F, 1000×; (C) 0% cold roll + 1800° F, 500×.

not sufficient to produce a ductile material. When cold worked Alloy No. 188 is annealed at 2000 and 2100° F, M_6C carbides appear in sizable quantity. Figure 14 compares the 30% cold reduced material with the unworked samples after 30-min exposures at 2000 and 2100° F. Exposures above 2100° F are necessary to dissolve most of these carbides.

Grain size control is an important consideration in metal processing. Therefore, time and temperature must be controlled to achieve optimum grain size as well as sufficient carbide solutioning. The solution annealed Alloy No. 188 had an average grain size of 34μ (ASTM 6.5) prior to cold

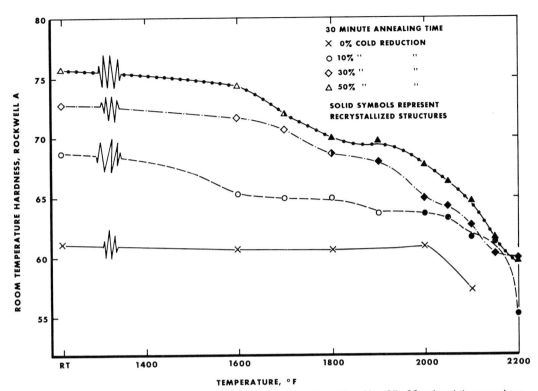

Fig. 10. Isochronal hardness recovery characteristics for Alloy No. 25, 30 min at temperature.

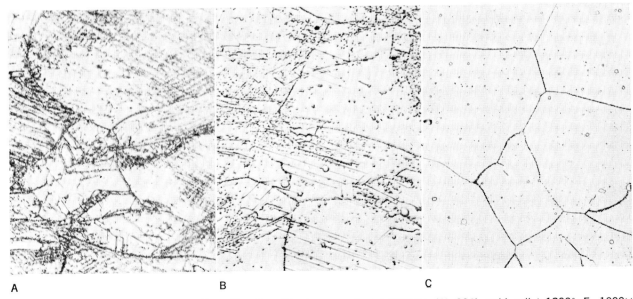

A   B   C

Fig. 11. Microstructures of Alloy No. 25 after 30-min annealing exposure: (A) 30% cold roll + 1800° F, 1000×; (B) 30% cold roll + 1900° F, 1000×; (C) 0% cold roll + 1800° F, 500×.

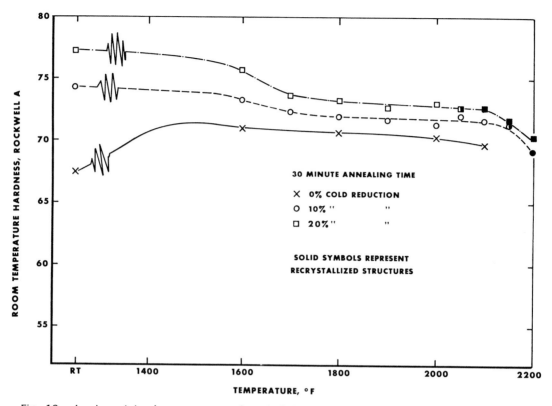

Fig. 12. Isochronal hardness recovery characteristics for Alloy No. 6B, 30 min at temperature.

rolling. The average recrystallized grain size for the 30 and 50% cold reduced material after exposures at 1700° F is approximately 4μ. Noticeable grain growth is evident after annealing at temperatures of 2000° F and above, the grain size being 6 to 12μ after 30 min at 2000° F. After 30 min at 2100° F, the 30% cold reduced material possesses an average grain size of approximately 41μ. Shorter annealing times at higher temperatures are therefore necessary to control the grain size to ASTM 6-6.5 and to generate sufficient carbide solutioning. Fifteen-minute exposures at 2100 and 2200° F yielded a grain sizes of 8μ and 68μ, respectively, for the 30% cold reduced material. Optimum solution annealing parameters appear to be less than 15 min at 2150° F.

Interpretations

Rolling of ideal frictionless, isotropic materials results in uniform deformation throughout the thickness of the material. Most metals, however, experience inhomogeneous deforma-

tion during cold rolling especially with light reductions. Shallow passes tend to work harden the surface layers relative to the interior. Consequently, to achieve a given reduction for Alloy No. 188 and No. 25, which exhibit high work hardening exponents, high separating forces would be required with several shallow passes relative to a single large reduction pass. This work hardening characteristic is also a factor in machining operations and implies that deep initial cuts could be made more readily than several shallow cuts.

Another obvious factor is that "warm working" Alloy No. 188 and No. 25 (below 1000° F) is not particularly advantageous. It is shown that 20% deformation produces high yield strengths and low ductility at 1000° F, a phenomenon which may be attributed to considerable aging in the deformation structure at this temperature.

It is also clear that relatively high temperatures (greater than 1600° F) must be employed before any appreciable restoration of annealed prop-

erties is observed. Furthermore, even with the onset of recrystallization of the matrix, a significant hardness persists due to the precipitation of secondary phases. The presence of secondary phases significantly reduces the ductility of the deformed and aged materials compared with the properties of solution annealed material.

There are certain possible advantages to using low-temperature annealing treatment (1600 to 2150° F). Sandrock and Leonard (4) showed that similar treatment were beneficial to the post-aging ductility of Alloy No. 25 after prolonged aging treatments. In their work, uniform precipitate distribution throughout the matrix could be accomplished by precipitation on the deformed structure, thus preventing appreciable grain and twin boundary precipitation. The work of Herchenroeder (5), showed that the formation of the embrittling Laves phase, Co_2W, in Alloy No. 25 could be retarded by utilizing solution annealing temperatures lower than 2250° F.

Obviously, low annealing tempera-

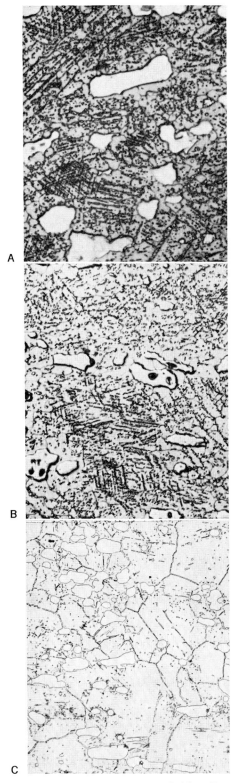

Fig. 13. Microstructure of Alloy No. 6B after 30-min annealing exposure: (A) 20% cold roll + 2000° F, 1000×; (B) 20% cold roll + 2050° F, 1000×; (C) 0% cold roll + 2000° F, 500×.

tures promote the formation and stabilization of fine grain size. Although fine-grained material improves yield strength and fatgiue life, reduced creep-rupture properties may be experienced at elevated temperatures. Large-grained material, on the other hand, is characterized by lower yield and ultimate strengths, but has high ductility.

Partial solutioning of primary carbides occurs above 2100° F for both Alloy No. 188 and No. 25. At these temperatures ($T/Tm \sim 0.90$), grain growth is extremely rapid and short annealing times are recommended for suitable grain size control.

The generalities of the effects of thermal treatments on the structure of Alloy No. 188 and No. 25 may be applied to Alloy No. 6B. This alloy behaves somewhat differently during mechanical deformations because of its lower work hardening exponent and low ductility. Solution annealed Haynes Alloy No. 6B undoubtedly has a higher carbon content in its supersaturated matrix as compared to alloy No. 188 and No. 25 because of its initial high carbon content and of the higher annealing temperature (2250° F) imposed on the alloy. Therefore, when the cold worked alloy is exposed to recrystallization temperatures near 2000° F, gross quantities of secondary carbides rapidly form on the substructure walls and, thus, restrict grain growth.

In summary, at least three important factors should be stressed as a result of this work on Haynes Alloy No. 188, and 25 and No. 6B. The first is that all three alloys are difficult to cold reduce. Alloys No. 188 and 25 exhibit high work hardening exponents while No. 6B exhibits a high yield strength and limited ductility. Exposing cold deformed material or working at temperatures above 800° F causes strain aging. Below 800° F, the yield strength of the cold deformed material remains quite high and, therefore, deformation processing at temperatures above room temperature would not be particularly beneficial.

The second factor is that heavy precipitation of carbides occurs in the

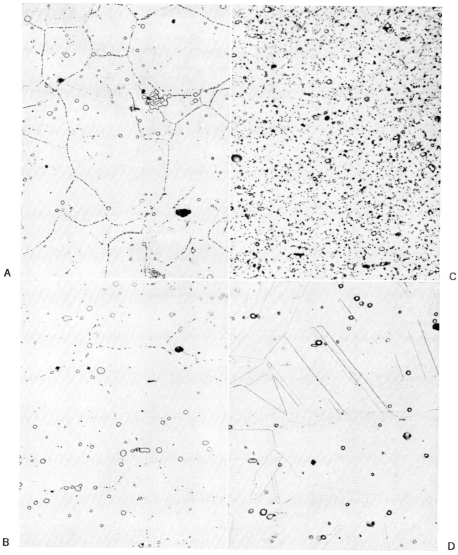

Fig. 14. Microstructures of Alloy No. 188 after 30-min annealing exposure: (A) 0% cold roll + 2000° F, (B) 30% cold roll + 2000° F, (C) 0% cold roll + 2100° F, (D) 30% cold roll + 2100° F, 500×.

range from 1500 to 2000° F for the cold worked alloys. This phenomenon curtails the use of recrystallization-annealing treatments in this range because the resultant structure will still cause high room temperature yield strengths.

The third point is that solutioning of carbides occurs above 2100° F for all three alloys. However, grain growth is quite rapid at these temperatures and short-time exposures are recommended to preserve a fine-grained structure.

ACKNOWLEDGMENTS

Appreciation is expressed to R. R. Mc-Graw, J. J. Simmons, J. W. Shutt and the personnel of Materials Systems Division Technology Laboratory for their assistance with this work.

REFERENCES

1. P. S. Kotval, Carbide Precipitation on Imperfections in Superalloy Matrices, Trans AIME 242 (1968) 1651.
2. B. R. Oliver and J. E. Bowers, The Determination of Yield Stress from Hardness Measurements, J Inst Metals, 94 (1966) 223.
3. H. E. Bleil, Dimensional Change of Stellite 6B During Phase Transformation from FCC to HCP Crystal Structure, Aerojet General Corp. Technical Memorandum 4923:65-4-357 (December 3, 1965).
4. G. D. Sandrock and L. Leonard, Cold Reduction as a Means of Reducing Embrittlement of Cobalt-Base Alloy (L-605), NASA TN D-3528, (August 1966).
5. R. B. Herchenroeder, Evaluation of Four Hours at 2100° F Heat Treatment of Haynes Alloy No. 25, Haynes Stellite Co., Project 816-41101-2 (January 4, 1962).

SECTION X:
Control of Structure and Properties

Controlling Microstructures and Properties of Superalloys Via Use of Precipitated Phases

Donald R. Muzyka

Historically, wrought precipitation-strengthened superalloys have been utilized in two basic heat treated states. Solution treatments well in excess of solvus temperatures of precipitating phases are used where relatively coarse grain sizes and optimum 1,400 to 1,800 F (760 to 980 C) creep/rupture properties are needed. Solution treating at or below solvus temperatures of the precipitating phases produces best tensile and low cycle fatigue properties up to about 1,200 F (650 C).

Microstructure and, in particular, grain size, are basic to the attainment of property requirements in superalloys. In a generalized concept of structure control, alloys considered are grouped into two basic types. Type I includes alloys such as Waspaloy, Astroloy, Rene' 95, and Pyromet 925, where the γ' phase is believed to be the primary phase useful for structure control. Type II covers alloys such as

Dr. Muzyka is supervisor, High Temperature Alloy Research, Research and Development Center, Carpenter Technology Corp., Reading, Pa.

A-286, 901, and Pyromet 860, where a secondary phase—in addition to the γ'—can be used for structure control. Extensions of these ideas to thermomechanical treatment and superplastic behavior of superalloys are also presented.

Introduction

In a recent paper, R. F. Decker (1) discussed our "state of enlightenment" on strengthening mechanisms of precipitation hardened superalloys. He considered the effects of alloying elements on the γ matrix, γ', phase stability, carbides and grain boundaries. Others, including Sullivan and Donachie (2, 3), Sabol and Stickler (4), and Sims (5) have written papers which describe the status of our knowledge of the precipitation strengthened superalloys. Currently, these alloys are grouped into three categories as follows: Fe-base, Fe-Ni base and Ni-base.

The papers listed above, for the most part, relate properties to microstructure in a general sense. However, no

TABLE I COMPOSITION OF WROUGHT SUPERALLOYS

Alloy	C	Cr	Ni	Fe	Co	Ti	Al	B	Mo	Cb	Others
					TYPE I						
Nimonic 80A	0.05	19.5	Bal	—	—	2.25	1.10	0.005	—	—	—
Waspaloy	0.05	19.5	Bal	—	13.5	3.00	1.30	0.005	4.3	—	0.05 Zr
M-252	0.15	20.0	Bal	—	10.0	2.50	1.00	0.005	10.0	—	0.10 Zr
Rene' 41	0.09	19.0	Bal	—	11.0	3.10	1.50	0.005	10.0	—	—
Astroloy	0.05	15.0	Bal	—	15.0	3.50	4.40	0.030	5.25	—	—
Pyromet 925	0.04	15.0	Bal	—	15.0	3.75	3.00	0.020	5.0	—	2.5 W
713C	0.12	12.5	Bal	—	—	0.80	6.10	0.012	4.2	2.00	0.10 Zr
IN 100	0.18	10.0	Bal	—	15.0	5.00	5.50	0.015	3.0	—	1.0 V, 0.05 Zr
Rene' 95	0.15	14.0	Bal	—	8.0	2.50	3.50	0.010	3.5	3.50	3.5 W, 0.05 Zr
TAZ-8A	0.125	6.0	Bal	—	—	—	6.00	0.004	4.0	2.50	4.0 W, 1.0 Zr, 8 Ta
NASA VI-A	0.13	6.1	Bal	—	7.5	1.00	5.40	0.020	2.0	0.50	5.5 W, 0.13 Zr, 9 Ta
NASA II-b	0.13	10.0	Bal	—	10.0	1.00	4.50	0.020	2.0	—	5.5 W, 8 Ta, 1 Hf, 1 V, .03 Zr
					TYPE II						
A-286	0.05	15.0	26.0	Bal	—	2.00	0.20	0.005	1.25	—	0.30 V
V-57	0.05	15.0	25.5	Bal	—	3.00	0.25	0.008	1.25	—	0.30 V
W-545	0.05	13.5	26.0	Bal	—	2.85	0.20	0.008	1.5	—	0.30 V
Unitemp 212	0.08	16.0	25.0	Bal	—	4.00	0.15	0.060	—	0.50	0.05 Zr
901	0.05	13.5	42.7	34.0	—	2.50	0.25	0.015	6.1	—	—
Pyromet 860	0.05	14.0	42.5	Bal	4.0	3.00	1.25	0.010	6.0	—	—
718	0.04	19.0	52.5	Bal	—	0.90	0.50	0.005	3.05	5.30	—
706	0.02	16.0	40.0	Bal	—	1.70	0.30	0.004	—	2.75	—
Rene' 62	0.05	15.0	Bal	22.0	—	2.50	1.25	0.005	9.0	2.25	—
Udimet 630	0.04	17.0	Bal	18.0	—	1.00	0.60	0.005	3.0	6.50	3.0 W
AF 1753	0.24	16.3	Bal	9.5	7.2	3.20	1.90	0.008	1.6	—	8.4 W, 0.05 Zr
D-979	0.05	15.0	45.0	Bal	—	3.00	1.00	0.010	4.0	—	4.0 W

general review of the fundamentals of processes which are needed to attain desired microstructures in wrought superalloys is available. One of the goals of this paper is to show that nearly identical properties can be achieved in alloys of various compositions, providing that processing is properly designed to control the microstructure.

There is increased emphasis on the attainment of high tensile and low cycle fatigue strength levels in wrought superalloys at temperatures below about 1,200 F (650 C) (6-9). Because high fatigue strengths are associated with fine grain sizes, this exposition on structure control should be useful for metallurgists interested in controlling grain size to achieve good tensile, fatigue, and low cycle fatigue strengths in superalloys.

This paper will review the state-of-the-art with regard to structure control in wrought precipitation strengthened superalloys. Table I lists compositions of alloys grouped in a new classification system with regard to types of processes necessary to achieve required microstructures and properties. This classification system, it is felt, will lead to efficient use of known information on a given wrought superalloy in developing useful processes for new alloys. All comments in this paper refer to heats of commercial alloys which have been balanced in composition to avoid "undesirable" phases such as sigma (10, 11). Where alloys containing Mu or Laves are mentioned, reference is made only to Mu or Laves which precipitate from solid solution during processing.

Some History

In the early days of forged nickel-base superalloys, the usual heat treatment included a solution treating step at a temperature well above the γ' solvus temperature. The purpose was to dissolve all γ', anneal the matrix to eliminate potentially mobile dislocations (12), and promote some grain growth in order to develop the maximum high temperature creep and stress rupture properties (13). Next, the importance of solution treatments just slightly above the γ' solvus was realized. Here, the idea was to solution treat at a temperature high enough to dissolve all γ' but low enough to prevent grain growth for best hardness, 1,000 to 1,400 F (540-760 C) tensile and creep/rupture properties, and fatigue properties.

The literature indicates that at least two heat treatments of "commercial" value were developed for each "commercial" nickel-base superalloy—one with a "low" and one with a "high" solution treating temperature (14). Aging sequences are extremely critical to the achievement of best property response in superalloys. For the purposes of this paper, it will be assumed that the "best" aging sequence is being used for the corresponding condition and alloy in each instance.

As time went on, forgers noted that the use of the "low" solution temperatures caused some problems in achieving uniform structural (and property) response in wrought superalloys (15, 16). For usual nickel-base superalloys, this was shown to be associated primarily with some degree of titanium segregation in the original ingot product. Various methods, including improved melting practices (17, 18) and thermal homogenization (19), were developed to improve homogeneity and the potential for effective structure control. But forgers also noted that "controlled deformation processing" (16), including carefully selected forging temperatures and forging sequences, and solution

treatments on the "low side" of the range allowed by various specifications, permitted additional opportunity to attain a uniform structural response. It is now recognized that, in many nickel-base superalloys, this "structure control" effect has a direct relationship with the γ' phase. As an example, Rehrer et al (20) have described in detail the structure controlling effects available due to γ' in a typical Type I nickel-base alloy, Waspaloy.

For the Fe-base and Fe-Ni base superalloys, on the other hand, phase relationships are such that other GCP (geometrically close packed) and TCP (topologically close packed) phases will precipitate in addition to γ' during elevated temperature exposures. For commercial Fe-base and Fe-Ni base alloys, these phases include Ni_3Cb, Ni_3Ti (or η), Laves, and Mu phases. The Ni_3Cb and Ni_3Ti phases are known to be the "equilibrium form" of the metastable FCC γ' occurring in various alloys. In 718 alloy, it is generally agreed that the strengthening precipitate has an equilibrium structural form of orthorhombic Ni_3Cb (21-26).

Solution heat treatment temperatures for Fe-Ni base superalloys generally exceed γ' solvus temperatures. But where 70 to 1,200 F (21 to 650 C) tensile properties are critical, solution temperatures are below the solvus temperature of the "third phase" (e.g., the HCP Ni_3Ti in A-286 or orthorhombic Ni_3Cb in 718). For 718 alloy, a 1,950 F (1,070 C) solution treatment produces high stress rupture strength, but stress rupture ductility is extremely low (22). Thus, this solution treating temperature is usually not used for 718 alloy where good stress rupture ductility is required.

The Classification System

Thus, as has been noted, it is possible to categorize precipitation hardening superalloys in two groups. First, Type I alloys of Table I have γ' as the major precipitating phase useful for structure control (that is, for ordinary hot worked structures, γ' controls recrystallization and grain growth). Type II alloys have a "third" phase (in addition to γ and γ') which is useful for structure control. For ordinary hot worked structures, the "third" phase controls both recrystallization and grain growth. However, under certain conditions for Type II alloys, γ' controls recrystallization and the "third" phase controls grain growth.

Typical phases that precipitate in superalloys are: GCP (γ', Ni_3Cb, and Ni_3Ti); TCP (Laves, Mu, and Sigma); and interstitial compounds (MC, $M_{23}C_6$, and M_6C). Phases of primary concern to this paper are the γ', Ni_3Cb and Ni_3Ti, all of which are common to superalloys.

In order to document a range of structure controlling effects in conventionally melted wrought superalloys, the following possibilities were considered:

Alloy Types	—Type I, Type II
Energy States	—high, low
Solution Temperatures	—below γ' solvus
	—between γ' and Ni_3X solvus (where applicable)
	—above γ' and Ni_3X solvus (where applicable)

Based upon extensive studies, the observations described in Fig. 1 are typical of structures attained. Figure 2 shows structures of a Type I alloy (Waspaloy) with a low initial energy starting state. It is believed to be a relatively strain free condition due to recrystallization late in the hot work-

Energy State	Solution Temperature	Recrystallization	Grains	Ni₃X
		TYPE I ALLOYS		
Low	Below γ'	No	As-Forged-Pinned by γ'	-
	Above γ'	Yes	Grow	-
High	Below γ'	Yes	Fine-Pinned by γ'	-
	Above γ'	Yes	Grow	-
		TYPE II ALLOYS		
Low	Below γ' +Ni₃X	No	As-Forged	Platelet/Globular
	Between γ' +Ni₃X	No	As-Forged	Platelet/Globular
	Above γ' +Ni₃X	Yes	Grow	-
High	Below γ' +Ni₃X	No	As-Forged	Spherical/Globular
	Between γ' +Ni₃X	Yes	Fine-Pinned by Ni₃X	Spherical/Globular
	Above γ' +Ni₃X	Yes	Grow	-

Fig. 1. The effects of solution temperatures vary with the type of alloy as well as with the temperature itself.

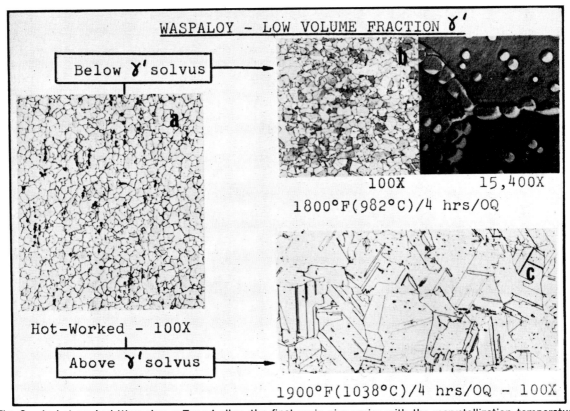

WASPALOY - LOW VOLUME FRACTION γ'

Below γ' solvus

100X   15,400X
1800°F(982°C)/4 hrs/OQ

Hot-Worked - 100X

Above γ' solvus

1900°F(1038°C)/4 hrs/OQ - 100X

Fig. 2. In hot worked Waspaloy, a Type I alloy, the final grain size varies with the recrystallization temperature. Compare with Fig. 3.

Source: *Metals Engineering Quarterly*, Nov 1971

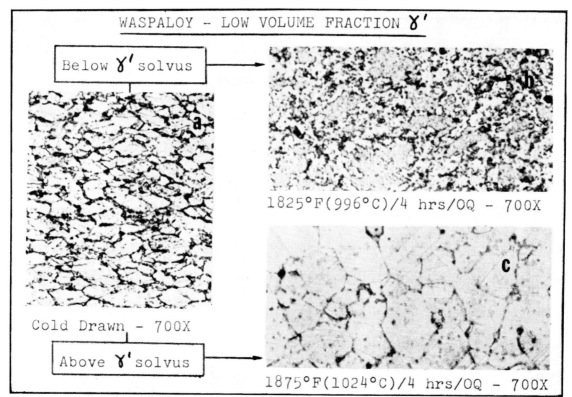

WASPALOY – LOW VOLUME FRACTION γ'

Below γ' solvus

Cold Drawn – 700X

Above γ' solvus

1825°F(996°C)/4 hrs/OQ – 700X

1875°F(1024°C)/4 hrs/OQ – 700X

Fig. 3. Cold worked Waspaloy develops different structures at different solution temperatures. Compare with Fig. 2.

ing cycle. ("Late" indicates recrystallization occurred, but insufficient temperature was available for grain growth. Also, insufficient processing occurred after recrystallization for further development of retained strain energy.) Grain size of the hot worked material was about ASTM 8. Solution treating below the γ' solvus, 1,800 F (980 C) for 4 hr and oil quenching, in this instance, maintains the fine, uniform as-forged grain structure. Relatively coarse "forging gamma prime" (20) pins the grain boundaries of the Waspaloy, thus inhibiting recrystallization and grain growth at 1,800 F (980 C). Solution treating at 1,900 F (1,040 C), above the γ' solvus, permits recrystallization and grain growth to occur.

Figure 3 shows structures of Waspaloy with an initial, high energy (strained) rate. The Waspaloy is from bar stock, experimentally annealed at 1,825 F (1,000 C) for 2 hr, water quenched, and cold drawn about 30%. Solution treating at 1,825 F, below the γ' solvus for the heat, leads to a very fine polygonized microstructure with grain boundaries apparently pinned by γ' particles. As will be shown later, this material has excellent tensile properties. Treating above the γ' solvus—at 1,875 F (1,025 C) permits recrystallization and some grain growth to occur.

These examples for a Type I alloy show that the γ' phase is useful for structure control. Type I alloys with high volume fractions of γ' give greatest flexibility in designing useful processes.

Figures 4 and 5 show structures for 718, a Type II alloy. As shown, heat treated 718 contains platelet or globular Ni_3Cb as a grain boundary pinning phase. (Pyromet 860 and 901 alloy can contain Ni_3Ti as a structure controlling phase.) The equiaxed as-hot-worked structure for 718 alloy is believed to have recrystallized relatively late in the

hot working cycle. Heating below the γ'' and Ni_3Cb solvus temperatures produces a structure of platelet-to-globular Ni_3Cb with a background of relatively coarse γ''. Heating to 1,750 F (950 C), above the γ'' solvus but below the Ni_3Cb solvus temperature, maintains the relatively fine as-hot-worked grain size with grain boundaries pinned by platelet-to-globular Ni_3Cb. It is believed that the platelet-to-globular Ni_3Cb in the grain boundaries contributes to good elevated temperature strength and ductility by inhibiting grain boundary sliding. (In fact, this type of strengthening mechanism may be applicable to all Type II alloys under proper conditions.) However, heating at 1,900 F (1,040 C), above the Ni_3Cb solvus permits grain growth, due to the loss of the structure controlling effects of the Ni_3Cb (27).

Figure 5 shows 718, as processed to a relatively higher energy (strained) state. Specimens probably did not re-

TABLE II PHASES FOR STRUCTURE CONTROL IN TYPE II ALLOYS

Alloy	Phase
A-286	Ni_3Ti
V-57 or W-545	Ni_3Ti
Unitemp 212	Ni_3Ti + Laves
901	Ni_3Ti
Pyromet 860	Ni_3Ti
718	Ni_3Cb
706	Ni_3Cb + Ni_3Ti
Rene' 62	Laves
Udimet 630	Laves
AF 1753	Laves
D-979	Mu

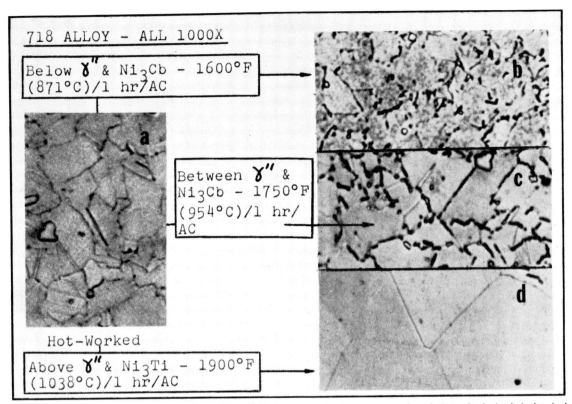

Fig. 4. A type II alloy, 718 is strengthened by Ni₃Cb in the grain boundaries when hot worked stock is heated at 1,750 F for 1 hr and cooled in air (right center). Higher temperatures can coarsen grains (right bottom). Compare with Fig. 5.

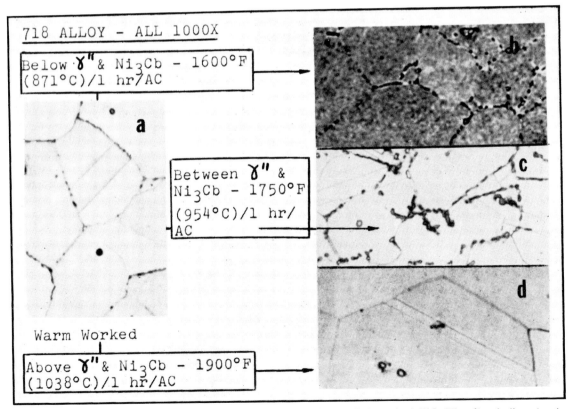

Fig. 5. Though warm worked 718 has a higher energy level than hot worked 718 (Fig. 4), similar structures develop when specimens are solution treated at similar temperatures.

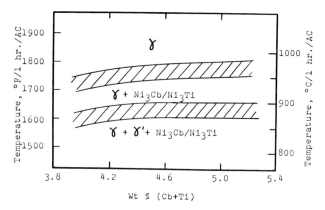

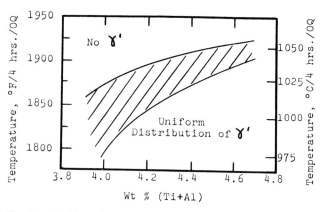

Fig. 6. In 706, phase relations change with temperature and composition. Hot forged material was heated at the indicated temperatures for 1 hr and air cooled.

Fig. 7. In Waspaloy, phase relations of γ' vary with solution temperature and amount of titanium plus aluminum.

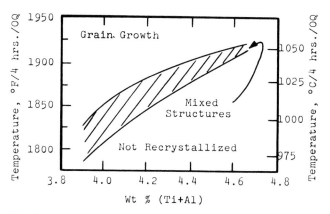

Fig. 8. The extent of recrystallization in Waspaloy is affected by the solution temperature and the amount of titanium plus aluminum. Compare with Fig. 6.

crystallize late in their respective hot/warm working cycles. Heating below the γ'' and Ni_3Cb solvus temperatures produces a structure consisting of grains with boundaries pinned by fine-globular Ni_3Cb and a background of coarse γ''. Heating at 1,750 F (950 C), above the γ'' solvus but below the Ni_3Cb solvus, causes recrystallization to occur due to loss of the γ'' phase. Also, the prior grain boundaries are apparently pinned by globular to platelet Ni_3Cb, which appears to be inhibiting grain growth. Finally, heating at 1,900 F (1,040 C), above the Ni_3Cb solvus temperature, permits grain growth, apparently due to the loss of the structure controlling effects of the Ni_3Cb.

These examples for a Type II alloy show that, in addition to the γ', a phase (in the examples given—Ni_3Cb) can be useful for structure control. As Table II shows, other Type II alloys have other phases, such as Laves and Mu, available for structure control. A relatively new alloy, 706 (an Fe-Ni-Cr alloy containing Ti and Cb) is interesting in that both Ni_3Cb and Ni_3Ti are apparently involved in the structure control process (28, 29).

Phase Boundaries

In order to select temperatures for processing superalloys for best moderate temperature properties, it is necessary to understand and establish phase relationships. Considerable shifting of phase reaction boundaries can occur due to compositional variations, as is shown for 706 alloy in Fig. 6. Our work demonstrates that the best approach is to design process and heat treatment sequences around knowledge of phase boundaries in pseudo-equilibrium diagrams of superalloys where the composition variable is related to "hardener content."

Figure 7 shows the γ' phase boundary in Waspaloy, and Fig. 8 shows how grain structure for conventionally hot worked bars varies with the γ' phase relationships (20). For this Type I alloy, fine structures are obtained in conventionally hot worked parts by essentially "forging-in" the desired structure, and solution treating at temperatures just below the γ' solvus temperature followed by aging (20, 30). In Waspaloy, the structure shown in Figure 3b was produced by this procedure. It correlates with excellent tensile properties for this alloy from 70 F (21 C) to at least 1,000 F (540 C). For each heat of a superalloy, there is an optimum solution heat treatment to give the particular structure and property response desired.

Figure 9 shows grain size data for another Type II alloy, 901. Here, and for other commercial Type II alloys (such as A-286 and 718), solution treatment of conventionally hot worked bars leads to essentially no grain coarsening until the η or Ni_3Cb solvus temperatures are exceeded.

These data point to a significant and useful observation about precipitation strengthened superalloys. The metallurgist can take two approaches to property optimization. He can (1) choose a composition and optimize the heat treatment, or (2) choose the heat treatment and optimize the composition.

The Type I and Type II alloys have varying degrees of flexibility related to composition and phase boundary relationships. For Type I alloys, as noted above, greatest flexibility in structure control is available for alloys with a high volume fraction of γ' available. However, most Type II alloys have relatively low volume fractions of γ' available. Thus, "solution treating" these alloys below their γ' solvus temperatures leads to relatively weak structures. Type II alloys, fortunately, have the "third phase." There is also a temperature separation between the γ' and "third phase" solvus temperatures. A final consideration in Type II alloys is the temperature of the "third phase" solvus. If this temperature is too low, it becomes difficult to employ a "structure control" process.

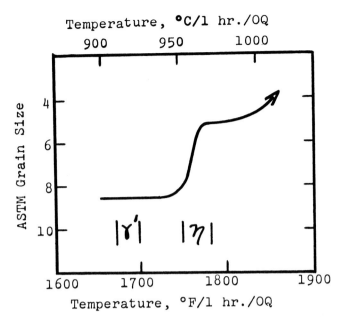

Fig. 9. When 901, a Type II alloy, is heated above the γ' solvus, the grains grow, then level off in size until the η solvus is passed.

Thermomechanical Treatment

There is recent increased interest on the part of jet engine manufacturers in thermomechanical treatment (TMT) or thermomechanical processing (TMP) of superalloys. In thermomechanical treatment, the express intent is to use the strain energy from the forming process to provide strength to the finished forged and heat treated part. For the most part, TMT benefits are lost when recrystallization occurs (31). This is where structure control effects, via precipitated phases to prevent recrystallization, can be very important.

Figure 10 shows recent data developed on Waspaloy. Here pre-annealed ~½ in. diam. bars were cold drawn 30%, and heat treated using "sub-solution" treatments to anneal the material. (The Russians call this process "polygonal annealing.") This state of stress relaxation without recrystallization is supposed to lead to an increase in strength and ductility over an annealed state. Subsolution treatments were followed by conventional aging at 1,550 F (840 C) for 4 hr, air cooling, reheating at 1,400 F (760 C) 16 hr, and air cooling. By an apparent combination of the pinning effects of the gamma prime in inhibiting recrystallization, as well as the apparent dislocation substructure—created by the cold work strain energy—a strong-ductile condition is created. Below 1,000 F (540 C), a significant improvement in the strength capability of Waspaloy has been achieved. The tensile strength data, shown in Fig. 10, are in the range expected for alloys much richer in alloy content than Waspaloy. Microstructure studies of these samples showed recrystallization occurred only at solution treatment temperatures of 1,875 F (1,025 C) or above. Microstructures for several of the conditions represented by Fig. 10 are shown in Fig. 3.

Preliminary data show that there is a practical limit to the time at elevated temperature that can be permitted during TMT if adequate final age hardening response is to be achieved in a superalloy. Other work at Carpenter, un-

published to date, shows that cooling rate after TMT should be as rapid as possible for best properties at 70 F to 1,200 F. These data re-emphasize the all-important fact that control of microstructure by TMT, or by precipitated phases (or both) can lead to equivalent properties for many grades of superalloys. Once adequate processes are devised, the user can select the most economical base alloy to do his job.

Superplasticity

The necessary conditions for a reasonable, useful degree of micrograin superplasticity are:

1. An extremely fine grain microstructure (5μ grain size or less) with "interconnected" grain boundaries to permit easy grain boundary sliding.
2. A second phase to obstruct grain growth, and maintain the fine structure during the forming process.
3. In the relationship $\sigma = K\dot{\epsilon}^m$ — a value of m = strain rate sensitivity, greater than about 0.30.

Deformation should take place at about half the absolute melting temperature.

On the premise that nearly any two phase alloy system may be capable of superplastic behavior, a brief study was conducted in the Carpenter Technology Research and Development Center to study this phenomenon in conventionally melted superalloys. Results indicated that Rene' 95 can be superplastic. Here, it is believed, the γ' phase acts

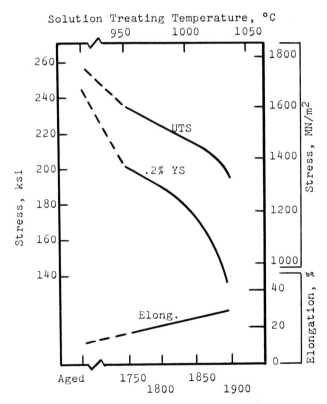

Fig. 10. Thermomechanical treatment of Waspaloy improves mechanical properties, particularly below 1,000 F. Specimens were annealed at 1,825 F for 2 hr, water quenched, cold drawn 30%, solution treated at the indicated temperatures for 4 hr, oil quenched, aged at 1,550 F for 4 hr, air cooled, reheated at 1,400 F for 16 hr, and air cooled.

as the structure controlling phase to maintain the micrograin austenite structure throughout the process. For Pyromet 860, excellent ductility was achieved (138% elongation), but the uniform neck free elongation typical of superplasticity was not observed. It is felt, however, that Pyromet 860 could be superplastic somewhere in the 1,550 to 1,800 F range at the proper deformation rate.

Metallographic studies of the Rene' 95 specimens indicated no apparent changes in austenitic grain size due to the superplastic deformation. The only apparent change in structure was a slight increase in the apparent γ' particle size due to exposures at 1,700 and 1,800 F (925 and 980 C). These results appear to support a grain boundary sliding mechanism of superplasticity in superalloys. Alloys such as Rene' 95 with a relatively high volume fraction of γ' will have a wide range of superplastic conditions (temperature and strain rate) available. In general, it is believed that alloys with greater volume fractions of precipitated phases which are stable over wide temperature ranges will be most easily handled by commercial superplastic forming techniques. Thus, of the superalloys listed in Table I, Astroloy, Pyromet 925, 713C, IN 100, Rene' 95, TAZ-8A, NASA VI-A and NASA II-b would be expected to have a wide latitude of useful superplasticity.

Summary

This review has been directed at some fundamental aspects of the processing of wrought superalloys—with major emphasis on structure control effects available from precipitating phases. Basic methods are suggested to process various superalloy compositions to nearly identical microstructures and property levels by utilizing solvus relationships of precipitating phases in selecting working and heat treating cycles. Structure control procedures are most useful to produce best tensile, impact, fatigue and low cycle fatigue properties from 70 F (21 C) to about 1,200 F (650 C) for most superalloys.

For Type I alloys, the γ' phase is paramount as a structure controlling phase—and attendant characteristics involving use of processing and heat treating temperatures in the vicinity of the γ' solvus are critical. Most flexibility in processing Type I alloys is achieved with alloys containing high volume fractions (>40) of γ' in the alloy.

For Type II alloys, a "third phase" (such as Ni_3Ti or Ni_3Cb) in addition to the γ' can be used to "pin" grain boundaries. Most flexibility in processing is achieved in alloys with a wide separation between γ' and Ni_3X solvus temperatures, a high volume fraction of γ' and Ni_3X, and the highest possible temperature for an Ni_3X solvus.

It is suggested that superalloy metallurgists seriously consider utilizing a division of precipitation hardening alloys into the Type I and Type II categories suggested here. For the future, it is believed that structure control will grow in importance as metallurgists strive to attain best properties in the most economical alloys.

Acknowledgements

The author is grateful to the management of the Carpenter Technology Corp. for permission to publish these results. The author would also like to thank many colleagues and associates who have provided useful comments and constructive criticism. Special thanks is extended to the members of the High Temperature Alloy Research Group for their contributions to the data presented in this paper.

REFERENCES

1. R. F. Decker: Strengthening Mechanisms in Nickel-Base Superalloys, "Symposium: Steel Strengthening Mechanisms," Zurich, 1969, Climax Molybdenum Co., Greenwich, Conn., 1970, pp. 147-170.
2. C. P. Sullivan and M. J. Donachie Jr.: Some Effects of Microstructure on the Mechanical Properties of Nickel-Base Superalloys, Met. Eng. Quart., Vol. 7, No. 1, Feb., 1967, pp. 36-45.
3. C. P. Sullivan and M. J. Donachie Jr.: Properties and Microstructure of Iron-Base Superalloys, Met. Eng. Quart. November 1971.
4. G. P. Sabol and R. Stickler: Microstructure of Nickel-Base Superalloys, Phys. Status Solidi, vol. 35, No. 1, 1968, pp. 11-52 (Eng.).
5. C. T. Sims: A Contemporary View of Nickel-Base Superalloys, Jnl. of Metals, vol. 18, No. 10, October, 1966, pp. 1,119-1,130.
6. W. F. Simmons and H. J. Wagner: Current and Future Usage of Materials in Aircraft Gas Turbine Engines, DMIC Memorandum 245, Battelle Memorial Institute, Columbus, Ohio, February 1, 1970.
7. M. J. Donachie Jr. and E. F. Bradley: Jet Engine Materials for the 1970's, Metal Progress, vol. 95, No. 3, March, 1969, pp. 61-64.
8. E. F. Bradley, D. G. Phinney and M. J. Donachie Jr.: The Pratt & Whitney Gas Turbine Story, Metal Progress, vol. 97, No. 3, March, 1970, pp. 68-75.
9. E. F. Bradley and M. J. Donachie Jr.: The Role of Materials in Flight Propulsion Systems, Jnl. of Metals, vol. 22, No. 10, October, 1970, pp. 25-30.
10. W. J. Boesch and J. S. Slaney: Presenting Sigma Phase Embrittlement in Nickel Base Superalloys, Metal Progress, vol. 86, No. 1, July, 1964, pp. 109-111.
11. L. R. Woodyatt, C. T. Sims, and H. J. Beattie Jr.: Prediction of Sigma-Type Phase Occurrence from Compositions in Austenitic Superalloys, Trans. TMS-AIME, vol. 236, No. 4, April, 1966, pp. 519-527.
12. N. J. Grant: Choice of High-Temperature Alloys—Influence of Fabrication History, Metal Progress, vol. 69, No. 5, May, 1965, pp. 81-86.
13. W. Betteridge, "The Nimonic Alloys," Edward Arnold, Ltd., London, 1959, Chapter 6.
14. J. A. Burger and D. K. Hanink: Heat Treating Nickel-Base Superalloys, Metal Progress, vol. 92, No. 1, July, 1967, pp. 61-66.
15. A. J. DeRidder and R. W. Koch: Controlling Variations in Mechanical Properties of Heat Resistant Alloys During Forging, Met. Eng. Quart., vol. 5, No. 3, August, 1965, pp. 61-64.
16. A. J. DeRidder and R. J. Noel: Deformation Processing of Superalloy Gas Turbine Components, presented at the SAE International Automotive Engineering Congress, January 13-17, 1969, Detroit, Mich., SAE Paper No. 690101.
17. F. N. Darmara: Vacuum Induction Melting: The Revolutionary Influence in Steelmaking, Jnl. of Metals, vol. 19, No. 12, December, 1967, pp. 42-48.
18. R. Schlatter: Vacuum Melting of Specialty Steels, Jnl. of Metals, vol. 22, No. 4, April, 1970, pp. 33-39.
19. R. P. DeVries and G. R. Mumau: Importance of a Relationship Between Dendrite Formation and Solidification in Highly Alloyed Materials, Jnl. of Metals, vol. 20, No. 11, November, 1968, pp. 33-36.
20. W. P. Rehrer, D. R. Muzyka, and G. B. Heydt: Solution Treatment and Al + Ti Effects on the Structure and Tensile Properties of Waspaloy, Jnl. of Metals, vol. 22, No. 2, February, 1970, pp. 32-38.
21. J. F. Barker: A Superalloy for Medium Temperatures, Metal Progress, Vol. 81, No. 5, May, 1962, pp. 72-76.
22. H. L. Eiselstein: Metallurgy of a Columbium Hardened Nickel-Chromium-Iron Alloy, ASTM STP 369, Phila., Penna., 1965, pp. 62-79.

23. J. F. Barker, E. W. Ross, and J. F. Radavich: Long Time Stability of Inconel 718, *Jnl. of Metals,* vol. 22, No. 1, January, 1970, pp. 31-41.

24. H. J. Wagner and A. M. Hall: Physical Metallurgy of Alloy 718, DMIC Report 217, Battelle Memorial Institute, Columbus, Ohio, June 1, 1965.

25. P. S. Kotval: Identification of the Strengthening Phase in "Inconel Alloy" 718, *Trans. TMS-AIME,* vol. 242, No. 8, August, 1968, pp. 1764-1765.

26. D. F. Paulonis, J. M. Oblak, and D. S. Duvall: Precipitation in Nickel-Base Alloy 718, *Trans. ASM,* vol. 62, No. 3, Sept., 1969, pp. 611-622.

27. D. R. Muzyka and G. N. Maniar: Effects of Solution Treating Temperature and Microstructure on the Properties of Hot Rolled 718 Alloy, *Met. Eng. Quart.,* vol. 9, No. 4, November, 1969, pp. 23-37.

28. J. H. Moll, G. N. Maniar, and D. R. Muzyka: The Microstructure of 706, a New Fe-Ni Base Superalloy, Met. Trans., vol. 2, No. 8, August 1971, p. 2143-2151.

29. J. H. Moll, G. N. Maniar, and D. R. Muzyka: Heat Treatment of 706 Alloy for Optimum 1200°F Stress-Rupture Properties, Met Trans. vol. 2, No. 8, August 1971, p. 2153-2160.

30. D. R. Muzyka, J. H. Moll, and H. M. James: Quantitative Measurements of the Effects of Gamma Prime Size and Distribution of the Properties of Waspaloy, presented at WESTEC '70, Los Angeles, Calif., March 10, 1970.

31. J. G. Dunleavy and J. W. Spretnak: "Soviet Technology on Thermal-Mechanical Treatment of Metals." DMIC Memorandum 244, Battelle Memorial Institute, Columbus, Ohio, November 15, 1969.

SECTION XI:
Role of Directional Solidification

The Development of Columnar Grain and Single Crystal High Temperature Materials Through Directional Solidification

FRANCIS L. VERSNYDER AND M. E. SHANK

Pratt and Whitney Aircraft, Advanced Materials Research and Development Laboratory, Middletown, Conn. (U.S.A.)

ABSTRACT

Something exciting has been happening during these last several years. Structural components of a major engineering device were first constructed so that each one consisted of an individual metal crystal. These were alloy single crystal turbine blades for advanced aircraft jet engines. A new precision casting technique, based on directional solidification, which imparts significantly improved ductility and thermal shock resistance to high temperature creep resistant, nickel-base superalloys, has been carried through from reserach to production. This controlled solidification technique has been used to produce both columnar grain and alloy single crystal gas turbine components. The improvement in physical properties is achieved by controlling the solidification process to produce either columnar grains throughout a cast-to-size part, or a complete single crystal throughout a cast-to-size part, with a preferred [001] crystallographic orientation. This orientation is established parallel to the major stress axis of the part without the use of separate "seeding". These parts have exhibited superior structural strength and stability in the severe operating environments associated with gas turbine engine operation. A comparison is made between the properties of superalloys having conventional equiaxed grains, directionally solidified columnar grains and [001] oriented single crystals. The evolution of this new process is traced from its beginning in columnar grain directional solidification experiments through the pilot-plant operation. The feasibility of producing parts using the "directional solidification process" has been demonstrated in production foundry facilities where several thousand gas turbine blades and vanes have been cast-to-size in various complex shapes.

INTRODUCTION

Twenty-odd years ago there was a flood of new data and information on high-temperature alloys. An ASTM symposium[1] was held wherein it was pointed out that during the war period the development of the turbo-supercharger and gas turbine for aircraft and other applications involved the invention and development of many new high-temperature alloys. Although most of the high-temperature testing facilities of the U.S.A. were assigned to this research work, little of the information was made public during the war. Although alloys for elevated temperature service can be dated to the invention of Nichrome in 1906, it was first the turbo-supercharger and then the gas turbine that spurred the substantial development of these materials. Since the development of these alloys has been so closely tied to the end application, it will be a purpose of this article to examine their specific characteristics in the context of their most demanding application, namely the first stage stator vane and first stage rotor blade of the gas turbine engine. The early, high-temperature, heat-resistant alloys were both cobalt-base and nickel-base used in both cast and wrought form. Over this twenty-year time period the trend has been to nickel-base alloys and cast parts. For example, between 1961 and 1967, nine new cobalt-base superalloys were introduced in the U.S.A., but twenty-three new nickel-base superalloys appeared during the same time period[2]. Further, with increased emphasis on higher turbine inlet temperatures and air-cooling to achieve these temperatures, complex castings of the strongest alloys have been required[3,4]. Consequently, the context for the development to be described here is cast nickel-base alloys utilized in both solid and air-cooled configurations. Since, in an engineering sense, an engine test of the material in its final configuration is the final arbiter of its success, we will here follow this development from the first pouring of metal through laboratory test results, and finally engine test results.

Alloys and alloy development

Although a fascinating topic in its own right, alloy development will not of itself be pursued in depth here. A host of good review papers, beginning in the mid-1950's up to the present, more than adequately cover the full complexities and excitement of the subject[5–8]. Here we must take the eclectic approach and deal only with those facets which were critical to the development outlined herein and touch only lightly on all other aspects.

At the risk of over-simplification, it is possible to characterize the major nickel-base superalloys as being simple structurally, albeit complex in chemistry (Table 1)*. Basically, nickel-base high-temperature superalloys depend for their strength on a precipitation of gamma-prime (Ni_3Al) in gamma (Ni solid solution) with a volume fraction of gamma-prime of the order of 60 vol. % or more. The trend towards increasing the volume percent of the gamma-prime precipitate also led to another trend less desirable, that is a reduction in chromium content until present-day alloys contain chromium in the range 6–10 wt. %. It is clear that to continue to increase strength by decreasing oxidation resistance would only lead to less than useful alloys since the gas turbine environment requires the simultaneous optimization of a number of alloy characteristics, i.e. creep strength, oxidation resistance, structural stability, thermal fatigue resistance and impact resistance to name but a few. It became apparent then that some alternative means needed to be found which would break with traditional alloy development with the aim of increasing the high-temperature utility of these alloys and perhaps at the same time creating a new avenue for future alloy development.

Creep – Intergranular fracture

There have been many researches conducted on grain boundary behavior and high-temperature deformation, and on the properties of grain boundaries affecting their fracture behavior. In all of these studies, the grain boundary has been singled out for particular attention, since that is where most of the untoward events are occurring. Grain boundary sliding has been noted, intergranular cavitation and similarly the nucleation of voids in metals during creep, precipitation at grain boundaries and adjacent denuded zones, all of these aspects reveal the complexities of deformation and fracture during creep in polycrystalline alloys. One significant fact, however, agreed upon by most investigators, arises from this research, and that is that the principal untoward events, i.e. excessive precipitation, inter-

---

* These alloys contain refractory metal additions as solid solution strengtheners, and C, B and Zr to promote ductility. The carbon reacts to form a variety of carbide phases (MC, $M_{23}C_6$, M_6C and M_7C_3) depending on composition and heat treatment.

TABLE 1: NOMINAL COMPOSITIONS OF TYPICAL NICKEL-BASE SUPERALLOYS

Alloy designation	Nominal chemical composition (wt. %)									
	Cr	Co	Mo	W	Ti	Al	C	B	Zr	Other
Hastelloy X	22.0	1.5	9.0	0.6	–	–	0.10	–	–	18.5 Fe
M–252	20.0	10.0	10.0	–	2.6	1.0	0.15	0.005	–	–
Nimonic 90	19.5	18.0	–	–	2.4	1.4	0.07	–	–	–
Waspaloy	19.5	13.5	4.3	–	3.0	1.3	0.08	0.006	0.06	–
René 41	19.0	11.0	10.0	–	3.1	1.5	0.09	0.005	–	–
U 700	15.0	18.5	5.0	–	3.3	4.3	0.07	0.03	–	–
IN 100	10.0	15.0	3.0	–	4.7	5.5	0.18	0.014	0.06	1.0 V
Mar–M200	9.0	10.0	–	12.5	2.0	5.0	0.15	0.015	0.05	1.0 Nb
B–1900	8.0	10.0	6.0	–	1.0	6.0	0.10	0.015	0.10	4.0 Ta
TRW–NASA–VIA	6.0	7.0	2.2	5.8	1.0	5.3	0.16	0.18	0.12	0.34 Nb
										8.3 Ta
										0.6 Re
										0.5 Hf
NX 188	–	–	18.0	–	–	8.0	0.04	–	–	–

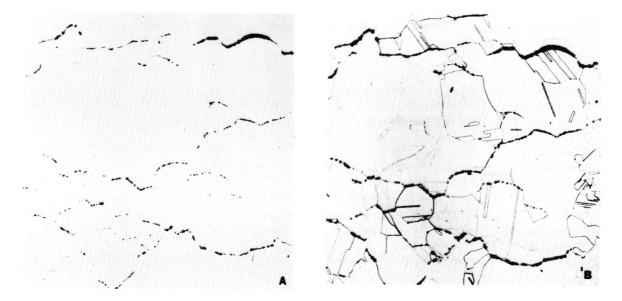

Fig. 1. (A) Voids on boundaries normal to the stress axis. Oxygen-free high-conductivity copper $\sigma = 2500$ p.s.i. at 752° F. As polished. Magnification 100 ×. (B) Same as (A) etched in potassium dichromate to reveal grain boundaries.

granular cavitation and void formation, all occur on the boundaries oriented normal to the principal stress axis[9] (see Fig. 1). Two courses of action then could be followed to deal with this problem: (a) gain sufficient understanding of the events occurring at these grain boundaries so as to control them in a way as to make the consequences benign or (b) eliminate these grain boundaries from the structure –and that is the development outlined in the remainder of this paper.

CASTING AND SOLIDIFICATION

Since the thrust of high-temperature alloy development, as outlined above, has been towards very high strength, cast, alloys, and since the state-of-the-art of precision casting of vacuum melted superalloys had reached a sufficient level of sophistication, and since directional solidification, as revealed by researchers in other areas, seemed a likely way of eliminating grain boundaries, this approach appeared to have a good chance of producing the desired results, *i.e.* the elimination of grain boundaries normal to the stress axis in a high-temperature alloy ingot or precision casting.

In cast metals and alloys there are two types of grains, described as equiaxed and columnar. The mechanism of formation of these two types of grains depends on many variables, including metal composition, pouring temperature, mold temperature and temperature gradient and mold size and shape.

Columnar structures can be produced in a casting provided that two conditions are met. First, the heat flow must be unidirectional causing the solid–liquid interface at the growing grains to move in one direction. Second, there must be no nucleation in the melt ahead of the advancing interface.

An equiaxed grain is usually described as having an axial ratio approximately equal to one and forms as a result of the uniform three-dimensional growth of a nucleus within the melt. A columnar grain, however, usually has an axial ratio much greater than one and is more often nucleated at a surface.

The crystallographic orientation of each type of grain in a casting may be random except under particular conditions when a columnar grain may have a preferred orientation. Confusion arises, therefore, when the definitions described previously are applied to the macrograin structure of a turbine blade. The structure produced by conventional casting in shell molds is often described as producing either a random grain or an equiaxed grain. The latter case usually refers to the control of grain size and shape by control of metal and mold temperature with or without melt or mold-surface inoculation. In the absence of control a random assembly of grains with large variations in size and shape is produced. The macroetched surface of a casting produced under controlled conditions gives the appearance of a uniform equiaxed grain structure both in the thin and thick sections. Sectioning of such a casting reveals, however, that the grains have grown normal to each mold wall, impinging at the center of the casting. Where the thickness of the casting is approximately twice the grain diameter, the grains have an axial ratio of one and could therefore be described as equiaxed. However, when the surface grain diameter is small and/or the thickness of the casting is great, then the axial ratio of the grains is greater than one. The mechanism of solidification of an alloy in a shell mold cast

under controlled conditions may therefore be described as the uniform nucleation of randomly oriented columnar grains at a mold surface and their growth perpendicular to all mold surfaces. In the majority of cases, the length of the columnar grain is governed only by the width of the mold cavity, such that the growth is restricted by impingement of the grains with similar grains growing from an opposite face. Exceptions occur in thicker sections where the nucleation of true equiaxed grains may occur in the melt. The net result is nevertheless a preponderance of transverse grain boundaries.

The grain structure of the material which we shall describe refers to a particular type of columnar grain that is preferentially oriented. It is found that under the conditions of a steep temperature gradient, such as may be imposed by means of a water-cooled copper chill, the growth of grains with a [001] orientation is preferred. The means for controlling the direction of the growth of grains and crystal orientation has been given by VerSnyder and Guard[10] for Ni–Cr–Al alloy (see Fig. 2), by Fisher and Walter[11] for silicon irons (see Fig. 3) and by Flemings[12] for steel (see Fig. 4).

In the technique for the Ni–Cr–Al alloy ingot castings*, we find that the directionally solidified ingots were cast into a 4-inch square mold (Fig. 2) consisting of a shell of molding sand with an inner liner of "Exomold," a commercial exothermic material. The inner liner is thicker at the top than at the bottom so as to provide a temperature gra-

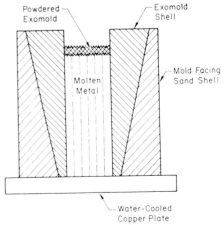

Fig. 2. Design of the mold for directional solidification of an ingot.

---

* The construction of the ingot mold was suggested by A. J. Keisler.

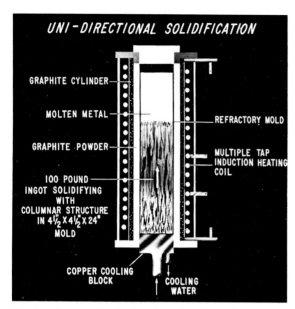

Fig. 3. Diagram of induction heated mold used to produce columnar structure.

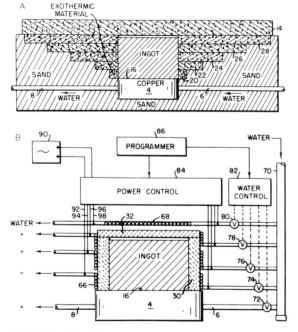

Fig. 4. (a) Mold configuration for obtaining directional solidification utilizing exothermic material. (b) Mold configuration for obtaining directional solidification utilizing a tapped induction coil heater.

dient along the height of the casting. The entire mold assembly rests on a water-cooled copper plate so that directional cooling is obtained. The top of the ingot is covered with exothermic material immediately after pouring to prevent freezing over. An increase in the diameter of the columnar grains was observed to occur varying directly with distance from the copper base plate. The upper one-quarter of the ingot was usually equiaxed. The variation of chemical composition from top to bottom was within the range of heterogeneity of conventional castings.

In the technique used by Fisher and Walter the ingot mold (Fig. 3) consisted of a $4\frac{1}{2}$ in. by $4\frac{1}{2}$ in. fused alumina tube, 24 in. long, encircled by a graphite cylinder open at both ends. The graphite was heated by an induction coil powered by a 50 kW, 9600-cycle source. Electrical shunt connections to the induction coil at full coil, two-thirds, and one-half coil length permitted control over the temperature in the solidifying melt. The base of the mold was a water-cooled copper block which served as a thermal sink to establish the temperature gradient required for maintaining columnar freezing conditions. With the equipment, singly- and doubly-oriented columnar ingot structures measuring up to 18 in. were obtained with the use of an exothermic "hot-topping" compound.

Flemings describes both an exothermic technique (Fig. 4a) and a tapped induction coil technique (Fig. 4b). In the exothermic technique the interior diameter of the mold is uniform; the exterior, however, is formed of a series of rings whose diameters increase with their distance away from the copper base plate. Thus, in the areas requiring temperature isolation for the longest period of time, a greater amount of exothermic material is provided. The casting operation is identical with that described above, that is, substantially unidirectional cooling is obtained, the heat being extracted through the surface of the water-cooled copper block. In the tapped induction coil technique, a programmer directs the power controller to terminate application of electric power to the induction heating coils below the solid-liquid zone as this zone progresses through the ingot. At the same time the programmer directs the water controller to increase the flow of cooling water through all coils surrounding the solid ingot. For example, if one-third of the ingot is solidified, the programmer will direct the power controller to terminate the application of power to the bottom induction heating coils. At the same time the programmer will direct the water controller to increase the flow of the water through this segment of the coil thus applying additional cooling to the segment through which induction heating is no longer being applied.

In all of these several techniques, the aim of course is to create the temperature gradient required for unidirectional solidification. Now with this background to hand, let us proceed to the specifics of turbine blade and vane directional solidification.

Directional solidification

In order to use laboratory equipment then available, it was decided that the initial directional solidification studies would be carried out using a simple resistance mold heater and a water-cooled copper chill base. A conventional vacuum induction melting furnace was modified for this purpose. Water and power connections were made through the furnace chamber wall to which a water-cooled copper chill and a resistance mold heater could be connected.

The vacuum pumping system consists of a 600 liters-per-second diffusion ejector pump backed by a 130 cubic feet-per-minute mechanical pump, which is capable of maintaining a pressure of approximately 1×10^{-2} torr during the melting and casting of a 25 lb. ingot. The pressure is measured with a Pirani gauge.

Ingot molds are prepared by winding 3-inch-diameter by 12-inch-long alumina tubes with two independently controlled molybdenum heating coils (Fig. 5). The windings are held in place on the outside of the tube with a refractory alumina cement. After drying, the mold-heater unit is cemented to the water-cooled copper chill plate. The assembly is placed inside a stainless steel flask. Thermocouple protection tubes are located at four levels between the windings on the alumina tube. The space between the tube and the sides of the flask is lightly

rammed with alumina grogg for insulation and the top of this rammed layer is capped off with the alumina cement. The complete assembly is air-dried and then placed in the vacuum chamber.

Platinum/platinum – 10% rhodium thermocouples are placed in the protection tubes. The water leads and power cables are connected to vacuum furnace-wall feedthrough connections. The furnace is then closed and evacuated and a radiation shield brought into position over the top of the mold. When the chamber pressure is less than 5×10^{-1} torr, power is applied to the coils to heat the mold. Two 30-ampere, 240-volt variable transformers are used for manual control of temperature. The mold is heated to 2700° F (1480° C) and is held at this temperature for 10 minutes before pouring. A vacuum-melted master ingot is melted and held at 2800° F (1540° C) for 10 minutes before pouring. The mold cover is then removed and a fast steady pour maintained to insure a uniformly chilled ingot bottom surface and to avoid cold shuts. The mold cover is replaced after the pour is completed.

The steep temperature gradient established in the growing ingot is controlled and maintained as the columnar grains grow from the chill by gradually reducing the power, first to the bottom winding, and then to the top. When solidification is complete, mold power is turned off and the ingot is cooled in the furnace to below 1800° F (980° C). The furnace is then opened and the ingot broken out of the mold. Grain structure is determined by macroetching, in a solution of hydrochloric acid and hydrogen peroxide. Test material is then machined from the ingot by conventional or electrical discharge techniques. Seven to eight inches of columnar material is regularly obtained in the ten- to eleven-inch ingots made by this method (Fig. 6).

This laboratory technique was then adapted to the fabrication of directionally solidified gas-turbine parts. The first parts unidirectionally solidified were turbine blades. They were cast in a four-blade in-line silica shell mold (Fig. 7). The blades were cast with their larger end closest to the chill. A rectangular bottom gate connecting the four blades was left open on the bottom and the edges of the mold cemented to the copper chill. This mold was wound with a two-zone resistance heater and was cast using the procedure developed for casting ingots.

A circular-clustered five-blade mold design was later developed for pilot plant casting of unshrouded turbine blades (Fig. 8). A vertical rectangular exten-

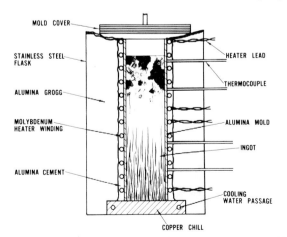

Fig. 5. Directional solidification 3-inch-diameter ingot mold design.

MOLD COVER

STAINLESS STEEL FLASK

ALUMINA GROGG

MOLYBDENUM HEATER WINDING

ALUMINA CEMENT

HEATER LEAD

THERMOCOUPLE

ALUMINA MOLD

INGOT

COOLING WATER PASSAGE

COPPER CHILL

Fig. 6. Three-inch-diameter directionally solidified ingot.

sion of the blade root was left open to the chill. The blade was cast with the largest section at the bottom so that the progressive growth of straight columnar grains upward through the airfoil would be unrestricted. The mold was heated and poured using the same procedures as before. Blades similar to those shown in Fig. 9 were produced for engine testing.

In the pilot plant development it was necessary to establish compatibility with existing casting techniques and, to provide increased production capacity. It was decided to adapt the process to use a permanent mold heater. A new 30-lb. capacity furnace (Fig. 10) was designed using an induction-heated graphite susceptor for mold heating. Hot rate-of-rise vacuum measurements during each stage of a melting and solidification cycle in the original furnace were used to calculate gas load data, to size the new pumping system. These data indicated that a 10-inch diffusion pump with 4400 liters per second of pumping speed at pressures in the low 10^{-3} torr range would be required. This enabled the system to operate under maximum gas load conditions for casting a 25 lb. ingot at approximately 1×10^{-3} torr. In order to decrease the time required to evacuate the system low enough to start the diffusion pump, a 260 cubic feet per minute Rootes-type blower matched to a 150 cubic feet per minute mechanical pump was chosen. Four-inch roughing and fore-pump lines were used. A top nozzle baffle minimizes backstreaming of the diffusion pump and a 12-inch gate valve and high vacuum line prevent throttling of the oversize pump inlet. This system provides a degree of vacuum refinement of the alloy which is not attainable in a booster-pumped system. A multistation Pirani gauge and a hot cathode ionization gauge are used to measure the system pressure. The solidification cycle is carried out in the low 10^{-4} torr range. After completion of the cycle and pumping overnight, the ultimate pressure in the furnace chamber is normally 1×10^{-6} torr. This yields a cold leakup rate of less than 5×10^{-4} torr per hour.

Separate 9600-cycle, 20 kW motor-generator induction power supplies for melting and for heating the mold provide controlled heating of the mold for minimum thermal shock and complete outgassing of the mold. After undesirable volatiles from the mold have been pumped out of the system, the other power supply is used for rapid melting of the alloy. This allows better control of mold preheat, and melt superheat can then be achieved with a single power supply used for both mold and crucible.

The mold heater that was developed consists of a 12-inch-diameter induction coil with two vertical zones 6 inches high (Fig. 11). The two zones are connected to the power supply with a switching arrangement that allows power to be applied to both zones or to the top zone only. A 10-inch-diameter by 14-inch-high vacuum-grade graphite susceptor, in-

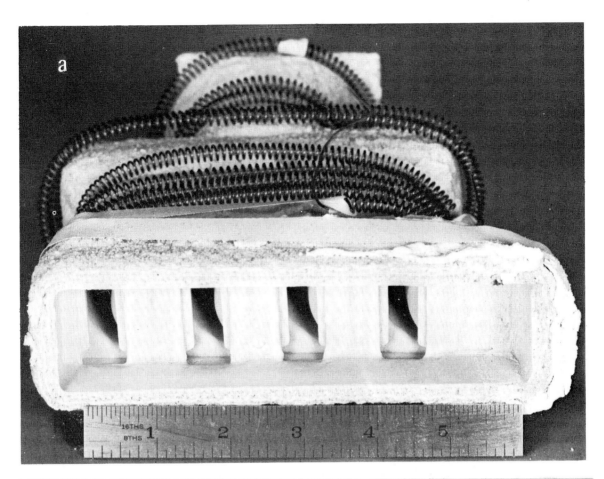

Fig. 7(a) Directional solidification mold showing open bottom and molybdenum heating elements. (b) Four-blade silica investment shell mold and first directionally solidified gas-turbine blade casting.

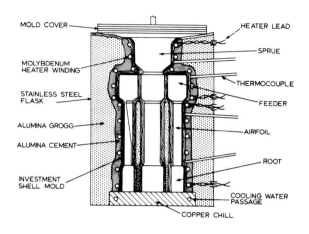

Fig. 8. Resistance-heated mold assembly for casting directionally solidified gas-turbine blades.

sulated on the outside with graphite felt, is positioned inside the coil. The top of the susceptor is adjusted to the same level as the top of the first turn of the induction coil. A 9-inch-diameter by 1-inch-thick copper chill, trace cooled on the bottom, is supported at the bottom of the susceptor by a stainless steel coaxial water shaft. The opening at the top of the susceptor is insulated with a graphite felt cover which can be lifted and turned out of the way during the pour. For measuring mold temperature, thermocouples in alumina protection tubes are inserted through a graphite felt radiation shield at the top of the mold heater.

The investment shell mold developed for use in the induction heated furnace is a symmetrical center-poured cluster with top gating. The large end

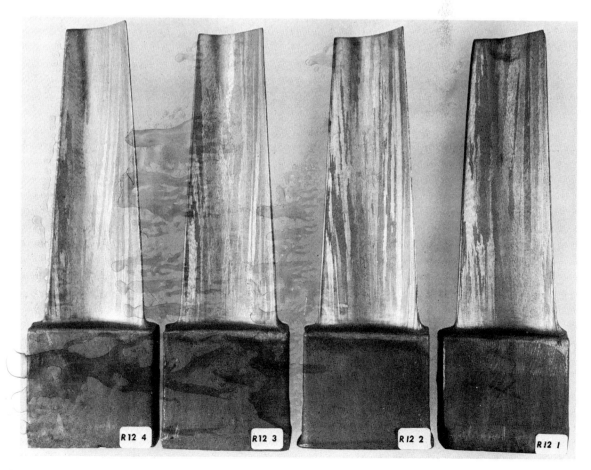

Fig. 9. Directionally solidified PWA 664 JT4 turbine blade castings in the macroetched condition showing the columnar grain structure.

Fig. 10. Vacuum induction melt furnace with independently controlled induction mold heater.

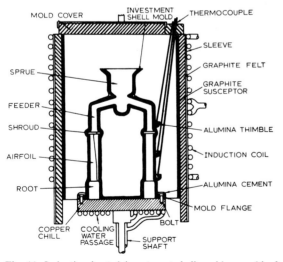

Fig. 11. Induction-heated investment-shell mold assembly for casting directionally solidified gas-turbine blades.

of the parts is turned down, and the bottom of each part is left open to the water-cooled copper chill. When an alloy with high superheat is poured into an open-bottom ceramic mold, large vertical forces are developed. To avoid having the mold "float" upward, thereby allowing the metal to run out, a mold was designed with a flat ceramic flange on the bottom. The flange is bolted to the chill to prevent the mold from lifting.

A short vertical extension of the maximum cross section of the part is added to the mold at the bottom of each root. This extension, called a starter[13], allows better heating of sharp corners in the root section of the mold, helping to prevent nucleation of grains on the surface of the mold. The extension also allows extra growth length for the development of properly oriented columnar grains. When this extra material is later cut off the part, any less favorably oriented grains, cold shuts and other chill-zone defects are removed.

Better grain orientation control is obtained when the starters and roots are filled rapidly and evenly. Therefore, gating is designed to accommodate a rapid, uninterrupted pour. Also fast, uniform growth of the columnar grains can be maintained best in a mold design which has constant or decreasing cross sections vertically. The growth rate decreases when solidification reaches an increased section in the mold. Thermal energy from the heat of solidification has to be conducted downward through the solid metal as growth progresses upward.

In order to provide adequate interdendritic feeding it appears necessary to have at least 1.5 vertical inches of feed metal over the top of the cast part. The mold design developed for casting unshrouded turbine blades has a 2-inch-high extension of the airfoil connected to a nearly horizontal round ingate. For small shrouded blades a 2-inch-long round feeder having a slightly larger cross-sectional area than that of the smallest section of the airfoil is found to give sound parts. The feeder is blended into the top of the shroud with a very large radius. For turbine vanes or blades with large top shrouds, additional feeding is provided by the use of short blind risers near the outer tips of the shroud. For most parts, the mold is positioned in the susceptor with the tops of the feeders just below the bottom of the upper coil zone.

It is found that better grain orientation and faster solidification can be achieved by the use of a coarse-machined surface (e.g., a diamond-shaped knurling) on the chill[14]. This is due to a combination of microscopic welding between the alloy and asperities on the machined surface of the chill, the increased surface area on the chill, and improved mechanical adherence.

Mold temperatures as high as 2850° F (1560° C) are used in this process. This temperature is above the recommended calibration temperature for platinum/platinum–rhodium thermocouples. The errors involved are not large enough to justify the inconvenience of using other available thermocouple materials. Temperature control is maintained through manual adjustment of the induction power control based on thermocouple readout on a multipoint millivolt temperature recorder. Therefore, since (1) temperature is measured outside the mold, rather than in the metal itself, (2) power control is manual, rather than automatic, and (3) relative temperature and rate of cooling are more important than accuracy of absolute temperature measurement, the small changes in calibration of the thermocouples caused by the high temperatures used in this process are found to be acceptable.

Usually six thermocouples are used on each mold. Three levels are monitored on two different parts on a mold cluster. The lowest level monitored is at the top of the root on the blade or vane, the second level is at mid-airfoil and the third is at the top of the blade or vane. Mold preheat temperatures are controlled to raise the temperature of the base of the mold as near as possible to the liquidus temperature of the alloy without heating the rest of the mold above the softening temperature of the mold material. For zircon shell molds, the top of the mold cannot be heated above 2800° F (1540° C) when parts are cast that require long solidification cycles. When short parts having simple cross sections are cast, slightly higher temperatures can be used for short periods of time without excessive sagging of the zircon shell. Alumina shell molds are dimensionally stable at temperatures up to 2900° F (1590° C) even for extended cycles.

To start a typical cycle, a mold is installed inside the heater and the furnace chamber evacuated. With the mold cover in place and cooling water flowing through the chill, power is started to both zones of the heater. Heating rates are controlled to prevent thermal shock of the mold and to prevent overloading of the vacuum pumps for system out-gassing. The mold is heated to the desired temperature and held for 15 to 20 minutes. This technique establishes equilibrium heat flow in the mold and further outgasses the ceramic shell. The melt is heated to 2850° F (1560° C) and held for approximately the same length of time for additional vacuum refinement of the alloy. An alloy superheat of 350 deg. F (195 deg C) gives sufficient fluidity for complete filling of the knurled grooves in the chill, and enhances the natural selection of the preferred [001] growth direction in the alloy. A smooth, rapid pour is used (as before) to insure even filling of the starters and to minimize heat loss out of the open mold cover.

As soon as the pour is complete and the mold cover is replaced, the power to the bottom section of the mold heater is switched off. The power to the top section of the heater is adjusted to hold a constant temperature at the top of the mold for 10 minutes. This establishes a vertical temperature gradient in the mold. By gradual power reductions to produce smooth thermocouple recorder curves through predetermined points, a controlled cooling rate is established in the alloy (Fig. 12). With this controlled gradient a [001] oriented continuous columnar structure is grown from the chill completely through the gas turbine part.

It is found that the best growth control is maintained when the mold heater power is controlled by monitoring the temperature of the alloy near the growing columnar grain structure. Each thermocouple registers a thermal arrest as the section of the casting near it is solidifying. Therefore, after the first 10 minutes of growth, the cooling rate can be controlled by monitoring the temperature of the

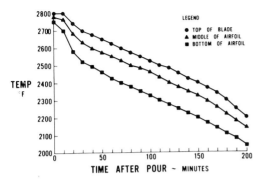

Fig. 12. Typical cooling curves resulting from casting directionally solidified gas-turbine blades.

mold just above the advancing solidification front. When the temperature in the control zone approaches the liquidus temperature of the alloy, control is changed to the next higher thermocouple. All thermocouples continue to print on the recorder chart until the end of the cycle. Cooling control is maintained until the top of the part is below the solidus temperature of the alloy. This provides adequate interdendritic feeding, and helps to prevent vertical grain-boundary hot tears. The casting is then cooled under vacuum or in an argon atmosphere to below 1000° F (540° C). Following solidification, the casting is cooled down, removed from the furnace, and broken out of the mold. The feeder head and risers are cut off and the components are sand blasted and etched for visual examination.

Monocrystals

This method, which depends in part on directional solidification, is reproducible, requires no "seeding" to establish the orientation and appears to be adaptable to commercial foundry practice. As many as 12 turbine blades have been cast at the same time using a slightly modified shell molding technique. The mold is preheated to a temperature exceeding the melting point of the alloy to be cast, and solidification is allowed to proceed by careful control of mold temperature and temperature gradients. This casting process results in preferred growth of the [001] orientation.

The molds designed for casting single crystals are very similar to those employed for producing columnar-grain directionally solidified parts. The only difference is the placement of a multiple-turn constriction in the path of the growing columnar grains to select one [001] oriented grain. A rectangular cavity, called a starter, is used to establish

columnar grain-growth competition near the water-cooled copper chill. Several properly oriented grains then grow through a series of small diameter upward-angled growth paths containing right-angle turns. This allows that grain which can grow the most rapidly in the directions of the perpendicular growth paths to competitively block out the growth of all other less favorably oriented grains.

Figure 13 shows the first constriction design which consistently produced good single crystal cast parts. The starter at the bottom of the mold is used to produce [001]-oriented columnar grains in the very steep gradient near the copper chill. A small diameter 30-degree upward-sloping constriction is used to select several of these directionally solidified columnar grains. The more favorably oriented grains grow along the full length of this constriction and encounter a sharp right-angle turn into a growth expansion zone. In this area each grain is allowed to grow most rapidly in the direction of one of its horizontal [001]-growth planes. The most favorably oriented of these grains gradually crowds out the other grains and reaches the other end of the growth expansion zone. A second right-angle turn is used for added assurance that only one grain continues to grow. This grain then grows into the

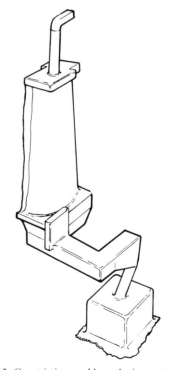

Fig. 13. Constriction mold producing cast single crystal gas-turbine blades.

bottom of the part and upward through the airfoil in the controlled gradient of the mold heater.

A more recently designed constriction allows bottom filling of the mold and the casting of two parts from each starter (Fig. 14). A rectangular bar serves as the downgate for filling the parts and as one of the constriction elements over the top of the starter. The larger cross section of this design produces better orientation control owing to the rapid, uniform filling of the starter. The bottom-filling feature produces a cleaner casting because of the reduced turbulence in the airfoil and the trapping of floating ceramic materials in the downgate as the molten alloy rises in the part.

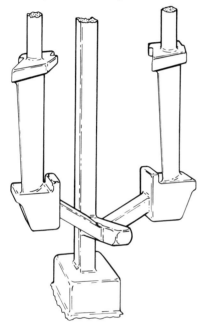

Fig. 14. Improved constriction mold design for bottom-filling of two blades per starter.

Distortion of the parts due to mold constraint has also been reduced with this design which has the parts connected to each other at the bottom only. After the alloy is poured, columnar grain growth starts upward from the chill. Columnar grains from the center of the starter, where orientation control is best, enter the first section of the constriction. Several of these grains are allowed to grow almost horizontally directly toward the susceptor in the first right-angle turn of the constriction.

Right-angle turns branching off to either side of this growth path then select a single preferentially oriented grain to grow into the base of either part. Approximately a 15 degree upward slope is maintain-

ed in the constriction to ensure a steep gradient. The increased cross section of this design gives improved thermal conductivity to promote preferred growth competition. This design produces single crystal blades for engine testing and single crystal bars for mechanical testing and alloy development programs.

A single crystal mold is set up in the heater the same as for a directional solidification casting (Fig. 15). The mold is bolted to the chill to close the bottom of the starters and prevent the mold from floating off the chill when the alloy is poured. The mold is positioned inside the heater with the top of the parts just below the separation plane between the top and bottom induction coils. Best starter preheat is obtained when the top of the chill is positioned approximately even with the bottom of the susceptor.

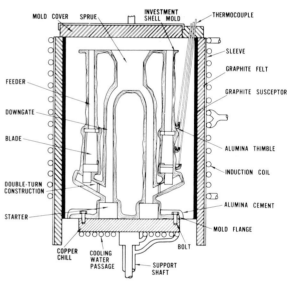

Fig. 15. Induction heated investment shell mold assembly for casting alloy single crystal gas-turbine blades.

Temperature is controlled by manual adjustments to the 20 kW power supply to produce smooth thermocouple recorder curves through planned control points. Platinum/platinum–10% rhodium thermocouples are connected to a multipoint millivolt recorder. The thermocouples are attached to the side of the mold with their tips inside short alumina thimbles. It has been found that the best growth rate control can be maintained when the temperature is monitored near the liquid metal just ahead of the advancing solidification front. Normally three levels from the bottom to the top of the part are monitored. Temperature control changes progressively from bottom to top as solidification front advances upward.

With the furnace under vacuum and a uniform power distribution between the top and bottom heater coils, the mold is heated at a controlled rate to promote ceramic outgassing without overloading the vacuum pumps. The mold is held for approximately 15 min. at 2850° F (1565° C) to achieve uniform temperature distribution through the thickness of the mold. A vacuum melted master alloy is heated to approximately 350 deg F (190 deg C) above its melting point. This high degree of superheat is found to enhance the natural selection of the preferred orientation by rapid growth competition near the chill. The high preheat temperature prevents the nucleation of unwanted grains on mold surfaces above the chill.

Immediately after the alloy is poured the power to the bottom zone of the mold heater is switched off. Columnar grain growth starts immediately in the very steep temperature gradient near the chill. The power to the top coil zone is adjusted to maintain the top of the part at a constant temperature for approximately 10 minutes. This establishes a natural gradient in the mold as the bottom zone of the susceptor cools down.

Induction power is then manually adjusted to produce smooth thermocouple recorder curves through predetermined points as the alloy solidifies from bottom to top (Fig. 16). Temperature control is maintained to below the solidus temperature of the alloy to promote interdendritic feeding and to

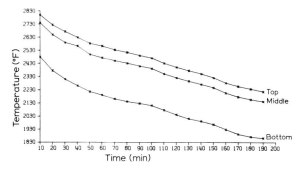

Fig. 16. Typical cooling curves resulting from casting alloy single crystals in a multiple constriction investment shell mold. The time indicated is the time after pour. Top, middle and bottom refer to positions on the airfoil of the casting.

Fig. 17. Multiple constriction competitive growth investment shell mold for producing eight alloy single crystal gas-turbine blades per casting.

prevent excessive mold constraint due to solidification shrinkage.

As a result of the progressive bottom-to-top solidification cycle, interdendritic shrinkage is kept to a minimum when the proper cooling rate is maintained. Inclusions tend to be floated upward out of the part since the solidification cycle is slower than in conventional casting. Dust or loose ceramic particles can be blown out of the mold before it is fastened to the chill since the mold is manufactured with both ends open. Gas porosity is minimized in the single crystal casting process on account of the thorough outgassing of the mold prior to the pour and of the zone freezing action which sweeps upward through the alloy. Figure 17 shows a bottom-poured multiple constriction investment shell mold which is used to produce eight cast single-crystal parts from four columnar-grained starters. A macroetched product of that mold is shown in Fig. 18.

Good filling of very small mold detail can be achieved because of the fluidity of the alloy when both the mold and the alloy are heated to well above the melting point of the metal. A large variety of blade sizes and shapes have been produced by this technique (Fig. 19). Parts have been cast solid and with cores to produce air-cooling passages.

Process control

The ideal structure of a directionally solidified casting consists of an assembly of columnar grains parallel to the major axis of the part and possessing a [001] preferred orientation. The ideal structure of a monocrystal casting consists of an unblemished polyphase monocrystal root, air-foil and shroud. The crystallographic orientation should be such that there is not more than a 10° deviation between the major axis of the part and the closest [001] as measured on a standard stereographic projection. Adherence to the previously described techniques usually results in the desired structure. However, defects are sometimes encountered, which, when properly identified (see Fig. 20), are relatively easily remedied. If defects are encountered they can generally be attributed to two main causes. They result either from a variation in the established time–temperature relationship necessary for the control of directional growth or from improper materials or equipment. Loss of temperature control may result in a loss of the preferred orientation, the formation of equiaxed grains, "freckle" segregation, longitudinal grain-boundary cracking in the case

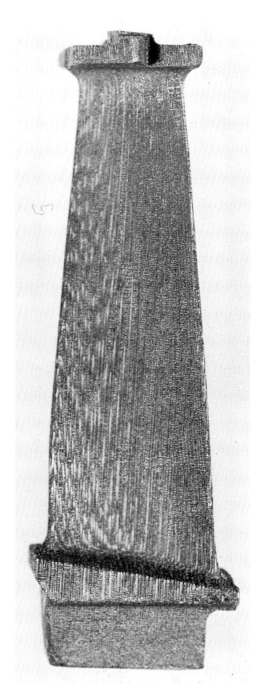

Fig. 18. Macroetched alloy single crystal JT–12 gas-turbine blade showing parallel dendritic structure and complete absence of grain boundaries.

of columnar grains, or shrinkage porosity. Improper materials or equipment may result in gas porosity, metal–mold reaction, inclusions, loss in dimensional precision, or grain divergence in the case of columnar grains.

Fig. 19. Comparison of the variety of sizes of blades that have been cast by the single crystal process.

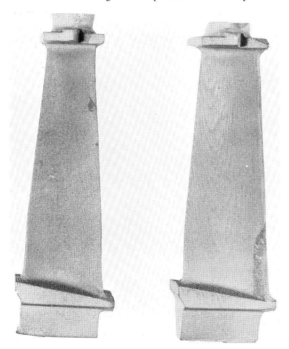

Fig. 20. Macroetched JT–12 blades showing freckle-chain in trailing edge (right) and equiaxed grain near top of airfoil (left).

Let us now discuss those aspects of process control which are pertinent to directional solidification and monocrystals and set aside those defects resulting from loss of process control, mentioned above, that are common to all investment casting techniques. Since it is clear that the cycle times are long (refer to Figs. 12, 16) compared with conventional casting, special attention must be paid to vacuum practice, and to structural defects arising from directional solidification behavior.

Gas porosity may be readily distinguished from shrinkage porosity in directionally solidified or monocrystal castings since it appears as almost perfectly spherical holes situated in the interdendritic interstices. When seen at all in a casting, this type of porosity is usually absent at the bottom of the casting but increases in amount with increase in distance from the water-cooled base. The mechanism of pore formation is believed to be due to an excessive concentration of the gas dissolved in the molten metal ahead of the advancing solid–liquid interface. The presence of gas porosity is, therefore, indicative of a gas content greater than can be accommodated in the solid metal.

A comparison was made of directionally solidified test bars in three furnaces having different vacuum capabilities (Table 2). The analyzed gas content of the material solidified in a vacuum of 5×10^{-4} torr shows decreased nitrogen and hydrogen owing to additional vacuum refinement. The increase in oxygen over that of the master ingot is probably due to the mold material used. In the furnaces with higher pressures and higher leakup rates, all three gases analyzed showed an increase over those of the starting material. The volume percentage of voids (determined metallographically) increased with higher working pressures. It has been shown that with good vacuum practice and proper solidification control, gas porosity can be eliminated in directionally solidified castings. The increased creep life and ductility shown at the lower pressures emphasize the importance of a properly designed vacuum system capable of producing working pressures less than 1×10^{-3} torr with a very low furnace leakup rate.

Structural defects, because of their potentially detrimental effect on mechanical properties, are of primary importance in setting standards for acceptance of directionally solidified or monocrystal castings during inspection. Typical defects are the extent of divergence of the columnar grains with respect to the major stress axis of the part, the nucleation of columnar grains with other than the desired [001] orientation, the presence of equiaxed grains, and in the case of monocrystals an orientation with respect to the part axis outside the limit mentioned above.

It has been shown[15] that the transverse properties of a directionally solidified casting are equivalent to the properties of a conventional casting made in the same material. Therefore, one may expect a continuous deterioration in properties proportional to the degree of grain divergence from the longitudinal axis. Grain divergence is usually avoided by maintaining the surface of the copper chill in a clean and uniform condition.

The nucleation of misoriented grains, that is columnar grains with other than the [001] preferred crystallographic orientation, has a marked effect on mechanical properties. A recent investigation[16] of the effect of crystallographic orientation on the mechanical properties of Mar–M200 single crystals showed that at 1400° F (760° C) and 100,000 p.s.i. the rupture life varied from 5 to over 1900 hours for crystals of [011] and [001] orientations, respectively. Misoriented grains are the result of allowing the temperature of the mold to fall below the melting point of the alloy with the result that grains are nucleated at the mold wall with a random crystallographic orientation. Misoriented columnar grains which are not obviously nucleated at the mold wall may be detected by attention to the dendritic orientation within the grain. Their formation is prevented by strict control of the mold cooling cycle.

Equiaxed grains appear in several forms. A transition from columnar or monocrystal to equiaxed grains may be observed when either too small a temperature gradient exists to sustain unidirectional growth or when areas within the liquid zone cool to the point where nucleation of new grains ahead of the liquid–solid interface becomes energetically feasible. The columnar-equiaxed transition is often associated with extensive microporosity and often results in the formation of transverse cracks during cooling after solidification. Random nucleation of equiaxed grains may also occur at the mold surface, apparently at about the same time as

TABLE 2: EFFECT OF VACUUM PRACTICE ON DIRECTIONALLY SOLIDIFIED TEST BARS CAST FROM IDENTICAL MASTER INGOTS

Casting furnace vacuum conditions		Impurity content (p.p.m.)				Si (wt. %)	Voids (vol. %)	Creep test results (as-cast)			
								1400° F – 100 k.s.i.		1800° F – 30 k.s.i.	
Leakup rate (torr/min)	Working pressure (torr)	O_2	N_2	H_2	S			Rupture life (h)	Elong. (%)	Rupture life (h)	Elong. (%)
1×10^{-4}	5×10^{-4}	1.1	1.3	0.2	21	0.047	0.2	260.8	12	49.7	28
1.6×10^{-3}	1×10^{-3}	1.7	2.4	0.5	39	0.054	0.3	157.2	9	48.1	27
1.6×10^{-2}	$>2 \times 10^{-3}$	1.6	1.7	0.5	44	0.066	0.4	36.4	8	20.4	20
Master ingot		0.4	1.4	0.3	52	0.066					

columnar grains or a monocrystal are growing in the liquid metal, resulting in superimposed equiaxed grains. This type of defect is usually observed toward the top of the casting where the liquid metal is close to the freezing range. The small amount of energy required to cause nucleation under these conditions can be provided by a small inclusion or mold irregularity. The equiaxed grains are usually shallow and in many cases may be remoed by the usual polishing techniques.

Another form of equiaxed grain has been observed and is referred to as "freckle". Freckling is a casting defect that occurs in unidirectionally solidified castings and vacuum consumable-electrode ingots[17]. The defect may be described as extended trails of macro-segregation, enriched in the normally segregating elements and depleted in the inversely segregating elements. The name "freckles" is suggested by the spotted appearance of these trails when macro-etched because of the presence of excess eutectic material, second phase particles, porosity and small randomly oriented grains.

In unidirectionally solidified castings, freckle trails are normally found at the outer surface of the casting and are aligned parallel to the direction of gravity (Fig. 21). Copley et al.[18] have directly observed the formation of freckle trails in a unidirectionally solidified transparent model system, 30 weight percent NH_4Cl-H_2O. A unidirectionally solidified casting of 30 weight percent NH_4Cl-H_2O can be divided into three zones: the top- or liquid-zones, the mushy-zone containing both dendrites and liquid, and the solid-zone which contacts the chill. Copley et al. discovered upward flowing liquid jets in the mushy-zone. These jets erode the mushy-zone causing localized segregation and the formation of dendritic debris which produces residual vertical chains or trails of equiaxed grains. It was further shown that the jets in 30 weight percent NH_4Cl-H_2O are free convection resulting from a density inver-

Fig. 21. (a) Freckle chains (as revealed by etching) formed during the solidification of a 4-in. diameter Mar–M200 alloy single crystal (Red. approx. $\frac{1}{2}$). (b) Enlarged view of a freckle chain (as revealed by etching). Note the large variation in reflectivity of the individual grains in the chain indicative of substantial orientation mismatches. (10 ×)

sion in the mushy-zone. The jets are initially comprised of eutectic fluid which is very corrosive with respect to the dendrites. An analysis of driving force, thermal transport and solute transport effects indicates that convective jets are likely in metallic alloys, where the light elements segregate normally and/or the heavy elements segregate inversely.

As mentioned previously, cooling curves are generated as a means of process control. These curves represent plots of temperature, T, *versus* time, t, and are measured from several thermocouples located at various distances (heights), z, from the chill plate, which provides a convenient reference plane. This information can be replotted in the form of isochronal thermal distribution profiles (curves of temperature *versus* height at various times). Assuming that the measured temperatures are representative of (or at least directly related to) those in the melt, the intersections of the thermal profiles with the liquidus and solidus temperatures give respectively a measure of the liquidus and solidus positions as a function of time. The time interval required for both the liquidus and solidus surfaces to pass a given point represents the local solidification time, Δt or t_1. Furthermore, the slope $(\partial T/\partial z)$ at these intersection points represents the thermal gradient, G, which can be measured at the liquidus and solidus interfaces. By taking the slope of an interface position *versus* time curve, the growth rate, $R = (\partial z/\partial t)$, can be determined as a function of time or distance for either the liquidus or the solidus interface.

To describe the effect of growth rate on freckling, Copley *et al.* considered the special case of steady-state solidification (solidification at constant thermal gradient, G, and growth rate, R). Under these conditions, the amount of erosion produced by an incipient jet is proportional to the local solidification time

$$\Delta t = \frac{T_L - T_S}{RG}$$

which is the time required for the mushy-zone to pass a point in the casting. If a critical time Δt^* is required to produce enough damage to stabilize the jet and produce dendritic debris, then the equation above predicts for a given melting range and thermal gradient a critical growth rate below which freckling will occur.

Figure 22 summarizes the effect of growth rate and thermal gradient on freckling. For $G > G^*$, freckling is suppressed because there is insufficient

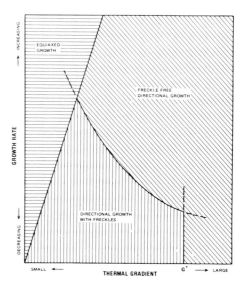

Fig. 22. Freckle free growth regime is established as the growth rate and thermal gradient are varied.

potential energy available to drive the convective jets. For $G < G^*$, freckling is predicted for

$$R < \frac{T_L - T_S}{\Delta t^* G}$$

where Δt^* has been taken as constant. Finally, steady-state unidirectional solidification is restricted to the region

$$R < \frac{K_T G_S}{\Delta H}$$

where K_T is the thermal conductivity, G_S is the gradient in the solid at the solidus, and ΔH is the heat released when 1 cm^3 of liquid at the liquidus temperature is transformed to solid at the solidus temperature. For growth rates greater than this amount, more heat is released in the mushy-zone than can be conducted away through the solid. At steady-state the thermal gradient in the liquid-zone must become negative in order to preserve a heat balance so that unidirectional solidification cannot be maintained.

Once the boundaries of the desired regime have been empirically established, one would simply choose process variables which influence growth rate (R) and thermal gradient (G) in such a way as to maximize the probability of producing defect-free castings.

ENGINEERING PROPERTIES

When directionally solidified, Mar–M200 alloy is

one of the strongest, if not the strongest, high temperature nickel-base superalloy. Since it is also representative of the nickel-base superalloys, and since a considerable background of data and information was already available on the alloy conventionally cast, it was chosen as the principal material for the development being described.

Structure, orientation and heat treatment

The main structural features of the alloy are as follows: The as-cast material is heavily cored, owing to pronounced dendritic segregation during solidification; the dendrites are rich in tungsten and cobalt whereas the interdendritic regions are rich in chromium, titanium, nickel and carbon. The structure consists of ~ 60 vol. % coherent precipitate of γ', basically $Ni_3(Al, Ti)$, in a matrix of γ (nickel-base solid solution), interspersed in the interdendritic regions with minor MC carbides and γ/γ' eutectic. The γ' particles, and γ/γ' eutectic, contain more titanium, aluminum and nickel, and less tungsten, cobalt and chromium than the γ matrix. Heat treatment partially removes segregation, eliminates the eutectic, refines the γ' dispersion in γ, and gives additional partially coherent $M_{23}C_6$ carbide. An as-cast crystal, therefore, is composed primarily of two oriented phases $(\gamma + \gamma')$, whereas the normally heat-treated crystal consists of three oriented phases $(\gamma + \gamma' + M_{23}C_6)$. Typical electron transmission micrographs of the $\gamma + \gamma'$ structure are shown in Fig. 23.

Crystals grown in the $\langle 100 \rangle$ orientation develop a simple unidirectional dendritic structure, Fig. 24, since $\langle 100 \rangle$ happens to be the preferred direction of growth for dendrites in this material. Crystals grown in the $\langle 110 \rangle$ and $\langle 111 \rangle$ orientations, how-

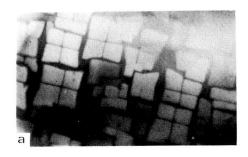

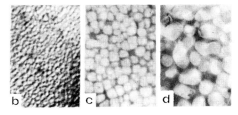

Fig. 23. Coherent precipitate of γ' in γ in the Ni-base superalloy Mar–M200: (a) as-cast; (b) after solutionizing at 2250° F for 4 hours, followed by air quenching; (c) and (d) after aging treatments at 1500° F and 1700° F for 64 hours, respectively.

ever, tend to promote equal growth, generally in separate colonies, in the two and three geometrically favored $\langle 100 \rangle$ growth directions, respectively. In other orientations of growth, a single $\langle 100 \rangle$ dendrite direction generally prevails, which appears to be the one that is most steeply inclined to the actual growth direction of the crystal. In such crystals the continuous primary dendrite arms are arranged in regular rows or plantations, rather than as a random forest. This is indicative of a common origin for each dendrite row, probably by sideways growth of a secondary arm of the original dendrite. In accord with this idea the rows of dendrites define a cube plane, which can be revealed by macroetching as elliptical surface markings.

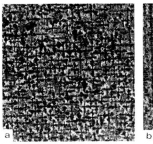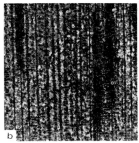

Fig. 24. Macrostructure of (a) transverse and (b) longitudinal sections of a crystal of Mar–M200 grown in the cube direction. Note the continuity of dendrite arms in the direction of crystal growth. (4 ×)

As mentioned previously, directionally solidified Mar–M200 grows with a ⟨100⟩ preferred orientation, as do most cubic alloys[19]. Nature was remarkably kind since when this orientation is parallel to the major stress axis of a turbine part it provides the best combination of properties for such turbine blades and vanes. A detailed study[16] of the effect of orientation on the properties of monocrystal Mar–M200 confirmed the viability of nature's choice. Consequently, all properties to be subsequently presented, unless specifically noted otherwise, refer to data taken with the [001] direction parallel to the stress axis.

The mechanical properties of Mar–M200 are particularly susceptible to heat treatment. It is not surprising, therefore, that the consistency of the "as-cast" properties, which depend upon the thermal history of the casting following solidification, may be improved by heat treatment. In the conventionally cast polycrystalline alloy the effect of heat treatment on the relative strength of grain boundaries with respect to the strength of the grains is of considerable importance. It follows therefore that conventionally cast Mar–M200 is heat treated in such a manner that grain-boundary strength and ductility are optimized. This heat treatment usually involves an intermediate temperature anneal of 20 to 60 hours at 1600–1650° F. In the absence of any grain boundaries, the heat treatment may be designed solely to increase the bulk strength[20]. In this case, solution heat treatment of the alloy followed by an intermediate temperature age is shown to be effective (Fig. 25).

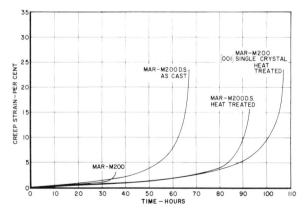

Fig. 25. Effect of structure and heat treatment on the creep behavior of Mar–M200 at 1800° F and 30 k.s.i. Mar–M200 D.S. heat treated and the Mar–M200 [001] single crystal heat treated both received 2250° F for one hour followed by an aging treatment of 1600° F for 32 hours.

The following comparative properties of conventionally cast, directionally solidified and monocrystal are presented with each material having received a heat treatment designed to optimize its properties. The heat treatments, unless specifically noted otherwise, are as follows:

Conventionally cast	– Furnace cooled, 1600° F for 50 hours
Directionally cast	– Furnace cooled
Monocrystal (tensile properties)	– Furnace cooled
Monocrystal (creep properties)	– Furnace cooled, 2250° F for 1 hour, 1600° F for 32 hours.

Tensile properties, static and dynamic modulus

The tensile properties of the three cast forms of Mar–M200 are compared in Figs. 26a, b, c and d. The Figures show that the tensile and yield strengths of monocrystal Mar–M200 are equivalent to the corresponding values of directionally solidified and conventionally cast Mar–M200. The tensile ductility of the monocrystal material is superior at all temperatures. The temperature dependence of the tensile ductility displays a minimum at approximately 1400–1500° F for all the materials. These data indicate that the phenomenon is an intrinsic bulk property of the nickel-base superalloy and is not associated with the presence of grain boundaries, although the prominence of this phenomenon may be masked by fracture at transverse boundaries as in conventionally cast Mar–M200. The peak in tensile strength observed in all three cast forms coincides with the minimum in tensile ductility, an effect observed in most nickel-base superalloys[6,9].

The low modulus of elasticity of directionally-solidified and [001]-oriented monocrystal materials is largely responsible for their improved resistance to cyclic thermal strain. The elastic constants of a Mar–M200 crystal are slightly higher than those of nickel, but in the same ratio. The elastic constants of directionally-solidified material can be derived from those of the single crystal[21]; however, for bending of a blade or for twisting about its span the stiffnesses of both materials are the same. Whereas the room temperature moduli of wrought or conventionally-cast alloys are 32×10^6 p.s.i. and 12×10^6 p.s.i. in bending and torsion, respectively, the corresponding values for both directionally-

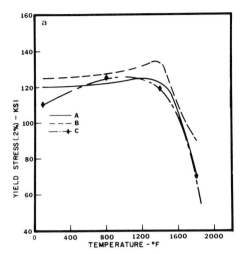

Fig. 26a

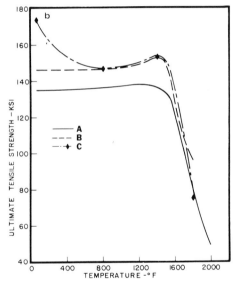

Fig. 26b

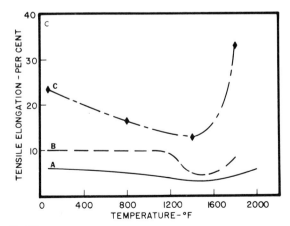

Fig. 26c

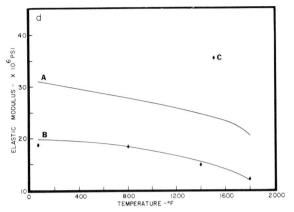

Fig. 26d

Fig. 26. (a) Variation of the 0.2% yield stress with temperature for (A) conventionally cast, (B) directionally solidified, and (C) monocrystal Mar–M200. (b) Variation of ultimate tensile strength with temperature for (A) conventionally cast, (B) directionally solidified and (C) monocrystal Mar–M200. (c) Variation of tensile elongation with temperature for (A) conventionally cast, (B) directionally solidified and (C) monocrystal Mar–M200. (d) Variation of the static modulus of elasticity with temperature for (A) conventionally cast, (B) directionally solidified and (C) monocrystal Mar–M200.

solidified and monocrystal materials are 19×10^6 and 18×10^6 p.s.i.

The variation of the dynamic longitudinal modulus of directionally-solidified material with temperature is shown in Fig. 27. The resonant bending frequency of directionally-solidified JT4 turbine blades at room temperature was found to be approximately 82% of that of the conventionally cast blades, in agreement with the ratio of the square roots of their moduli.

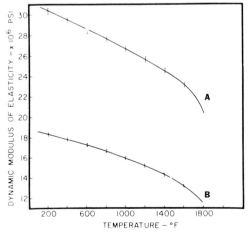

Fig. 27. Variation of the dynamic modulus of elasticity with temperature for (A) conventionally cast and (B) directionally solidified Mar–M200.

Creep and stress-rupture properties

A significant advantage of directionally solidified and monocrystal materials is revealed through a study of their creep and stress-rupture properties. From Fig. 28 one can see that stress to produce rupture in 100 hours is higher for directionally solidified Mar–M200 than for the conventionally cast alloy and an advanced wrought alloy (Udimet 700). At all temperatures minimum rupture ductilities of directionally solidified Mar–M200 were greater than average values of conventionally cast Mar–M200. This most dramatic effect of directional solidification is illustrated for a number of high temperature alloys in Fig. 29. Figure 29 shows two scatterbands; the lower band refers to the conventionally cast materials, the elongation values varying from less than 5% at 1400° F to less than 10% at 2000° F. The rupture elongation of the unidirectionally solidified materials displays a minimum in excess of 5% at 1400° F, increasing to over 15% at 2000° F.

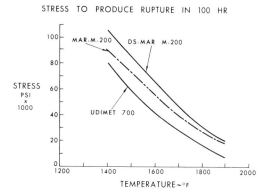

STRESS TO PRODUCE RUPTURE IN 100 HR

Fig. 28. Comparison of stress-rupture properties at various temperatures of wrought U–700, conventionally cast and directionally solidified Mar–M200.

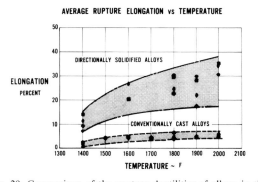

AVERAGE RUPTURE ELONGATION vs TEMPERATURE

Fig. 29. Comparison of the rupture ductilities of alloys in the conventionally cast and unidirectionally solidified conditions.

In order to document more fully the properties of the columnar-grained material, several tests were carried out on specimens which were cut from transverse sections of unidirectionally solidified ingots. In these specimens the columnar grains were oriented normal to the axis of the specimens. As might be expected, the transverse properties of columnar grained Mar–M200 were found to be equivalent to the conventionally cast material as shown in Fig. 30.

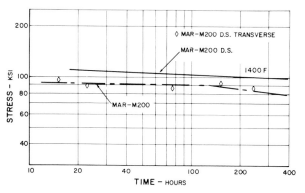

Fig. 30. Transverse and longitudinal stress-rupture properties of directionally solidified Mar–M200 compared with conventionally cast Mar–M200 at 1400° F.

The creep and stress rupture property comparison of conventionally cast, directionally solidified and monocrystal Mar–M200 is given in Table 3 and in Fig. 31. The monocrystal demonstrates longer rupture life and a lower minimum creep rate. In addition, the high values of rupture ductility obtained with the directionally solidified material are also observed with the monocrystal. Clearly the difference in the three cast forms is a function of the grain boundary effect, the effect of a preferred orientation, and the difference in thermal history.

A further note on the creep behavior of directionally solidified materials is relevant here. The amount of primary creep at 1400° F is highly stress

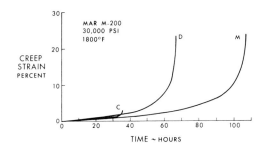

Fig. 31. Comparison of the creep properties at 1800° F of conventional (C), directional (D) and monocrystal (M) Mar–M200.

TABLE 3: CREEP AND STRESS-RUPTURE PROPERTIES OF CONVENTIONALLY CAST, DIRECTIONALLY SOLIDIFIED AND MONOCRYSTAL Mar–M200

	1400° F/100 k.s.i.			1600° F/50 k.s.i.			1800° F/30 k.s.i.		
	Rupture life (h)	Elongation (%)	Min. creep rate (in./in./h)	Rupture life (h)	Elongation (%)	Min. creep rate (in./in./h)	Rupture life (h)	Elongation (%)	Min. creep rate (in./in./h)
Conventionally cast	4.9	0.45	70.0×10^{-5}	245.9	2.2	3.4×10^{-5}	35.6	2.6	23.8×10^{-5}
Directionally solidified	366.0	12.6	14.5×10^{-5}	280.0	35.8	7.7×10^{-5}	67.0	23.6	25.6×10^{-5}
Monocrystal	1914.0	14.5	2.2×10^{-5}	848.0	18.1	1.4×10^{-5}	107.0	23.6	16.1×10^{-5}

dependent (Fig. 32). At the stress levels found in turbine parts the amount of primary creep should be less than 0.5%. At the lower stress levels it is no longer proper to consider primary creep as a rapidly occurring amount of plastic deformation. Rapid primary creep only occurs at the higher stresses. The time to completion of primary creep increases exponentially as the stress is lowered, Fig. 33. At a stress of 50 k.s.i. it should take 10,000 hours to reach the completion of primary creep in directionally solidified material at 1400° F.

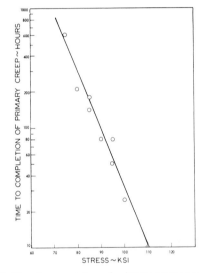

Fig. 33. Time to completion of primary creep as a function of stress at 1400° F for solutionized and aged D.S. Mar–M200.

A comparison of the fatigue life data for conventional, directional and monocrystal Mar–M200 is given in Fig. 34. Extrapolation of the curves to $\frac{1}{2}$-cycle gives good agreement with the tensile reduction in area measured on columnar-grained specimens tested at the same strain rate.

Columnar-grained and single crystal materials exhibit exactly similar fatigue lives at both temperatures. However, as shown in Fig. 34, the polycrystalline Mar–M200 has a fatigue life that is one to two orders of magnitude lower than that of the directionally solidified materials. At elevated temperatures, crack initiation and propagation occur at grain boundaries, and the rate of intergranular crack initiation and propagation is considerably faster than transgranular cracking for the same conditions of testing. (A detailed description of the effect of carbides, grain boundaries and other structural features on elevated temperature fatigue is given elsewhere[22].) At these temperatures, single

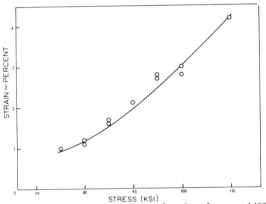

Fig. 32. Amount of primary creep as a function of stress at 1400° F for solutionized and aged D.S. Mar–M200.

Fatigue properties

Although good fatigue resistance in materials for aircraft power plants has always been considered essential, the last decade has been characterized by a new awareness of the importance of fatigue behavior, especially the low cycle and thermal fatigue characteristics of high temperature materials. This section will deal with the low-cycle fatigue character of directional and monocrystal materials; a later section on engine testing will describe the thermal fatigue behavior.

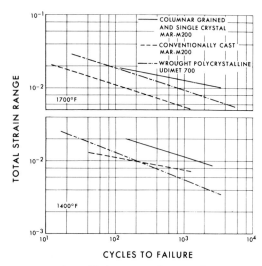

Fig. 34. Comparison of fatigue properties at 1400° F and 1700° F for a typical wrought nickel-base alloy (U–700) with conventionally cast, directionally solidified and monocrystal Mar–M200.

crystals have substantially longer high strain fatigue lives than those of polycrystals.

Nickel-base superalloys exhibit an extremely heterogeneous mode of deformation both at low temperatures and under conditions of high-strain rates at elevated temperatures because of a high volume fraction of γ' in the microstructure[23]. Crack initiation in the nickel-base superalloys usually occurs at defects in the microstructure. These include microporosity in cast alloys and brittle phases such as carbides, nitrides, sulfides and borides in both wrought and cast alloys. The rate of crack initiation is a sensitive function of defect size for planar slip conditions[24].

Even with single crystals microstructural defects play an important role (Fig. 35) with implications to alloy design and development. In Fig. 35 it is apparent that slip and cracking are preferentially initiated at precracked MC carbides (photo inset). The absence of MC carbides results in retardation of crack initiation. In addition, the occurrence of premature failure at large carbide particles at high strains is eliminated.

Fracture–Creep, fatigue and impact

Intergranular initiation and propagation of cracks are the principal mode of progress to complete fracture in high temperature creep. As already noted (see earlier section on Creep–Intergranular fracture) void formation, cracking and other malignant events occur principally on grain boundaries normal to the major stress axis of the specimen or part. Even with directionally solidified materials crack initiation may be intergranular beginning at longitudinal grain boundary segments that are transverse to the major stress axis (Fig. 36). Examination of this creep-tested Mar–M200 directionally cast material, Fig. 36, shows that the columnar grain boundaries are not truly linear but follow an irregular path between interlocking dendrites. Small segments of the boundary are almost transverse to the columnar axis. These small transverse sections

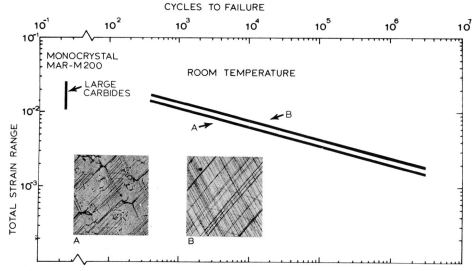

Fig. 35. Comparison of fatigue properties at room temperature for monocrystal Mar–M200 with carbide size varying from "large carbides" to "moderate" (A) to "carbide free" (B).

358

Fig. 36. Crack formation during creep in directionally solidified Mar–M200. The longitudinal grain boundary, A, contains transverse segments, B, at which cracks form preferentially. Propagation of the cracks occurs transverse to the applied stress along interdendritic interstices, C. (105 ×)

act in the same way as transverse grain boundaries in conventionally cast material.

Low temperature crack initiation in nickel-base superalloys occurs preferentially at microstructural defects such as pores and brittle second phases, *i.e.* carbides, but propagates in fatigue transgranularly as described by Gell and Leverant[25]. At elevated temperatures crack initiation is intergranular for polycrystalline conventionally cast nickel-base alloys even in fatigue but subsequent propagation is a function of temperature, frequency and mean stress and is usually a combination depending on these factors[26]. The significant aspect of intergranular cracking is that the rate of crack initiation and propagation in grain boundaries is much greater than that for transgranular cracking. As a result, the high-strain fatigue life of directional and monocrystal materials is significantly improved as compared with their equiaxed counterpart.

Impact loading (encountered in jet engines as occasional foreign object damage) involves very

high stress rates where the time available for plastic flow is very limited. An experiment was devised by Gilman[27] to demonstrate that grain boundaries are a source of failure even at low temperatures. The alloy used was Mar–M200 in the three cast forms, first conventionally cast with large equiaxed grains; second, directionally solidified in the direction of the cylindrical axis; and third, a monocrystal with [001] orientation parallel to the cylindrical axis. The specimens were (cracked) in impact tension, produced by a steel driver plate whose velocity was sufficient to produce a peak impact pressure of about 80 kbars and comparable tension at the plane where the rarefaction waves met (magnitude at fracture about 20 kbars = 290,000 p.s.i. = 2×10^{10} d/cm^2) with quite different crack patterns, as shown in Fig. 37. At (A) in this Figure it may be seen that the fracture pattern follows the grain boundaries in the conventionally cast structure. Figure 37B shows longitudinal cracks that follow the boundaries of the elongated grains that result from directional solidification plus some transverse cracks through the grains. The last photograph (Fig. 37C) shows a random pattern of cracks through the grains. These cracks commonly began at carbide matrix interfaces. If these interfaces could be eliminated, as well

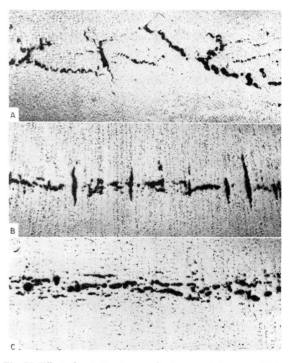

Fig. 37. Effect of cast structure on fracture mode in a very high stress rate test of Mar–M200. (A) Conventionally cast. (B) Directionally solidified. (C) Single crystal.

as the grain boundaries, still further change in the fracture pattern would be expected.

It may be concluded that the way to optimize all of the properties of nickel-base high temperature materials, exemplified by Mar–M200, is to eliminate all internal interfaces.

ENGINE AND ENGINE-SIMULATION EVALUATION

The final proof of any materials development occurs when the material is first tested in an experimental engine, and then either flight-service-tested and/or incorporated as a bill-of-materials part. We will here describe some of the major types of tests, and results that provided the demonstration of technical feasibility for directional and monocrystal materials.

The life-limiting characteristics of turbine blades and vanes have no doubt become clear to the reader by now. However, to highlight the purpose of the tests to be described, a brief review is in order. A typical blade shape and vane shape are given in Fig. 38. The columnar grain directional structure shown in Fig. 38, however, is the structure obtained by directional solidification after VerSnyder[28]. Turbine blades (rotating airfoil rows) and turbine vanes (stationary airfoil rows) are primarily life-limited by one of three failure modes: creep, thermal fatigue or corrosion. A secondary failure mode of turbine blades is high-frequency fatigue. There is also occasional foreign object damage to contend with. All three primary failure modes are dependent upon the blade metal temperature and stress state (either the absolute value or its rate of change).

Fig. 38. Typical turbine blade and vane shape. These parts have been etched to show the columnar grain directional structure.

Creep is the elongation of a blade due to the centrifugal loads at temperature. The rate of elongation or growth is largely a function of the average blade metal temperature but will also be affected by cyclic conditions. The creep phenomenon also occurs in turbine vanes but the phenomenon is quite complex. It is generally termed "bowing" and is a result of a combination of the pressure differential between the concave and convex surfaces of the vane and of cyclic conditions. The second failure mode, thermal fatigue, is a thermally induced low-cycle fatigue failure resulting from alternating strains caused by restrained expansions and contractions of the airfoil leading and trailing edges during engine transients. As already implied, this cyclic phenomenon does affect the creep behavior. Corrosion (also referred to as hot corrosion) will not be dealt with here since this would introduce a volume of subject matter not relevant to the comparisons to be made. Suffice it to say that directional and monocrystal materials behave much the same as conventionally cast material.

Engine-simulation test

It is often desirable to perform simple screening tests on new materials prior to taking the more complex and expensive step of engine testing. A number of such tests are commonly employed in the gas turbine industry. They include ballistic impact, rig "bowing" tests, rig, "thermal shock" tests and others. The test of most significance in this instance is the rig "thermal shock" test. The first tests to be run were comparative tests of conventionally cast and directionally solidified Mar–M200.

All tests were conducted in a rig in which the temperature of the test material was cycled by turning the fuel on and off to a modified gas-turbine combustion chamber. A cycle consisted of 1.5 min hot and 0.5 min cold. During the test the materials were subjected to 30 min at temperature prior to each 100 cycles.

The results of the direct comparison of conventional and directional material are given in the histogram of Fig. 39. Note that at the end of the test no directional solidified Mar—M200 specimens had failed. The condition of typical test specimens, which were first stage turbine blades, is shown in Fig. 40a and 40b. Additional tests were run which included monocrystal Mar–M200 and these results are presented in Table 4. The monocrystal blades were run to a total of 2400 cycles without a failure. This was a

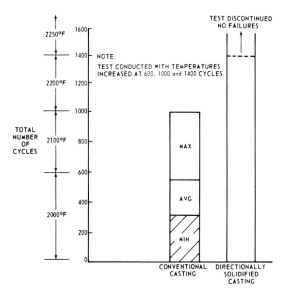

Fig. 39. Histogram of thermal-fatigue test results of conventionally and directionally solidified Mar–M200.

total hot time of 72 hours and an endurance time of 92 hours, far in excess of the values for conventionally cast materials, as shown in the table. Directional and monocrystal materials, both coated and uncoated, possess excellent thermal-shock resistance.

Engine test–Blades

The first engine test of directionally solidified Mar–M200 turbine blades was performed in a first-stage JT4 turbine and was a comparative test, the base line material being conventionally cast Mar–M200. It had been previously established in this type of accelerated engine endurance test that an aluminide coating extended the time to first cracking from 17 hours to in excess of 50 hours for Mar–M200 blades. Therefore, it was decided to coat the conventionally cast blades for the test. It was also decided, on the basis of laboratory data, that the most revealing test of the directionally solidified material could be achieved by leaving this material uncoated. Consequently this engine test was run with coated conventionally cast and uncoated directional Mar–M200 first turbine blades. All blades were etched and examined prior to final processing for the test, typical examples of which are shown in Fig. 41.

The blades after final processing were assembled in alternate positions in the first stage rotor of the JT4 experimental engine as shown in Fig. 42.

The test consisted of 15 endurance cycles prior to each hot section inspection. Each endurance

Fig. 40a. Uncoated conventionally cast Mar–M200 first stage turbine blade, tested in a thermal shock rig, after 600 cycles at 2000° F, 400 cycles at 2100° F and 400 cycles at 2200° F. One crack was visible after the forst 800 cycles.

Fig. 40b. Uncoated directionally solidified Mar–M200 first stage turbine blade, tested in a thermal shock rig, after 600 cycles at 2000° F, 400 cycles at 2100° F and 400 cycles at 2200° F. The blade shows no cracks.

TABLE 4: SUMMARY OF TYPICAL THERMAL SHOCK DATA—Mar–M200 ALLOY

Material	Specimen	Cycles at 2000° F	2100° F	2200° F	Total cycles	Condition
Conventionally cast	Gas turbine blade	300	–	–	300	Cracked
Conventionally cast	Gas turbine blade	600	400	–	1000	Cracked
Conventionally cast	Gas turbine blade	600	200	–	800	Cracked
Directionally solidified	Gas turbine blade	600	400	400	1400	No damage
Directionally solidified	Gas turbine vane	600	400	400	1400	No damage
Directionally solidified	Gas turbine blade	600	400	400 + 115 cycles at 2250° F + 51 cycles at 2300° F	1566	No damage
Monocrystaloy	Gas turbine blade	600	400	1400	2400	No damage

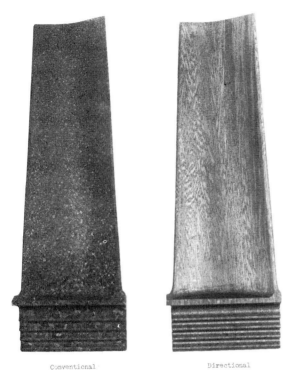

Conventional   Directional

Fig. 41. Conventionally cast Mar–M200, shroudless JT4 test blade in macroetched condition showing normal grain structure and directionally solidified Mar–M200, shroudless JT4 test blade in macroetched condition showing columnar grain structure.

Fig. 42. Coated conventionally cast and uncoated directionally solidified turbine blades assembled in the rotor. The alternate blades, which appear darker, are the coated conventionally cast blades.

cycle consisted of six 5-minute periods at maximum turbine inlet temperature alternated with 2-minute idle periods, followed by 30 minutes at maximum inlet temperature steady state. That is approximately one hour total hot time and 12 thermal cycles per "endurance" cycle. The maximum turbine inlet temperature was varied from 1865° F to 2020° F. The test was terminated when approximately

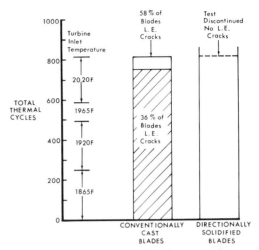

Fig. 43. Histogram of JT4 engine test results.

one-half of the conventionally cast blades had significant leading edge cracks.

The results of this engine test are summarized in the histogram of Fig. 43. Note that there was no failure of the directionally solidified blades. The presence of cracks in more than 50% of the conventionally cast blades, when the test was terminated after 70 hours of endurance running, indicates that the true life of this material under these conditions had been determined. The directionally solidified material on the other hand was still considered serviceable and available for continued engine testing. The condition of the blades at the termination of the test is shown by two typical blades in Fig. 44.

Since it is difficult to get meaningful blade growth measurements in the relatively short and very severe cyclic endurance test another experimental engine test was run under steady-state conditions. This test was, in turn, compared with a similar test run with conventionally cast blades. The test was conducted with a full set of directionally solidified Mar–M200 blades at a turbine blade metal temperature of 1622° F and an average blade stress of 31,200 p.s.i. The results of this test are presented in Fig. 45. Note that the average blade growth is one-half that of the conventional material. Note also that the maximum growth of the directional material falls approximately on the average growth curve of the conventional material. The test of the directional material was finally concluded at 660 hours.

Fig. 44. Coated conventionally cast and uncoated directionally solidified turbine blades after 70.0 hours of engine test. Note the transverse crack in the leading edge of the conventionally cast blade (left). Oxidation of the directionally solidified blade (right) is due to the absence of a coating.

The first engine test of monocrystal Mar–M200 turbine blades was performed in a first stage JT12 turbine and was a comparative test, the baseline material being forged U–700. Both materials were coated with an aluminide coating to limit the extent of oxidation-erosion. All blades were etched and examined prior to final processing. Two typical monocrystal Mar–M200 blades that were prepared in the pilot lots for this test are shown in Fig. 46. An etched cross-section of one of the pilot lot blades is shown in Fig. 47. The regular alignment of the dendrites, characteristic of [001] oriented monocrystal Mar–M200, can be seen in this macrophoto. The blades were subsequently machined, coated and assembled in the first stage rotor of the JT12 engine. These monocrystal parts were made according to the invention of Piearcey[29].

The operating parameters for the steady state engine test were a turbine inlet temperature of 1515° F, a rotor speed of 15,000 r.p.m. and a burner pressure of 68 p.s.i.g. The total running time of the test was in excess of 230 hours. All blades survived

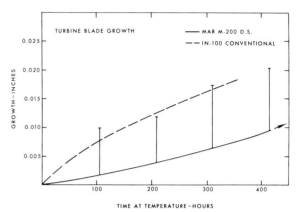

Fig. 45. Turbine blade growth results from experimental military engine tests of conventionally cast IN–100 and directionally solidified Mar–M200. The two curves present a comparison of average growth while the bars attached to the curve of the Mar–M200 D. S. indicate the maximum blade growth values for this material.

Fig. 46. Typical monocrystal Mar–M200 JT12 turbine blade castings macroetched for inspection prior to being prepared for testing in an experimental engine.

the test in satisfactory condition. Growth measurements were made at the end of 147 hours of running time; typical data are presented in the histogram of Fig. 48. The growth of the monocrystal Mar–M200 varied between 0.0002 in. and 0.0005 in. while the "standard" U-700 blades grew an average of 0.0018 in. This represents a difference in creep growth of approximately a factor of 3. These engine tests confirmed the salient features, defined in the laboratory testing of directional and monocrystal material, i.e., superb thermal fatigue resistance and longer creep lifetimes.

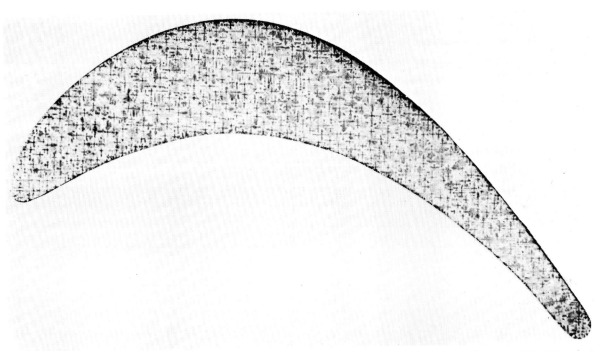

Fig. 47. Etched transverse section of a monocrystal Mar–M200 JT12 turbine blade showing regular alignment of dendrites. (×8; reduced in reproduction ×⅔)

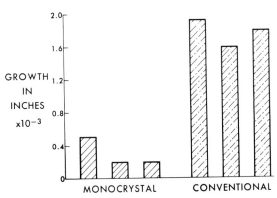

Fig. 48. Histogram comparison of the blade growth results of running monocrystal Mar–M200 and conventional U–700 turbine blades in a JT12 engine test.

Engine test–Vanes

The first engine test of directionally solidified Mar–M200 turbine vanes was performed in a first stage JT8D uncooled turbine and was a comparative test, the base-line material being the bill-of materials W152* cobalt-base alloy. Since all vane materials for this engine are coated, the directional material was also coated using an aluminide coating. The directionally solidified Mar–M200 vanes, together with the WI52 conventionally cast vanes, were assembled in alternate positions in an experimental JT8D engine run in a sea level test stand. The test was an endurance test conducted in two phases at maximum turbine inlet temperatures between 1850° F and 1900° F.

The first phase of the endurance test conformed to the requirements of the standard FAA cycling procedures. Each cycle encompassed various power setting changes during 12 accelerations from idle to take-off power were made. A total engine time of 78 hours was accumulated during this phase with seven hours being at maximum turbine inlet temperature.

The second phase of this test was a 1000-cycle endurance test. Each cycle was a 5-minute cycle consisting of approximately one minute at take-off power, two minutes at idle, 10 seconds at 1.3 EPR (an engine pressure ratio of 1.3 simulates conditions during the application of thrust reversers), and the remainder of the time at idle. A total engine time of 83 hours was accumulated during this phase with 17 hours being at maximum turbine inlet temperature.

The test was terminated at the end of phase II with a total engine time of approximately 161 hours.

*Nominal chemical composition (wt. %)

Cr	Fe	W	Cb + Ta	C	Co
21.0	1.75	11.0	2.0	0.45	Bal.

The post-test condition of the vanes was generally good, with the exception of the few located in the normally hotter burner positions. These did show a degree of distress. Measurement of trailing edge bow resulted in an average value of 0.015 in., a bow resistance far superior to that of cobalt base W152, which under similar conditions would be expected to bow an average of 0.24 in. No transverse cracks were observed in any of the Mar–M200 directionally solidified vanes, indicating that the material possessed the desired combination of resistance to bow and thermal shock.

Comparison vanes from an engine test that dramatically demonstrates the advantage in bow resistance that can be obtained by directional solidification are shown in Fig. 49.

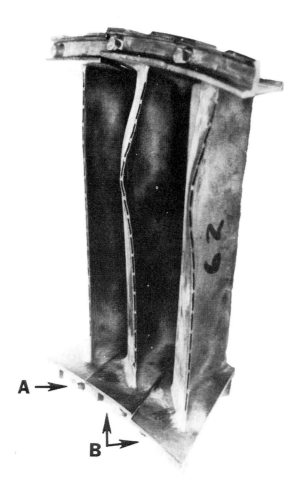

Fig. 49. Bowed conventionally cast cobalt-base alloy turbine vanes (B) and unbowed directionally solidified Mar–M200 turbine vane (A) were run at the same time and the same conditions in an experimental military engine.

Flight service testing of directionally solidified Mar–M200 was begun in late 1964. Sixty-thousand hours of flight time were accumulated on 40 half-sets of parts. Half-set combinations of directionally solidified Mar–M200 and bill-of-materials conventionally cast WI52 were placed in airline service test as JT8D first-stage turbine nozzle vanes. By October of 1966 enough information was available to make a statistical analysis of the various service observations and measurements. On the basis of this analysis, directionally solidified material proved to have higher crack resistance and substantially improved bow resistance.

CONTINUING AND FUTURE DEVELOPMENT

The benefits arising from directional solidification and monocrystals to high temperature properties are now becoming more widely recognized. As a consequence, additional developments in this field can be expected and indeed are to some degree coming to view now. Developments in process technology will almost certainly proceed in the areas of new techniques, new apparatus, novel mold design and new methods of establishing the desired rate and gradient during solidification. New methods will be advanced which will be of importance to both research and to production. In the research area, Kear et al.[30] have described a novel method of obtaining oriented crystals with various compositions, and in the production area Chandley[31,32] and Fleck[33] have described techniques for obtaining or maintaining directional solidification in turbine airfoils. Almost certainly articles other than gas turbine blades and vanes will also benefit from the application of these ideas. Bolling[34], for example, has described directional solidification to obtain integral turbine disks and blades while Copley et al.[35] have invented a high-temperature single crystal spring. Since directional solidification and monocrystals open a new approach to alloy design, new alloys specifically tailored to their particular characteristics can be expected. In fact, one such alloy has already been announced by Pratt and Whitney Aircraft. The alloy designated as NX 188 and described by Maxwell[36] possesses an improved melting point and exceptional creep rupture strength and structural stability. These few examples give some foretaste of the future.

SUMMARY

A new precision casting technique, based on directional solidification, which imparts significantly improved ductility and thermal shock resistance to high-temperature creep resistant, nickel-base superalloys has been carried through from research to production. This controlled solidification technique has been used to produce both columnar grain and alloy single crystal gas turbine components. The benefits, either directly demonstrated or implied for gas turbine service, are:

Superior thermal shock resistance

Longer cyclic strain life

Longer creep life

Better intermediate temperature ductility

Good thin-wall properties

The feasibility of producing parts using the "directional solidification process" has been demonstrated in production foundry facilities where several thousand gas turbine blades and vanes have been cast-to-size in various complex shapes.

ACKNOWLEDGEMENTS

The work described in this review covers seven years of research, invention and innovation, also laboratory, rig and engine testing, carried out in about that order, many of these phases, of course, progressing simultaneously. Literally hundreds of persons were involved, and it is impossible, therefore, to make individual acknowledgements. Many of those to whom credit is due are listed in the references appended to this paper either as authors of technical publications or inventors to whom patents have been issued. There is, unfortunately, no way to list those people who took part in the rig and engine testing phases, but we are deeply indebted to everyone involved.

Much of the material contained herein was assembled by Dr. E. J. Felten. Drs. A. F. Giamei, C. P. Sullivan, and C. H. Wells read and criticized portions of the manuscript for which we extend our thanks.

We wish to give special thanks for permission to use material and illustrations, some of which is unpublished work, to the following: Drs. J. L. Walters (Fig. 3), M. C. Flemings (Fig. 4), J. J. Gilman (Fig. 37) and D. N. Duhl (Figs. 32 and 33).

Last, but most importantly, we wish to thank the management of Pratt and Whitney Aircraft for the long-range vision and support which make this type of invention, development and innovation a continuing possibility.

REFERENCES

1 *Symposium on Materials for Gas Turbines*, Am. Soc. Testing Mater., Philadelphia, 1946.
2 W. F. SIMMONS AND H. J. WAGNER, *Metal Progr.*, 91 (1967) 86.
3 W. R. MARTENS AND W. A. RAABE, *Am. Soc. Mech. Engrs.*, 67GT17, 1967.
4 W. H. SENS AND R. M. MEYER, *Soc. Auto. Engrs.*, 680278, 1968.
5 F. L. VERSNYDER, *J. Metals*, 8 (1956) 1445.
6 C. G. BIEBER, *Metals Eng. Quart.*, 1 (1961) 92.
7 C. P. SULLIVAN AND M. J. DONACHIE, JR., *Metals Eng. Quart.*, 7 (1967) 36.
8 G. D. SABOL AND R. STICKLER, *Phys. Status Solidi*, 35 (1969) 11.
9 F. L. VERSNYDER, in HEHEMANN AND AULT (eds.), *High Temperature Materials*, John Wiley and Sons, New York and London, 1959.
10 F. L. VERSNYDER AND R. W. GUARD, *Trans. Am. Soc. Metals*, 52 (1960) 485.
11 H. J. FISHER AND J. L. WALTER, *Trans. AIME*, 224 (1962) 1271.
12 M. C. FLEMINGS AND H. F. TAYLOR, *U. S. Pat. 3,204,301*, 1965; also M. C. FLEMINGS, R. V. BARONE, S. Z. URARNAND AND H. F. TAYLOR, *Trans. Am. Foundrynen's Soc.*, 69 (1961) 422.
13 L. W. SINK, *U. S. Pat. 3,417,809*, 1968.
14 B. J. PIEARCEY, *U. S. Pat. 3,485,291*, 1969.
15 B. J. PIEARCEY AND B. E. TERKELSEN, *Trans. AIME*, 239 (1967) 1143.
16 B. H. KEAR AND B. J. PIEARCEY, *Trans. AIME*, 239 (1967) 1209.
17 J. PRESTON, in E. L. FOSTER (ed.), *Trans. Intern. Vacuum Metal. Conf.*, Am. Vacuum Soc., New York, 1967.
18 S. M. COPLEY, A. F. GIAMEI, S. M. JOHNSON AND M. F. HORNBECKER, *Trans. AIME*, to be published.
19 W. A. TILLER, *Trans. AIME*, 9 (1957) 847.
20 B. J. PIEARCEY, *U. S. Pat. 3,310,440*, 1967.
21 C. H. WELLS, *Trans. Am. Soc. Metals*, 60 (1967) 270.
22 G. R. LEVERANT AND M. GELL, *Trans. AIME*, 245 (1969) 1167.
23 M. GELL AND G. R. LEVERANT, *Proc. Third Bolton Landing Conf. on Ordered Alloys*, 1969, to be published.
24 M. GELL AND G. R. LEVERANT, *Trans. AIME*, 242 (1968) 1869.
25 M. GELL AND G. R. LEVERANT, *Acta Met.*, 16 (1968) 553.
26 M. GELL AND G. R. LEVERANT, *Fracture 1969*, Chapman and Hall, 1969.
27 J. J. GILMAN, *Trans. Am. Soc. Metals*, 59 (1966) 597.
28 F. L. VERSNYDER, *U. S. Pat. 3,260,505*, 1966.
29 B. J. PIEARCEY, *U. S. Pat. 3,494,709*, 1970.
30 B. H. KEAR, S. M. COPLEY, M. F. HORNBECKER AND L. W. SINK, *Trans. AIME*, 245 (1969) 1361.
31 G. D. CHANDLEY, *U. S. Pat. 3,376,915*, 1968.
32 G. D. CHANDLEY, *U. S. Pat. 3,441,078*, 1969.
33 D. G. FLECK, *U. S. Pat. 3,411,563*, 1968.
34 G. F. BOLLING, Seminar *Solidification*, Am. Soc. Metals, 1969.
35 S. M. COPLEY, D. N. DUHL AND B. H. KEAR, U. S. Pat. Pending, 1970.
36 D. H. MAXWELL, Pratt and Whitney Aircraft, FRDC, *Report GP69-186*, 1969, to be published in *Metals Eng. Quart.*

BIBLIOGRAPHY*

B. J. Piearcey and F. L. VerSnyder, *S.A.E. J.*, *74–6* (1966) 84.

B. J. Piearcey and F. L. VerSnyder, *S.A.E. J.*, *74–8* (1966) 36.

F. L. VerSnyder and B. J. Piearcey, *A.I.A.A. J.*, *Aircraft, 3–5* (1966) 390.

B. J. Piearcey and F. L. VerSnyder, *Metal Progr.*, *90* (1966) 66.

G. A. Webster, *Phil. Mag.*, *14* (1966) 775.

G. A. Webster and B. J. Piearcey, *Trans. Am. Soc. Metals, 59* (1966) 847.

F. L. VerSnyder, R. B. Barrow, B. J. Piearcey and L. W. Sink, *Mod. Castings, 50–12* (1967) 360.

B. J. Piearcey and F. L. VerSnyder, *Trans. S.A.E., 75* (1967) 247.

B. J. Piearcey and R. W. Smashey, *Trans. AIME, 239* (1967) 451.

---

* Supplemental list of publications directly related to the topic of directional solidification and monocrystals.

S. M. Copley and B. H. Kear, *Trans. AIME, 239* (1967) 984.

G. A. Webster and B. J. Piearcey, *Metal Sci. J., 1* (1967) 97.

B. J. Piearcey, B. H. Kear and R. W. Smashey, *Trans. Am. Soc. Metals, 60* (1967) 634.

F. L. VerSnyder, R. B. Barrow, B. J. Piearcey and L. W. Sink, in E. L. Foster (ed.), *Trans. Intern. Vacuum Metal. Conf.*, Am. Vacuum Soc., New York. 1967.

E. J. Lyons, *U. S. Pat. 3,346,039*, 1967.

S. M. Copley, B. H. Kear and F. L. VerSnyder, in Burke, Reed and Weiss (eds.), *Surfaces and Interfaces II*, Syracuse University Press, 1968.

F. L. VerSnyder, in *Yearbook of Science and Technology*, McGraw-Hill Book Company, New York, 1968.

B. H. Kear, S. M. Copley and F. L. VerSnyder, *Trans. Japan Inst. Metals, 9* (1968) 672.

D. R. Parille, *U. S. Pat. 3,401,738*, 1968.

R. B. Barrow, *U. S. Pat. 3,405,220*, 1968.

F. L. VerSnyder, R. B. Barrow, B. J. Piearcey and L. W. Sink, *Trans. Am. Foundrymen's Soc., 55* (1969) 10.

G. R. Leverant and B. H. Kear, *Met. Trans., 1* (1970) 491.

SECTION XII:
Protective Coatings

PROTECTIVE COATINGS FOR HIGH TEMPERATURE ALLOYS

STATE OF TECHNOLOGY

G. William Goward
Materials Engineering and Research Laboratory
Pratt & Whitney Aircraft
400 Main Street, East Hartford, Connecticut 06108

ABSTRACT

Coatings used on nickel- and cobalt-base superalloy blades and vanes in gas turbine engines typify the state of coating technology for high temperature alloys.

Coatings formed by interdiffusion of aluminum with the alloys to form layers consisting mainly of intermetallic compounds, such as NiAl and CoAl, were the first systems used for protection of gas turbine airfoils. The protectivity of these systems is derived from the formation of protective alumina scales. In a general way, coating degradation occurs by cyclic oxidation, molten salt hot corrosion and, at higher temperatures, interdiffusion with the substrate. Thermal fatigue properties are governed by the brittle-ductile transition behavior of the intermetallic compounds NiAl and CoAl. Both positive and negative effects occur, depending on the shapes of thermal strain-temperature curves for particular applications. Significant increases in hot corrosion and oxidation resistance have been obtained by the incorporation of noble metals, such as platinum, in aluminide coatings.

The so-called MCrAlY overlay coatings, based on nickel, cobalt, iron and combinations thereof with chromium, aluminum and yttrium can be formulated over a wide range of compositions nominally independent of those of substrate alloys. Improved oxidation resistance and, in part, hot corrosion resistance is derived from yttrium which enhances protective oxide adherence. Mechanical properties, principally ductility, and therefore thermal fatigue resistance, can be adjusted to the requirements of specific applications.

Incremental improvements in performance of the MCrAlY coatings are expected as research programs define degradation mechanisms in greater detail and more complex compositions are devised. More basic evaluations of mixed metal-ceramic insulative coatings have been initiated to determine if these systems are capable of effecting further increases in airfoil durability.

INTRODUCTION

For many applications of high temperature alloys, the use of protective coatings is required to achieve practical lives for components which are designed to operate at high temperatures in oxidizing environments. Coatings used for surface protection of nickel- and cobalt-base superalloy airfoils in gas turbine engines typify the state of coating technology for such applications.

In presenting a review of this technology, this paper will qualitatively describe the turbine environment and thus define the requirements of protective coatings for superalloy airfoils. Available coating systems will be discussed in terms of structure, composition, processing methods, oxidation and hot corrosion resistance and effects on the mechanical behavior of coating-substrate system. Finally, typical development approaches which may lead to improved coatings will be described.

THE GAS TURBINE ENVIRONMENT - COATING REQUIREMENTS

The major conditions which may affect coating behavior in the gas turbine are as follows:

(a) gas and metal temperature;
(b) gas composition including extraneous contaminants;
(c) gas and metal temperature cycling;
(d) gas pressure and velocity.

In a variety of applications, gas stream temperatures can range from 1200°F for power generation turbines to 2500°F for high performance aircraft gas turbines. Corresponding metal temperatures range from 1100°F to 2000°F with projections to about 2200°F for the most advanced alloys, for example, directionally solidified eutectics, under current development. Increasingly sophisticated methods of cooling airfoils are employed to maintain metal temperatures within the range of practical use with regard to strength and surface stability.

Nominally, the gas stream in the turbine is in an oxidizing state with the major gaseous constituents being oxygen, nitrogen, carbon oxides, water and sulfur oxides. While other deleterious species can be present, and will be dealt with appropriately, it is clear that at the higher temperatures cited, simple oxidation must play a major role in degradation of the surfaces of superalloy airfoils. Although a moderate degree of oxidation resistance can be derived from the cobalt-chromium base of cobalt alloys and the nickel-chromium or nickel-chromium-aluminum base of nickel alloys, the overriding importance of high creep strength imposes unsatisfactory limits on the practical oxidation resistance of these alloys. Thus, a most important requirement for protective coatings, particularly in the higher temperature ranges, is to provide sufficient oxidation resistance to maintain the airfoil in its designed shape for a period at least nominally equal to its practical mechanical life.

In many applications of gas turbines, and in particular in those used for aircraft propulsion, the gas temperature, and therefore the metal temperature, is cycled over quite wide ranges as the engine is operated, for example, at idle, takeoff, climb, cruise and thrust reverse power requirements. Such thermal cycling imposes a major effect on oxidation behavior of alloys and coatings, that of protective oxide spallation. This phenomenon is normally attributed to stresses arising from the different coefficients of thermal expansion of metal and protective oxide. Because the amounts of those critical elements, such as chromium and aluminum, which are selectively oxidized to form protective oxides on superalloys, are limited by mechanical property considerations, cyclic oxidation properties of these alloys are inadequate for many applications, particularly those which require metal temperatures above approximately 1400-1500°F. Early coating technology was based on simply increasing the surface concentration, to a depth of a few mils, of, for example, chromium or aluminum to form the corresponding protective oxides for a longer time under cyclic conditions. Contemporary coating technology now recognizes the desirability of improved oxide adherence to further prolong the effective use of these elements, particularly aluminum. Thus the definition of enhanced oxidation resistance as a major requirement of protective coatings can be refined to include, first, the requirement for an increased reservoir of those elements which form protective oxides and, second, enhanced adherence of those oxides under thermal cycling conditions.

Under service conditions, a variety of extraneous contaminants can be ingested along with the intake air, or introduced with the fuel, into the turbine. A possible list of these contaminants is quite large; some of the more important are sea salt, runway dust or sand and fuel contaminants such as vanadium and lead. While the exact mechanisms by which these contaminants produce harmful deposits on airfoil surfaces have not been completely established, it is clear that deposition does universally occur and that the deposited contaminants, usually in the form of molten sulfates or oxides, cause greater or lesser degrees of acceleration of the oxidation process. This accelerated oxidation falls under the general term hot corrosion; in specific instances where metallic sulfides are one of the products of reaction, the term sulfidation has been widely used. Again, some resistance to hot corrosion can be obtained in nickel- and cobalt-base superalloys but for similar reasons of strength requirements, this resistance is inadequate for many practical applications. This is particularly true for the nickel-base alloys used for rotating blades, where the needs for higher strength have necessitated reduction in chromium concentrations, this element being perhaps the most effective in obtaining resistance to hot corrosion. Thus, another important requirement for protective coatings is substantially superior resistance to hot corrosion compared to that of the base alloys.

In some gas turbine applications, abrasive particles such as sand can be ingested with the intake air. In addition, although not well documented, so-called "hard carbon" can be formed during combustion of fuel in turbine burners. Such particles, accelerated to high velocity in the turbine gas stream can cause erosive removal of metal from turbine airfoils. Because this phenomenon appears to occur in only a few specialized applications, the technology base for coating or materials solutions is quite limited. It is anticipated, however, that the problem may be more serious with the use of coal derived fuels in gas turbines and programs have been initiated to describe erosion mechanisms to lay the base for possible materials solutions[1].

Cyclic operation of turbines, mentioned above, can also produce significant thermal and mechanical strains on superalloy airfoils. Thermally induced strains, caused by unequal rates of heating and cooling of the surface and interior of the metal are generally the more important of these two with respect to the behavior of the coating-alloy system[2,3,4]. Typical thermal stresses for various parts of airfoils are schematically illustrated in Figure 1. As shown, depending on the particular design, location on the airfoil and mode of engine operation, thermally induced tensile strains can reach a maximum at either relatively low or high metal temperatures. These cyclic thermal strains cause what is commonly referred to as thermal fatigue cracking of turbine airfoils. If the airfoils are coated, cracking almost always originates in the coating, and it must be recognized that strains large enough to cause coating cracks are also sufficiently large to cause propagation of these cracks into the base alloy[2,3,4]. Thus, another highly important requirement for protective coatings is maximum possible resistance to thermal stress cracking.

It is also required that coatings have minimal influence on other mechanical properties such as high temperature creep and high frequency fatigue strength. Influences on creep behavior are relatively straightforward provided that high temperature coating processing is matched to the heat treatment requirements of the alloy and due account is taken of reduction of load bearing area, it being assumed that the coating is incapable of carrying any significant load. Possible positive and negative effects on high frequency fatigue behavior are less well understood and must be measured and accounted for in specific applications where problems are anticipated.

Insofar as is currently understood, the effects of high gas pressure and velocity on surface stability can be related only in a secondary manner to those factors already discussed. High gas pressures and velocities (e.g. up to 25 atmospheres, 1650 feet per second) cause relatively high values of the heat transfer coefficient, the net effect of which is the generation of the previously mentioned thermal strains, which in an advanced high performance turbine can be of the order of 0.2 to 0.3% at the airfoil surface. Although Hancock[5] has demonstrated that cyclic mechanical strains can damage protective oxides on steels, no definitive work has been done to demonstrate that strains significantly greater than those caused by the thermal expansion mismatch between the oxide and metal occur and cause additional mechanical damage to the oxide under practical turbine conditions. Dils[6] has demonstrated the possibility that gas temperature and pressure fluctuations which occur in the turbine at relatively high frequency under nominally steady state (isothermal) conditions can also damage protective oxides but again the damage arises because of the thermal expansion mismatch between the oxide and underlying metal.

A second possible effect of gas pressure and velocity is that related to the deposition, evaporation and composition of corrosive salts on turbine airfoils. Again, insufficient information is available to make completely definitive statements on these effects but first principle reasoning indicates that increased pressure will cause, for example, increased condensation of salts from the vapor phase and will also retard re-evaporation of such salts[7]. In addition, both pressure and velocity must also influence the composition of salts which deposit on airfoils[8].

Summarizing, the most important requirements for protective coatings for gas turbine superalloy airfoils are:

(a) resistance to thermal cyclic oxidation;
(b) resistance to hot corrosion;
(c) resistance to thermal fatigue cracking.

Secondary factors which should be considered for practical applications are:

(a) effect on airfoil creep behavior;
(b) effect on high frequency fatigue resistance;
(c) resistance to particulate erosion.

CONTEMPORARY COATINGS FOR SUPERALLOYS

DIFFUSION COATINGS

Although the gains to be realized by aluminizing various alloys, particularly steels, were probably initially recognized early in this century, it was not until the middle fifties that investigations began on the possibility of extending the life of gas turbine airfoils by such coatings. The first practical applications of such coatings were during the late fifties on cobalt-base first stage vanes, typically on alloys such as X-40 and WI-52. In the early to middle sixties, similar coatings were applied to nickel-base superalloy rotating blades.

The two most important processing methods for diffusion coatings are slurry-fusion and pack cementation, with the latter now being in predominant use. In the slurry-fusion process, a suspension of aluminum or an aluminum alloy is sprayed or dipped onto the alloy to a controlled thickness and the system is then heat treated in the range of 1600-2000°F to cause diffusional formation of the coating. Depending on the aluminum alloy used, part of the coating formation process may involve fusion and subsequent further solid state diffusion[9]. In the pack cementation process, parts are immersed in a powder mixture of aluminum or an aluminum alloy, an activator such as an ammonium halide, and an inert diluent, commonly alumina. Depending on the nature of the aluminum alloy, the system is then heated from 1200°F to 2000°F for times ranging from two to 24 hours. Subsequent heat treatments may be required to further diffuse the coating and to develop the proper mechanical properties of the superalloy. Coating thickness is usually in the range of one to four mils. Two archetypical types of coating structures formed on nickel-base superalloys are shown in Figure 2. The structures are governed primarily by the nature of diffusional formation of two principle intermetallic compounds in the nickel-aluminum system, Ni_2Al_3 and NiAl as described by Goward and Boone[8]. Other types of structures can be formed but these can usually be related to the two basic structures. The above described slurry coatings have similar structures but may also show some remnants of fusion of the original slurry.

A great deal of more or less empirical work has been done on the modes of degradation or wear-out of these diffusion coatings. Depending on variations of the environmental conditions described above, it is now generally accepted that various combinations of the following contribute to coating degradation.

(a) Cyclic Oxidation - The principle protective oxide which causes enhanced oxidation of the diffusion aluminide coatings is Al_2O_3, although other less protective oxides can form, primarily because the diffusion coatings contain greater or lesser amounts of the substrate alloying elements either in solution or as precipitated phases. Repeated spallation and reformation of alumina depletes the coating of aluminum to a concentration where less protective oxides such as Cr_2O_3 and NiO form and internal oxidation of residual aluminum occurs.

(b) Hot Corrosion - Molten salts, such as Na_2SO_4, deposited on airfoil surfaces cause more rapid coating degradation. Although detailed mechanistic studies have not been performed on coatings, it is reasonable to extrapolate the hot corrosion mechanisms for simpler nickel-base systems described by Pettit and others[10,11,12] to these more complex coatings. Thus, degradation could be initiated by basic fluxing of the protective Al_2O_3 and subsequently propagated by a combination of basic fluxing and sulfide oxidation. If the coatings contain sufficient amounts of deleterious elements such as molybdenum or tungsten, derived from the substrate, acidic fluxing of protective oxides is another possible degradation mode. Acidic fluxing can also occur by deposition of acidic oxides such as V_2O_5 when low grade fuels contained vanadium are burned in the turbine.

(c) Interdiffusion - At sufficiently high temperatures and times, nominally above 1700-1800°F for practical purposes, the diffusional process which occurs during coating formation continues. This decreases the aluminum concentration and causes a further tendency to inability to form protective Al_2O_3. Limited work tends to indicate that interdiffusion rates are quite dependent on the superalloy composition[13] but completely definitive quantitative relationships have not yet been developed.

(d) <u>Erosion</u> - Although this effect is widely quoted as a degradation mode very few clear cut cases have been documented in the open literature and studies are just beginning to define the mechanisms and severity of particulate erosion at high temperatures[1]. The possibility must, however, be recognized that high velocity particulates such as sand and "hard carbon" from turbine burners can contribute significantly to coating removal.

(e) <u>Mechanical Effects</u> - As mentioned, the most important mechanical property of a coating is resistance to cracking by thermally induced stresses. Figure 3 illustrates the ductility, or strain-to-crack, of a typical NiAl-based diffusion aluminide coating. As expected from prior work on NiAl, the coating has quite low ductility up to 1200-1400°F where a transition from brittle to ductile behavior occurs. Much of the older empirical work done on the thermal fatigue resistance of coated superalloys, for example, consisted of cycling usually arbitrary specimen shapes in and out of poorly characterized hot gas streams or flames from a variety of burner rigs. While such testing may empirically rank the performance of coatings in the particular test chosen, it is of limited use for specific practical applications because the shape and magnitude of the strain-temperature cycle is usually unknown. Duplication of analytically defined turbine airfoil strain-temperature cycles in laboratory scale burners is difficult or impossible and to obtain more quantitative data on thermal fatigue behavior resort must be made to so-called thermomechanical fatigue testing.[3,4] This involves the use of a testing machine which imposes controlled strain-temperature cycles, similar to those shown in Figure 1, on an appropriate specimen of the coated superalloy. Typical data indicate, as might be expected, that if sufficiently large strains are imposed below the transition temperature, coating cracking can occur in a few cycles or ultimately one cycle. Such behavior is obviously then defined as a negative effect of the coating on the base alloy, since such cracks will propagate into the alloy with continued cycling. Conversely, if the peak strain is imposed above the transition temperature of the coating, several thousands of cycles may be required to cause fatigue cracks to initiate. Under these conditions, cracking may initiate in the uncoated alloy prior to initiation in the coated alloy. Such behavior can be defined as a positive effect of the coating on the alloy.

For many advanced air cooled airfoil designs, thermal strains can peak at relatively low temperatures[2,3,4] and brittle coatings can prove to be inadequate for practical applications. Since there is no known method of significantly altering the brittle behavior of intermetallic compounds such as NiAl (and CoAl), it is now generally recognized that coatings containing only these phases can have limited applicability in advanced high performance engines.

With regard to secondary effects of coatings on mechanical properties, those of influence on high temperature creep and high frequency fatigue deserve brief mention. First principle considerations tend to indicate that coatings should have little influence on creep behavior other than that obviously caused by reduction in load bearing area. This assumes, of course, that due attention is given to assuring that the thermal aspects of coating processing are consistent with the heat treatment requirements of the base alloy. Indeed, some studies have shown beneficial effects of coatings on creep behavior[14], the effects being only partially explainable on the basis of inhibition of creep cracking on otherwise exposed grain boundaries.

Paskiet, Boone and Sullivan[15], in a limited study on the effects of a diffusion aluminide coating on the high frequency fatigue behavior of Udimet 700, found that the coating increased the fatigue strength from room temperature to about 900°F and thereafter decreased the fatigue strength up to about 1300°F at which point no influence was observed.

During the past decade, numerous attempts have been made to improve the oxidation and hot corrosion resistance of diffusion aluminide coatings. A moderate increase in resistance to hot corrosion at intermediate (1500-1800°F) temperatures can be achieved by increasing the chromium concentration over that normally derived from the superalloy substrate. This can be accomplished by a chromizing process, such as pack cementation or electroplating, prior to aluminizing. Because chromium has only limited solubility, about 6-8%, in NiAl, additional amounts of this element exist in such coatings as a second phase which can be deleterious to high temperature (1800-2000°F) oxidation resistance.

Perhaps the most important improvement is that which has been achieved by incorporation of noble metals, such as platinum, in diffusion aluminide coatings[16]. The coatings are synthesized by first electroplating thin layers, about 0.25 mil, of these metals on the superalloy surface and then aluminizing with a process adjusted in aluminum activity such that the outer layer of the coatings contain most of the noble metal as an aluminum containing intermetallic. These coatings show significant improvements in hot corrosion resistance in the temperature range of 1750°F[17] and moderate improvements in oxidation resistance at higher temperatures. Detailed mechanisms to explain these improvements have not been established but investigations by Felten and Pettit[18] on platinum-aluminum alloys suggest that improved oxide adherence may play an important role. It is also reasonable to hypothesize that the high stability of, for example, platinum-aluminum intermetallic compounds may cause increased coating stability with respect to degradation caused by coating-alloy interdiffusion. With regard to thermal fatigue resistance, it is important to recognize that these coatings are still based on intermetallic compounds, including NiAl, which exhibit brittle fracture behavior in the temperature range up to 1200-1400°F, and, therefore no significant improvements, in comparison to the simple diffusion aluminide coatings, are to be anticipated.

OVERLAY COATINGS

Recognition of the possible fundamental limitations of diffusion coatings prompted the initiation of exploratory programs on alternate coating concepts in the middle sixties. Major emphasis was placed on the possibility of cladding the superalloys with alloys more resistant to cyclic oxidation and hot corrosion. It had been recognized for many years that enhanced cyclic oxidation resistance of nickel-chromium[19] and iron-chromium-aluminum[20] alloys was caused by improved oxide adherence derived from the addition of trace amounts of alkaline earth and rare-earth elements. The development of the iron-chromium-aluminum-yttrium alloys by Wukusik and Collins[21] served to bring increased attention to the possible practical utility of these effects for turbine airfoil coatings. At that time, practical processing methods for the application of such coatings were not available. Many methods of processing, including diffusion bonding, powder sintering or melting, plasma spraying, and physical vapor deposition, including sputtering, were evaluated at their then state-of-art level. Of these, physical vapor deposition, involving vaporization of a continuously fed ingot with a high voltage electron beam[17], and condensation of the vapor on parts appropriately rotated in the vapor cloud, emerged as the most promising method of application of a variety of experimental coating alloys. After considerable refinement, the process now operates on a large scale production basis. Sputtering[22] and plasma spraying are still undergoing active evaluation as promising alternate processing methods.

The first of a series of the so-called MCrAlY coatings was an iron-chromium-aluminum-yttrium alloy[23], containing aluminum in the range of 12% in contrast to the 3-5% range in the structural alloys developed by Wukusik and Collins[21]. This FeCrAlY coating, applied in the range of five mils in thickness to nickel-base superalloys, showed significantly enhanced durability in comparison to diffusion aluminide coatings, particularly with respect to high temperature hot corrosion resistance. It was later recognized that because of the greater stability of NiAl in comparison to that of iron-aluminum compounds, coating-alloy interdiffusion was causing loss of aluminum by formation of NiAl at the FeCrAlY-nickel-base superalloy interface. Subsequent development work resulted in more stable systems such as CoCrAlY[24], NiCrAlY[25], and a mixed nickel-cobalt-base composition, NiCoCrAlY[26].

Since the specific practical applications of these coating systems are proprietary in nature, it is not possible, or even particularly useful, to cite performance or life improvements to be derived from their use. Rather, it is more useful to focus, in what follows, on the principles involved in the performance of the coatings and their various possible modes of degradation.

(a) Microstructure - A typical MCrAlY coating microstructure, specifically a CoCrAlY composition, is shown in Figure 4. The coating composition is adjusted to give a two-phase structure, in this case the intermetallic compound CoAl in a ductile cobalt solid solution matrix. Chromium partitions to a major extent to the solid solution phase as expected from available phase diagram information. Yttrium has very low solubility in the alloy, probably less than 0.01% and the fractional percentage of this element present appears as finely distributed precipitate of a cobalt-yttrium intermetallic compound.

(b) Oxidation Resistance and Degradation - Oxidation resistance of the MCrAlY coatings is derived from formation of protective Al_2O_3. Chromium assists in Al_2O_3 formation during the transient oxidation period as described by Pettit and Giggins[27]. Improved resistance to Al_2O_3 spallation under thermal cycling conditions is caused by the presence of yttrium which results in significant improvements in oxide adherence. Of the various mechanisms which have been investigated to explain this effect, Pettit and Giggins[28] have recently concluded that yttrium improves the adherence of Al_2O_3 to NiCrAlY and CoCrAlY alloys by essentially two processes, namely, by provision of vacancy sinks which prevent void formation at the alloy-Al_2O_3 interface, and perhaps more importantly, by forming macro- and micro-pegs of yttrium-rich oxides, respectively at grain boundaries and within grains, which mechanically key the Al_2O_3 to the alloy surface.

In a very general way, degradation under cyclic oxidation conditions still proceeds mainly by limited but significant spalling of Al_2O_3 and resultant loss of aluminum from the coating to the point where less protective oxides such as Cr_2O_3 and NiO or CoO form. At quite high temperatures (2000-2200°F), coating-substrate interdiffusion may play a significant role in coating degradation with regard to both dilution of aluminum concentration and diffusion of other elements from the substrate into the coatings. As would be expected, NiCrAlY coatings are more stable on nickel-base alloys than the corresponding CoCrAlY compositions.

(c) Hot Corrosion Resistance and Degradation - In a rather extensive study of sodium sulfate induced hot corrosion of model NiCrAl(Y) and CoCrAl(Y) alloys, Goebel and Pettit[12] have concluded that hot corrosion resistance of these alloys is derived from two major factors, namely, the presence of adherent Al_2O_3 which retards the basic fluxing initiation stage, and the presence of significant amounts of chromium which retards the basic fluxing-sulfide oxidation propagation stage. With regard to more specific compositional effects, it has been empirically observed that the cobalt-chromium-aluminum alloys are more resistant to molten sulfate attack than the corresponding nickel-based alloys. This difference in behavior is attributed at least partially to the greater tendency toward more reducing conditions at the sulfate-metal interface for the nickel-base alloys, which leads to higher oxide ion concentration in the sulfate and the formation of liquid nickel sulfides.

Thus, selection of MCrAlY coating compositions for specific practical applications is governed to a large extent by the anticipated environmental conditions. For example, where severe hot corrosion conditions are anticipated, as in industrial or marine turbine applications, CoCrAlY compositions are the logical choice. Selection for other applications follows the same line of reasoning with a tendency to use of more nickel-rich compositions for less corrosive but higher temperature conditions.

(d) <u>Erosion Resistance and Degradation</u> - This aspect of degradation is mentioned only for the sake of completeness. As previously discussed, little can be said with respect to the particulate erosive removal of protective coatings other than that the possibility exists. Whether or not the MCrAlY coatings offer any advantages over the diffusion coatings is still an open question.

(e) <u>Mechanical Property Effects</u> - The electron beam physical vapor deposition process, and other possible processes for fabricating the MCrAlY type of coatings, allow the preparation of a wide range of coating compositions. As would be anticipated, aluminum concentration plays a major role in establishing coating ductility. Those compositions which result in either a matrix phase of CoAl or NiAl with dispersed solid solution phases, or phase pure CoAl or NiAl, exhibit brittle-ductile transition behavior similar to that for the diffusion aluminide coatings, as shown in Figure 5[3]. Conversely, as shown in Figure 5, significant ductility, of the order of 1-3%, can be obtained in the lower temperature ranges from ambient to 1200°F, with ductility continuously increasing as the volume fraction of CoAl or NiAl (and Ni_3Al) is decreased. This increased ductility is, of course, obtained at a sacrifice of aluminum content and hence of oxidation and hot corrosion resistance. For practical gas turbine applications, each specific case must be analyzed in terms of the expected magnitude and shape of the critical thermal strain-temperature curves schematically illustrated in Figure 1. For example, when the thermal strain achieves its maximum value at relatively low temperatures, substantial improvement in thermal fatigue behavior can be gained with a ductile MCrAlY coating compared to a diffusion aluminide coating as illustrated in Figure 6. Conversely, if the thermal strain achieves its maximum value at intermediate to high temperatures where all coatings exhibit ductile behavior, the use of coatings with high ductility at low temperatures is not necessary; for such cases, it is more logical to maximize oxidation and hot corrosion resistance by using coatings with higher aluminum contents. For a more comprehensive treatment of thermal fatigue behavior of coatings, the reader is referred to recent papers on this subject by Leverant, Strangman and Langer[3] and Strangman and Hopkins[4].

With regard to the effects of the MCrAlY coatings on creep behavior of the superalloys, the major factor which must be considered in blade design is the added centrifugal stress caused by the mass of the coating which for some specific designs can be significant. The initial interdiffusion affected zone is usually about 0.25 to 0.5 mil and is for most practical purposes negligible. For very high temperature applications, for example, above 2000°F, significant coating-substrate interdiffusion can occur. For blade designs where creep is the limiting life factor, this effect may have to be accounted for during the initial design using creep data obtained from appropriate laboratory testing.

DEVELOPMENT OF IMPROVED COATINGS

Notwithstanding the coating life extensions achieved with MCrAlY and noble metal aluminides, further improvements in protective and mechanical capabilities are required to assure effective technical and economic use of alloys with higher temperature capabilities currently under development. In addition, it can be logically projected that the achievement of significant increases in the durability of airfoils in gas turbines used for power generation and marine propulsion will rely in large part on the availability of coatings with improved hot corrosion resistance. This is particularly true if the trend to use of lower grade fuels continues and if it develops that economic considerations dictate the use of lower grade synthetic fuels for power generation.

Current development programs are concentrated in approximately four general areas, as follows:

(a) increased high temperature-life capabilities by modification of MCrAlY compositions;
(b) improved, lower cost methods of coating processing;
(c) increased resistance to hot corrosion for power generation and marine applications;
(d) ceramic-based "thermal barrier" coatings.

Based on a combination of a moderate amount of fundamental understanding resulting from research on oxidation and hot corrosion mechanisms, and practical experience, development of improved coatings proceeds by a process of enlightened empiricism. A typical example is the development, under NASA contract,[29] of a coating system for the protection of an advanced directionally solidified eutectic alloy, γ / γ'-δ (Ni_3Cb in a Ni-Ni_3Al matrix). The program first screened a number of compositions based on NiCrAlY by furnace cyclic oxidation at 2000°F to 2200°F and hot corrosion (Na_2SO_4) at 1600°F of coated specimens. The most promising coating was one synthesized by sputtering a thin layer of platinum over a Ni-18%Cr-12%Al-0.3%Y coating applied by electron beam physical vapor deposition. Further evaluation of the coating was accomplished by burner rig oxidation testing at 2000°F, using a coated conventional superalloy as a control to assess coating performance for projected practical applications. Thermomechanical fatigue testing was then performed with a strain temperature cycle typical of that of an advanced air-cooled airfoil. It was concluded that thermal fatigue resistance of the coating-alloy system could be a limiting factor in practical applications. Thermal fatigue properties were significantly influenced by relatively high thermal expansion mismatch strains caused by an appreciable disparity between the thermal expansion coefficients of the coating and base alloy; such strains are additive to thermally induced strains. It is planned to ultimately test the coating-alloy system in an experimental engine run to fully evaluate system performance. Recommendations for future development include efforts to lower the expansion mismatch strains to effect improved thermal fatigue behavior.

Many, if not most, efforts to develop improved coatings are limited by the restricted capabilities of processing methods to apply a variety of coating compositions and structures on an experimental, and ultimately, a production basis. The most important limitations of the contemporary electron beam physical vapor deposition process lie in its line-of-sight nature and its inability to deposit complex compositions containing low vapor pressure elements or compounds. Current development programs involving vapor cloud ionization and biasing, gas scattering, and dual source evaporation are aimed at minimizing these limitations. At the same time, alternate processes such as sputtering[22] and plasma spraying are being evaluated to afford a greater degree of compositional and structural flexibility.

The reality of more or less continuous ingestion of sea salt into gas turbines used for marine propulsion results in perhaps the most hot corrosive environment of all applications. While significant airfoil life improvements have been accomplished by replacement of diffusion aluminide coatings with certain CoCrAlY compositions[22,30], logistic and economic considerations indicate that significant improvements, on the order of factors of four to five, are still required. A coordinated multi-task contract program under the direction of the Naval Ship Engineering Center[30] has been initiated to effect such life improvements. Since it is known[17] that sodium chloride, possibly deposited along with sulfate salts during lower power (low metal temperature) operation of marine gas turbines, significantly accelerates the hot corrosion attack of CoCrAlY, one of the tasks of the program is to devise a laboratory test which reproduces the corrosion microstructure of real airfoils to allow more accurate ranking of developmental coatings[31]. Parallel and subsequent tasks will involve semi-empirical development of coatings with improved resistance to sulfate-chloride induced hot corrosion, using principles and methods previously discussed.

So-called "thermal barrier coatings" are based on the relatively old idea[32] of insulating metals from the thermal effects of high temperature gaseous environments. Coating based on more or less stabilized zirconium oxide are utilized to increase the durability of sheet metal components, such as burner and afterburner liners in gas turbines. The coatings, applied by plasma spraying to thicknesses of 5 to 20 mils, are at least two layered, consisting of a "bond coat" of an oxidation resistant alloy and a top coat of stabilized zirconia. To minimize the effects of thermal expansion mismatch stresses between the zirconia and the base metal, coating systems consisting of gradation of metal, through metal-zirconia mixtures, to pure zirconia[33] have been applied with some degree of success. Because of the difficulty of developing major improvements in temperature capabilities of metallic alloy systems and the significant losses in turbine efficiency resulting from increased airfoil cooling, serious re-consideration is being given to the potential of thermal barrier coatings for protection of turbine airfoils. The potential benefits from the use of such coatings are quite large and include reduction of metal temperatures by about 100 to 200°F, or reduction in cooling air requirements for increased turbine efficiency, and reduction of thermal stresses from temperature cycling. The major potential problem is thermal stress induced spallation of the insulating ceramic layer. Whether or not this problem can be solved by a combination of improvement in coating properties and advanced airfoil design concepts is as yet unknown. If the problem can be solved, thermal barrier coatings may well provide the transition between metallic systems and the apparent ultimate solution of bulk ceramic airfoils.

REFERENCES

1. EPRI Contract RP 543-1, Study of Materials for Use Under Corrosion-Erosion Conditions in Coal Gasification Systems, to Pratt & Whitney Aircraft.

2. Linask, I. and Dierberger, J., A Fracture Mechanics Approach to Turbine Airfoil Design, ASME Gas Turbine Conference, Houston, Texas, March 2-6, 1975, Paper 75-GT-79.

3. Leverant, G. R., Strangman, T. E., and Langer, B. S., Parameters Controlling the Thermal Fatigue Properties of Conventionally Cast and Directionally Solidified Turbine Alloys, TMS-AIME Symposium on Superalloys, Seven Springs, PA, September 12-15, 1976.

4. Strangman, T. E and Hopkins, S. W., Thermal Fatigue of Coated Superalloys, American Ceramic Society, Annual Meeting, Washington, D.C., May 6, 1975.

5. Hancock, P., Proceedings of 1974 Gas Turbine Materials in the Marine Environment Conference, MCIC Report 75-27, pp. 225-236.

6. Dils, R. R., Fatigue of Protective Metal Oxides in Combustion Chamber Exhaust Gases; in Fatigue at Elevated Temperatures, ASTM

STP 520, American Society for Testing and Materials, 1973 pp. 102-111.

7. Fryxell, R. E. and Bessen, I. I., Coating Life Assessment in Gas Turbines Operated for Ship Propulsion, Proceedings of 1974 Gas Turbine Materials on the Marine Environment Conference, MCIC Report 75-27, pp. 259-276.

8. Hanby, V. I., Sodium Sulfate Formation and Deposition in Marine Gas Turbines, Journal of Engineering for Power, $\underline{96}$ 129 (1974).

9. Goward, G. W. and Boone, D. H., Mechanisms of Formation of Diffusion Aluminide Coatings on Nickel-Base Superalloys, Oxidation of Metals $\underline{3}$ 475 (1971).

10. Bornstein, N. S. and DeCrescenti, M. A., Met. Trans. $\underline{2}$ 2875 (1971).

11. Goebel, J. A., Pettit, F. S., and Goward, G. W., Mechanisms for the Hot Corrosion of Nickel-Base Alloys, Met. Trans. $\underline{4}$ 261 (1973).

12. Goebel, J. A. and Pettit, F. S., Hot Corrosion of Cobalt-Base Alloys, Report ARL TR75-0235, Air Force Contract F33615-72-C-1757, June, 1975.

13. Fleetwood, M. J., Influence of Nickel-Base Alloy Composition on the Behavior of Protective Aluminide Coatings, J. Inst. Met. $\underline{98}$ 1 (1970).

14. Field, T. T., Henricks, R. J., and Gell, M., The Effect of Oxidation Resistant Coatings on Creep Properties of Several Nickel-Base Superalloys, Sixth Annual Spring Meeting of AIME, Pittsburgh, PA, May, 1974.

15. Paskiet, G. F., Boone, D. H., and Sullivan, C. P., Effects of Aluminide Coating on the High Cycle Fatigue Behavior of a Nickel-Base High Temperature Alloy, J. Inst. Met. $\underline{100}$ 58 (1972).

16. Bungart, K., Protective Diffusion Layer on Nickel- and/or Cobalt-Base Alloys, U.S. Patent 3,677,789, 1972.

17. Goward, G. W., Coatings and Coating Processing for Gas Turbine Airfoils Operating in a Marine Environment, Proceedings of 1974 Gas Turbine Materials in the Marine Environment Conference, MCIC Report 75-27, pp. 277-296.

18. Felten, E. J. and Pettit, F. S., Development, Growth and Adhesion of Alumina on Platinum-Aluminum Alloys, Oxidation of Metals $\underline{10}$ (1976).

19. Hessenbruck, W., Metallen and Legierangen fur Hohe Temperature, Teil I, Springer, Berlin, 1940.

20. Pfeiffer, H., Werkst. und Korros. $\underline{8}$ 574 (1975).

21. Wukusik, C. S. and Collins, J. F., An Iron-Chromium-Aluminum Alloy Containing Yttrium, Materials Research and Standards $\underline{4}$ 637 (1964).

22. Fairbanks, J., High Rate Sputter Deposition of Protective Coatings on Marine Gas Turbine Hot Section Superalloys, Proceedings of 1974 Gas Turbine Materials in the Marine Environment Conference, MCIC Report 75-27, pp. 429-456.

23. Talboom, F. P. and Grafwallner, J., Nickel- or Cobalt-Base Superalloy with a Coating of Iron, Chromium and Aluminum, U.S. Patent 3,542,530.

24. Evans, D. J. and Elam, R. C., Cobalt-Base Coating for the Superalloys, U.S. Patent 3,676,085.

25. Goward, G. W., Boone, D. H., and Pettit, F. S., High Temperature Oxidation Resistant Coating Alloy, U.S. Patent 3,754,903.

26. Hecht, R. J., Goward, G. W., and Elam, R. C., High Temperature NiCoCrAlY Coatings, U.S. Patent 3,928,026.

27. Pettit, F. S. and Giggins, C. S., Oxidation of Ni-Cr-Al Alloys Between 1000-1200°C, J. Electrochem. Soc. $\underline{118}$ 1782 (1971).

28. Giggins, C. S. and Pettit, F. S., Oxide Scale Adherence Mechanisms and the Effects of Yttrium, Oxide Particles and Externally Applied Loads on the Oxidation of NiCrAl and CoCrAl Alloys, Report ARL TR 75-0234, Air Force Contract F33615-72-C-1702, June, 1975.

29. Strangman, T. E., Felten, E. J., and Ulion, N. E., High Temperature Oxidation Resistant Coatings for the Directionally Solidified Ni-Cb-Cr-Al Eutectic Superalloy, American Ceramic Society, 78th Annual Meeting, Cincinnati, Ohio, May 2-5, 1976.

30. Fairbanks, J. W., Coating Technology for the Marine Gas Turbine Airfoil Application, ASME Gas Turbine and Fluids Engineering Conference, Paper 78-GT-111, New Orleans, LA, March 2-5, 1976.

31. Pettit, F. S., Initiation and Propagation of Sulfate-Induced Hot Corrosion and Effects Produced by Sodium Chloride, Third Conference on Materials for Marine Gas Turbines, Bath, England, September 20-23, 1976.

32. Svirskii, L. D., Effect of a Thermally Insulated Heat Resistant Coating on the Performance of Gas Turbine Blades, Zashita Metallov, $\underline{6}$ 733 (1970).

33. Boone, D. H., Peroulakis, A., and Wilkins, C. R., Ceramic-Metal Thermal Barrier Coatings for Gas Turbine Engines, American Ceramic Society, Williamsburg Materials and Applications Forum, September 1974.

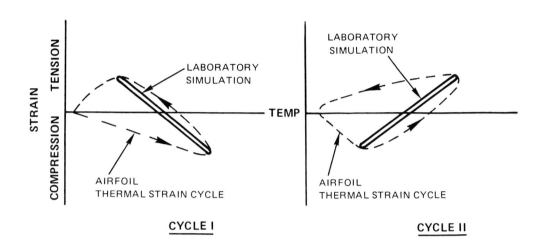

Figure 1.   Typical airfoil thermal strain cycles and laboratory thermal-mechanical simulation.

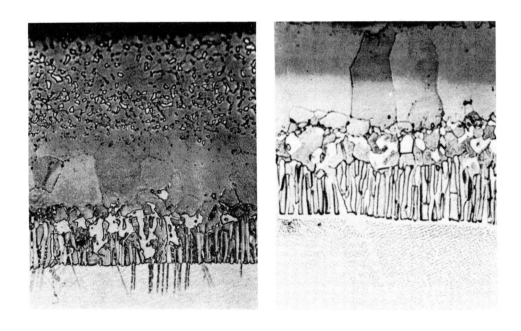

Figure 2.   Archetypical inward (left) and outward (right) diffusion aluminide coatings on nickel-base superalloys (400X).

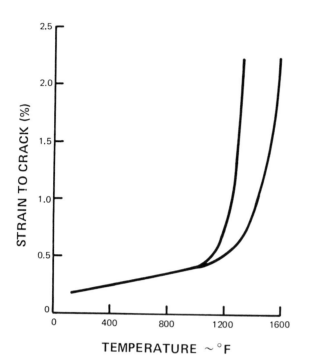

Figure 3. Typical brittle-ductile behavior of diffusion aluminide coatings.

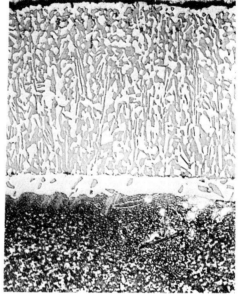

Figure 4. Microstructure of two phase CoCrAlY overlay coating.

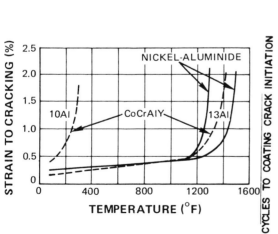

Figure 5. Ductility of CoCrAlY overlay and diffusion aluminide coatings.

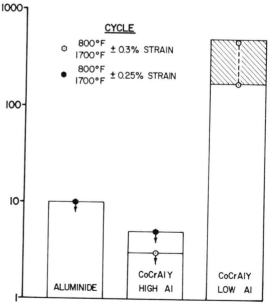

Figure 6. Thermal-mechanical fatigue behavior of brittle and ductile coatings.

SECTION XIII:
Welding

Welding the nickel alloys : Dissimilar metals, high-temperature service

Successful welding of dissimilar metals depends on filler metal choice and control of dilution in the weld metal

by HAROLD R. CONAWAY, *senior technical representative, Huntington Alloys, Inc.*

Welding dissimilar metals

In welding of dissimilar metals the composition of the weld metal depends on electrode or filler metal composition and the amount of dilution from the two base metals. The amount of dilution varies with the welding process, operator technique, and joint design.

Among the nickel-alloy welding products commonly used for dissimilar joints between nickel alloys and between nickel alloys and other materials, the nickel-chromium electrodes and filler metals (Inco-Weld A electrode, Inconel welding electrodes 112 and 182, Inconel filler metals 82, 92, and 625) are particularly versatile. They tolerate dilution from a variety of base metals without becoming crack sensitive.

In cases where more than one welding consumable is metallurgically compatible with the base metals, the selection is based on the required joint strength, service environment, and on the cost of the welding product. For example, either Inconel welding electrode 112 or Inco-Weld A electrode is suitable for a joint between Monel alloy K-500 and Inconel alloy 718. If maximum

ALLOWABLE DILUTION OF NICKEL WELDING CONSUMABLES, PERCENT [a]

	Cu	Cr	Fe	Si[c]
Nickel Filler Metal 61	Unlimited	30	25	
Nickel Welding Electrode 141	Unlimited	30	40	
Monel Filler Metal 60	Unlimited	8	22 (SAW); 15[b]	
Monel Filler Metal 67	Unlimited	5	5	
Monel Welding Electrode 187	Unlimited	5	5	
Monel Welding Electrode 190	Unlimited	8	30 (SMA)	
Inconel Filler Metal 62	15	15	25	0.75
Inconel Filler Metal 82	15	15	25	0.75
Inconel Filler Metal 92	15	15	25	0.75
Inconel Filler Metal 601	15	15	25	0.75
Inconel Filler Metal 625	15	15	25	0.75
Inconel Welding Electrode 112	15	15	40	0.75
Inconel Welding Electrode 132	15	15	40	0.75
Inconel Welding Electrode 182	15	15	40	0.75

a Use as guideline only
b For gas shielded welding, 5 percent for welds to be stress relieved; 15 percent not stress relieved.
c Total

WELDING CONSUMABLES FOR DISSIMILAR JOINTS BETWEEN NICKEL ALLOYS AND OTHER MATERIALS

Nickel Alloy	Electrode or Filler Metal	Dissimilar Material					
		Stainless Steel	Low Alloy and Carbon Steels	5-9 Nickel Steels	Cast Iron	Copper	Copper-Nickel
Nickel 200,201	Electrode	Inco-Weld A	Inco-Weld A	Inco-Weld A	Ni-Rod 55	Monel 190	Monel 190
	Filler Metal	Inconel 82	Inconel 82	Inconel 82	—	Monel 60	Monel 60
Monel alloys 400, K-500, 502	Electrode	Inco-Weld A	Monel 190	Monel 190	Ni-Rod 55	Monel 190	Monel 190
	Filler Metal	Inconel 82	Nickel 61	Nickel 61	—	Monel 60	Monel 60
Inconel alloys 600, 601	Electrode	Inco-Weld A	Inco-Weld A	Inco-Weld A	Ni-Rod 55	Nickel 141	Nickel 141
	Filler Metal	Inconel 82	Inconel 82	Inconel 82	—	Nickel 61	Nickel 61
Inconel alloys 625, 706, 718, X-750	Electrode	Inco-Weld A	Inco-Weld A	Inconel 112	Ni-Rod 55	Nickel 141	Nickel 141
	Filler Metal	Inconel 82	Inconel 82	Inconel 625	—	Nickel 61	Nickel 61
Incoloy alloy 800	Electrode	Inco-Weld A	Inco-Weld A	Inco-Weld A	Ni-Rod 55	Nickel 141	Nickel 141
	Filler Metal	Inconel 82	Inconel 82	Inconel 82	—	Nickel 61	Nickel 61
Incoloy alloy 825	Electrode	Inconel 112	Inconel 112	Inconel 112	Ni-Rod 55	Nickel 141	Nickel 141
	Filler Metal	Inconel 625	Inconel 625	Inconel 625	—	Nickel 61	Nickel 61

Electrode and filler metal selection for normal conditions.

joint strength is required, Inconel welding electrode 112 is the better choice. If maximum strength is not required, the joint is welded more economically with Inco-Weld A electrode.

For base metal combinations not covered in the table, evaluate electrodes and filler metals on the basis of their ability to accept dilution from the two base metals without forming a crack-sensitive composition or developing other undesirable characteristics. For a joint between a nickel alloy and a non-nickel material, a nickel-alloy welding product gives the best results. Consumables normally used on austenitic stainless steels usually give crack-sensitive welds in nickel alloys. The information given here is a guideline for welding dissimilar joints. The effects of interalloying dissimilar materials are too complex to predict results in all cases.

Dilution

Chemical analysis of the deposited bead most accurately determines the dilution produced by a set of welding conditions — welding process, technique, joint design. A close approximation of dilution can be made using the area of a joint cross section: Dilution is the area of the original base metal divided by the final weld-metal cross-sectional area.

For flat position SMA butt welding, a dilution of 30 percent is used to calculate weld deposit compositions: the electrode supplies 70 percent of the completed weld bead, and the base metal 30 percent. With SMAW, operator technique has less effect on dilution than with other welding processes. Variations in technique can cause the rate to range from 20 to 40 percent.

The dilution rate for SMAW is well known. For other processes, dilution depends more on technique. Dilution with GMAW usually ranges from 10 to 50 percent, depending on torch manipulation and type of metal transfer. Spray transfer gives the highest dilution, and short circuit the lowest. The greatest variation in dilution is with GTA. Depending on welding technique, dilution rates can range from 20 to over 80 percent. For a joint made without filler metal, dilu-

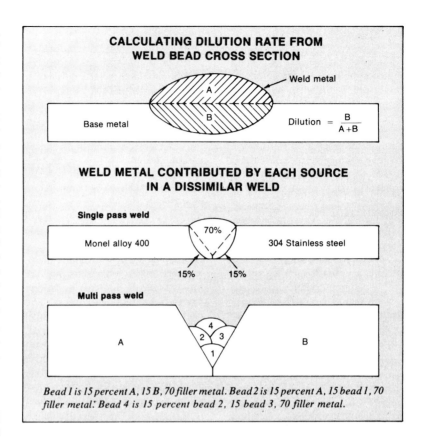

CALCULATING DILUTION RATE FROM WELD BEAD CROSS SECTION

Weld metal

Base metal   Dilution = $\dfrac{B}{A+B}$

WELD METAL CONTRIBUTED BY EACH SOURCE IN A DISSIMILAR WELD

Single pass weld

Monel alloy 400   70%   304 Stainless steel

15%   15%

Multi pass weld

A   B

Bead 1 is 15 percent A, 15 B, 70 filler metal. Bead 2 is 15 percent A, 15 bead 1, 70 filler metal. Bead 4 is 15 percent bead 2, 15 bead 3, 70 filler metal.

tion would be 100 percent, since all of the weld metal is supplied by the base materials.

With a known dilution rate, the approximate composition of the final weld metal can be determined. For example, assuming 30 percent dilution, the weld metal in a joint between Monel alloy 400 (67 Ni, 32 Cu) and Type 304 stainless steel (8 Ni, 18 Cr, 74 Fe) made with Inco-Weld A electrode (70 Ni, 15 Cr, 8 Fe) would be 15 percent Monel alloy 400, 15 percent type 304 stainless steel, and 70 percent Inco-Weld A electrode. The amount of weld metal contributed by each source is calculated as follows:

Contribution of Inco-Weld A electrode:

70 × 70 Ni = 49 Nickel
70 × 15 Cr = 10.5 Chromium
70 × 8 Fe   = 5.6 Iron

Contribution of Monel alloy 400:

15 × 67 Ni = 10 Nickel
15 × 32 Cu = 4.8 Copper

Contribution of Type 304 stainless steel:

15 × 8 Ni   = 1.2 Nickel
15 × 18 Cr = 2.7 Chromium
15 × 74 Fe = 11.1 Iron

The electrode contribution plus base metal contribution is the composition of the weld deposit: 60.2 nickel (49 + 10 + 1.2), 13.2 chromium (10.5 + 2.7), 16.7 iron (5.6 + 11.1), and 4.8 copper.

In a multipass weld, composition remains constant along each bead but varies with bead location. The root bead is diluted equally by the two base metals. Subsequent beads may be diluted by a base metal and by a previous bead or entirely by previous beads.

Dilution limits

Copper, chromium, and iron are the elements normally of concern in considering dilution of nickel welding products. All of the products can accept unlimited dilution by nickel without detriment.

Dilution limits given in the following discussion apply only to commonly used welding materials that have solid-solution compositions. Precipitation hardening weld metals are not usually recommended for dissimilar joining. The values mentioned here are guidelines only. Borderline cases may require a trial joint

WELDING NICKEL ALLOYS

evaluation. When weld metal will be diluted by more than one potentially detrimental element, allowance should be made for possible additive effects. A joint in steel to be welded by a gas-shielded process with filler metal 60 requires a barrier layer of Monel welding electrode 190, nickel filler metal 61, or nickel welding electrode 141.

Nickel consumables

Nickel welding electrode 141 and Nickel filler metal 61 are excellent for dissimilar welding, but have lower strength than other nickel-alloy welding products.

Ni-Rod and Ni-Rod 55 welding electrodes are specially formulated for the welding of cast irons. These dilution limits do not apply.

Inconel

Inconel (nickel-chromium) consumables are the most widely used materials for dissimilar welding. The products produce high strength weld deposits that can be diluted by many other materials with no reduction of mechanical properties. Inco-Weld A electrode has exceptional dissimilar welding capability.

Besides weld metal dilution, differences in thermal expansion and melting point may influence the selection of filler metal for dissimilar joints, especially if the joints will be exposed to high service temperatures.

Unequal expansion of joint members stresses the joint and reduces fatigue strength. If one of the base metals is of lower strength, select a filler metal that has an expansion rate near that of the weaker base metal. The stress of unequal expansion will then concentrate on the stronger side of the joint.

A joint between austenitic stainless steel and mild steel illustrates the importance of considering differences in thermal expansion. The coefficient of expansion of mild steel is lower than that of stainless steel. From the standpoint of dilution, either a stainless steel electrode or an Inconel nickel-chromium electrode would be suitable. The stainless electrode has an expansion coefficient near that of the stainless base metal; Inconel is closer to that of mild steel. If the joint is welded with the stainless electrode, both the weld metal and the stainless base metal will expand more than the mild steel, placing the line of differential expansion

along the weaker (mild steel) side of the joint. If the joint is welded with an Inconel welding product, the stress resulting from unequal expansion is confined to the stronger, stainless steel side of the joint

Differences in melting point between the two base metals, or between the weld metal and base metal can, during welding, result in rupture of the material having the lower melting point. Solidification and contraction of the material with the higher melting point places stress on the lower melting point material, which is in a weak, incompletely solidified condition. Applying a layer of weld metal on the low-melting-point base metal before the joint is welded often eliminates this problem. Stress level is lower during application of the weld-metal layer. During completion of the joint, the previously applied weld metal reduces the melting-point differential across the joint.

Carbon migration is sometimes important in the selection of a filler metal for a dissimilar joint involving carbon steel. Nickel alloy weld metals are effective barriers to carbon migration. In joining carbon steels to stainless steels, high-nickel weld metals are sometimes used to prevent undesirable carbon migration.

Welding precipitation-hardening alloys

Precipitation hardening (age hardening) alloys are usually GTA welded.

Precipitation hardening nickel alloys fall into two systems: the nickel-aluminum-titanium system, which includes Duranickel alloy 301, Monel alloy K-500, and Inconel alloy X-750; and the nickel-columbium-aluminum-titanium system, which includes Inconel alloys 706 and 718.

Both systems have good weldability. The significant difference between the two is the rate at which precipitation occurs. The aluminum-titanium-hardened alloys respond quickly to precipitation-hardening temperatures. The columbium-aluminum-titanium-hardening alloys respond more slowly. The delayed precipitation reaction enables the alloys to be welded and directly aged with less possibility of cracking.

Cracking can occur when high residual welding stresses are present during the aging treatment. It occurs in the area of highest stress, in the base metal near the heat-affected zone. Little stress relief occurs at aging temperatures, and high residual welding stresses may exceed the rupture strength of the base metal at aging temperatures. Stresses introduced by precipitation add to the residual stresses.

Heat input during welding should

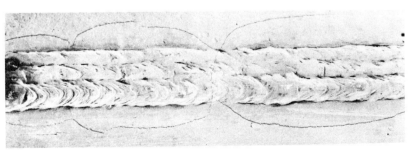

This failure in Inconel alloy X-750 was caused by residual welding stress. Joint is 2 inches (50 mm) thick, welded in the age-hardened condition and re-aged at 1,300°F/20h (700°C/ 20h) after welding.

be moderately low to obtain the highest joint efficiency. For multiple-bead or multiple-layer welds, use several small beads instead of a few large, heavy beads.

In gas-shielded welding of multiple-pass joints, clean between layers to remove oxide films as they accumulate. The oxide films should be removed when they are heavy enough to be visible on the weld surface. They must be removed by abrasive blasting or grinding. Power wire brushing only polishes the oxide surface. Oxide films inhibit fusion and result in laminar oxide stringers. Inclusions of this type act as mechanical stress raisers and can significantly reduce joint efficiency and service life.

Rigid or complex structures must be assembled and welded with care to avoid excessively high stress levels. Units or subassemblies should be given sufficient annealing treatments to ensure a low level of residual stress when they are precipitation-hardened. Any part that has been subjected to severe bending, drawing, or other forming operations should be annealed before it is welded. The alloy manufacturer will provide appropriate thermal treatments. Heat in a controlled-atmosphere furnace to limit oxidation, and minimize subsequent cleaning operations. If material containing partially filled weld grooves has been thermally treated,

remove the oxide from the weld by grinding or abrasive blasting before welding is resumed.

Welding the Al-Ti system

The strong tendency to base-metal cracking of welded and directly aged aluminum-titanium-hardened alloys makes special handling of welded structures necessary. The alloys must be given the proper thermal treatment after welding and before they are precipitation-hardened. Thermal treatment must be carried out with a fast, uniform rate of heating to avoid prolonged exposure to temperatures in the precipitation-hardening range. The best way to attain this high heating rate is to charge the weldment directly into a furnace heated to temperature. If the part mass is large in relation to furnace area, heat the furnace to 200-500°F (110-275°C) above the normal heat-treat temperature, then reset the furnace controls when the parts reach temperature. The stress created by repair or alteration welding must be relieved in the same way by rapid heating to annealing temperature before re-aging.

Sometimes, particularly with complicated structures, postweld stress relief is not practical. Preweld heat treatment may be helpful in such cases. Two procedures used successfully for Inconel alloy X-750 are:

1. Heat at 1550°F (840°C) for 16 hours, air-cool, and weld.

2. Heat at 1950°F (1065°C) for 1 hour, furnace-cool at 25-100°F (15-55°C) per hr to 1200°F (650°C), air-cool, and weld.

All of the aluminum-titanium-hardened alloys can be welded in the aged condition. Because of the problem of base metal cracking, however, the weld and heat-affected zone must not be exposed to age-hardening temperatures after welding. If service temperatures are in the age-hardening range, anneal and re-age the weldment before putting it in service.

Welding the Nb-Al-Ti system

Inconel alloys 706 and 718 resist postweld cracking. Alloy 718 annealed sheet can be welded and directly aged without cracking. However, repair welding or welding of sheet in the aged condition and re-aging after welding (highly restrained condition) can result in base-metal cracking. Anneal highly restrained or complicated structures after welding and before age-hardening to avoid base metal cracking. Use the rapid heating procedure described for aluminum-titanium-hardened alloys.

Generally, Inconel alloys 706 and 718 are welded in the annealed or solution-treated condition. If complex units must be annealed in conjunction with welding or forming operations, the annealing temperature should be consistent with the specification and end-use requirements.

Fabrication for high-temperature service

Equipment for high-temperature service is often of nickel alloy. The design and welding of the equipment must consider the severe demands imposed by high temperatures.

Design factors

Nickel weld metals have high creep-rupture strength but usually have lower stress-rupture ductility and lower mechanical- and thermal-fatigue strength than wrought material. Proper design of welded structures minimizes the effects of these lower properties.

Design equipment so that welds

are located where low fatigue strength and stress-rupture ductility will not be detrimental. Generally, this means placing the welds where least high temperature deformation occurs. For example, a horizontal pipe having a longitudinal weld should be turned to locate the weld at the top rather than at the bottom, to reduce weld metal elongation when the pipe sags during high-temperature service.

To minimize the effects of thermal or mechanical fatigue, locate welds in low stress areas. Corners and areas where shape or dimensional

Thermal fatigue cracks start at incomplete welds.

changes occur are points of stress concentration and should not contain

WELDING NICKEL ALLOYS

welded joints. Butt joints are best, because the stresses act axially rather than eccentrically, as they do in corner and lap joints.

Welding procedures

The weld should completely penetrate or close the joint. If the design permits, allow no unfused area. Thermal fatigue failures often start in incomplete welds that create stress concentrations.

Welds that must be placed where changes in sectional size or direction occur require careful welding procedures to minimize stress concentrations. Avoid undercutting, lack of penetration, weld craters, and excessive weld reinforcement. If complete penetration in a joint in rod or bar stock is not possible, the weld metal should be continuous and seal the joint so that none of the process atmosphere can enter.

Welds in heat-treating fixtures fabricated of round or flat stock should flare smoothly into the base metal without undercut. When fixtures are to be subjected to heating and quenching cycles, wrap-around or loosely riveted joints are sometimes desirable, because they allow some freedom of movement.

All welding slag must be removed from completed joints. Slag that melts during high-temperature service can be corrosive. In oxidizing environments, the slag becomes increasingly fluid and aggressively attacks the metal. In reducing atmospheres, the slag picks up sulfur and causes failure by sulfidation in atmospheres that are otherwise low in sulfur. In one case, with only 0.01 per-

Attack by molten welding flux.

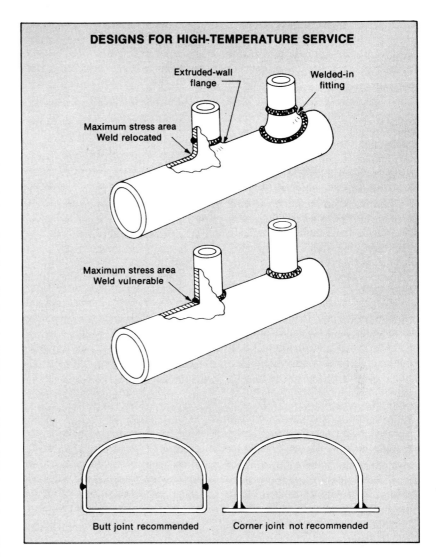

DESIGNS FOR HIGH-TEMPERATURE SERVICE

Extruded-wall flange

Welded-in fitting

Maximum stress area
Weld relocated

Maximum stress area
Weld vulnerable

Butt joint recommended   Corner joint not recommended

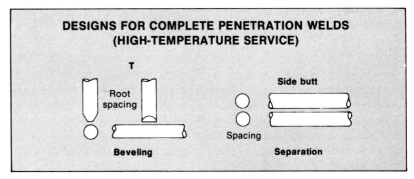

DESIGNS FOR COMPLETE PENETRATION WELDS
(HIGH-TEMPERATURE SERVICE)

T

Root spacing

Side butt

Spacing

Beveling

Separation

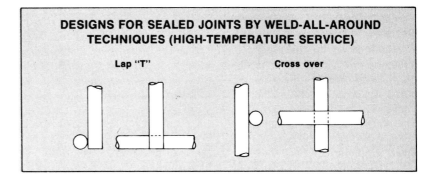

DESIGNS FOR SEALED JOINTS BY WELD-ALL-AROUND
TECHNIQUES (HIGH-TEMPERATURE SERVICE)

Lap "T"   Cross over

cent sulfur in the atmosphere, the sulfur content of slag was found to increase from 0.05 to 1.6 percent in 1 month. Also, sulfur pickup can depress the melting temperature of the slag, causing the slag to become corrosive at lower temperatures.

If all slag cannot be removed from tight areas, like the root of a lap or crossover joint in round stock, subsequent welding passes must completely enclose the joint to prevent contact of the slag with the atmosphere.

Testing and inspection

In transverse face-bend specimen (above, left) the Nickel 200 side of the joint absorbed almost all the elongation. Average elongation is 20 percent, if gauge length spans the entire joint. Elongation on the Nickel 200 side is 40 percent. In longitudinal bend-test specimen (above, right) the weld runs central and parallel to the long edge. All parts of the joint elongate at the same rate.

Bend testing

Free-bend or guided-bend tests are useful to evaluate strength and ductility of welded joints. Specimens for bend testing may have the weld in the transverse or longitudinal direction.

Transverse specimens, commonly used for qualification, give misleading ratings of weld metal ductility. The bending area includes five zones: weld metal, two heat-affected zones, and two areas of unaffected base metal. During bending of the specimen, greater elongation occurs in the unaffected base metal and heat-affected zones if those areas are softer than the weld metal. Low elongation of the weld metal could be erroneously interpreted as an indication of low ductility. This effect is more pronounced in dissimilar joints in which one member is of lower strength than the other.

A longitudinal test specimen having the weld in the center, parallel to the long edge, is preferred. All parts of the joint are forced to elongate at the same rate regardless of strength. Such a test is more realistic.

Misleading ductility values can also be obtained from transverse tensile tests. Longitudinal tests are better for determining weld quality. ■

Welds in nickel alloys are tested and inspected by the same methods used for welds in steel. Some of them: visual inspection, radiography, various mechanical tests, and sectioning and etching for micro- and macro-examination. Magnetic particle inspection is not useful, since only a few alloys, such as Nickel 200, are sufficiently magnetic to produce a recording. Fluorescent and nonfluorescent dye-penetrant inspections are useful for detection of small surface defects.

INDEX

393

Date D